Conversion Table between Millielectron Volts and Wave Numbers

meV	cm^{-1}	meV	cm^{-1}	meV	cm^{-1}	meV	cm^{-1}	meV	cm^{-1}
1	8	51	411	101	815	151	1218	201	1621
2	16	52	419	102	823	152	1226	202	1629
3	24	53	427	103	831	153	1234	203	1637
4	32	54	436	104	839	154	1242	204	1645
5	40	55	444	105	847	155	1250	205	1653
6	48	56	452	106	855	156	1258	206	1661
7	56	57	460	107	863	157	1266	207	1670
8	65	58	468	108	871	158	1274	208	1678
9	73	59	476	109	879	159	1282	209	1686
10	81	60	484	110	887	160	1290	210	1694
11	89	61	492	111	895	161	1299	211	1702
12	97	62	500	112	903	162	1307	212	1710
13	105	63	508	113	911	163	1315	213	1718
14	113	64	516	114	919	164	1323	214	1726
15	121	65	524	115	928	165	1331	215	1734
16	129	66	532	116	936	166	1339	216	1742
17	137	67	540	117	944	167	1347	217	1750
18	145	68	548	118	952	168	1355	218	1758
19	153	69	557	119	960	169	1363	219	1766
20	161	70	565	120	968	170	1371	220	1774
21	169	71	573	121	976	171	1379	221	1782
22	177	72	581	122	984	172	1387	222	1791
23	186	73	589	123	992	173	1395	223	1799
24	194	74	597	124	1000	174	1403	224	1807
25	202	75	605	125	1008	175	1411	225	1815
26	210	76	613	126	1016	176	1420	226	1823
27	218	77	621	127	1024	177	1428	227	1831
28	226	78	629	128	1032	178	1436	228	1839
29	234	79	637	129	1040	179	1444	229	1847
30	242	80	645	130	1049	180	1452	230	1855
31	250	81	653	131	1057	181	1460	231	1863
32	258	82	661	132	1065	182	1468	232	1871
33	266	83	669	133	1073	183	1476	233	1879
34	274	84	677	134	1081	184	1484	234	1887
35	282	85	686	135	1089	185	1492	235	1895
36	290	86	694	136	1097	186	1500	236	1903
37	298	87	702	137	1105	187	1508	237	1912
38	306	88	710	138	1113	188	1516	238	1920
39	315	89	718	139	1121	189	1524	239	1928
40	323	90	726	140	1129	190	1532	240	1936
41	331	91	734	141	1137	191	1541	241	1944
42	339	92	742	142	1145	192	1549	242	1952
43	347	93	750	143	1153	193	1557	243	1960
44	355	94	758	144	1161	194	1565	244	1968
45	363	95	766	145	1169	195	1573	245	1976
46	371	96	774	146	1178	196	1581	246	1984
47	379	97	782	147	1186	197	1589	247	1992
48	387	98	790	148	1194	198	1597	248	2000
49	395	99	798	149	1202	199	1605	249	2008
50	403	100	807	150	1210	200	1613	250	2016

meV	cm^{-1}	meV	cm^{-1}	meV	cm^{-1}	meV	cm^{-1}	meV	cm^{-1}
251	2024	301	2428	351	2831	401	3234	451	3638
252	2032	302	2436	352	2839	402	3242	452	3646
253	2041	303	2444	353	2847	403	3250	453	3654
254	2049	304	2452	354	2855	404	3258	454	3662
255	2057	305	2460	355	2863	405	3267	455	3670
256	2065	306	2468	356	2871	406	3275	456	3678
257	2073	307	2476	357	2879	407	3283	457	3686
258	2081	308	2484	358	2887	408	3291	458	3694
259	2089	309	2492	359	2896	409	3299	459	3702
260	2097	310	2500	360	2904	410	3307	460	3710
261	2105	311	2508	361	2912	411	3315	461	3718
262	2113	312	2516	362	2920	412	3323	462	3726
263	2121	313	2524	363	2928	413	3331	463	3734
264	2129	314	2533	364	2936	414	3339	464	3742
265	2137	315	2541	365	2944	415	3347	465	3750
266	2145	316	2549	366	2952	416	3355	466	3759
267	2153	317	2557	367	2960	417	3363	467	3767
268	2162	318	2565	368	2968	418	3371	468	3775
269	2170	319	2573	369	2976	419	3379	469	3783
270	2178	320	2581	370	2984	420	3387	470	3791
271	2186	321	2589	371	2992	421	3396	471	3799
272	2194	322	2597	372	3000	422	3404	472	3807
273	2202	323	2605	373	3008	423	3412	473	3815
274	2210	324	2613	374	3016	424	3420	474	3823
275	2218	325	2621	375	3025	425	3428	475	3831
276	2226	326	2629	376	3033	426	3436	476	3839
277	2234	327	2637	377	3041	427	3444	477	3847
278	2242	328	2645	378	3049	428	3452	478	3855
279	2250	329	2654	379	3057	429	3460	479	3863
280	2258	330	2662	380	3065	430	3468	480	3871
281	2266	331	2670	381	3073	431	3476	481	3879
282	2274	332	2678	382	3081	432	3484	482	3888
283	2283	333	2686	383	3089	433	3492	483	3896
284	2291	334	2694	384	3097	434	3500	484	3904
285	2299	335	2702	385	3105	435	3508	485	3912
286	2307	336	2710	386	3113	436	3517	486	3920
287	2315	337	2718	387	3121	437	3525	487	3928
288	2323	338	2726	388	3129	438	3533	488	3936
289	2331	339	2734	389	3137	439	3541	489	3944
290	2339	340	2742	390	3146	440	3549	490	3952
291	2347	341	2750	391	3154	441	3557	491	3960
292	2355	342	2758	392	3162	442	3565	492	3968
293	2363	343	2766	393	3170	443	3573	493	3976
294	2371	344	2775	394	3178	444	3581	494	3984
295	2379	345	2783	395	3186	445	3589	495	3992
296	2387	346	2791	396	3194	446	3597	496	4000
297	2395	347	2799	397	3202	447	3605	497	4009
298	2404	348	2807	398	3210	448	3613	498	4017
299	2412	349	2815	399	3218	449	3621	499	4025
300	2420	350	2823	400	3226	450	3629	500	4033

Tunneling Spectroscopy

Capabilities, Applications, and New Techniques

Tunneling Spectroscopy

Capabilities, Applications, and New Techniques

Edited by

Paul K. Hansma

University of California at Santa Barbara
Santa Barbara, California

PLENUM PRESS • NEW YORK AND LONDON

Library of Congress Cataloging in Publication Data

Main entry under title:

Tunneling spectroscopy.

Includes bibliographical references and index.
1. Tunneling spectroscopy. I. Hansma, Paul K., 1946–
QC454.T75T86 1982 539.7′2112 82-16161
ISBN 0-306-41070-2

A Division of Plenum Publishing Corporation
233 Spring Street, New York, N.Y. 10013

Printed in the United States of America

This book
is dedicated to
Bob Jaklevic
and
John Lambe

Contributors

J. G. Adler, Department of Physics, University of Alberta, Edmonton, Alberta T6G 2J1, Canada

S. de Cheveigné, Groupe de Physique des Solides de L'Ecole Normale Supérieure, Université Paris VII, Tour 23, 2 place Jussieu, 75251 Paris Cedex 05, France

Robert V. Coleman, Department of Physics, University of Virginia, Charlottesville, Virginia 22901.

L. H. Dubois, Bell Laboratories, Murray Hill, New Jersey 07974.

H. G. Hansma, Department of Biological Sciences, University of California, Santa Barbara, California 93106.

P. K. Hansma, Department of Physics, University of California, Santa Barbara, California 93106.

K. W. Hipps, Department of Chemistry and Chemical Physics Program, Washington State University, Pullman, Washington 99164.

R. C. Jaklevic, Physics Department, Ford Scientific Laboratory, P.O. Box 2053 SL3012, Dearborn, Michigan 48121.

John Kirtley, IBM T. J. Watson Research Center, Yorktown Heights, New York, 10598.

J. Klein, Groupe de Physique des Solides de l'Ecole Normale Supérieure, Université Paris VII, Tour 23, 2 place Jussieu, 75251 Paris Cedex 05, France.

R. M. Kroeker, IBM San Jose Research Laboratory, 5600 Cottle Road, San Jose, California 95193.

Bernardo Laks, Instituto de Fisica, DFA, Universidade Estadual de Campinas, Campinas, San Paulo, Brazil.

A. Léger, Groupe de Physique des Solides de l'Ecole Normale Supérieure, Université Paris VII, Tour 23, 2 place Jussieu, 75251 Paris Cedex 05, France.

Ursula Mazur, Department of Chemistry and Chemical Physics Program, Washington State University, Pullman, Washington 99164.

D. L. Mills, Department of Physics, University of California, Irvine, California 92717.

W. J. Nelson, School of Physical Sciences, New University of Ulster, Coleraine BT52 1SA, Northern Ireland.

M. Parikh, Hewlett-Packard Laboratories, 3500 Deer Creek Road, Palo Alto, California 94304.

D. G. Walmsley, School of Physical Sciences, New University of Ulster, Coleraine BT52 1SA, Northern Ireland.

M. Weber, Department of Physics, University of California, Irvine, California 92717.

W. Henry Weinberg, Division of Chemistry and Chemical Engineering, California Institute of Technology, Pasadena, California 91125.

Henry W. White, Physics Department, University of Missouri, Columbia, Missouri 65211.

E. L. Wolf, Ames Laboratory—USDOE and Department of Physics, Iowa State University, Ames, Iowa 50011.

Preface

This book has been compiled to give specialists, in areas that could be helped by tunneling spectroscopy, a rounded and relatively painless introduction to the field. Why relatively painless? Because this book is filled with figures—A quick glance through these figures can give one a good idea of the types of systems that can be studied and the quality of results that can be obtained.

To date, it has been somewhat difficult to learn about tunneling spectroscopy, as papers in this field have appeared in a diversity of scientific journals: for example, *The Journal of Adhesion*, *Journal of Catalysis*, *Surface and Interface Analysis*, *Science*, *Journal of the American Chemical Society*, *Physical Review*—over 45 different ones in all, plus numerous conference proceedings. This diversity is, however, undoubtedly healthy. It indicates that the findings of tunneling spectroscopy are of interest and potential benefit to a wide audience. This book can help people who have seen a few papers or heard a talk on tunneling spectroscopy and want to learn more about what it can do for their field.

Tunneling spectroscopy is presently in a transitional state. Its experimental methods and theoretical basis have been reasonably well developed. Its continued vitality will depend on the success of its applications. Crucial to that success, as pointed out by Ward Plummer, is the adoption of tunneling spectroscopy by specialists in the areas of application.

At present, tunneling spectroscopy is still usually done by tunneling spectroscopists—those who specialize in this field. Will tunneling spectroscopy successfully pass from the hands of specialists in tunneling spectroscopy into the hands of specialists in applications? Will it be used by specialists in lubrication, corrosion, catalysis, adhesion, and surface science to help solve their problems?

There has already been some progress in this direction. For example, Henry Weinberg, author of Chapter 12, is a chemical engineer who specializes in catalysis and surface science; Kerry Hipps and Ursula Mazur, authors of Chapter 8, are chemists who specialize in the study of inorganic ions. That these groups use tunneling spectroscopy as a tool in their research areas is one of the brightest signs of hope for the future of tunneling spectroscopy.

I am grateful to Ellis Rosenberg of Plenum Publishing Corporation for inviting me to edit this book; to Arnold Adams, John Adler, Atiye Bayman, Kerry Hipps, Peter Love, Mark Lowenstine, Ursula Mazur, John Moreland, John Rowell, Douglas Scalapino, Qi Qing Shu, and Henry Weinberg, who gave unselfishly of their time in suggesting improvements in various chapters; to Joan Jacobs for efficiently and cheerfully handling the secretarial work involved in editing a book; to Helen Hansma for indexing the book with help from the chapter authors; to Robert Coleman, Gene Rochlin, and Douglas Scalapino for introducing me to tunneling spectroscopy; and finally to the University of California, the Division of Materials Research of the National Science Foundation, and the Office of Naval Research for supporting tunneling spectroscopy at the University of California, Santa Barbara.

University of California
Santa Barbara

Paul K. Hansma

Contents

1. Introduction **1**

Paul K. Hansma

1. Why? Why? Why? 1
   1.1. Why Do Vibrational Spectroscopy? 1
   1.2. Why Do Tunneling Spectroscopy? 4
   1.3. Why Not Do Infrared, Raman, or Electron Loss Spectroscopy Instead? 4
2. A Water Analogy for Tunneling Spectroscopy 4
   2.1. Water Flow 4
   2.2. Tunneling 5
3. Strengths of Tunneling Spectroscopy 8
   3.1. Spectral Range 8
   3.2. Sensitivity 10
   3.3. Resolution 14
   3.4. Selection Rules 17
4. Weaknesses of Tunneling Spectroscopy 20
   4.1. Junction Geometry 20
   4.2. Top Metal Electrode 20
   4.3. Cryogenic Temperatures 25
5. General Experimental Techniques 26
   5.1. Introduction 26
   5.2. Sample Preparation 26
      5.2.1. The Substrates 26
      5.2.2. The Vacuum Evaporator 27
      5.2.3. Junction Fabrication 28
      5.2.4. Care and Handling of Completed Junctions 33
      5.2.5. Cryogenics 34
      5.2.6. Characterizing Junctions and Obtaining Spectra 36
      5.2.7. Calibration and Measuring Peak Positions 38
6. Conclusions 39
   References 39

2. The Interaction of Tunneling Electrons with Molecular Vibrations . . . 43
John Kirtley

1. Introduction . . . 43
2. Elastic Tunneling . . . 44
3. Inelastic Tunneling . . . 47
   3.1. Simple Long-Range Models . . . 47
   3.2. Complex Long-Range Models . . . 52
   3.3. Short-Range Models . . . 62
4. Conclusions . . . 68
   References . . . 69

3. Tunneling Spectroscopies of Metal and Semiconductor Phonons . . . 71
E. L. Wolf

1. Introduction . . . 71
2. Threshold Spectroscopy of Normal State Phonons . . . 77
   2.1. Semiconductors . . . 77
   2.2. Metals . . . 86
3. Superconductive Tunneling: The Effective Phonon Spectrum $\alpha^2F(\omega)$ . . . 88
   3.1. Superconductivity . . . 88
   3.2. The Tunneling Density of States in $C-I-S$ Junctions . . . 89
   3.3. McMillan–Rowell Inversion for $\alpha^2F(\omega)$ . . . 91
4. Proximity Tunneling Methods . . . 95
   4.1. $C-I-NS$ Junctions in the Thin-N Limit . . . 97
   4.2. Phonons in the Superconductor S . . . 99
   4.3. Phonons in the Proximity Layer N . . . 104
5. Conclusions . . . 106
   References . . . 107

4. Electronic Transitions Studied by Tunneling Spectroscopy . . . 109
S. de Cheveigné, J. Klein, and A. Léger

1. Introduction . . . 109
2. Experimental . . . 110
3. Results . . . 111
   3.1. Rare Earth Oxides . . . 111
   3.2. Large Molecules . . . 113
4. What Are Not Electronic Transitions? . . . 118
5. Conclusions . . . 118
   References . . . 119

5. Light Emission from Tunnel Junctions . . . 121
D. L. Mills, M. Weber, and Bernardo Laks

1. Introduction . . . 121
2. Planar Tunnel Junctions and Surface Polaritons . . . 126

3. Light Emission from Tunnel Junctions: The Theoretical Picture and Examples . . . 136
3.1. General Remarks . . . 136
3.2. Light Emission from Slightly Roughened Junctions . . . 138
3.3. Light Emission from Junctions Grown on Holographic Gratings . . . 142
3.4. Light Emission from Small-Particle Junctions . . . 146
3.5. Summary . . . 147
4. Conclusions . . . 148
References . . . 151

6. Comparisons of Tunneling Spectroscopy with Other Surface Analytical Techniques . . . 153

L. H. Dubois

1. Introduction . . . 153
2. Major Surface Analytical Techniques: A Brief Survey . . . 154
2.1. Techniques for Studying Surface Chemical Composition . . . 155
2.1.1. X-Ray Photoelectron Spectroscopy . . . 155
2.1.2. Auger Electron Spectroscopy . . . 155
2.1.3. Secondary Ion Mass Spectrometry . . . 158
2.2. Determination of Surface Electronic Structure . . . 159
2.2.1. Ultraviolet Photoelectron Spectroscopy . . . 159
2.2.2. Electron Energy Loss Spectroscopy . . . 161
2.3. Techniques for Surface Structural Analysis . . . 163
2.3.1. Surface Extended X-Ray Absorption Fine Structure . . . 163
2.3.2. Low-Energy Electron Diffraction . . . 165
2.3.3. Transmission Electron Microscopy . . . 166
2.3.4. Gas Adsorption . . . 167
2.3.5. Scanning Electron Microscopy . . . 167
2.4. Observation of Surface Vibrational Modes . . . 168
2.4.1. Infrared Spectroscopy . . . 168
2.4.2. Surface Raman Spectroscopy . . . 174
2.4.3. High-Resolution Electron Energy Loss Spectroscopy . . . 177
2.4.4. Inelastic Neutron Scattering Spectroscopy . . . 179
3. The Application of Modern Surface Analytical Techniques to the Characterization of Carbon Monoxide Adsorbed on Alumina Supported Rhodium . . . 182
3.1. Sample Preparation and Morphology . . . 182
3.1.1. High-Surface-Area Samples . . . 182
3.1.2. Low-Surface-Area Samples . . . 184
3.2. Vibrational Spectroscopic Analysis . . . 185
3.3. ^{13}C Nuclear Magnetic Resonance Studies . . . 191
3.4. Adsorbate Structure and Bonding from Studies of Model Systems . . 192
4. Conclusions . . . 194
References . . . 195

7. The Detection and Identification of Biochemicals 201
Robert V. Coleman

1. Introduction 201
2. IET Spectra of Biological Compounds 203
   2.1. Amino Acids 203
   2.2. Pyrimidine and Purine Bases 209
   2.3. Nucleotides and Nucleosides 214
3. Surface Adsorption and Orientation Effects on the IETS of Nucleotides 221
4. uv Radiation Damage Studies with IETS 223
5. Conclusions 226
   References 227

8. The Study of Inorganic Ions 229
K. W. Hipps and Ursula Mazur

1. Introduction 229
2. Why Study Inorganic Ions by Tunneling Spectroscopy? 230
   2.1. Direct Observation of Transitions Forbidden in Photon Spectroscopy 231
      2.1.1. Vibrational Transitions 231
      2.1.2. Electronic Transitions 238
   2.2. Impregnation Catalysts 241
   2.3. Speciation of Metal Ions in Natural Waters 243
3. Doping Techniques and Insulator Surfaces 245
   3.1. Solution Phase Doping of Alumina Barriers 245
   3.2. AlO_x and MgO Supported O_ySiH_x Barriers 248
4. Solution Phase versus Gas Phase Adsorption 250
5. Representative Spectra 252
   5.1. Metal Cyanide Complexes 252
   5.2. Metal Glycinates 257
   5.3. Other Inorganic Systems 259
6. The Role of Counterions 262
7. Oxidation and Reduction Processes 263
8. What's Next? 266
9. Conclusions 267
   References 267

9. Studies of Electron-Irradiation-Induced Changes to Monomolecular Structure 271
M. Parikh

1. Introduction 271
   1.1. Why Study Irradiation-Induced Molecular Structure Changes? 271
   1.2. Why Use Tunneling Spectroscopy? 272
   1.3. Scope of this Chapter 272

2. Present State-of-the-Art Experiments 273
2.1. Electron Irradiation Experiments 273
2.2. Underlying Assumptions 275
2.3. Determination of "Damage" Cross-Sections 276
2.4. General Trends 278
3. Suggestions for Future Experiments 280
3.1. Review of Zeroth-Order Experiments 280
3.2. First-Order Experiments 281
3.3. Second-Order Experiments 282
4. Conclusions 284
References 284

10. Study of Corrosion and Corrosion Inhibitor Species on Aluminum Surfaces 287
Henry W. White

1. Introduction 287
1.1. General Remarks 287
1.2. Corrosion of Aluminum in Organic Media 288
1.3. Corrosion by Chlorinated Hydrocarbons 290
1.4. Corrosion Inhibitors for Aluminum in Chlorinated Solvents 291
1.5. Corrosion and Inhibitor Surface Species 291
2. Corrosion of Aluminum by Carbon Tetrachloride 293
2.1. Proposed Reactions 293
2.2. Tunneling Spectroscopy Studies 293
2.2.1. Experimental Procedure 293
2.2.2. Surface Species 294
3. Inhibition of Corrosion by Formamide 296
3.1. Surface Species 296
3.2. Inhibition Mechanism 299
4. Corrosion of Aluminum by Trichloroethylene 300
4.1. Reaction with Aluminum 300
4.2. Surface Species and Reactions 300
4.3. Corrosion Mechanism 304
5. Corrosion Inhibitors for Aluminum in Hydrochloric Acid 304
5.1. Acridine Surface Species 304
5.2. Orientation of Thiourea on Aluminum Oxide 306
6. Conclusions 307
References 308

11. Adsorption and Reaction on Aluminum and Magnesium Oxides 311
D. G. Walmsley and W. J. Nelson

1. Introduction 311
2. Clean Aluminum Oxide 311
3. Dirty Aluminum Oxide 313

4. Doped Aluminum Oxide 315
4.1. Formic Acid 316
4.2. Acetic Acid and Closely Related Molecules 318
4.3. Higher Acids 321
4.4. Unsaturated Acids 326
4.5. Unsaturated Hydrocarbons 326
4.6. Phenols 331
4.7. Aromatic Alcohols and Amines 333
4.8. Bifunctional Molecular Species 338
4.9. Chemical Mixtures 339
5. Clean Magnesium Oxide 341
6. Doped Magnesium Oxide 343
6.1. Benzaldehyde 343
6.2. Formic, Acetic, and Propionic Acids .344
6.3. Phenol 348
6.4. Carboxylate Mode Shift 349
6.5. Benzyl Alcohol 349
6.6. Unsaturated Hydrocarbons 350
6.7. Diketone 353
7. Technical Postscript 353
8. Conclusions 355
References 356

12. The Structure and Catalytic Reactivity of Supported Homogeneous Cluster Compounds 359

W. Henry Weinberg

1. Introduction 359
2. Experimental Procedures 361
3. Results and Discussion 363
3.1. $Zr(BH_4)_4$ on Al_2O_3 at 300 K 363
3.2. $Zr(BH_4)_4$ on Al_2O_3 at 475 K 367
3.3. The Interaction of $Zr(BH_4)_4$ on Al_2O_3 with D_2, D_2O, and H_2O . . .368
3.4. The Interaction of $Zr(BH_4)_4$ on Al_2O_3 with C_2H_4, C_3H_6, and C_2H_2 368
3.5. The Interaction of $Zr(BH_4)_4$ on Al_2O_3 with Cyclohexene, 1,3-Cyclohexadiene and Benzene 373
3.6. $Ru_3(CO)_{12}$ on Al_2O_3 375
3.7. $[RhCl(CO)_2]_2$ on Al_2O_3 379
3.8. $Fe_3(CO)_{12}$ on Al_2O_3 383
4. Conclusions 386
References 388

13. Model Supported Metal Catalysts 393

R. M. Kroeker

1. Introduction 393

2. Special Techniques 394
3. Experimental Results 397
3.1. Carbon Monoxide on Rhodium 397
3.1.1. Chemisorption of CO on Rhodium 397
3.1.2. Hydrogenation of CO on Rhodium 402
3.2. Carbon Monoxide on Iron 405
3.3. Carbon Monoxide and Hydrogen on Nickel 407
3.4. Carbon Monoxide on Cobalt 412
3.5. Ethanol on Silver 415
4. Future Areas of Study 415
4.1. Acetylene on Palladium 415
4.2. Carbon Monoxide on Ruthenium 418
4.3. Carbon Monoxide on Platinum 418
4.4. Other Molecules; Other Reactions 419
4.5. Low-Temperature Adsorption 419
5. Conclusions 420
References 421

14. Computer-Assisted Determination of Peak Profiles, Intensities, and Positions 423

J. G. Adler

1. Introduction 423
2. Measurement of Tunneling Conductance and Its Derivatives 424
2.1. Modulation Spectroscopy 424
2.2. A Survey of Measuring Circuits 426
2.3. Calibration of Tunnel Conductance and Its Derivatives 428
3. Interfacing with a Computer 430
3.1. General Considerations 430
3.2. Analog-to-Digital Conversion 430
3.3. Digital Data Transmission from Analog Instrumentation 431
3.3.1. IEEE 488 Standard Interface 431
3.3.2. The BCD Interface 432
3.3.3. The Serial Interface (RS-232C) 433
3.3.4. The Parallel Interface 433
4. Peak Profile Determination 434
4.1. General Remarks 434
4.2. Factors Affecting Peak Profile 435
4.3. Peak Profiles of Junctions with Composite Barriers 441
4.4. Peak Intensities 444
4.5. Peak Positions 445
5. Data Handling 445
5.1. General Comments 445
5.2. Data Calibration 446
5.3. Data Storage 447
5.4. Data Analysis 448
References 449

15. Infusion Doping of Tunnel Junctions **451**

R. C. Jaklevic

1. Introduction 451
   1.1. Doping Requirements 451
   1.2. Review of Other Doping Methods 452
        1.2.1. Vapor Phase Doping 452
        1.2.2. Liquid Phase Doping 453
        1.2.3. Infusion Doping 453
2. Experimental Description of Infusion 454
   2.1. Junction and Film Preparation 454
   2.2. Infusion Techniques 454
   2.3. Infusion Monitoring—Resistance and Capacitance 456
3. Experiments Relating to Physical Mechanisms of Infusion 457
   3.1. Resistance and Capacitance Behavior 457
   3.2. Film Porosity 458
   3.3. Water Infusion and Organic Molecules 459
   3.4. Masking Experiments 461
   3.5. Sn and Au Overlay Films 462
4. Examples of Molecules Infused 462
   4.1. Acids and Bases 463
   4.2. Solvents and Alcohols 465
   4.3. Solid Phase Molecules 467
5. Applications of Infusion 468
   5.1. Hydrogenation and Deuteration of Propiolic Acid 468
   5.2. Solid-State Anodization of Aluminum 469
   5.3. Other Applications 470
6. Conclusions 471

   References 472

16. Vibrational Spectroscopy of Subnanogram Samples with Tunneling Spectroscopy **475**

P. K. Hansma and H. G. Hansma

References 481

1

Introduction

Paul K. Hansma

1. Why? Why? Why?

Tunneling spectroscopy is a sensitive technique for measuring the *vibrational spectra* of molecules. It was discovered by Jaklevic and Lambe[1,2] in 1966.

1.1. Why Do Vibrational Spectroscopy?

Vibrational spectroscopy is a powerful way of identifying molecules and molecular fragments. Since a functional group, say $—CH_3$, has roughly the same vibrational frequencies wherever it appears in a molecule, researchers can deduce the presence or absence of $—CH_3$ by the presence or absence of vibrations at its characteristic frequencies. Table 1 shows the characteristic frequencies of many functional groups.

From this knowledge of the presence or absence of functional groups, together with whatever other information is available, researchers can guess the structure of their unknown molecules. After they guess, they can use vibrational spectroscopy to see if they are right or wrong.

They do this by comparing the vibrational spectrum of their unknown molecule to the vibrational spectrum of what they guess it is. There are many excellent, extensive collections of vibrational spectra[3–7] for use in this comparison. If the spectrum of the guessed molecule is not in an accessible collection, the researcher can consult the literature or, if worse comes to worst, measure it himself.

Paul K. Hansma • Department of Physics, University of California, Santa Barbara, California 93106.

Table 1. Spectra-Structure Correlations. (Probable Positions of Characteristic Infrared Absorption Bands)[a]

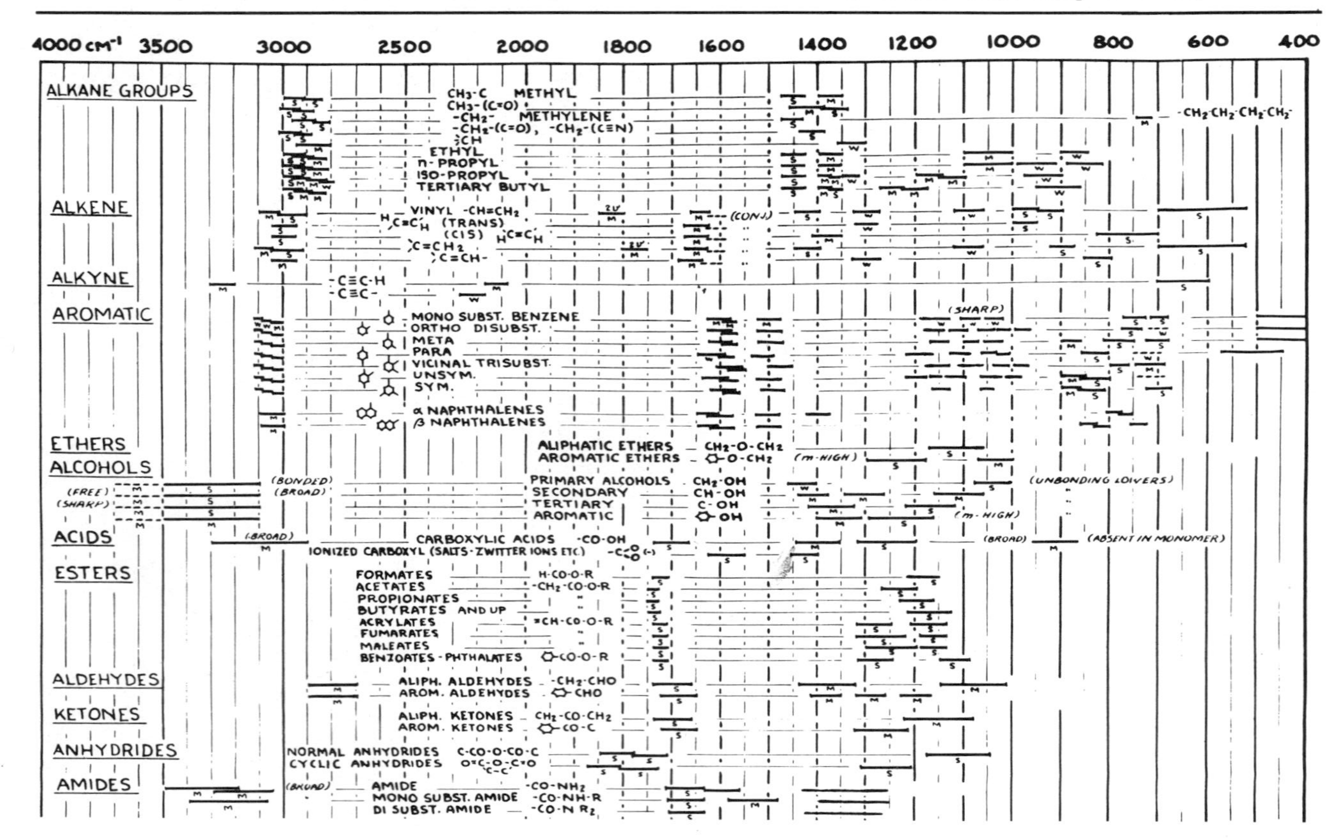

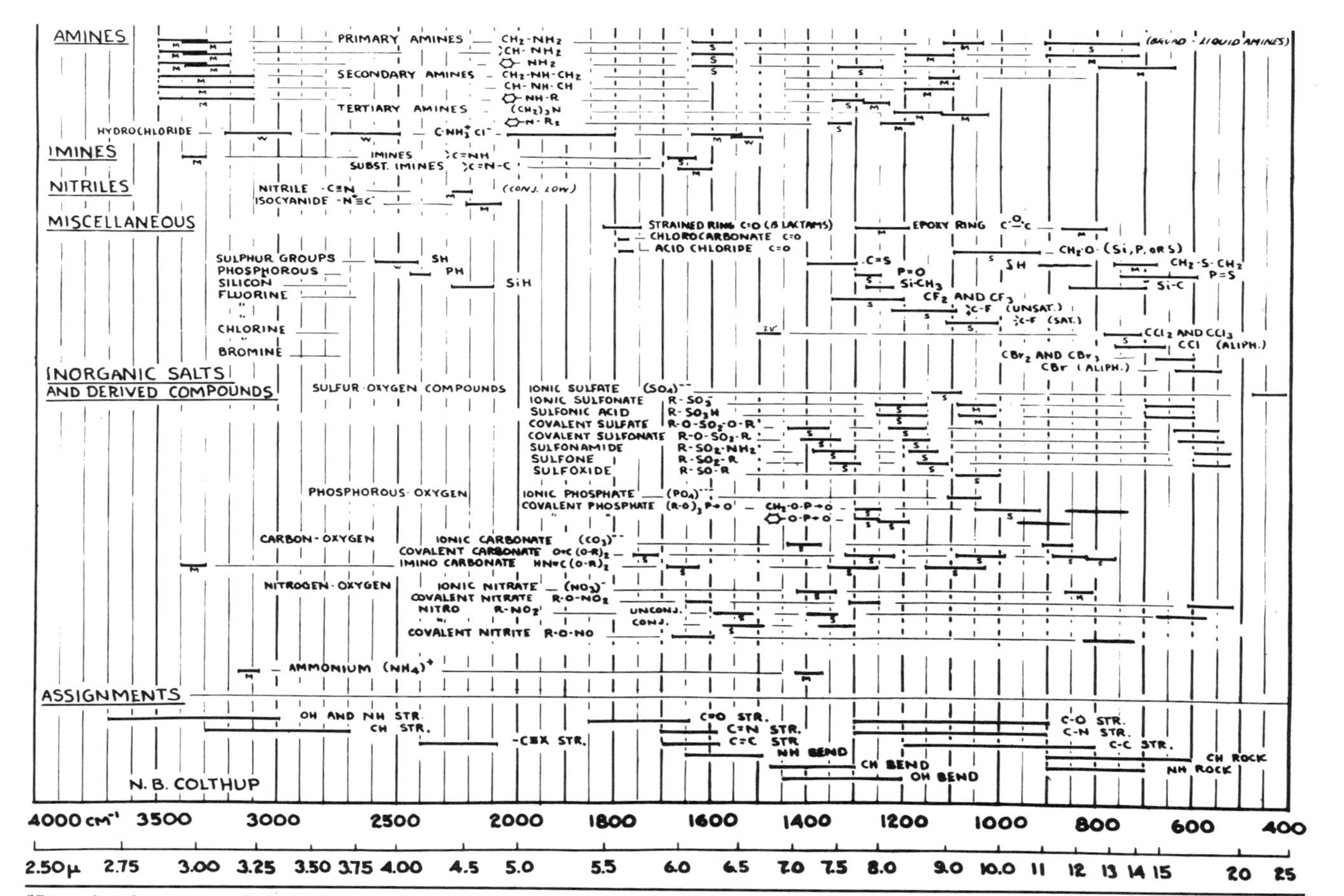

[a]Reproduced, courtesy of N. B. Colthup, Stamford Research Laboratories, American Cyanamid Company, and the editor of the *Journal of the Optical Society of America*.

It is worth noting that even if the first guess is wrong, the collections of spectra can be of great help. Usually they are grouped with similar molecules near each other. Thus, if the first guess is wrong, the correct spectrum may be nearby.

1.2. Why Do Tunneling Spectroscopy?

Section 3 will show that tunneling spectroscopy has a unique combination of spectral range, sensitivity, resolution, and selection rules. It is specially suited to the vibrational spectroscopy of adsorbed monolayers on surfaces.

1.3. Why Not Do Infrared, Raman, or Electron Energy Loss Spectroscopy Instead?

Why not indeed. For most applications these are the spectroscopies of choice. For bulk samples, the resolution and availability of commercial apparatus make infrared or Raman best. For surface absorbed species on real, high-surface-area catalysts, infrared and Raman spectroscopy have problems, but tunneling spectroscopy will not work at all. For surface adsorbed species on single-crystal surfaces, electron-energy-loss spectroscopy or, possibly, advanced infrared or Raman techniques are best.

But, as will be shown in Chapters 7–13, tunneling spectroscopy excels for some important tasks; for example, obtaining complete vibrational spectra of molecules and complexes on alumina or magnesia or on supported metal particles.

Furthermore, the problem of identifying adsorbed surface species is sufficiently complex that any technique that can give additional information is of value. In this way, tunneling spectroscopy complements not only the other vibrational spectroscopies but other surface analytical techniques as well. Chapter 6 discusses this point in depth.

2. A Water Analogy for Tunneling Spectroscopy

2.1. Water Flow

Consider the tube shown in Figure 1. It has two open channels separated by a height h. If we measure the steady-state flow as a function of pressure, we will find two components: (1) a steadily increasing flow through the bottom channel and (2) a flow through the top channel, which has a threshold pressure ρgh, and steadily increases thereafter.

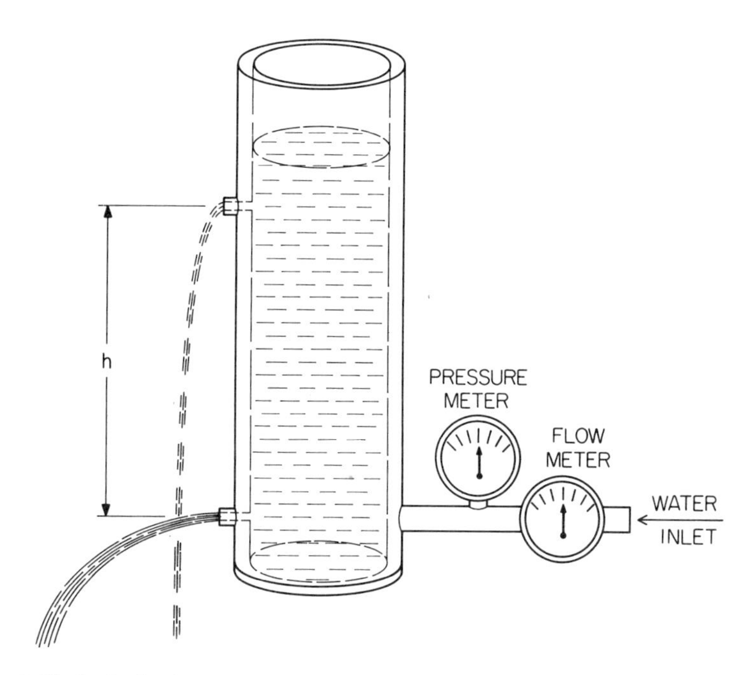

Figure 1. The basic physics behind tunneling spectroscopy can be understood with the help of a water analogy.

Thus, the total flow F has a kink at the characteristic pressure $P = \rho gh$ as shown in Figure 2. This kink might be difficult to observe. It would be easier to see the step in dF/dP and even easier to see the peak in d^2F/dP^2 if we could find a way to measure them.

A key point is that, even if the tube were covered by a basket, we could determine the height of the opening by looking at the d^2F/dP^2 versus P curve: an opening at height h is revealed by a peak at pressure $P = \rho gh$.

2.2. Tunneling

Figure 3 shows an idealized view of a tunnel junction. Here, we measure the current as a function of voltage. Again, we will find two components: (1) a steadily increasing current due to *elastic* electron tunneling and (2) a current, which has a threshold voltage $h\nu/e$ and increases steadily thereafter, due to *inelastic* electron tunneling. This threshold is set by the requirement that the electrons must give up an energy $h\nu$ to excite the molecular vibration. Since their tunneling energy is eV, we must have $eV \gtrsim h\nu$.

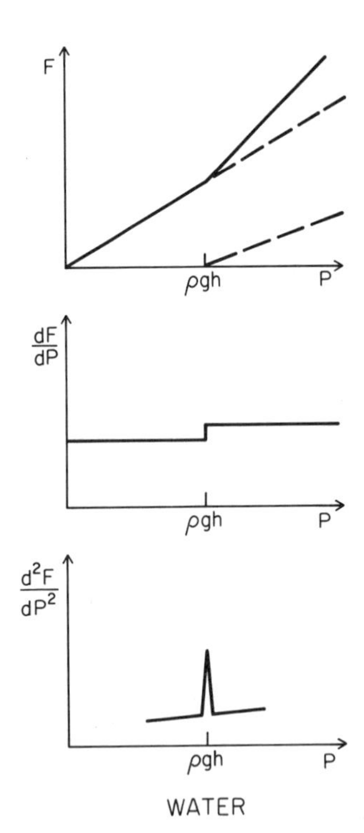

Figure 2. The flow-versus-pressure curve has a kink when it becomes possible for water to flow out of the upper channel. This kink becomes a step in the first derivative and a peak in the second derivative.

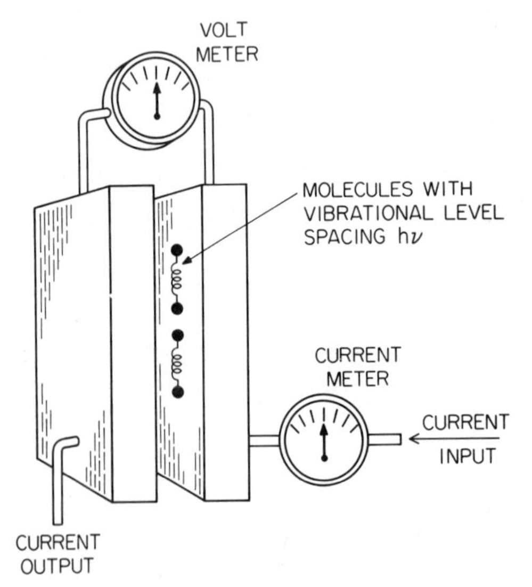

Figure 3. In this schematic view of a tunneling junction, the molecules with vibrational level spacing $h\nu$ are sandwiched between two metal electrodes. In an actual tunnel junction the metal electrodes would be separated by less than 100 Å.

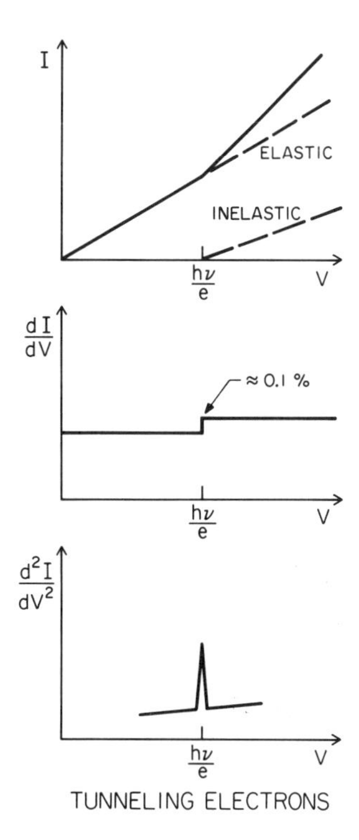

Figure 4. Current versus voltage has a kink when the inelastic electron tunneling channel opens up. This kink becomes a step in the first derivative and a peak in the second derivative.

Figure 4 shows the total current I, which is the sum of the current through the elastic and inelastic tunneling channels. It has a kink at $V = h\nu/e$ which becomes a step in dI/dV versus V and a peak in d^2I/dV^2, just as for the water analogy. A plot of d^2I/dV^2 versus V* is called a tunneling spectrum.

Thus, a tunneling spectrum reveals the vibrational energies of molecules included between the metal electrodes, since a vibrational energy of $h\nu$ results in a peak at $V = h\nu/e$. A real tunneling spectrum has many peaks since the typical molecules that are studied have many vibrational modes. For example, Figure 5 shows the tunneling spectrum of a monolayer of benzoate ions on alumina. Each of the sharp peaks in this spectrum corresponds to the opening of an inelastic electron tunneling channel. Each channel corresponds to a particular molecular vibration.

*More generally, d^2V/dI^2 is actually measured. It has peaks in the same locations since $|d^2V/dI^2| = |d^2I/dV^2(dV/dI)^3|$ and $(dV/dI)^3$ is a smooth, relatively slowly varying junction.

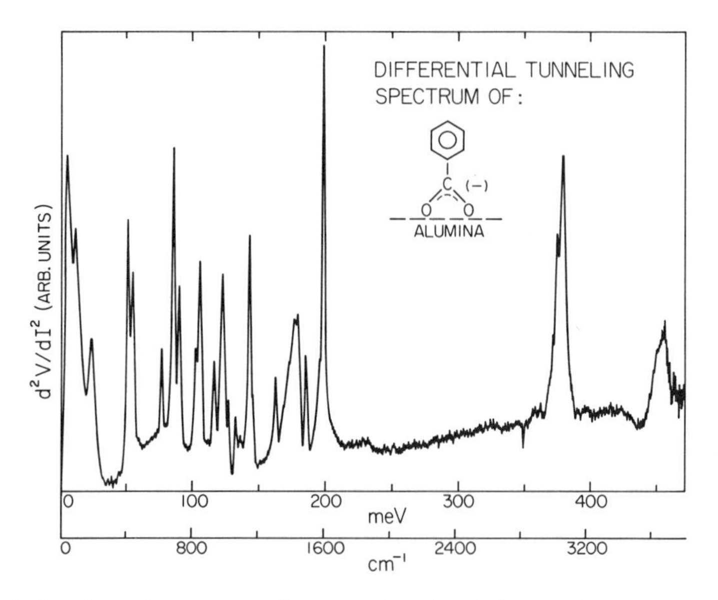

Figure 5. A real tunneling spectrum has a number of peaks because there are a number of vibrations in the typical organic molecule. In this case, we seek peaks corresponding to all of the vibrations of the benzoate ion. This spectrum resembles and can be interpreted like an infrared-absorption-versus-energy or Raman-intensity-versus-energy spectrum.

3. Strengths of Tunneling Spectroscopy

3.1. Spectral Range

The spectral range of tunneling spectroscopy includes all molecular vibrations. It extends from 0 to beyond 4000 cm^{-1}. For example, the spectrum shown in Figure 5 includes all the molecular vibrations of benzoate ions on alumina. The easiest part of this range to work in is the part from 400 to 4000 cm^{-1}. This is consequently the most common range for published tunneling spectra: it is the spectral range covered in most of the spectra in this book.

Tunneling spectra that include the range below 400 cm^{-1} are shown in Figure 6. The problem with working below 400 cm^{-1} is that tunnel junctions for tunneling spectroscopy are usually made with a lead top electrode and run at temperatures for which lead is superconducting. The structure in a tunneling spectrum due to phonons of lead is very intense relative to the structures due to molecular vibrations.

Structure due to the electron–phonon interaction has, of course, been a very interesting study in itself and has contributed a great deal of funda-

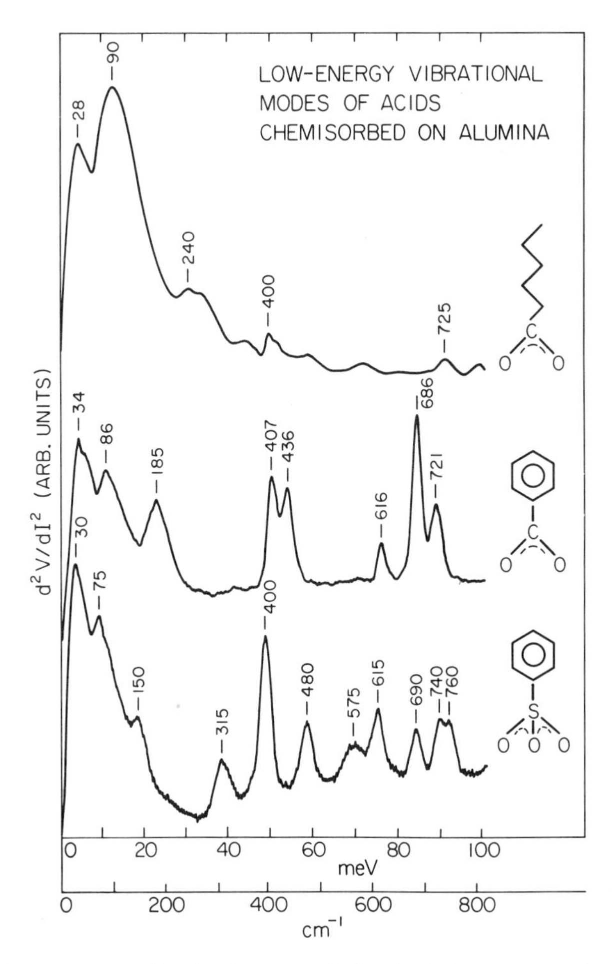

Figure 6. The low-energy vibrational modes of chemisorbed species can be studied with tunneling spectroscopy. These modes, which are inaccessible to most vibrational spectroscopies, remain largely unexplored.

mental knowledge about the nature of the superconducting state, as will be discussed in more detail in Chapter 3. This wonderful structure is, however, a nuisance if one wants to see the molecular vibrations in this range.

There are two ways to minimize this structure. The first is to quench the superconductivity in the lead electrode with a magnetic field. Fortunately, a field of only about 700 G is required. This field can be supplied with permanent magnets. Thus, in our laboratory, to quench the field due to the superconductivity of the lead electrode, we tape onto the back of the slide a small ($1/8 \times 1/8 \times 1/2$-in) samarium–cobalt magnet.

The second way to minimize the structure is to plot a difference spectrum between a doped and an undoped junction.[8, 9] It turns out that both of these methods must be used together to see the structure below 400 cm^{-1}. For even when the superconductivity of the lead electrode is quenched, the metallic phonon structure is comparable to or a little stronger than the structure due to molecular vibrations.[8]

Thus, to study the region below 400 cm^{-1}, one can plot a differential tunneling spectrum with the superconductivity of the lead electrode quenched by a permanent magnet. This is not particularly difficult, but it has probably not been done more because of the difficulty in interpreting the features found below 400 cm^{-1}. For example, the features in the three spectra shown in Figure 6 have not yet been interpreted. Part of the reason for this is that the typical spectral range of infrared spectrometers is 400–4000 wave numbers (by a happy coincidence). Thus, a great deal of work has been done on the assignment of features in this spectral range and we can make assignments in tunneling spectroscopy by analogy. Much less work has been done in the range below 400 cm^{-1}, especially for chemisorbed molecules on surfaces.

The range above 4000 wave numbers can also be studied. In fact, as will be seen in Chapter 4, tunneling spectra have been taken out to beyond 16,000 cm^{-1} (beyond two electron volts). In this range electronic transitions can be seen.

The main point is that tunneling spectroscopy can cover a wide range or, with a little extra work, the entire range of molecular vibrational frequencies. This becomes more impressive when one realizes that this range is being covered for a single monolayer over a small area sample, which brings us to the next section.

3.2. Sensitivity

The sensitivity of tunneling spectroscopy is sufficient to see a fraction of a monolayer of material over a junction area less than 1 mm on a side. In fact, almost all of the tunneling spectra in this book are taken for approximate monolayer coverage: one monolayer is just about the right amount of stuff for a tunneling spectrum.

How is it known that most of the tunneling spectra in this book are for one monolayer? Because most of the spectra turn out to be the spectra of adsorbed molecules on alumina. This alumina is the barrier of the tunnel junction, as will be discussed later. There is only room for one monolayer of such adsorbed species. Above that monolayer could be unreacted molecules, but these are typically not observed.

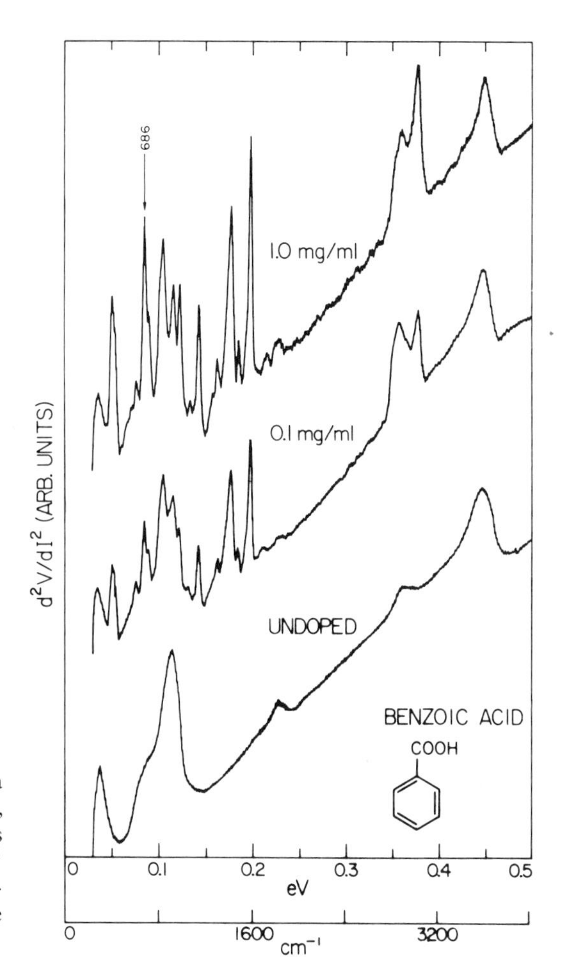

Figure 7. As the concentration of a liquid doping solution is increased, the intensity of the tunneling peaks increases. The peak at 686 cm^{-1} was chosen to study this dependence, because it is in a region free from background interference.

A more direct way of determining what the coverage is, is by radioisotope studies[10, 11] such as the one illustrated in Figure 7. In this experiment, a tunnel junction was doped with benzoic acid molecules by dropping onto the oxidized aluminum aqueous solutions of benzoic acid of various concentrations (a more detailed discussion of experimental techniques in tunneling spectroscopy will be given in Section 5 of this chapter). Note that, as could be expected, as the solution concentration was decreased, the intensity of the peaks in the tunneling spectra decreased, indicating that there is less stuff on the surface. Figure 8 shows the intensity of a particular peak in the tunneling spectra as a function of solution concentration. Again, we see the peak height decreasing with the solution concentration. We also see, however, a saturation of peak height above solution concentrations of order 0.2 mg/ml. This saturation can be understood with the aid of the other curve in Figure 8. This is a curve of the actual surface concentration that results from doping with given solution concentrations.

The surface concentration was measured by doping oxidized aluminum strips with radioactively labeled benzoic acid and then counting the strips in a scintillation counter to determine the absolute number of molecules on the surface. As can be seen, the saturation in the tunneling peak intensity correlates with the saturation in the surface coverage. That saturation in surface coverage occurs at approximately 1 molecule/15^2 Å: monolayer coverage.

In this experiment, satisfactory tunneling spectra were observed down to approximately 3% of a monolayer. No particular care, however, was

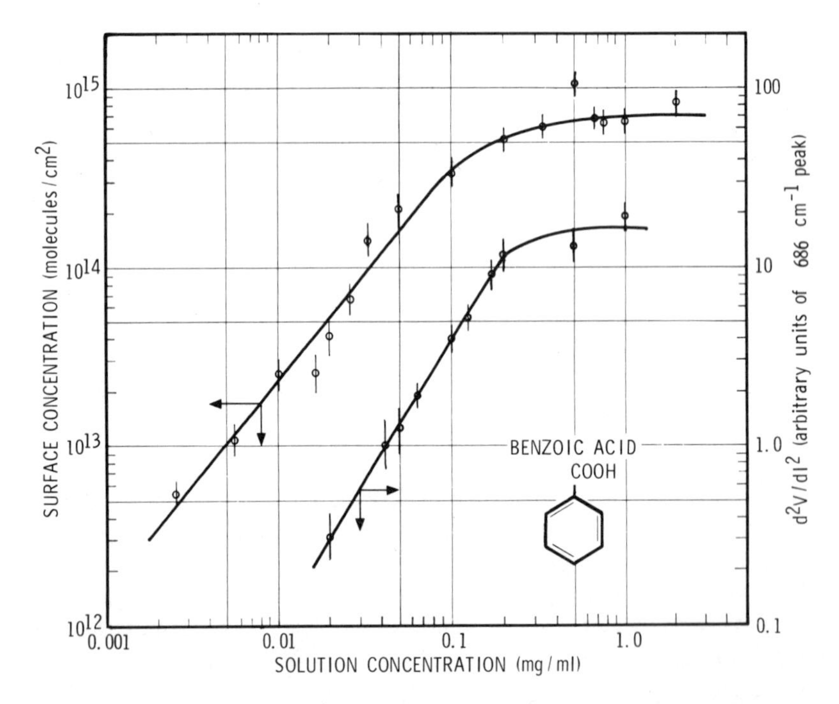

Figure 8. Intensity of the 686-cm^{-1} peak of the previous spectrum is shown as a function of a solution concentration. Also shown is the surface concentration as a function of solution concentration. Note that the tunneling peak height increases and the saturates following the increase and then saturation of the surface concentration at approximately one molecule per 15 Å^2: monolayer coverage.

taken to determine the lower limit to sensitivity. A more recent experiment was aimed specifically at determining the sensitivity.[12] In this experiment, a small quantity of *p*-deutero benzoic acid was mixed with benzoic acid and the resultant mixture was doped onto the tunneling junction to give a monolayer. Tunneling spectra were then measured as the *p*-deutero benzoic acid was progressively diluted in order to determine the lower limit at which the *p*-deutero acid could be detected.

Figure 9 shows the tunneling spectrum of benzoic acid and *p*-deutero benzoic acid. Note that as expected the *p*-deutero benzoic acid has a characteristic C–D stretching peak near 2300 cm^{-1}. Figure 10 shows a blowup on both the *x* and *y* axis of that region of mixtures of *p*-deutero benzoic acid and benzoic acid of various concentrations. Note that the peak due to *p*-deutero benzoic acid can still be observed for a solution concentration of 1%. The *p*-deutero benzoic acid and benzoic acid should be roughly equal in their competition for surface sites. Thus, it is reasonable to assume that a 1% solution of *p*-deutero benzoic acid and benzoic acid gives a

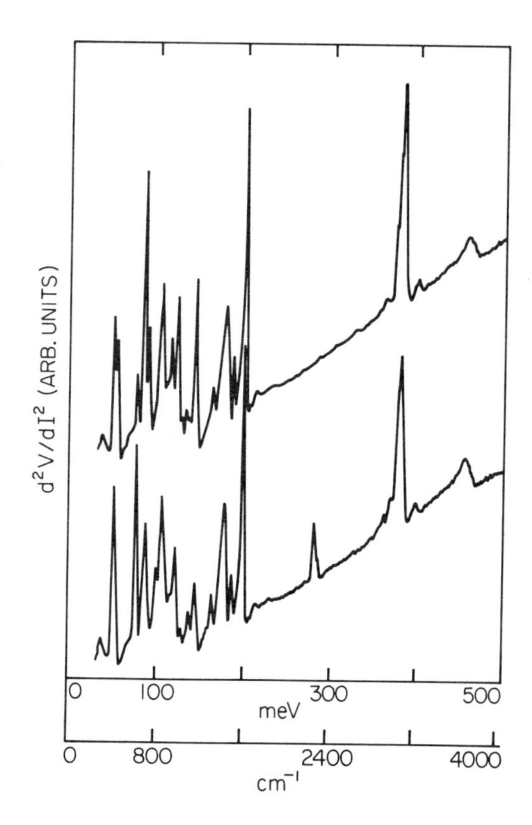

Figure 9. The upper trace is the tunneling spectrum of benzoic acid, the lower trace that of *p*-deutero benzoic acid. Richard Kroeker made mixtures of these to measure the sensitivity of tunneling spectroscopy. See reference 12.

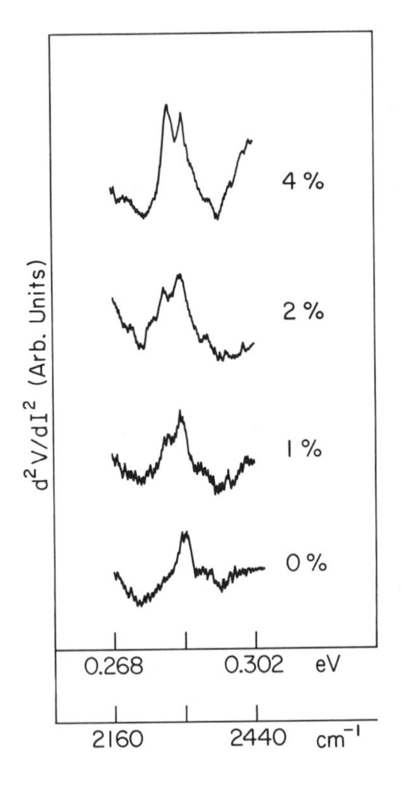

Figure 10. This is a blowup of the region around 2300 cm^{-1} of the spectrum for dilute mixtures of *p*-deutero benzoic acid in benzoic acid. Note that the characteristic C–D stretching peak near 2300 wave numbers can still be detected for 1% *p*-deutero benzoic in benzoic acid. This corresponds to detecting one deuterium atom per 1500 Å^2.

surface concentration of 1% *p*-deutero benzoic acid in a benzoic acid monolayer. Hence, this experiment detected 1% of a monolayer.

This number becomes more impressive when one realizes that only one of the hydrogen atoms in the benzoic acid was replaced with deuterium and that each benzoic acid molecule occupies 15 Å^2. Thus, in this experiment, one deuterium atom/1500 Å^2 was detected in a sample less than 1 mm on a side!

3.3. Resolution

Vibrational modes of surface species have characteristic widths of order 1 meV (8 cm^{-1}) or greater even when measured with vibrational spectroscopies with intrinsically high resolution such as Fourier transform infrared spectroscopy. In addition to this characteristic width, tunneling spectra contain two broadening contributions: modulation voltage broadening and thermal broadening. Figure 11 shows the peaks that would be found in a tunneling spectrum for a delta function vibrational mode (that is a vibra-

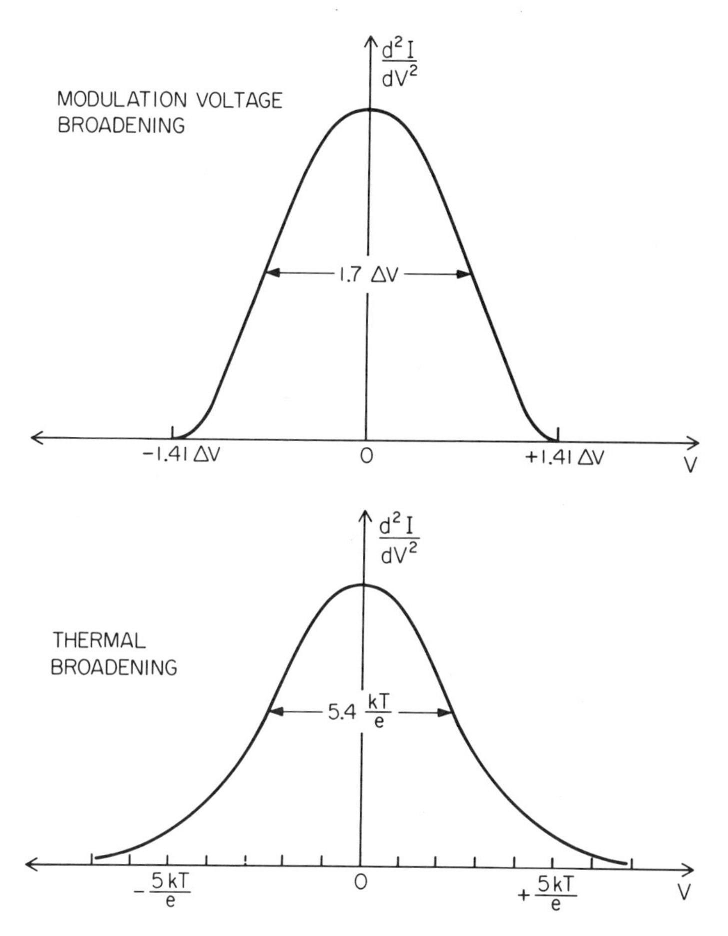

Figure 11. If there were a vibrational mode with negligible intrinsic width, in a tunneling spectrum it would appear to have width from two contributions: (1) modulation voltage broadening that results from the modulation voltage applied to measure the derivative and (2) thermal broadening that results from the smearing of the electrons around the Fermi energy at nonzero temperature.

tional mode with negligible intrinsic width). The modulation voltage broadening comes from the modulation current that is applied to measure the second derivative. The thermal broadening comes from the thermal smearing of the electron distribution in each metal electrode around the Fermi energy; the maximum energy of tunneling electrons is not exactly eV (V is the applied voltage). Table 2 shows some practical numbers for these contributions.

Table 2. The Effect of Temperature and Modulation Voltage on the Resolution, Peak Shift, and Time to Trace a Tunneling Spectrum[a]

		Resolution		Trace time			Peak shift		Relative trace time
Temp.	Modulation	meV	cm^{-1}	1 sec	3 sec	10 sec	meV	cm^{-1}	(same signal noise)
4.2	2	3.9	32	15 min	40 min	2 hr	0.7	6	1
	1	2.6	21	20 min	1 hr	3 hr	0.9	7	25
	0.7	2.1	17	25 min	$1\frac{1}{4}$ hr	$3\frac{1}{2}$ hr	1.0	8	125
1	2	3.4	28	15 min	40 min	2 hr	0.7	6	1.1
	1	1.8	14	30 min	$1\frac{1}{2}$ hr	5 hr	0.9	7	35
	0.7	1.3	10	40 min	2 hr	6 hr	1.0	8	200

[a]At commonly used temperatures and modulation voltages, the resolution of tunneling spectroscopy varies from 10 to 32 cm^{-1}. The time to take a complete tunneling spectrum varies depending on the time constant of the lock-in amplifier that is used to measure the second-harmonic signal. Superconductivity in the top lead electrode makes the actual resolution slightly better than listed in this table. It also, however, shifts the peak up by a small amount. This peak shift, which varies from 6 to 8 cm^{-1}, under the conditions of this table should be subtracted from measured peak positions before comparison with infrared and Raman spectra. Tunneling spectra are generally measured at the poorest resolution sufficient to obtain the desired information, because the price paid in relative trace times for higher resolution is high.

Thus, the resolution in the tunneling spectra presented in this book typically ranges between 1 and 4 meV (10 and 32 cm^{-1}). This is not, however, the maximum resolution that can be achieved. Professor Walmsley and co-workers, by using a helium dilution refrigerator and small modulation voltages, have been able to set the current world record[(13)] in the resolution of tunneling spectroscopy at roughly 2 cm^{-1}. This high resolution is not commonly used (even by Professor Walmsley) because the relative trace time, that is the time to obtain a spectrum with a same signal-to-noise ratio rises rapidly as the resolution improves, as is also shown in Table 2. Thus, most spectra are taken at the poorest resolution sufficient to obtain the information desired. Fortunately, since the vibrational modes of surface adsorbed molecules typically have an intrinsic width of a meV or greater, the ultimate in resolution is not required.

The relative trace time for same signal-to-noise ratio rises so rapidly because the second harmonic signal that is proportional to d^2V/dI^2 is also proportional to the square of the modulation voltage. Thus, as the modulation voltage is decreased in an attempt to get better resolution, the signal falls as the square of the modulation voltage. If this were not bad enough, the improvement in signal to noise that can be obtained by averaging with a time constant on a lock-in amplifier increases only as the square root of the averaging time. Thus, if a signal started out at a strength of one and we cut the modulation voltage by a factor of 2, the signal would fall to 1/4. In order to quiet the noise down by a factor of 4 to obtain the same signal-to-noise ratio would require a time constant 16 times longer. Things get even a little worse, because, in order to obtain a spectrum that is not blurred by too rapid a sweep, it is a good rule of thumb to spend greater than six time constants in each resolution element of the spectrum. As the resolution becomes better, there are more resolution elements per tunneling spectrum. In short, there is a big price to pay for improvements in resolution as summarized in Table 2.

The trace times, shown in this table, give an idea of how long it takes to plot a complete tunneling spectrum with the given time constants on the lock-in amplifier that measures the second-harmonic voltage. These times are based on spending six time constants per resolution element. In our laboratory we have found these to be lower limits on acceptable trace times. If traces are swept faster than these times, shifts in peaks will be observed.

The table also includes peak shifts due to superconductivity in the top metal electrode. These will be discussed later.

3.4. Selection Rules

Basically there are no strong selection rules in tunneling spectroscopy. Selection rules are the general consequence of symmetry. In the case of optical spectroscopy, selection rules arise because the wavelength of light is very long relative to the size of molecules. Thus, the electric field of the light is uniform over the size of molecules. This is not true for tunneling spectroscopy.

It is a commonly shared belief among tunneling spectroscopists that Raman and infrared modes appear with comparable intensities in tunneling spectra. It would be highly desirable to have a tunneling spectrum of a simple molecule that had separate infrared and Raman modes.[14] Unfortunately, the chemisorbed molecules, which are typically studied by tunneling spectroscopy, do not typically have enough symmetry to have separate infrared and Raman modes. Simple molecules with a lot of symmetry such as benzene present real problems to the tunneling spectroscopist, because they will leave an aluminum-oxide surface at room temperature. N. I. Bogatina *et al.*[15] have obtained some results using a clever liquid doping technique with cooled substrates. It would be desirable to obtain a spectrum by vapor doping onto a cooled substrate that could be kept cold until measurements were complete; such an *in situ* tunneling spectrometer would also open many other possibilities. For example, the pioneering work of O. I. Shklyarevskii *et al.*[16] on the transition from physical adsorption to chemisorption of carboxylic acids could be extended to many other systems.

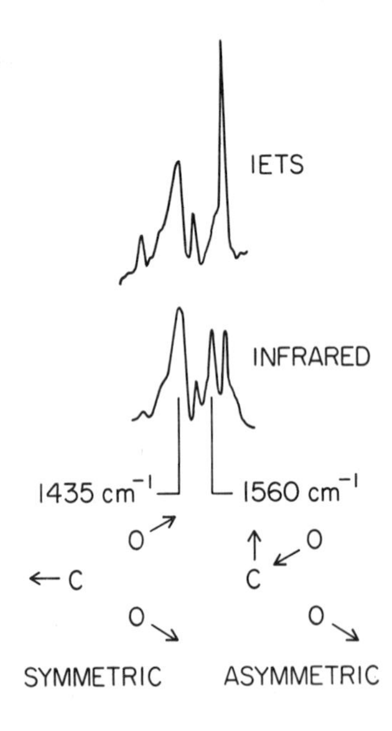

Figure 12. The asymmetric carboxylate group vibration is weak in tunneling spectra, perhaps because of the orientation effect. It involves an oscillating dipole moment parallel to the oxide surface.

Some of the best information available at present comes from the work of Hipps and Mazur on inorganic ions. See, for example, their discussion of $K_4Fe(CN)_6$ data that is presented in Figure 9 of Chapter 8. This type of data is also beginning to answer questions about the coupling of tunneling electrons to vibrational modes that are neither Raman nor infrared active. Theory predicts that these modes should be observed.[17]

There is an orientational selection preference. Theory[17] and experiment agree that tunneling electrons interact somewhat more strongly with vibrational modes that involve oscillating dipoles parallel to the direction of electron flow (perpendicular to the surface plane). For example, Figure 12 shows the tunneling and infrared spectra of the spectral range that includes the COO^- vibrations for a carboxylic ion on alumina.[15] In this case, the symmetric vibration of the COO^- ion is perpendicular to the surface while the antisymmetric vibration is parallel to the surface. Note that though these occur with comparable intensities in the infrared spectra, in the tunneling spectra the symmetric mode is much stronger.* This same trend is followed for all adsorbed carboxylic acids in tunneling spectra.[17–19]

*Though, as noted by Walmsley,[18] the peak associated with the symmetric mode probably also contains some contribution from CH vibrations. The shape certainly indicates an unresolved multiple peak.

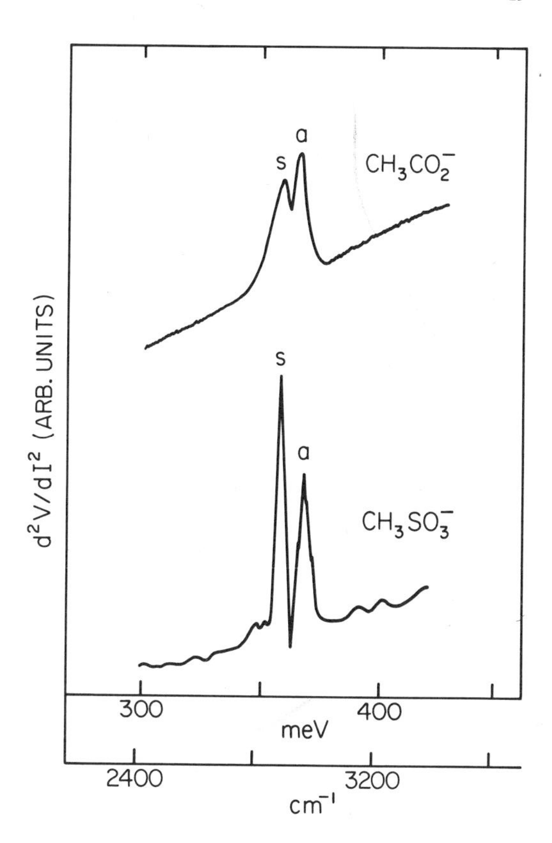

Figure 13. Note that the relative intensity of the symmetric and antisymmetric methyl group vibrations changes if the methyl group is attached to aluminum oxide by a carboxylate ion or a sulfonate ion.

Another example of this effect can be seen in Figure 13. Careful experiments by Jim Hall showed that the SO_3^- ion sits like a tripod on the oxide surface.[20] Thus, if a CH_3 group is attached to the SO_3^- tripod, the CH_3 symmetric vibration will be perpendicular to the oxide while the CH_3 antisymmetric vibration will be parallel to the oxide. Inspection of Figure 13 shows that indeed the symmetric vibration is more intense than the antisymmetric vibration even though the antisymmetric vibration is degenerate.

However, as noted by Jim Hall, for the same CH_3 group attached to the oxide with a CO_2^- ion, the relative intensities are reversed. We know from other tunneling studies that the two oxygen atoms of the CO_2^- ion are roughly equidistant from the surface.[17–19] But, with only two point attachment to the oxide, it is possible for the molecule to cant relative to the surface. The higher intensity of the antisymmetric CH_3 vibration relative to the symmetric CH_3 vibration suggests that it is indeed canted. Thus, we get a detailed picture of the orientation of these ions from the preference of tunneling electrons to couple to vibrational modes oscillating perpendicular to the surface.

This will be discussed in more detail in Chapter 2.

4. Weaknesses of Tunneling Spectroscopy

4.1. Junction Geometry

The most serious limitation of tunneling spectroscopy is that it can only be performed with tunnel junctions! Figure 14 shows that the unknown layer, the layer that can be identified with tunneling spectroscopy, must be sandwiched between two metal electrodes together with an excellent quality tunneling barrier. The only systems that can be studied are systems that can be modeled within this constraint.

Thus, it is impossible, at least at present, to study real catalysts or industrially produced surfaces. In the petroleum industry, for example, it is often desired to study chemisorption on real catalyst samples. This is impossible with tunneling spectroscopy at present: no real catalyst has ever been studied.

Fortunately, however, the two best tunneling barriers, aluminum oxide and magnesium oxide, are of interest in themselves as catalysts and as catalyst supports. Also, studies have been done by modifying the alumina on tunneling barriers.[21-23] Certainly this type of research will continue. Gradually, tunneling spectroscopists can be expected to increase the scope and diversity of model surfaces that can be studied. Chapters 8–13 contain a survey of the present state of the modeling art.

4.2. Top Metal Electrode

It is somewhat surprising that tunneling spectra can be observed at all. As Figure 14 shows, the unknown layer is covered with an evaporated top metal electrode. Why does the top metal electrode not destroy the unknown layer?

It turns out that many top metal electrodes will indeed destroy the unknown layer. For example, junctions with a top electrode of aluminum, chromium, or most of the transition metals will have tunneling spectra that bear little or no resemblance to the original molecules that were doped into the junction.[24] Presumably, the top metal electrode has damaged the layer of molecules. Fortunately, this does not seem to be the case for lead, thallium, and, under some conditions, indium, tin, silver, and gold. For these metals, tunneling spectra are obtained that closely resemble infrared

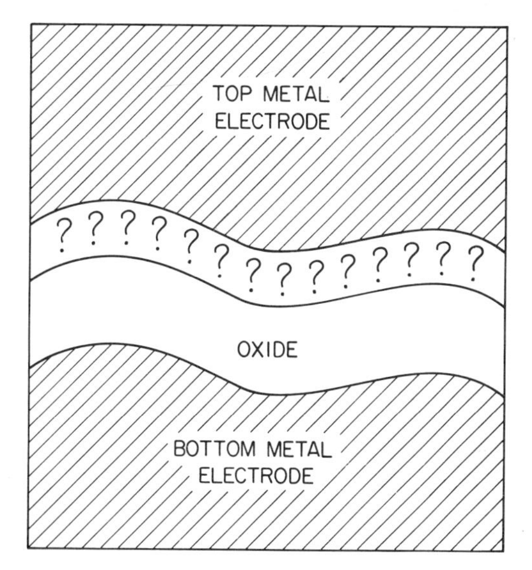

Figure 14. This schematic view of the barrier region of a tunneling junction illustrates a constraint of tunneling spectroscopy. The unknown layer, symbolized by question marks, must be part of an excellent tunneling barrier and covered by top metal electrode.

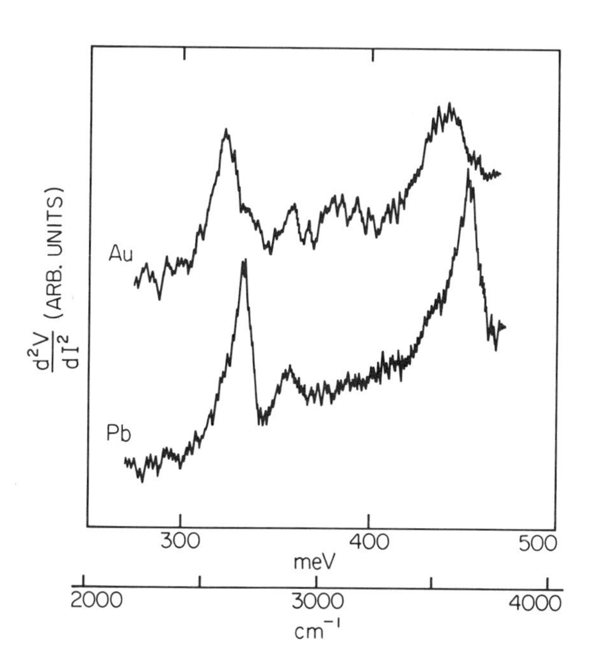

Figure 15. The top metal electrode can produce peak shifts. Here, we see the OH vibration that occurs near 460 meV with a lead top electrode that is downshifted and broadened for a gold top electrode.

spectra or Raman spectra of the surface adsorbed molecules without a top electrode. There are, however, some shifts due to the top metal electrode.[25–28]

Figure 15 shows shifts in the OH stretching frequency for a gold top metal electrode as opposed to a lead top metal electrode.[25] Notice that for the gold top electrode the OH peak is broadened and its center is shifted downward.

Figure 16 shows the very much smaller shift in a hydrocarbon vibrational mode.[26] This peak is one of the strongest in the benzoic acid spectrum (see Figure 5). Its shift was measured by John Kirtley for several different metal electrodes and for lead in the superconducting and normal

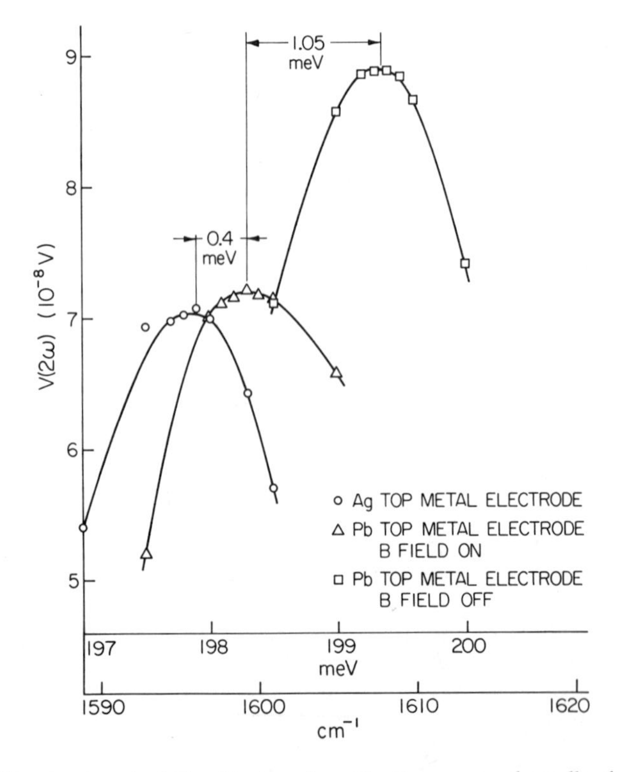

Figure 16. Vibrational mode shifts of hydrocarbon vibrations are much smaller than the shifts in Figure 15. Here we have a blown-up view of the top of the sharp peak that occurs near 200 meV in the spectrum of benzoate ions (see Figure 5). There is a shift of 1.05 meV due to superconductivity of the lead electrode. This is easily corrected for, as shown in Table 2. There is a smaller shift of 0.4 meV depending on whether a normal lead or (normal) silver electrode is used. This shift can be understood within the framework of an image dipole model but is not easily estimated or corrected for.

state. Notice first that, when lead is made normal, the peak shifts downward in energy. This is because the superconducting energy gap of lead shifts peaks by Δ/e for zero modulation voltage and by $\lesssim\Delta/e$ for non-zero-modulation voltage, where Δ is the superconducting energy gap. This shift can be calculated in a straightforward manner: calculated values[26] are included in Table 2. It should be subtracted from the measured tunneling peak positions before comparison with infrared, Raman, or electron energy loss results.

Figure 17 shows the measured shifts for a number of vibrational modes of benzoate ions and the OH groups for several top metal electrodes.[26] The x axis of the curve is scaled according to $1/R^3$, where R is the atomic radius

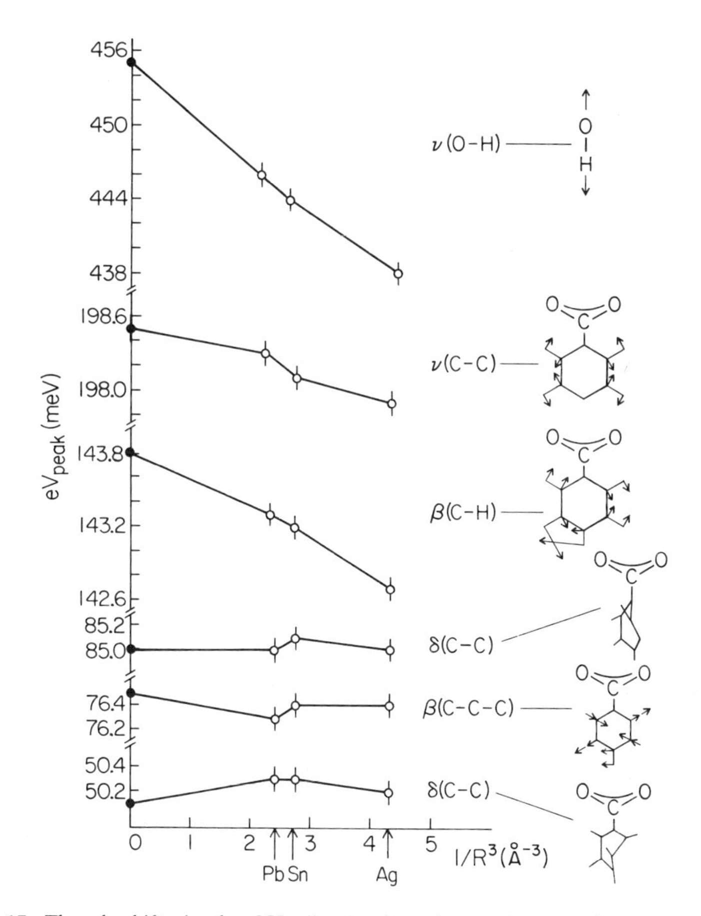

Figure 17. Though shifts in the OH vibrational mode can be several percent, shifts in hydrocarbon modes are typically less than one percent.

of the metal, because a simple theory for the effect is based on an image dipole effect in the top metal electrode which has a contribution that scales as the cube of the distance to the effective plane of the top metal electrode. These shifts are fairly benign. Their magnitudes are small: less than 1% for hydrocarbon modes and a few percent for the OH group. They can be approximately calculated within the framework of the simple image dipole theory.[26]

More recently, however, some more disturbing results were obtained by Atiye Bayman.[27] Figure 18 shows spectra for CO on iron particles supported by the aluminum oxide tunneling barrier. Note that, with different top metal electrodes, the low-energy vibrational mode structure changed qualitatively. The changes were so dramatic that it is not even clear that they can be interpreted as peak shifts. It does remain true that metals with similar atomic radii give similar spectra. The radii of Pd, Tl, Sn, In, Ag, and Au are 1.745, 1.71, 1.582, 1.57, 1.442, and 1.439 Å, respectively. Thus, within the pairs of comparable radii the spectra are comparable, but between pairs the spectra changed dramatically.

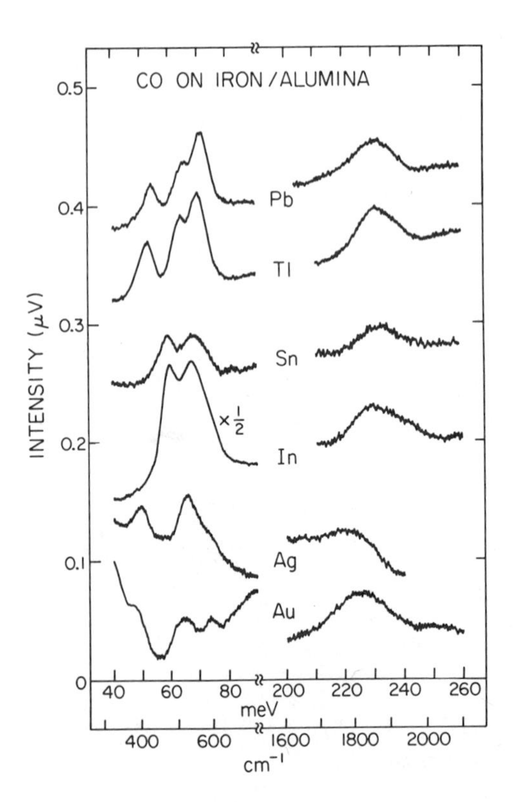

Figure 18. The effect of the top metal electrode can be much more dramatic in exotic systems. For CO, which has a very large dipole derivative, on small iron particles on alumina, the low-frequency vibrations change their character dramatically in going between pairs of metals with similar atomic radii.

Fortunately, the hydrocarbon modes she studied in these systems were very little shifted from top metal electrode to top metal electrode and from the tunneling results to optical results.

Another strange result that turned up in these studies was a *dip* in a tunneling spectrum at the position of the strongest *peak* in the corresponding infrared spectrum. These results serve as a warning that though tunneling spectra of adsorbed hydrocarbons on oxides seem to be very little perturbed by the top metal electrode and easily understood, tunneling spectra in more exotic systems—in particular those containing metal particles or molecules with very large dipole derivatives—must be treated with caution.[27, 28]

4.3. Cryogenic Temperatures

The final major disadvantage of tunneling spectroscopy is that tunneling spectra must be run at cryogenic temperatures to obtain resolutions sufficient for vibrational spectroscopy.[29] Figure 19 shows the thermal broadening in cm^{-1} and meV that occur at various temperatures. Note that by room temperature the broadening is so severe that it would be pointless to do ordinary vibrational spectroscopy. This is a real disadvantage if one wanted to study, say, chemical reactions proceeding at room temperature or elevated temperatures. The best that can be done is to let the reaction proceed and then chill the sample to cryogenic temperatures for measuring the spectra. This technique has been used and will be discussed in Chapter 13.

As a day-to-day matter, running spectra at cryogenic temperatures is only a minor nuisance. For most spectra it is not necessary to run below 4.2 K. Thus, liquid helium does not have to be transferred into research Dewars. Probes containing the sample can simply be slipped down the neck of the Dewars that liquid helium is shipped in. This will be discussed in more detail in Section 5.2.

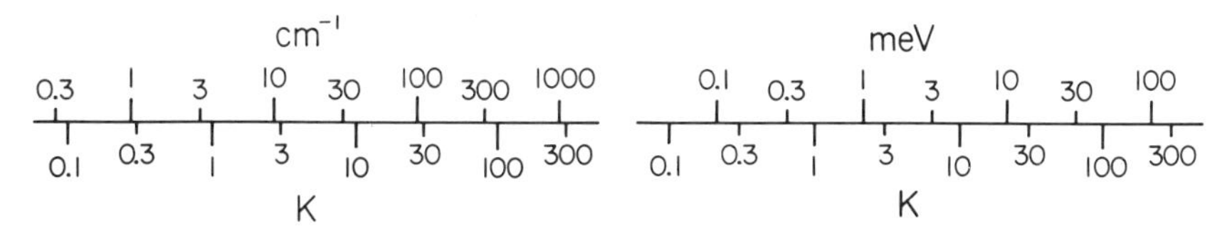

Figure 19. The thermal broadening mechanism illustrated in Figure 11 becomes a serious problem at elevated temperatures. For example, peak widths would be of order 1000 cm^{-1} or 100 meV at room temperature.

5. General Experimental Techniques

5.1. Introduction

The experimental techniques for tunneling spectroscopy have at their heart the experimental techniques for electron tunneling. Three articles on general techniques of electron tunneling that I have found particularly readable and helpful are by Giaver,[30] McMillan and Rowell,[31] and Coleman.[32]

It should be understood that the closest classical analog of tunneling is cooking.* Every chef has his own recipes and procedures based on experience and superstition. It is not practical to test every superstition in detail; when the chef finds a good recipe he follows it.

The purpose of this section is to give, in some detail, a recipe for making aluminum–insulator–lead tunnel junctions doped with organic molecules and measuring their tunneling spectra. This is certainly not the only recipe and probably not the best recipe, but it has been tested and does work.

5.2. Sample Preparation

Sample preparation is the most critical part of tunneling spectroscopy research. After the apparatus for tunneling spectroscopy is assembled, researchers typically find most of their experimental time and energy is spent in sample preparation. Though measuring the spectrum involves cryogenics and low-noise electronics, it is relatively straightforward once the necessary equipment is assembled. Sample preparation remains a challenge.

5.2.1. The Substrates

Many types of smooth insulating substrates have been used for tunneling spectroscopy. These include fused alumina substrates, sapphire substrates, silicon substrates, ordinary glass microscope slides, and cover glasses. It really makes no difference to the experimental results what substrate is used. Therefore, choice of substrates is based on convenience to the experimenter. In our group, we use ordinary glass microscope slides because these are readily available and cheap. We originally used Corning slides but, at the suggestion of R. V. Coleman, tried Clay Adam's Gold Seal slides and found them easier to clean.

*I am grateful to Professor G. I. Rochlin for this observation.

Cleaning slides is an important part of sample preparation. Good junctions for tunneling spectroscopy cannot be obtained on dirty substrates. Many failures to obtain good tunneling junctions can be traced to dirty substrates.

There are roughly as many cleaning procedures as there are experimental groups. Our own cleaning procedure involves washing staining racks full of microscope slides with a high-pressure water spray and detergent solution, rinsing thoroughly ($\gtrsim 5$ min/staining rack) with high-pressure water, rinsing with deionized water, rinsing with ethanol, and then storing under ethanol. We clean roughly six staining racks at a time.

When we are ready to use a rack of slides, we give it a last-minute rinse in fresh ethanol, then blow it dry with a heat gun for one minute and store the clean, dried slides in a dust-proof container in a clean section of the lab. We withdraw the slides one or two at a time for making tunnel junctions. The remaining racks can be stored under ethanol for months.

Slides can be tested for cleanliness by breathing on them and holding them up to a bright light. The pattern of fog on the slide will show up small traces of remaining dirt and detergent residue.* If the slides are found to be dirty, our superstition (based on a little experience) is that they should be discarded. It is better to start again with new slides than to try to rewash the almost clean slides.

5.2.2. The Vacuum Evaporator

Many types of vacuum evaporators have been successfully used for making electron tunnel junctions for tunneling spectroscopy. The basic requirement is that it be relatively clean from organic contaminents. Experience has shown, however, that even inexpensive oil diffusion pump systems with mechanical pumps as backing pumps can be maintained at the desired degree of cleanliness. Most groups have, however, found that common facilities cannot be maintained at the desired degree of cleanliness. In other words, regarding the vacuum evaporator, "Be it ever so simple, it should be your own."

The cleanliness of the vacuum system can be initially established by washing all interior parts carefully with detergent and water and thoroughly rinsing and drying them with a heat gun. Thick deposits can be cleaned off with glass beads in a sandblaster. It is important, however, to wash parts after they have been sandblasted unless the sandblaster is maintained

*We do not use the slide that has been breathed on because of possible contamination. We assume it is representative of other slides in the same staining rack.

specifically for critical cleaning. All cleaned parts for the vacuum system should be handled only with gloves. Though plastic gloves are probably the best, we use nylon gloves because they are much more comfortable and can be washed and reused.

Final cleanings of the vacuum system can be obtained by baking, the case of a bakable system, or by glow discharge cleaning with an ac glow discharge in argon gas between the chamber walls and an aluminum electrode. We use currents of order 0.2 A, argon pressures of order 100 mTorr, and times of order 1–10 min. It is important to have argon flowing through the system, either through the cracked high-vacuum valve into a diffusion pump or into the roughing system during the glow discharge cleaning. It is also important not to continue the glow discharge cleaning for too long. Our experience is that glow discharge cleaning for longer than 10 min does more harm than good by uncovering old organic deposits.

It also helps to do several evaporations all over the walls to clean up those surfaces and outgas the immediate vicinity of the evaporation sources. The real test of system cleanliness is freedom from hydrocarbon contamination in the spectrum of undoped junctions.

5.2.3. Junction Fabrication

Figure 20 shows a schematic view of junction fabrication. The aluminum for the bottom metal electrode is most commonly evaporated from a three-stranded tungsten filament (typically 3×0.040 in.). The aluminum should be of high purity (we use 0.58-mm-diam wire of 99.999% purity from Alfa Products). The aluminum is loosely wrapped around the tungsten filament and evaporated. We generally get about 20 evaporations from each tungsten filament by wrapping on new aluminum wire before each evaporation.

The substrate is placed roughly 8 in. from the source—the compromise here is: the farther from the source, the more uniform will be the film, but the more aluminum will be required, and there is a real limit to how much aluminum can be evaporated from the tungsten filament. If it is evaporated to "dryness," it then cannot be reused because Al wire will not rewet it.

The mask is placed close to the slide; in our case the slide rests directly on the top surface of the mask. The masks can be made from aluminum or stainless steel; brass should be avoided because of possible zinc sublimation. The thickness of the aluminum is not at all critical. In one evaporator we simply watch the aluminum source through the glass slide and evaporate for 5 sec after the aluminum film is completely opaque. In other systems, we monitor the thickness and aim for roughly 800 Å.

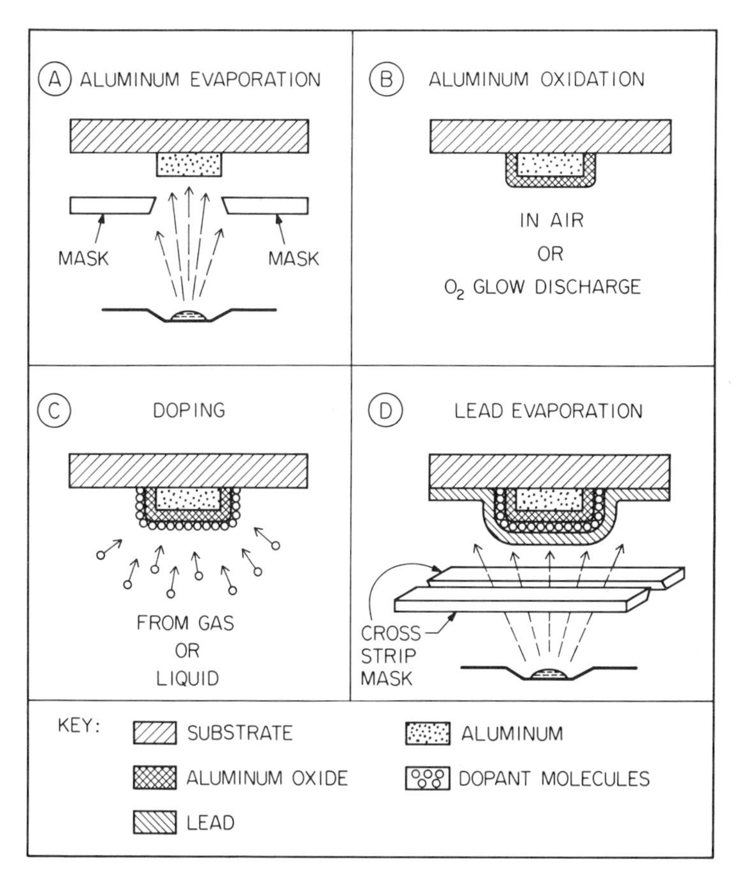

Figure 20. A schematic view of the four steps involved in making a conventional crossed film junction for tunneling spectroscopy. A schematic view of the completed junctions is shown in Figure 23. A photograph of real junctions is shown in Figure 22. The masks are shown away from the substrate for clarity. In practice, the aluminum mask touches the substrate and the lead mask is held of order 0.1 mm from the substrate by dimpling it with a center punch.

Next, the aluminum is oxidized in air or in an oxygen glow discharge. We have obtained more reproducible results by oxidizing in air after venting the chamber to pure, dry oxygen. A few minutes at room temperature is sufficient for a liquid doped tunneling junction, while up to 10 or 20 min at up to 200°C is required for supported metal junctions and some light-emitting junctions. Cleanliness of the air is a key to obtaining good air-oxidized tunnel junctions. For example, in our laboratory, we allow no organic solvents or other volatile organic materials outside of a fume hood, allow no smoking (this is very important), have charcoal filters on all incoming air into the laboratory, and maintain the laboratory at a positive pressure relative to outside air. These steps are difficult to initiate, but once established give good results for the oxidation and aid in keeping the vacuum evaporator clean.

Other researchers, especially those in polluted urban areas, have found it better to do oxygen glow discharge oxidation. A typical geometry is to have an aluminum ring below the junctions and to operate a dc glow discharge in an oxygen pressure of order 50 mTorr with currents of order 10 mA and times of order 10 min.[33,34] Experimentation is necessary because each vacuum system has a different geometry and the glow discharge pattern is dependent on the geometry of the chamber.

The next step is doping the oxidized aluminum strip. What is desired is to obtain roughly a monolayer of chemisorbed molecules on the oxidized aluminum. There are two basic techniques for doing this—gas phase and liquid phase doping.

In gas phase doping the oxidized aluminum strip is exposed to a vapor of the dopant in a vacuum chamber.[1,2] In the case of volatile substances, the vapor can be introduced from a bulb that is isolated from the vacuum chamber by a stopcock. Glass stopcocks with Teflon or metal plugs and Viton "O"-ring seals work well for this purpose. We (1) have a glass bulb of order 3 cm in diameter blown onto the end of the stopcock, (2) fill the bulb with a few cubic centimeters of the dopant, (3) degas it by freezing it in liquid nitrogen, pumping on it, thawing it, refreezing it, etc. for perhaps three cycles, then (4) fractionally distill it by evaporating off a little bit into the liquid nitrogen trap of the degassing pumping system and finally attach it to the vacuum chamber for doping. If a small amount of the dopant is evaporated off this way before use and if all of the dopant is not used in exposing junctions, then essentially the "heart cut" of a distillation of the dopant is used in doping the junction.

Typical exposures are 10 mTorr of dopant for one minute, though exposures vary widely with dopant. The correct exposure can be determined by trial and error. The ultimate test is, of course, the quality of the tunneling spectrum that is obtained. A good rule of thumb is, however, that the junction resistance of doped junctions should increase by roughly two orders of magnitude over the resistance of undoped junctions.

Nonvolatile substances can be vapor doped by placing them in a resistively heated boat below the oxidized Al strips.[35] The amount of power to the boat and time of heating can be varied to control exposure. For difficult cases, a thermocouple attached to the boat can help in determining proper conditions.

A second major technique for doping junctions is liquid doping.[36–38] In this case, the dopant molecules are applied to the oxidized aluminum from a liquid solution. The concentration of molecules on the surface is controlled by varying the concentration of the dopant molecules in the solution. Some researchers dip the entire slide into a solution containing the molecules and shake or blow off the excess solution. Others place a drop of

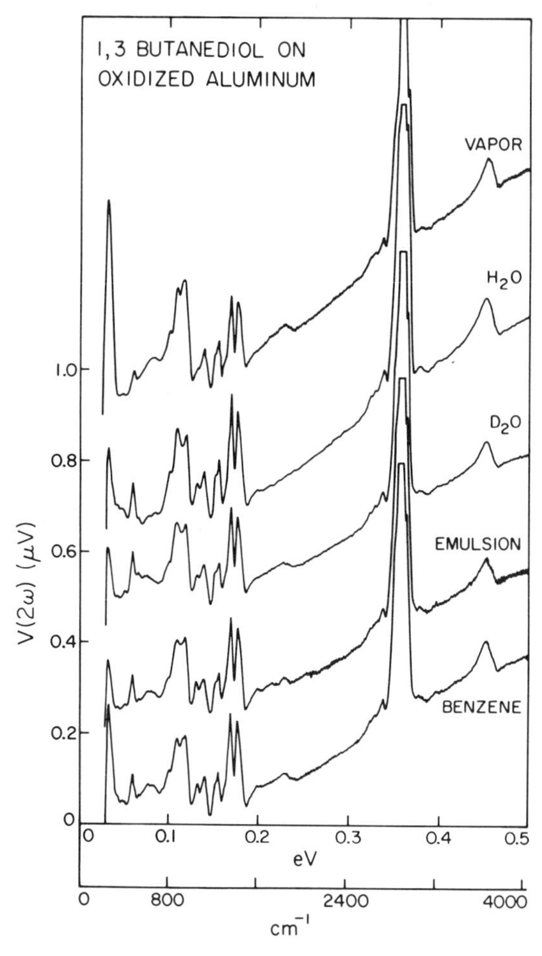

Figure 21. A comparison of 1,3-butanediol doped onto oxidized aluminum with various techniques. Note the similarity of the vapor-doped junction with those doped from various liquid solutions. The similarity of the liquid-doped junctions suggests that the solvent neither modifies the chemisorbed layer nor remains after doping (see reference 40).

the dopant solution on the junction area and spin it off. In our laboratory, we now do liquid doping using pipettors with disposable tips and a small spinner that spins the slides at 3000 rpm (made from a salvaged blower motor). We typically use drop volumes of order 10 μl and spin the slide immediately after placing the drop onto the junction area. By using small volumes of dopant solution, it is possible to dope each junction on the slide separately, with different concentrations. This aids in finding the optimal concentration for obtaining tunneling spectra with high signal-to-noise ratio.

Figure 21 shows results obtained by Atiye Bayman for 1,3-butanediol doped in various ways.(39) The upper trace shows the results from gas phase doping; the lower four traces show results of liquid doping from various solutions. Note that the spectra are very similar, though there are intriguing differences, in particular in the intensity of the peak near 500 cm^{-1}. The use of other solvents can, however, give markedly different spectra. In our experience water, D_2O, benzene, and hexane are all good solvents for liquid doping; they leave no trace. Solvents such as acetone, ethanol, and acetonitrile can leave traces. Methanol is an intermediate case. It generally seems to leave no trace but we do not trust it fully.

Two relatively new techniques for doping junctions, infusion doping(40, 41) and picomole doping, will be discussed in Chapters 15 and 16.

Finally, the junction is completed with an evaporated lead electrode. This can be done in the same vacuum system in which the aluminum is evaporated, though luxuriously equipped labs may find it convenient to use a separate evaporator so that the aluminum evaporator is always maintained clean—no dopants are ever introduced into it. The lead is typically evaporated from tantalum or molybdenum boats at rates up to 40 Å/sec for a total thickness of 2000–4000 Å. It is a superstition in our laboratory that it is good to begin the lead evaporation without a shutter so that the lead rate starts from zero. Lead electrodes thinner than 2000 Å will work but experience has shown us that the thicker the lead electrode, the longer the junctions can be stored. Presumably, some degradation processes involve infusion down through the lead electrode.(40, 41) Thicker lead electrodes slow this diffusion.

Though the evaporator can be vented immediately after the aluminum evaporation, it is good to wait 3–5 min after the lead evaporation, because it takes a long time for lead to cool from the evaporation temperature until it is no longer molten. If the evaporator is vented while the lead is still molten, it will oxidize.

For a researcher new to the area, who wants a good first sample for checking out his apparatus and procedures, I would suggest an aluminum film ≈ 0.5 mm wide by 800 Å thick, an air exposure of long enough to walk to a spinner for oxidation, liquid doping with a solution of 0.7 g/l benzoic acid in benzene (in a fume hood), and a lead film ≈ 0.5 mm wide by 3000 Å thick. If a fume hood is not available, 0.5 g/l benzoic acid in water works almost as well. For vapor doping, I would join G. Walmsley (see Chapter 11) in suggesting phenol.

5.2.4. *Care and Handling of Completed Junctions*

After the junctions are removed from the vacuum evaporator, their resistance can be measured immediately with a low-voltage ohmmeter. Though only a few years ago such ohmmeters had to be home built,[42] now they are available commercially.* Low-voltage ohmmeters made to check in-circuit resistance without turning on diodes are ideal.

To measure the junction resistance, first it is good to measure the strip resistances of both strips and take this into account, since strip resistances are often an appreciable fraction of junction resistances. As stated in the previous section, a good rule of thumb is that doped junctions should have resistances about two orders of magnitude larger than undoped junctions. This is easily checked for in the case of liquid doped junctions, since it is possible to leave the center junction on the slide undoped and directly compare this undoped junction to the doped junctions.

The ideal resistance for tunnel junctions depends upon the junction size. It is of order 50–200 Ω for a 0.5×0.5-mm junction. Basically, what is desired is to have the junction resistance low because noise in junctions, in particular shot noise, increases with junction resistance. If the junction resistance is too low, however, ohmic heating will cause an unesthetic burst of noise at the voltage where the lead electrode goes normal. Since the ohmic heating varies as V^2/R, for larger resistances, this occurs at larger voltages. The goal is to have it happen beyond the voltages of interest in a given tunneling spectrum. This sets the lower limit on junction resistance.

The next step is to cool the junctions to liquid-helium temperature. Though it is possible to store some junctions for months in a dry atmosphere, in general it is desirable to get the junctions into liquid helium within minutes after removing them from the vacuum system. In our laboratory, we walk rapidly from the evaporator room to the room with the helium Dewars and accept only urgent interruptions.

*For example, the Fluke 8024A.

5.2.5. *Cryogenics*

Figure 22 shows, on the left, a 1×3-in. glass microscope slide with five aluminum strips and, on the right, a microscope slide with both aluminum and lead strips. Note that the junction areas vary in size. This is an aid in getting junctions of the ideal resistance range. The middle view shows two junctions that have been cut out from the slide and mounted in a sample holder. The junctions are cut from the microscope slide using a carbide scriber to scribe the glass while it is held in a specially made vice that holds

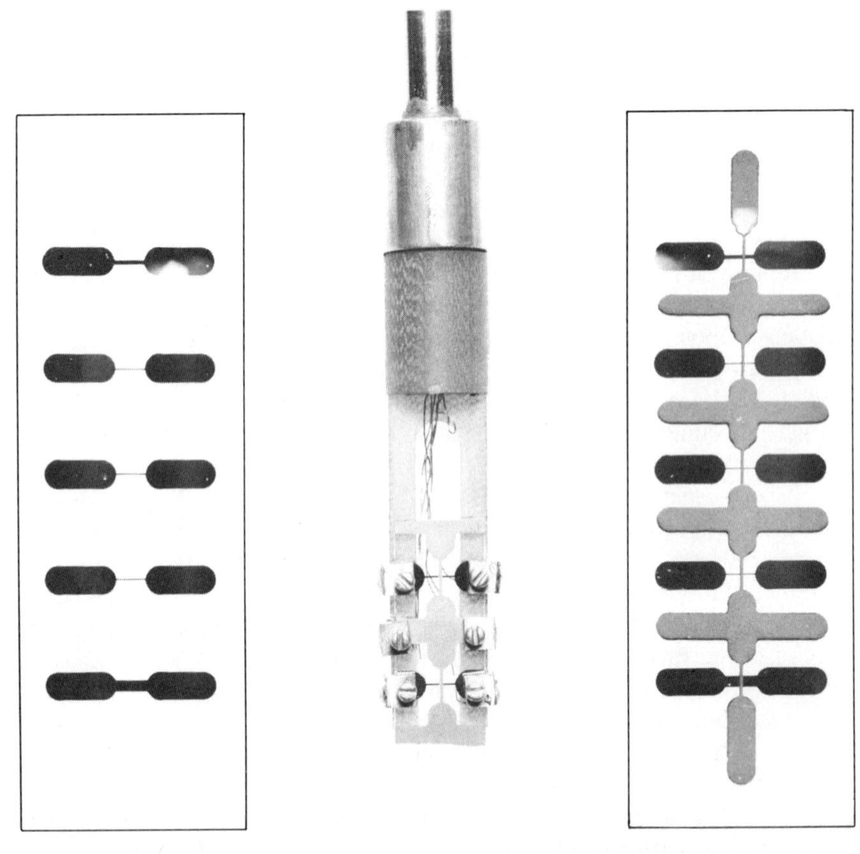

Figure 22. On the left is a 1×3-in. glass microscope slide with five aluminum strips ready for oxidation and doping. On the right five completed junctions are formed at the intersections of the aluminum bottom metal electrode and the lead top metal electrode. After measuring the resistances with an ohmmeter we cut out the pair that we most want to run and mount it with brass screw clamps on a probe as shown in the center. The probe is then inserted down the neck of a helium storage Dewar for measurements.

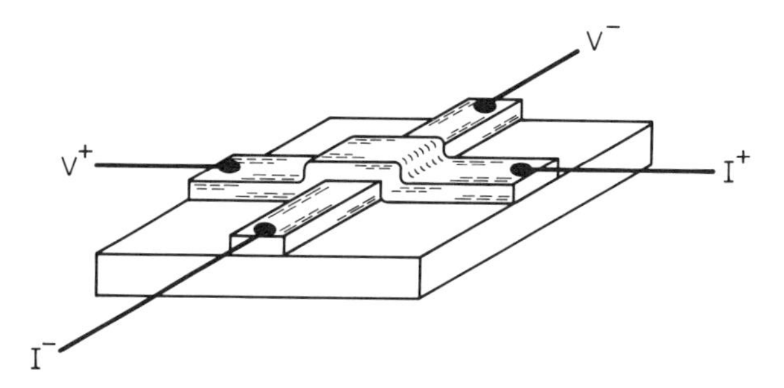

Figure 23. A schematic view of the four terminal measurement used in tunneling spectroscopy to minimize the effect of strip resistance.

it firmly between two flat, hard jaws. The upper jaw is used as a guide in making the scratch; then the slide is bent until it breaks along the scratch (about 95% of the time).

The sampler holder is designed to fit down the 5/8-in. i.d. neck of vendor-supplied liquid helium storage Dewars. With this type sample holder, it is unnecessary to transfer liquid helium into a research Dewar.

Note that six contacts are made to the two junctions using six small screw clamps that are machined from brass and use 0-80 brass screws with rounded tips. In our experience such screw clamps are preferable to silver paint or indium solder for attaching the leads to the junctions, though each technique has its own advantages.*

Figure 23 shows a schematic view of a four terminal measurement on a tunneling junction. The purpose of the four terminal measurement is to minimize the effect of the strip resistance on the measured junction characteristics. Thus, the voltage drops along the current-carrying strips do not contribute to the measured voltage.† Note that the sample holder of Figure 23 apparently violates the four-terminal geometry. However, since it is the lead strip for which the contact is made on the same side of the junction, there are no voltage drops because the lead strip is superconducting. In cases where the lead strip is normal and the strip resistance is nonnegligible, it is important to make the connections to the lead strip on opposite sides of the tunnel junction.

*We are grateful to R. V. Coleman for introducing us to these screw clamps.

†This is only true for the strip resistances on the leads coming into the junction. The strip resistances over the area of the junction can contribute to peak shifts. This contribution has not yet been studied in any detail and would certainly be worthy of a careful investigation.

5.2.6. *Characterizing Junctions and Obtaining Spectra*

After the junctions are inserted into liquid helium, it is important to measure the $I-V$ characteristics within a few millivolts of zero voltage to establish that the junctions are good tunnel junctions. This is done by looking at the structure due to the superconducting lead energy gap. (The origin of this structure is discussed in detail by, for example, Giaever[30] and McMillan and Rowell[31]). For our purposes, it is only important to note that the conductance at zero voltage should be approximately $1/7$ the conductance far away from zero voltage. Thus, the slope of the $I-V$ curve should be about $1/7$ as large at zero voltage as it is for voltages far from zero voltage. If the ratio is closer to unity, it indicates current flow in channels other than tunneling channels or excessive noise that is smearing the characteristic.

We look at the $I-V$ characteristic of every tunnel junction. It only takes a minute or so. Usually, we hook up the electrical leads and watch the $I-V$ characteristic as we lower the junction down into the Dewar. One can quickly learn to estimate by eye the desired $1/7$ resistance ratio. Routine use of this diagnostic tool will identify problems of (1) bad electrical connections, (2) excessive noise, and (3) poor-quality junctions.

After the $I-V$ characteristics are measured, it is time to measure tunneling spectra. Though it is possible, and for some experiments desirable, to build elaborate measuring circuits including bridge circuits,[43-46] a very simple measurement apparatus will suffice for most experiments—if high-quality components are used.

As shown in Figure 24, all that is really necessary is to apply a slowly varying current to the junction with a slowly varying voltage in series with a resistor and to apply an ac modulation current from a low-distortion oscillator in series with a large resistor. It is, of course, important to have only one ground in the circuit. Two common choices for where to put the ground are (1) at one side of the slowly varying voltage source or (2) at one side of the voltage measurement apparatus. Either way, the ac source must be isolated from ground. This can be done with either a transformer or two capacitors.

Though the resistors are shown as variable resistors, our experience is that a decade switch with metal film resistors is preferable to a potentiometer. A potentiometer has inductance and becomes electrically noisy with time.

The dc and ac voltages across the sample must be measured. The ac voltage is measured with a lock-in amplifier. A modern lock-in amplifier

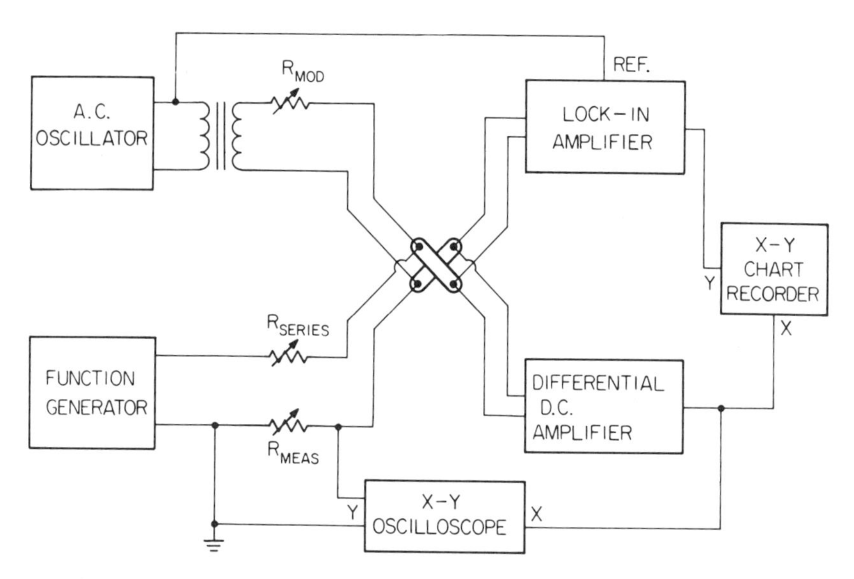

Figure 24. If high-quality components are used, tunneling spectra can be taken with a minimal amount of noncommercial equipment: just resistors and an isolation transformer. Careful attention to shielding and grounding is necessary, however, because of the low-level signals.

with a notch filter at the first harmonic frequency (or bandpass filter at the second harmonic frequency) can measure the second harmonic voltage across tunnel junctions with no external filtering. However, resonant circuits, such as used originally by Lambe and Jaklevic,[1] can give an appreciable boost in signal-to-noise ratio. The price one pays is uncertainty in the absolute magnitude and phase of the second harmonic signal. As Adler details in Chapter 14, the y axis can still be calibrated, but the procedure is not trivial. Unfortunately, details of the best resonant circuits have not been published. I would suggest a letter to Walmsley (coauthor of Chapter 11) for up-to-date information.

The dc voltage should be applied through a buffer amplifier to the x axis of a chart recorder. We have found it important to use a buffer amplifier since inputs of chart recorders are usually very noisy electrically and this noise is directly across the input to the lock-in amplifier.

One final note is that it is convenient to have the slowly varying dc sweeper also function as a low-frequency ac sweeper for tracing out the $I-V$ characteristics of junctions. Schematic diagrams for the combination sweeper that we use in our group are available upon request.

5.2.7. *Calibration and Measuring Peak Positions*

After the second harmonic voltage is plotted as a function of the dc voltage, this tunneling spectrum's x axis should be calibrated. A convenient means for doing this is to measure the voltage across the junctions with a digital voltmeter and, at a few selected voltages, make lines on the chart by twirling the y axis offset. Thus, the completed spectrum has a few lines, for example, one at zero volts, one at 0.2, and one at 0.4 V. Since amplifier gains can change and chart recorder x axes can easily become uncalibrated, it is good practice to make calibration marks on every usable spectrum. Making of marks all the way from the bottom to the top of the paper will help identify tilts of the chart paper.

Figure 25 shows a simple construction for determining peak positions in the case of a significantly sloping background. The peak positions can be measured conveniently with a ruler from the zero-voltage calibration line. After the positions are measured and converted to meV or cm^{-1} by using the calibration marks, it is important to subtract off a correction due to the superconducting energy gap of the lead before comparing to infrared or Raman results. The size of this correction is listed in Table 2 for common measuring conditions.

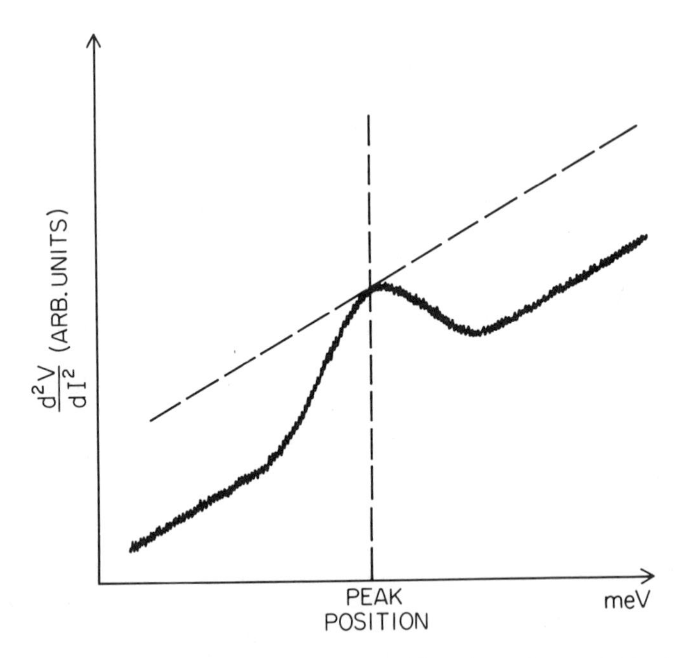

Figure 25. A simple construction for finding the position of a peak on a sloping background.

6. Conclusions

(1) Tunneling spectroscopy is a technique for measuring the *vibrational* spectrum of minute quanitites of material.

(2) The *spectral range* of tunneling spectroscopy includes all molecular vibrations.

(3) Its *sensitivity* is sufficient to see a small fraction of a monolayer over an area of order 1 mm on a side.

(4) Its *resolution* has no theoretical limits, but is generally set at 10 to 40 cm^{-1} by practical considerations.

(5) There are *no selection rules*. There is a slight selection preference for vibrations oscillating perpendicular to the plane of the junction's barrier.

(6) Problems include the limitations set by the *junction geometry*, the *cryogenic temperatures* necessary for good resolution, and the *top metal electrode* which can perturb the molecular vibrations being measured.

(7) *Sample preparation* occupies most of the time in tunneling spectroscopy research. Measuring spectra is straightforward once the necessary low-noise electronics has been assembled.

Acknowledgements

I thank Atiye Bayman for developing the water analogy and for suggesting revisions for this chapter; John Kirtley, Jim Hall, John Langan, Richard Kroeker, and Atiye Bayman for discussions and data on which this chapter is based; and the National Science Foundation (current grant # DMR79-25430) and the Office of Naval Research for their support of tunneling spectroscopy at the University of California.

References

1. R. C. Jaklevic and J. Lambe, Molecular vibration spectra by electron tunneling, *Phys. Rev. Lett.* **17**, 1139–1140 (1966).
2. J. Lambe and R. C. Jaklevic, Molecular vibration spectra by inelastic electron tunneling, *Phys. Rev.* **165**, 821–832.
3. See, for example, C. J. Pouchert, *The Aldrich Library of Infrared Spectra*, Aldrich Chemical Co., Milwaukee (1970).
4. I. Shimanouchi, *Tables of Molecular Vibrational Frequencies*, *Consolidated Volume I.* National Standard Reference Data Series, National Bureau of Standards, U.S., No. 39 (1972).
5. M. Avram and GH. D. Mateescu, *Infrared Spectroscopy*, Wiley-Interscience, New York (1972).

6. L. J. Bellamy, *The Infrared Spectra of Complex Molecules*, John Wiley and Sons, New York (1975).
7. E. Maslowsky, Jr., *Vibrational Spectra of Organometallic Compounds*, Wiley-Interscience, New York (1977).
8. S. Colley and P. Hansma, Bridge for differential tunneling spectroscopy, *Rev. Sci. Instrum.* **48**, 1192–1195 (1977).
9. J. Adler and R. Magno, unpublished.
10. J. D. Langan and P. K. Hansma, Can the concentration of surface species be measured with inelastic electron tunneling?, *Surf. Sci.* **52**, 211–216 (1975).
11. A. A. Cederberg, Inelastic electron tunneling spectroscopy intensity as a function of surface coverage, *Surf Sci.* **103**, 148–176 (1981).
12. R. M. Kroeker and P. K. Hansma, A measurement of the sensitivity of inelastic electron tunneling spectroscopy, *Surf. Sci.* **67**, 362–366 (1977).
13. D. G. Walmsley, R. B. Floyd, and S. F. J. Read, Inelastic electron tunneling spectra lineshapes below 100 mK, *J. Phys. C* **11**, L107–L110 (1978).
14. N. I. Bogatina, Selection rules in tunnel spectroscopy for highly symmetrical molecules, *Opt. Spectrosc.* **38**, 43–44 (1975).
15. N. I. Bogatina, I. K. Yanson, B. I. Verkin, and A. G. Batrak, Tunnel spectra of organic solvents, *Sov. Phys.—JETP* **38**, 1162–1165 (1974).
16. O. I. Shklyarevskii, A. A. Lysykh, and I. K. Yanson, Tunnel spectra of carboxylic acids: The transition from physical absorption to chemisorption, *Sov. J. Low Temp. Phys.* **2**, 328–333 (1976).
17. J. Kirtley, D. J. Scalapino, and P. K. Hansma, Theory of vibrational mode intensities in inelastic electron tunneling spectroscopy, *Phys. Rev. B* **14**, 3177–3184 (1976).
18. N. M. D. Brown, R. B. Floyd, and D. G. Walmsley, Inelastic electron tunneling spectroscopy (IETS) of carboxylic acids and related systems chemisorbed on plasma-grown aluminum oxide—Part 1, *J. Chem. Soc., Faraday Trans. 2* **75**, 17–31 (1979); N. M. D. Brown, W. J. Nelson, and D. G. Walmsley, Inelastic electron tunneling spectroscopy (IETS) of carboxylic acids and related systems chemisorbed on plasma-grown aluminum oxide—Part 2, *J. Chem. Soc., Faraday Trans. 2* **75**, 32–37 (1979).
19. J. T. Hall and P. K. Hansma, Chemisorption of monocarboxylic acids on alumina: A tunneling spectroscopy study, *Surf. Sci.* **76**, 61–76 (1978).
20. J. T. Hall and P. K. Hansma, Adsorption and orientation of sulfonic acids on aluminum oxide: A tunneling spectroscopy study, *Surf. Sci.* **71**, 1–14 (1978).
21. P. K. Hansma, D. A. Hickson, and J. A. Schwartz, Chemisorption and catalysis on oxidized aluminum metal, *J. Catal.* **48**, 237–242 (1977).
22. A. F. Diaz, U. Hetler, and E. Kay, Inelastic electron tunneling spectroscopy of a chemically modified surface, *J. Am. Chem. Soc.* **99**, 6780–6781 (1977).
23. N. K. Eib, A. N. Gent, and P. N. Henriksen, Formation of SiH bonds when SiO is deposited on alumina, *J. Chem. Phys.* **70**, 4288–4290 (1979).
24. R. Kroeker, previously unpublished observation.
25. J. R. Kirtley and P. K. Hansma, Effect of the second metal electrode on vibrational spectra in inelastic-electron-tunneling spectroscopy, *Phys. Rev. B* **12**, 531–536 (1975).
26. J. R. Kirtley and P. K. Hansma, Vibrational-mode shifts in inelastic electron tunneling spectroscopy: Effects due to superconductivity and surface interactions, *Phys. Rev B* **13**, 2910–2917 (1976).
27. A. Bayman and P. K. Hansma, Shifts and dips in inelastic electron tunneling spectra due to the tunnel junction environment, *Phys. Rev. Abst.* **12**, (1981).
28. K. W. Hipps and U. Mazur, An inelastic electron tunneling spectroscopy study of some iron cyanide complexes, *J. Phys. Chem.* **84**, 3162–3172 (1980).

29. R. J. Jennings and J. R. Merrill, The temperature dependence of impurity-assisted tunneling, *J. Phys. Chem. Solids* **33**, 1261 (1972).
30. J. Giaever, Electron tunneling and superconductivity, *Rev. Mod. Phys.* **46**, 245–250 (1974) (his Nobel Prize acceptance speech).
31. W. L. McMilland and J. Rowell, in *Superconductivity* (R. D. Parks, ed.), p. 561, Marcel Dekker, New York (1969).
32. R. V. Coleman, R. C. Morris, and J. E. Christopher, *Methods of Experimental Physics VII. Solid State Physics* (R. V. Coleman, ed.), Academic Press (1974).
33. J. L. Miles and P. H. Smith, The formation of metal oxide films using gaseous and solid electrolytes, *J. Electrochem. Soc.* **110**, 1240–1245 (1963).
34. R. Magno and J. G. Adler, Inelastic electron-tunneling study of barriers grown on aluminum, *Phys. Rev. B* **13**, 2262–2269 (1976).
35. M. G. Simonsen and R. V. Coleman, Inelastic-tunneling spectra of organic compounds, *Phys. Rev. B* **8**, 5875–5887 (1973).
36. P. K. Hansma and R. V. Coleman, Spectroscopy of biological compounds with inelastic electron tunneling, *Science*, **184**, 1369–1371 (1974).
37. M. G. Simonsen, R. V. Coleman, and P. K. Hansma, High-resolution inelastic tunneling spectroscopy of macromolecules and adsorbed species with liquid-phase doping, *J. Chem. Phys.* **61**, 3789–3799 (1974).
38. Y. Skarlatos, R. C. Barker, G. L. Haller, and A. Yelon, Detection of dilute organic acids in water by inelastic tunneling spectroscopy, *Surf. Sci.* **43**, 353–368 (1974).
39. A. Bayman and P. K. Hansma, Inelastic electron tunneling spectroscopic study of lubrication, *Nature* **285**, 97–99 (1980).
40. R. C. Jaklevic and M. R. Gaerttner, Electron tunneling spectroscopy—external doping with organic molecules, *Appl. Phys. Lett.* **30**, 646–648 (1977).
41. R. C. Jaklevic and M. R. Gaerttner, Inelastic electron tunneling spectroscopy. Experiments on external doping of tunnel junctions by an infusion technique, *Appl. Surf. Sci.* **1**, 479–502 (1978).
42. B. D. Wallace, Low power at ohmmeter's probes allows safe usage on most sensitive components, *Electron. Des.* **14**, 110 (1974).
43. D. E. Thomas and J. M. Rowell, Low-level second-harmonic detection system, *Rev. Sci. Instrum.* **36**, 1301–1306 (1965).
44. J. G. Adler and J. E. Jackson, System for observing small nonlinearities in tunnel junctions. *Rev. Sci. Instrum.* **37**, 1049–1054 (1966).
45. A. F. Hebard and P. W. Shumate, A new approach to high resolution measurements of structure in superconducting tunneling currents, *Rev. Sci. Instrum.* **45**, 529–533 (1974).
46. S. Colley and P. K. Hansma, Bridge for differential tunneling spectroscopy, *Rev. Sci. Instrum.* **48**, 1192–1195 (1977).

2

The Interaction of Tunneling Electrons with Molecular Vibrations

John Kirtley

1. Introduction

Electrons can tunnel with appreciable probability from one metal electrode to the other in metal–insulator–metal tunneling junctions if the insulating layers are sufficiently thin (~ 30 Å). If no energy is lost by the electrons in the transition, the tunneling is called elastic. Inelastic electron tunneling, in which electrons in filled states on one side of the metal lose energy to some excitation in the barrier region, but still have enough energy to finish up in a previously empty final state on the other side of the barrier, can also occur. The inelastic tunneling process occurs, for the low temperatures at which tunneling spectra are run, only if the bias energy eV is greater than the excitation energy $\hbar\omega$. Since the inelastic tunneling process represents an additional tunneling channel, the total conductance of the junction is greater for biases above the onset voltage $V = \hbar\omega/e$ than below it. The conductance increases are quite small, so that second derivatives of the current–voltage characteristics of the junctions are taken: the steps in conductance then appear as peaks, each peak corresponding to the excitation of a particular vibrational mode of the barrier region.

Although there is also structure in the current–voltage characteristics of the junctions due to superconductivity,[1] metal phonons,[2] and oxide

John Kirtley • IBM T. J. Watson Research Center, Yorktown Heights, New York 10598.

phonons,[3] this chapter will deal exclusively with the amplitudes of the conduction steps due to the excitation of the vibrational modes of the molecular species in the barrier region.

As will be demonstrated extensively in this volume, the positions in energy of the vibrational modes reveal a great deal about the identity and chemical configuration of the molecular species included in the barrier region. What can we learn in addition from the intensities of the modes? Early work in the field was aimed at delineating a few well-defined universal rules governing tunneling spectroscopy intensities, similar to the selection rules observed in infrared,[4] Raman,[4] and to some extent, high-resolution inelastic electron loss[5,6] spectroscopies. It was hoped that these rules could be used to determine the orientation of the molecules on the surface, assist in the assignment of the molecular vibrations, and help determine to what extent the molecule was perturbed by its environment. This chapter should make it clear that the rules in inelastic electron tunneling are not simple. However, while the analysis of tunneling spectroscopy intensities is more complex than might be hoped for, and we are a long way away from a complete understanding, it is still possible to obtain useful information from intensities.

Succeeding sections will briefly review (1) elastic tunneling theory in order to establish the very simple ideas and terminology to be used later; (2) the simple long-range models of Scalapino and Marcus,[7] and Lambe and Jaklevic[8]; and (3) the somewhat more complex partial charge model of Kirtley, Scalapino, and Hansma,[9] and the variations thereof by Kirtley and Hall.[10] The last will show that, at least for a molecule of high symmetry, orientation information can be obtained from intensity data. We will argue that short-range interactions need not be considered in analyzing tunneling spectroscopy intensities.[11] We will then conclude with an estimate of work that remains to be done in developing the theory of inelastic electron tunneling.

2. Elastic Tunneling

The most direct approach to calculating tunneling currents involves the solution of the full-time independent Schrödinger equation for a given barrier potential.[12] The desired solution has the form of incoming and reflected plane wave states on one side of the barrier, as well as transmitted states on the other side (Figure 1a). Transmission probabilities $D(\bar{k}_i, \bar{k}_f)$ are obtained by taking the ratio of the transmitted to the incident fluxes. Tunneling currents are obtained by summing over incoming and outgoing momenta. If it is assumed that momentum parallel to the surface is

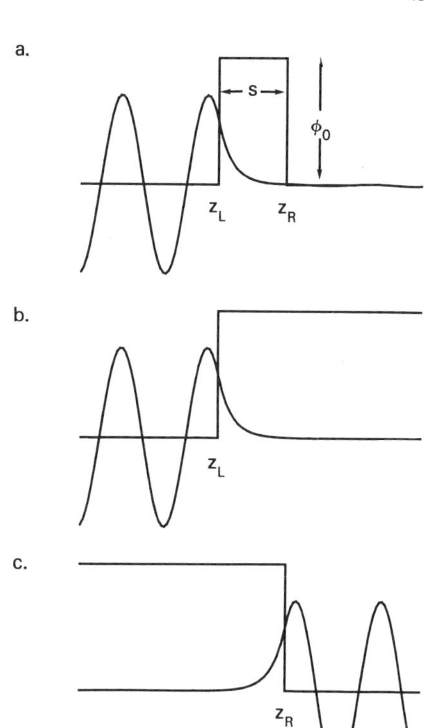

Figure 1. (a) In an "exact" calculation, incident and reflected plane waves on the left are matched through the barrier to a transmitted wave on the right. The barrier potential is taken to be square, with a width s and height ϕ_0. The transmission probability is simply the square of the ratio of the incident to the transmitted amplitudes. This figure was prepared with a transmission probability of $\sim 10^{-4}$. Transmission probabilities for tunneling spectroscopy junctions are typically 6 orders of magnitude smaller. In the transfer Hamiltonian formalism, z_R is extended to infinity for the calculation of the initial-state wave function (b) and z_L is extended to negative infinity for the calculation of the final-state wave function (c). An interaction potential is used to transfer an electron from the initial to the final state.

conserved, the total elastic current through the junction is given by

$$j_e = \frac{2e}{\hbar} \int dE [f(E) - f(E + eV)] \int \frac{d^2 k_\parallel}{(2\pi)^2} D(\bar{k}_i, \bar{k}_f)_{k_{i\parallel} = k_{f\parallel}} \qquad (1)$$

where $f(E)$ is the Fermi occupation function $f(E) = (1 + e^{E/k_B T})^{-1}$.

This approach is exact within the limits of the approximations to the potentials that are made. It has been used for many calculations of elastic tunneling processes, but becomes increasingly cumbersome when applied to problems with complex potentials. Applications of the "direct" approach to the problem of inelastic electron tunneling can be found in references 13–16.

A somewhat more flexible technique was first introduced by Bardeen[17] to explain Giaever's observation of an energy gap in the current–voltage characteristics of tunneling junctions with superconducting films.[1] Bardeen's approach calculated tunneling rates using first-order time-dependent perturbation theory. In the Bardeen picture the tunneling barrier is conceptually divided into two parts. The initial electronic state is localized on the left side of the tunneling barrier by extending the barrier such that $z_R = \infty$. (See Figure 1b.) The final state is similarly localized by setting $z_L = -\infty$ (Figure 1c). Both initial and final state wave functions take the form, parallel to the interface, of sinusoidally varying wave functions, while normal to the interface, the wave functions are sinusoidal outside the tunneling region, and exponentially attenuated into the tunneling region. The Bardeen picture ignores the effects of multiple scattering: the exponentially decaying wave from the left is not included when matching wave functions on the right-hand surface. If the WKB approximation[18] for the wave functions is used, the initial wave functions inside and outside the barrier are given by

$$\psi_i^{\text{out}} \sim k_z^{-1/2} e^{i(k_x x + k_y y)} \sin(k_z z + \gamma) \tag{2}$$

$$\psi_i^{\text{in}} \sim K_z^{-1/2} e^{i(k_x x + k_y y)} \exp\left(-\int_{z_L}^{z_R} K_z \, dz\right) \tag{3}$$

where $K_z = \{2m[\phi(z) - E_z]/\hbar^2\}^{1/2}$, $\phi(z)$ is the one-electron barrier potential, and $E_z = E - \hbar^2(k_x^2 + k_y^2)/2m$ is the kinetic energy of the electronic initial state in the z direction (normal to the interface). Similar results are obtained for the final-state wave functions.

To treat elastic tunneling, we take the solution of the time-dependent Schrödinger equation to be a linear combination of initial- and final-state wave functions:

$$\psi(t) = a(t)\psi_i e^{-i\omega_0 t} + \sum_f b_f(t)\psi_f e^{-i\omega_f t} \tag{4}$$

Inserting this into Schrödinger's wave equation with the Hamiltonian given by the kinetic energy term plus the barrier potential term as outlined above, and solving to first order for $b_f(t)$, leads to the standard Fermi's Golden Rule result for the transition rate/unit time:

$$\omega_{if} = \frac{2\pi}{\hbar} |M_{if}|^2 \delta(\omega_0 - \omega_f) \tag{5}$$

with

$$M_{if} = -\frac{\hbar^2}{2m} \int dx \int dy \left(\psi_i^* \frac{\partial}{\partial z}\psi_f - \psi_f \frac{\partial}{\partial z}\psi_i^*\right)_{z = \text{const}} \tag{6}$$

where the integral can be evaluated anywhere within the barrier. In the simplest case of $\phi(z)=\phi=\text{const}$, the WKB-type wave functions [Eqs. (2) and (3)] give matrix elements

$$M_{if} \sim e^{-K(z_L - z_R)} \tag{7}$$

with $K=[2m(\phi - E_z)/\hbar^2]^{1/2}$. For a typical barrier height of 2 eV above the Fermi level, $K=0.724$ Å^{-1}: the tunneling rate decreases by a factor of 4 for every additional angstrom of oxide thickness. To obtain the total elastic current we sum over all possible initial and final momentum states (ignoring superconductivity for simplicity):

$$j_e = \frac{4\pi e}{\hbar} \sum_{k_i} \sum_{k_f} |M_{if}|^2 \left[f(\varepsilon_i) - f(\varepsilon_f + eV) \right] \delta(\varepsilon_i - \varepsilon_f) \tag{8}$$

This approach, which is called the transfer Hamiltonian formalism, has been used in all of the applications to inelastic tunneling which will be discussed here. Two drawbacks of the transfer Hamiltonian method are[15, 16] (1) it defines two separate states for the left and right electrodes which are not orthogonal and hence are not complete sets of states for the full system; (2) the interaction region must be within the potential barrier, as opposed to the metal electrodes. The first objection becomes less important as the tunneling barrier becomes thicker, and the two sets of states become more weakly coupled. Tunneling spectroscopy junctions have relatively thick tunneling barriers, making the coupling very weak: The neglect of higher-order terms is fully justified. The second objection, while valid in certain special cases,[13, 14] should not be important for most of the effects we are interested in, since the molecular potentials are strongly screened within the first few angstroms of the metal electrodes. Further, the transfer Hamiltonian approach is sufficiently simple that fairly realistic models for the molecular potential can be used. The transfer Hamiltonian method has been used with great success to describe, for example, such widely varying effects as gap structure in the $I-V$ characteristics of tunnel junctions with superconducting electrodes,[19] Josephson effects,[20] and phonon structure[3] in metal–insulator–metal tunneling junctions.

3. Inelastic Tunneling

3.1. Simple Long-Range Models

The first treatment of inelastic tunneling was presented by Scalapino and Marcus (SM).[7] They assumed that the tunneling electron–molecular potential could be represented by the component of a point dipole normal

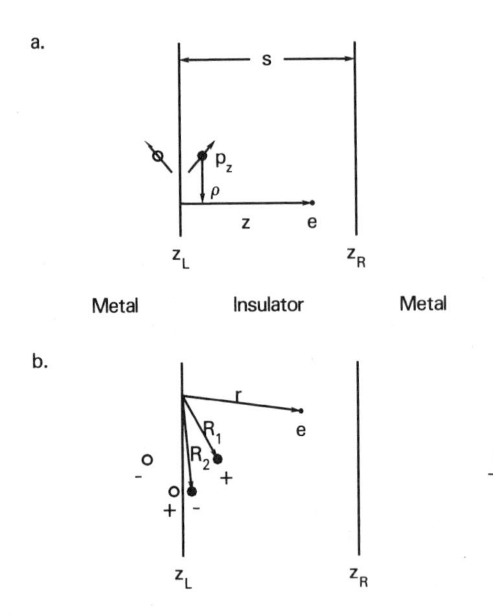

Figure 2. (a) In the theory of Scalapino and Marcus, a point dipole and its image are located close to one metal electrode. The component of the dipole normal to the surface adds to its image; the component parallel to the surface cancels. The tunneling electron penetrates the barrier a distance ρ from the dipole and is a distance z from the imaging plane. (b) In the theory of Kirtley, Scalapino, and Hansma, a series of point charges have images in both metal electrodes.

to the interface p_z and its image in the closest metal surface (Figure 2a):

$$V_{\text{int}}(z) = \frac{2ep_z}{(z^2+\rho^2)^{3/2}} \tag{9}$$

where $\rho \equiv (x^2+y^2)^{1/2}$.

They added (9) to the barrier potential and calculated the change in the tunneling current due to this potential using the transfer Hamiltonian formalism. Following the calculation outlined above for the elastic current, and using WKB-type wave functions [Eqs. (2) and (3)], they found that the matrix element M_{if} was given by Eq. (7) as before, but with

$$K = \left[2m(\phi + V_{\text{int}} - E_z)/\hbar^2\right]^{1/2} \tag{10}$$

Setting $\phi - E_z = \phi_0$ a constant, and expanding to lowest order in V_{int}/ϕ_0, they found

$$M_{if} \sim \left[1 + \left(\frac{2m}{\phi_0}\right)^{1/2} \frac{ep_z}{\hbar s} g\left(\frac{\rho}{s}\right)\right] e^{-(2m\phi/\hbar^2)^{1/2}s} \tag{11}$$

where $s=|z_R - z_L|$, and

$$g(x)=\frac{1}{x}-\frac{1}{(1+x^2)^{1/2}}$$

In this expression p_z is taken to be the expectation value of the dipole moment operator normal to the surface between the ground and the first excited state of the vibrational mode. At low temperatures (tunneling spectra are run at or below 4.2 K) all of the molecules are initially in their ground state and the inelastic tunneling electrons only flow in one direction for a particular bias polarity. The inelastic current equation analogous to Eq. (8) becomes

$$j_i^k=\frac{4\pi e}{\hbar}\sum_{k_i}\sum_{k_f}|M_{if}|^2 f(\varepsilon_i)[1-f(\varepsilon_f+eV)]\delta(\varepsilon_i-\varepsilon_f-\hbar\omega_k) \qquad (12)$$

where j_i^k is the inelastic current due to the excitation of the kth vibrational mode of the molecule with frequency ω_k. If we assume that the tunneling electron conserves its momentum parallel to the surface, the ratio of the increase in conductance as an inelastic channel opens up, divided by the elastic junction conductance, is given simply by the square of the ratio of the change in the elastic matrix element due to the vibrational potential, divided by the total elastic matrix element:

$$\frac{\Delta\sigma_i^k(\rho)}{\sigma_e}=\frac{2me}{\phi_0}\left(\frac{e}{\hbar s}\right)^2 p_z^2 g^2\left(\frac{\rho}{s}\right)\Theta(\mathrm{eV}-\hbar\omega_k) \qquad (13)$$

where $\Theta(x)=1$ for $x>0$ and 0 otherwise. In this expression the Θ function allows the inelastic channel to open up only when the electrons have enough energy to excite the vibration ($eV>\hbar\omega_k$). The factor ρ is interpreted as the distance from the molecular impurity to where the tunneling electron penetrates the barrier. The theory of Scalapino and Marcus therefore takes a very local view of the tunneling electron. To obtain the total inelastic conductance they integrated over ρ, summed over all vibrations of each impurity, and multiplied by n molecules per unit area.

Scalapino and Marcus assumed in multiplying by n that the molecular vibrational mode cross sections were independent of the molecular surface density. Langan and Hansma[21] have shown, using radioactively labeled molecules, that tunneling spectroscopy intensities increased slightly more rapidly than surface coverage as the surface density was increased.

Cunningham, Weinberg, and Hardy[22] suggested that this nonlinearity was due to cooperative behavior. For this to be true, however, there must be some coupling mechanism between the molecules to keep their vibrations in phase: There is no other evidence of such coupling. Following a suggestion by Langen and Hansma, Cederberg[23] has developed an alternate explanation for the nonlinearity: Tunneling electrons are more likely to penetrate the barrier region where there are no molecular dopants. Therefore the elastic current decreases rapidly as the molecular concentration increases, so even if the inelastic conductance step is proportional to the density of molecular scatterers, the ratio of the conductance step to the total junction conductance increases more rapidly than the surface concentration. In this chapter we are concerned with absolute intensities only to the extent that we insist that a "good" theory predicts them approximately correctly. We will concentrate instead on relative intensities for different vibrational modes of a particular molecule. For this purpose multiplying by n is good enough.

Scalapino and Marcus obtained finally

$$\frac{\Delta\sigma_i}{\sigma_e} = \frac{4\pi nme^2}{\phi_0\hbar^2}\ln\left(\frac{s}{\rho_0}\right)\sum_k |\langle k|p_z|0\rangle|^2\Theta(eV-\hbar\omega_k) \tag{14}$$

To derive Eq. (14) it was necessary to take both a lower and upper cutoff on the integration of $g^2(\rho/s)$. Since these cutoffs appeared in a logarithm, their values were not critical. Nevertheless they represented a weakness of the theory. For reasonable values of s, ρ_0, ϕ_0, n, m, and p_z the calculated magnitude of the conductance jump due to a monolayer of OH ions was $\sim 1\%$, in agreement with experiment. Since point dipoles oriented parallel to the interface cancel with their images, while dipoles oriented perpendicular to the surface add, the theory predicted that only vibrational modes with net oscillating dipole moments normal to the surface will cause inelastic transitions. Further, it predicted that only vibrational modes involving a change in the dipole moment of the molecule should be observed using tunneling spectroscopy, with intensities proportional to p_z^2. This is the same proportionality factor that appears in infrared absorption, so one might expect a strong correlation between infrared and tunneling intensities. Unfortunately, while the modes that are strong in ir spectroscopy tend to be strong in tunneling spectroscopy also, the correlation is not exact.

One possible explanation for this incomplete correlation is that the tunneling electron interacts with the induced as well as the intrinsic oscillating dipole of the molecule. Lambe and Jaklevic[8] extended the theory of Scalapino and Marcus to include the potential due to the bond polarizability. This term couples to Raman active modes, which have a change in the polarizability associated with a given vibration, but not necessarily a change in the net dipole moment. The interaction, including the effect of the nearest image of the dipole, is

$$V_i^R = -\frac{4\pi e \alpha z^2}{\left(z^2+\rho^2\right)^2} \tag{15}$$

where α is the molecular bond polarizability. Following the procedure of Scalapino and Marcus, they found

$$\frac{\Delta\sigma_i}{\sigma_e} = \frac{4\pi mne^2}{\phi_0 \hbar^2}\frac{e^2}{16s^2}\int_{r_0}^{s} t^2(\rho/s)\rho^2\,d\rho \sum_k |\langle k|\alpha|0\rangle|^2 \Theta(eV-\hbar\omega_k) \tag{16}$$

where

$$t(x) = \frac{1}{x^2}\left[\frac{1-x^2}{1+x^2}+\frac{1}{x}\tan^{-1}\left(\frac{1}{x}\right)\right]$$

In this case the function $t(x)$ is strongly divergent as x approaches 0, and the integral is sensitive to the value of the cutoff r_0. The reason for this divergence is that the polarizability α is taken to be independent of ρ, while as some point the induced dipole must saturate for small spacings. Nevertheless, estimates of the size of the conduction step for reasonable values of ρ_0, s, n, ϕ_0, and α gave magnitudes quite comparable to those for the electron–dipole interaction. Therefore one should expect to observe both Raman and infrared modes using tunneling spectroscopy. Indeed, observations of both the infrared and Raman active vibrational modes of benzene,[24] anthracene,[25] and ferrocyanide[26] have been reported.

The theory of Scalapino and Marcus, as extended by Jaklevic and Lambe, worked well for calculating intensities to within an order of magnitude. However, there were some difficulties with the theory: (1) It assumed that momentum parallel to the interface was conserved in the tunneling process. Strictly speaking this is not the case, and several interesting effects occur when this restriction is relaxed. (2) In calculating the tunneling current as a function of ρ it was implicitly assumed that the electronic wave function was localized on a small scale with respect to the interaction distance. This assumption led to divergent integrals unless cutoffs in ρ were introduced. (3) The molecular potential was taken for the molecule as a whole, although the electrons should be expected to "see" the molecular structure.

3.2. Complex Long-Range Models

In order to avoid these difficulties, Kirtley, Scalapino, and Hansma (KSH)[9] used a transfer Hamiltonian theory with a local, although still empirical potential, taken to be the sum of Coulomb potentials due to a set of partial charges localized on the atoms in the molecule:

$$V_i(\bar{r}) = -\sum_n \frac{e^2 Z_n}{|\bar{R}_n - \bar{r}|} \tag{17}$$

where $\bar{R}_n$ is the position of the nth atom in the molecule and $Z_n e$ is the partial charge associated with the nth atom. In first-order perturbation theory the part of the total interaction potential [Eq. (17)] which connects initial and final states with different energies is that which vibrates at the frequency associated with the energy difference. KSH expanded Eq. (17) to first order in the oscillating atomic positions:

$$\bar{R}_n^k(t) = \bar{R}_n^0 + \delta\bar{R}_n^k e^{i\omega_k t}$$

$$\bar{R}_n^0 = b_n\hat{x} + c_n\hat{y} + a_n\hat{z} \tag{18}$$

where $\bar{R}_n^k$ and $\delta\bar{R}_n^k$ are, respectively, the positions and vector displacements of the nth atom due to the kth vibrational mode. The interaction potential becomes

$$V_i^k(\bar{r}, \bar{R}) = \sum_n -e^2 Z_n^k \delta\bar{R}_n^k \cdot \nabla_n \left(\frac{1}{|\bar{r} - \bar{R}_n^0|} \right) e^{i\omega_k t} \tag{19}$$

This potential has the form of a sum of dipole potentials, each dipole located on an atom in the molecule:

$$V_i^k(\bar{r}, \bar{R}) = \sum_n -e\bar{\mu}_n^k \cdot \nabla_n \left(\frac{1}{|\bar{r} - \bar{R}_n^0|} \right) e^{i\omega_k t} \tag{20}$$

with the local dipole defined by

$$\bar{\mu}_n^k \equiv e Z_n^k \delta\bar{R}_n^k \tag{21}$$

Notice that in Eqs. (19)–(21) we have written the partial charge model in a form in which the partial charges associated with each atom are in general different for each vibrational mode. This avoids the difficulties pointed out by Rath and Wolfram[33]: The charge distributions associated with a given molecular vibration do not necessarily follow the nuclear motion, and can be different for each vibrational mode. However, it appears that in doing so we have sacrificed any predictive properties: The problem is overdefined. This overdefinition can be avoided if we take into account the symmetry properties of the vibrations of at least fairly symmetrical molecules, as we will see below.

In the tunneling junction the vibrating charges are sandwiched in a dielectric medium between two good metals (in the infrared). Therefore we should account for the image potentials formed. KSH took not only the image in the nearest surface, as in Scalapino and Marcus, but also the image in the farther metal surface, and its image in the near metal surface, and so on (see Figure 2b). Then the potential becomes

$$V_i^k(\bar{r}, \bar{R}) = \sum_{p=-\infty}^{\infty} \sum_n -\frac{e}{\varepsilon}\bar{\mu}_n^k \cdot \nabla_n \left[\frac{1}{|\bar{r} - \bar{R}_n - 2ps\hat{z}|} - \frac{1}{|\bar{r} - \bar{R}_n - (2ps - 2a_n)\hat{z}|} \right] \tag{22}$$

where s is the thickness of the oxide, and the dielectric constant characterizing the oxide and the molecular layer comprising the barrier region has been taken to be a constant ε.

Just as in the case of Scalapino and Marcus, KSH defined initial- and final-state wave functions in the WKB approximation. But contrary to Scalapino and Marcus, who inserted $V_i(\bar{r}, \omega)$ into the definition of K and used the full Hamiltonian to transfer electrons across the barrier, KSH used only the zeroth-order Hamiltonian

$$H_0(\bar{r}) = \frac{p^2}{2m} + \phi, \qquad z_R < z < z_L \tag{23}$$

0 otherwise, to calculate K. They inserted the full Hamiltonian $H(\bar{r}) = H_0(\bar{r}) + V_i^k(\bar{r}, \bar{R})e^{i\omega_0 t}$ into Eq. (4), and solved for $b_f(t)$ (taking $\omega_f = \omega_i - \omega_0$) to first order in the interaction potential. In elastic tunneling the full time-independent Hamiltonian transfers electrons across the barrier with no loss of energy. In inelastic tunneling the oscillating interaction potential $V_i^k(\bar{r}, \bar{R})$, transfers electrons across the barrier, with energy loss in the transition. Inclusion of the interaction potential as a transfer Hamiltonian (as in KSH), rather than causing a modification of the tunneling electron wave functions (as in Scalapino and Marcus) is much simpler for a com-

plicated potential. The matrix elements were then given by

$$M_{if}^{k} = \int_{-\infty}^{\infty} d^3x\, \psi_f^* V_i^k(\bar{r}, \bar{R}) \psi_i \tag{24}$$

The inelastic tunneling current equation was the same as given above for the theory of Scalapino and Marcus [Eq. (12)]. The evaluations of the matrix elements were somewhat complex but quite straightforward. The results for the matrix elements have been presented elsewhere several times[9, 10] and will not be reproduced here. The final expression for the step in conductance due to the opening of an inelastic tunneling channel was

$$\Delta\sigma^k = \frac{8\pi n e^2}{\hbar}\left(\frac{L}{\pi}\right)^6\left(\frac{m}{\hbar^2}\right)^3 (\varepsilon_f)^{1/2}(\varepsilon_f' - eV)^{1/2}$$

$$\times \int_0^{2\pi} d\phi \int_0^{2\pi} d\phi' \int_0^1 d(\cos\theta) \int_0^1 d(\cos\theta') |M_{if}^k|^2 \Theta(\hbar\omega_k - eV) \tag{25}$$

where L was a normalization length that canceled out with its counterpart in the matrix elements, ε_f and ε_f' were the Fermi energies of the initial and final state metals, and the matrix elements were evaluated at the Fermi surfaces (assumed spherical) of the two metals.

The integrals over θ and ϕ were integrals over possible initial and final angular directions for the tunneling electrons. The electrons are most likely to tunnel normal to the interface, since they then have to travel a shorter distance through the forbidden energy region, but off-axis scattering can be appreciable. When we take into account the extra momentum space accessible to larger scattering angles, electrons incident normally on the oxide interface are most likely to tunnel inelastically into a final state with momentum an angle of 7° off normal when the scattering dipole is perpendicular to the interface. The matrix elements have relative phases associated with them which are given by $e^{i\alpha_{\parallel}\rho_n}$, where $\alpha_{\parallel}$ is the change in momentum parallel to the interface between the initial and final electron states, and ρ_n specifies the position of the individual dipole scatterer in a plane parallel to the interfaces. These phase factors tend to add or cancel in the matrix element sum, depending on the spatial extent of the molecule, and the amount of off-axis scattering. Roughly speaking, since $k_f \sim 1$ Å^{-1}, this means that the matrix elements for scattering centers 8 Å apart will be 1 rad out of phase with each other (all other things being equal) and will constructively and destructively interfere, destroying the symmetries one might expect from a long-wavelength interaction. Off-axis scattering is even more important when the scattering dipole is oriented parallel to the interface. One of the angular integrations can always be done analytically[10]; the other three were done numerically.

When reasonable values for the parameters n, s, ϕ, a, and Z were taken, KSH found that the predicted absolute intensities for a monolayer of hydroxyl ions were in agreement with experiment. Just as in Scalapino and Marcus, the theoretical changes in conductance were normalized to the elastic conductance expected for the same model barrier potential. In all cases discussed in this chapter, the barrier potential has been assumed to be constant inside the oxide, and zero in the metal electrodes. Significantly better fits to the current–voltage characteristics of both undoped,[27] and intentionally doped[28–32] junctions can be obtained by assuming more complex barrier potential shapes. While these more complex barriers are easily incorporated into models of elastic tunneling, including them in inelastic tunneling models is very difficult. To date only square barriers have been treated in inelastic electron tunneling theories. While normalization of the inelastic tunneling current to the elastic current allows one to correct in a crude way for uncertainties in the detailed tunneling barrier characteristics, there are several interesting correlations of relative tunneling spectroscopy intensities with barrier potential parameters[30–32] that are worth exploring.

When the molecules were relatively close to one metal surface, KSH predicted, as did Scalapino and Marcus, that vibrations with net dipole moments normal to the surface should couple more strongly with the tunneling electrons than vibrations with net dipole moments parallel to the surface. However, for dipoles deep within the tunneling barrier, this is no longer true. There are two reasons for the relaxation of this intensity preference. First, off-axis scattering tends to allow constructive and destructive interference between the matrix elements of the individual scattering dipoles. Second, near the center of a parallel plate capacitor, the dipole potential, which is odd in the coordinate parallel to the dipole, tends to integrate to zero for dipoles normal to the interface, but not for dipoles oriented parallel to the interface. This effect is shown in Figure 3, where the angle-averaged matrix element squared (which has the dimensions of a cross section), is plotted as a function of the position of the molecule in the barrier, for point dipoles oriented perpendicular and parallel to the surface, for three different vibrational energy losses. This figure shows that for molecules located close to the interfaces, dipoles oriented normal to the surface have a higher probability of excitation, but that deep within the oxide, dipoles oriented parallel are favored, although at a lower scattering amplitude. Therefore the orientation intensity preference will depend sensitively on the position of the dopant molecule in the tunneling barrier.

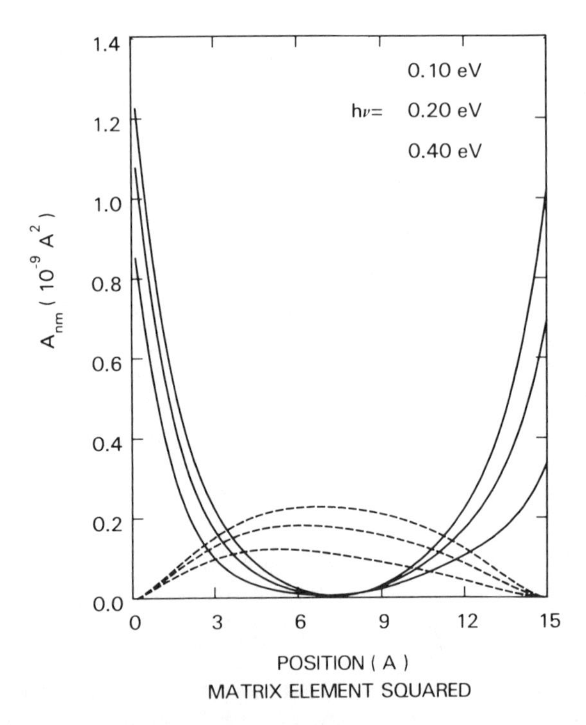

Figure 3. Plot of A_{nm}, the angle-averaged matrix element squared for an inelastic tunneling process in the theory of Kirtley, Scalapino, and Hansma, for three different vibrational mode energies, for a point dipole oriented parallel (dashed curves) or perpendicular (solid curves) to the interface, as a function of the position of the dipole in a square barrier of height 2 eV and width 15 Å. A_{nm} can be thought of as the cross section for an inelastic transition. Dipoles oriented normal to the interface are more likely to excite an inelastic transition if close to one of the metal electrodes, but dipoles oriented parallel to the interface are favored deep within the barrier. Reprinted, with permission, from reference 10.

The theory of Scalapino and Marcus, with its molecular dipole potential, only allowed vibrational modes with net oscillating dipole moments (infrared active modes). KSH with its sum of dipoles located at the atomic positions, relaxes these intensity preferences also. This relaxation of the rules again has two causes: Assume a molecule has an infrared inactive mode with two spatially separated oscillating dipoles that just cancel out to give no net molecular dipole moment. (1) If the initial and final electronic wave functions have different momenta parallel to the surface, the matrix elements for the two dipoles, if spaced the same distance from one metal electrode, have different phases, and will tend to incompletely cancel when the total matrix element for a particular vibrational mode is squared and angle averaged. (2) If the two dipoles are different distances from the metal electrode, they will have different weights (see Figure 3) in the matrix element sum, and will incompletely cancel. Numerical calculations of the modes of CO_2 indicate that the Raman active symmetric stretching mode, which has no net dipole moment, should have a tunneling spectroscopy intensity 4.2 times smaller than the infrared active antisymmetric stretching mode intensity, assuming an orientation with the molecular major axis normal to the interface, and only 20% smaller if the molecular axis lies parallel to the interface. Similarly, modes which are neither Raman nor infrared active should also be observable using tunneling spectroscopy. KSH drew these conclusions based on an interaction potential which contained only the intrinsic oscillating dipole potentials. Presumably if the polarizability of the molecule was included, Raman active modes would become even more "allowed." But as will be seen below, the KSH potential works reasonably well with infrared-derived dipole derivatives: it may not be necessary to include the polarizability interaction to model tunneling spectroscopy intensities.

Because of the complexity of the interaction of the tunneling electrons with even the long-range intrinsic dipole potential, any test of the theory of tunneling spectroscopy intensities with experiment must involve detailed calculations of all of the modes of a molecule with known orientation and position with respect to the metal electrodes. L. M. Godwin, H. W. White, and R. Ellialtioglu[34] applied the theory of KSH to the tunneling spectroscopy spectra of formic acid in an Al–Al_2O_3–Pb junction. Formic acid bonds to the surface of alumina as a formate ion with one oxygen bonded to the carbon and an Al atom, and with the second oxygen double bonded to the carbon. Godwin *et al.* argued that chemical intuition would indicate that the ion should bond to the surface with the single-bonded C–O bond normal to the interface. They restricted themselves to a simple partial charge model, setting $Z_n^k = Z_n$ independent of the vibrational mode k. The partial charges were assigned from infrared measurements. Godwin *et al.* found that the predicted intensity patterns fit experiment better if the molecule was assumed oriented with the C–O bond normal to the interface than if the C–O was assumed parallel to the interface. They also found that the carboxylate ion group intensities fit experiment well, but that the C–H bond mode intensities did not work as well. They attributed this discrepancy to the fact that the charge motion does not follow the nuclear motion in the deformation of a C–H bond. They stated that a more sophisticated model than the partial charge model would be required for accurate intensity calculations for C–H vibrational modes.

Such an improved model has recently been proposed by Kirtley and Hall[10] in an analysis of tunneling spectroscopy intensities of methyl-sulfonic acid on alumina. Hall and Hansma[35] analyzed the tunneling spectroscopy vibrational mode energies of this surface species. This analysis indicated that the SO_3H group bonds to the surface as a sulfonate ion (losing a proton). The sulfonate ion has the structure of two tetrahedrons aligned point to point along the C–S bond, with the methyl group hydrogens staggered with respect to the sulfonate group oxygens. The oxygen atoms are bound to the alumina in nearly equivalent chemical positions, and therefore the molecule is oriented with the C–S bond normal to the surface with the hydrogens closest to the Pb interface.

Kirtley and Hall[10] rewrote the sum over dipole potentials associated with each atom in the molecule [Eq. (19)] as a sum over dipole potentials associated with each symmetry coordinate. As an example of the distinction between atomic, bond, and symmetry coordinates, consider the C–H stretching modes of the methyl group in methyl sulfonic acid. This group has roughly tetrahedral structure with the three hydrogens and the carbon at the tips of the tetrahedron. In the symmetric stretching mode the atomic positions of the hydrogens move up relative to the center of mass of the molecule, and the atomic position of the carbon moves down with respect to the hydrogens. The relevant bond coordinates are the changes in the lengths of the C–H bonds. The symmetry coordinate associated with the symmetric C–H stretching mode has all three C–H bonds stretching or compressing in unison. In contrast, the symmetry coordinate associated with the antisymmetric stretch mode involves the compression of one C–H bond while the other two stretch. The vibrational normal modes can be expressed in terms of amplitudes of displacement of the symmetry coordinates A_{kj} where the matrix A_{kj} is nearly diagonal: each normal mode is primarily associated with the displacement of a particular symmetry coordinate.

In the symmetry group picture the interaction potential becomes

$$V_i^k(\bar{r}, \bar{R}) = -e \sum_j \bar{\mu}_j^k \cdot \nabla_j \left(\frac{1}{|\bar{r} - \bar{R}_j^0|} \right) e^{i\omega_k t} \tag{26}$$

where j is the symmetry coordinate number and $\bar{R}_j^0$ is the center position assigned to the jth symmetry coordinate.

This form of the potential had the advantages that direct comparison of the results could be made with comparable infrared measurements, and the orientation of the dipole associated with each symmetry coordinate could be unambiguously assigned (at least for an ion with as much symmetry as $CH_3SO_3^-$). It had the disadvantage that the positions of the oscillating charges had to be assigned to each symmetry coordinate. This meant that the sum was over groups of atoms rather than atoms, and was therefore less local. Kirtley and Hall found that the qualitative results obtained from the theory were the same for a given net dipole moment derivative with respect to symmetry coordinate whether the potential was cast in an atom-by-atom, bond-by-bond, or symmetry coordinate sum. This was the result of the long-range nature of the interaction and the relatively small size of the molecule.

Infrared intensity data are traditionally analyzed in terms of the dipole derivative $\partial\mu/\partial S_j$. The dipole moment in Eq. (26) was given by

$$\bar{\mu}_j^k = A_{kj}\frac{\partial\bar{\mu}}{\partial S_j} \tag{27}$$

where the orientation of the dipole associated with each symmetry coordinate was assigned by symmetry considerations, and A_{kj} was the amplitude of the jth symmetry coordinate due to the kth normal vibrational mode. The matrix A_{kj} was determined by standard normal mode analysis. The dipole derivatives $\partial\mu/\partial S_j$ were either taken from an analysis of infrared data or used as fitting parameters.

Since no complete analysis of the dipole derivatives $\partial\mu/\partial S_j$ existed for the $CH_3SO_3^-$ species, Kirtley and Hall used data for the methyl group vibrations of a number of similar species. They argued that the methyl group ligand dipole derivatives for ethane, which has a tetrahedral x-bonding structure and nonpolar C–C bond, should be most similar to $CH_3SO_3^-$.

Figure 4 compares experimental results with Kirtley and Hall's theoretical intensities for the methyl group vibrations, assuming infrared-derived ethane dipole derivatives, an orientation with the C–S bond normal to the interface, and the hydrogens at various distances from the Pb surface. As the Pb–hydrogen distance d was increased, the theoretical predictions for the intensities of the symmetric modes 2 and 4, which had net oscillating dipole moments normal to the interface, became weaker, while the antisymmetric modes 7, 9, and 11, which had net moments parallel to the surface, became stronger. This was just what was expected from Figure 3. The best fit occurred at $d = 1.5$ Å, although theory predicted a symmetric C–H bond mode No. 4 too large and its corresponding antisymmetric mode No. 11 too small in all cases.

If Kirtley and Hall assumed that the C–S bond was oriented parallel to the interface, the agreement between theory and experiment was poor for all assumed d's. Figure 5 compares the predictions of Kirtley and Hall for normal orientation versus parallel orientation for $d = 1.5$ Å with experiment. The better fit of theory to experiment for normal orientation supports the proposed orientation of Hall and Hansma.

Kirtley and Hall also derived molecular dipole derivatives from tunneling spectroscopy intensities using a nonlinear least-squares fit of the 11 dipole derivatives $\partial\mu/\partial S_j$ to the 11 measured tunneling spectroscopy inten-

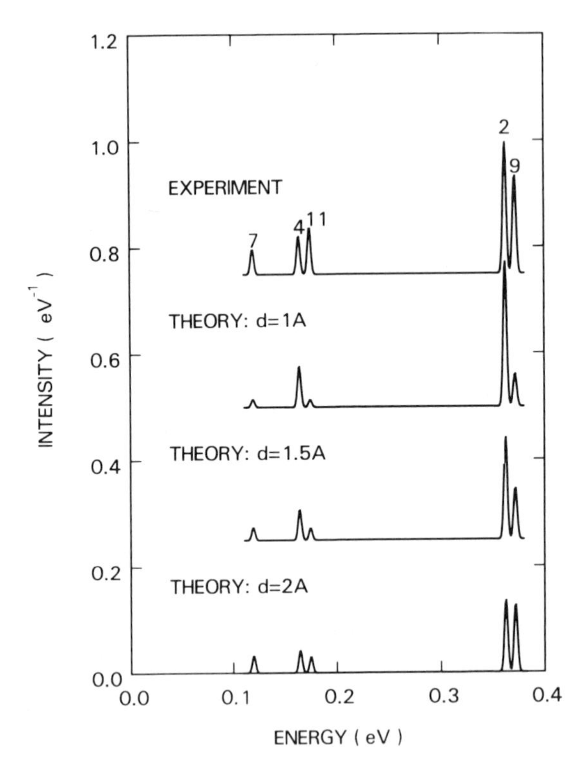

Figure 4. Comparison of the theory of Kirtley, Scalapino, and Hansma with experiment for the methyl group vibrations of methyl sulfonic acid on alumina, as measured with tunneling spectroscopy. The methyl group dipole derivatives were taken from infrared data for ethane. The methyl group was assumed oriented normal to the interface, with the hydrogens 1, 1.5, and 2 Å from the image plane of one metal electrode. As the assumed position moved further into the oxide, the symmetric C–H modes 2 and 4, which had net oscillating dipole moments normal to the interface, became weaker, but the antisymmetric C–H modes 4, 7, and 11, which had net oscillating dipole moments parallel to the interface, became stronger. This is just what was expected from Figure 3. The best fit of theory to experiment occurred at 1.5 Å. The spectra are offset from the baseline for clarity. Reprinted, with permission, from reference 10.

sities, assuming $d=1.5$ Å and the C–S bond normal to the oxide. The agreement between theory and experiment was good if the C–S bond was assumed normal to the interface and poor for parallel orientation.

These calculations showed that it was possible to obtain useful information from an analysis of tunneling spectroscopy intensities, at least in this highly symmetrical case. The relatively good agreement between experiment and theory was remarkable when one considers that (1) Kirtley and Hall only included the intrinsic molecular vibrational potential, neglecting the

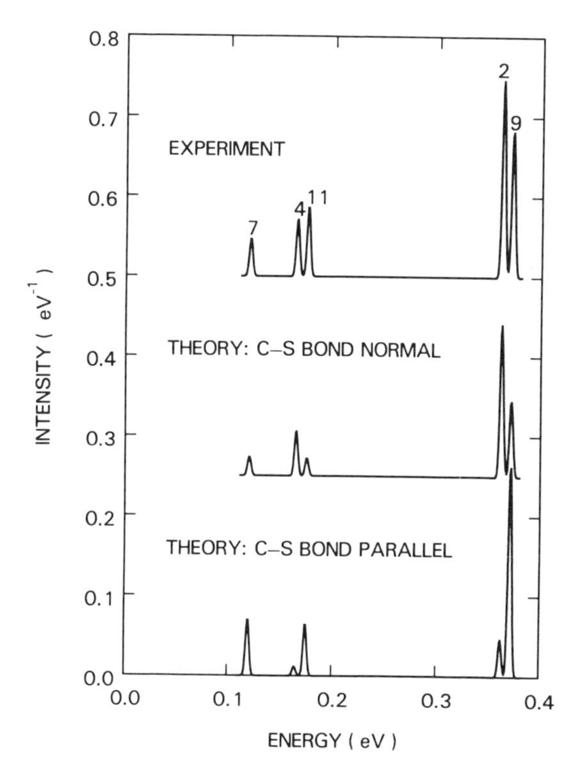

Figure 5. Comparison of the predictions of the theory of Kirtley, Scalapino, and Hansma with experiment for the methyl group vibrations of methyl-sulfonic acid on alumina, for assumed orientations with the C–S bond oriented normal and parallel to the interface. The better agreement between theory and experiment for normal orientation supports the proposed orientation of Hall and Hansma (reference 28). Reprinted, with permission, from reference 10.

induced dipole interaction potential; and (2) the dipole derivatives were taken from vapor state data. It is not unreasonable to expect modifications of the relative vibrational mode dipole derivatives in a tunneling junction geometry.

However, in at least one instance, the presence of a metal overlayer did not strongly affect the relative vibrational mode dipole derivatives of a molecular monolayer. Hartstein, Kirtley, and Tsang[36] measured the vibrational spectrum of 4-nitrobenzoic acid absorbed on the surface of a Si slab, using the multiple attenuated total reflection technique. They found that while the entire ir absorption spectrum increased in intensity when thin layers (<60 Å) of Ag were evaporated on the monolayer, the relative intensities did not change.

3.3. Short-Range Models

Another objection that can be made to the analysis of vibrational mode intensities using the KSH theory is that the interaction between the tunneling electron and the vibrating molecule is essentially long-range, falling off with molecule–tunneling-electron separation as $1/r^3$. In order to make confident comparisons of these theories with experiment, we must make sure that the short-range interactions are not important.

Kirtley and Soven[11] have calculated the short-range contribution to intensities in tunneling spectroscopy to explore this question. They chose the relatively simple molecule C–O, and employed the multiple scattering $X\alpha$ approach for their calculations.

In this approach, the exchange-correlation portion of the total molecular potential is approximated by a one-electron potential, similar to the Thomas–Fermi potential used in solid-state physics. This reduces the many-body Schrödinger equation to a semiempirical single-particle theory. The elastic tunneling current in the presence of the molecular potential was found for a given internuclear spacing. Then the spacing was varied, and the inelastic current was related to the change in the elastic current with respect to the vibrational normal coordinate in the first-harmonic approximation. The elastic tunneling current was found by solving Schrödinger's equation, in a manner very similar to that used for electron scattering in vacuum. The solution was found only to first Born approximation: the scattering from the right edge of the barrier was neglected when boundary conditions were matched at the left edge. The formulation of the $X\alpha$ procedure used by Kirtley and Soven involved spherical averaging of the molecular potential within "muffin tins" surrounding the atoms in the molecule, and outside of a "muffin tin" surrounding the molecule. This additional simplification of the problem eliminates any dipole or higher multipole fields, so in some sense the Kirtley and Soven theory is complementary to KSH: KSH includes the long-range interactions but has no short-range multiple scattering component; Kirtley and Soven includes the short-range but explicitly averages out the long-range fields. The inelastic scattering comes from a modulation of the wave functions transmitted through the barrier region by the molecular potential.

The molecular potential for fixed nuclear positions $\bar{R}_n$ was given by

$$V_{x\alpha}(\bar{r}, \bar{R}_n) = V_H(\bar{r}) + V_N(\bar{r}, \bar{R}_n) + V_{xc}(\bar{r}) \tag{28}$$

The static electron–electron (Hartree) interaction term was given by

$$V_H(\bar{r}) = \int d^3r' \frac{e^2 n(\bar{r}')}{|\bar{r} - \bar{r}'|} \tag{29}$$

where $n(\bar{r}')$ was the electron density at position $\bar{r}'$. The nuclear-attraction term was given by

$$V_n(\bar{r}, \bar{R}_n) = \sum_N \frac{-Z_N e^2}{|\bar{R}_N - \bar{r}|} \tag{30}$$

where Z_N was the screened nuclear charge. The exchange-correlation term was approximated by

$$V_{xc}(\bar{r}) = -3\alpha e^2[3n(\bar{r})/8\pi]^{1/3} \tag{31}$$

where α was a parameter of order 0.7. Using the muffin tin approximation, the molecular potential was spherically averaged within spheres centered on the individual atoms in the molecule, and outside an outer sphere surrounding the molecule. The intersphere region was volume averaged. An initial guess was made for the molecular potential; the wave functions were found by numerical integration and matching wave functions at the muffin-tin boundaries; and the potential was recalculated. This procedure was iterated to convergence. Finally, the molecular potential was exponentially attenuated at large distances and replaced by a polarizability potential as discussed by Davenport, Ho, and Schrieffer.[37]

The molecular potential was placed within the oxide barrier with one edge of the outer molecular sphere touching one counterelectrode edge. The barrier potential was taken to be a constant outside of the outer molecular sphere. The tunneling matrix element was calculated for a given internuclear spacing by matching wave functions across all boundaries. The wave functions in the oxide outside the molecular outer sphere was expanded in exponentially decaying plane waves. The wave functions on the molecular outer sphere boundary were expanded in spherical harmonics

$$\psi_{sp}(|\bar{r}| = R) = \sum_\lambda A_\lambda \left[k_\lambda^{(1)}(kR) Y_\lambda(\hat{r}) + \sum_L S_{\lambda L} k_L^2(kR) Y_L(\hat{r}) \right] \tag{32}$$

where $k = [2m(\phi - \varepsilon_f)/\hbar^2]^{1/2}$, $k_\lambda^{1,2}(kR) = -i^{-\lambda} h_\lambda^{1,2}(kR)$, $h_\lambda^{1,2}(kR)$ are spherical harmonics, and the sums λ and L are taken over all l and m indices. The scattering matrices $S_{\lambda L}$ were found by solving the Schrödinger equation for a scattering state of negative energy $-E$ in the presence of the molecular potential, following the same iterative procedure used for the molecular wave functions.

The boundary matching problem was simplified considerably by using the Bardeen picture: only waves decaying to the right were allowed for the initial-state wave function, and only states decaying to the left were allowed for the final-state wave function. The tunneling matrix element was then found using the Bardeen expression [Eq. (6)]. The total elastic matrix elements $M_{k'k''}$ were found in this way for a set of C–O internuclear spacings. The inelastic tunneling matrix $M^i_{k'k''}$ was defined by

$$M^i_{k'k''} = \langle \nu | M_{k'k''} | 0 \rangle \tag{33}$$

where $|0\rangle$ and $|\nu\rangle$ were, respectively, the ground and νth excited vibrational wave functions of the molecule (calculated in the harmonic oscillator approximation). Once the inelastic tunneling matrix elements were determined, the inelastic tunneling current was calculated as usual with Eq. (25).

A question arose concerning the assignment of the tunneling electronic energies relative to the molecular bound state energies. Kirtley and Soven argue that it was reasonable to assume that the one-electron potential in the vicinity of the molecule was given simply by the sum of the gas phase molecular potential and the oxide barrier potential. This assumption would place the tunneling electron energy at $-(\phi - \varepsilon_f) \sim -2$ eV relative to the molecular vacuum level potential. However, they did not limit themselves to this assumption. Figure 6 shows the predicted inelastic cross sections

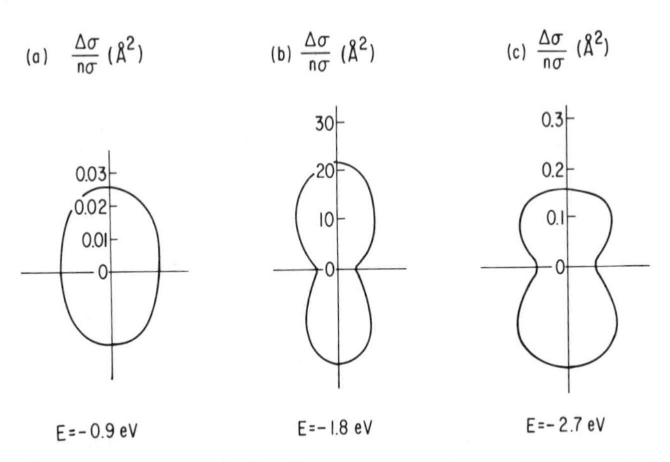

Figure 6. Prediction of the short-range multiple scattering theory of Kirtley and Soven for the first-harmonic vibrational excitation cross section of the carbon–oxygen stretch mode of CO in a tunneling junction geometry. In these polar plots 0^0 (vertically upward) corresponds to the CO oriented with its major axis normal to the interface, with the carbon closest to the oxide. The labeled energies correspond to the energy of the tunneling electron relative to the molecular vacuum level. The orientation dependence of the scattering cross-sections does not depend strongly on the tunneling electron energy, but the absolute values of the cross sections go through a strong resonance at -1.8 eV, corresponding to a bound-state energy of the molecule. Reprinted, with permission, from reference 11.

($\Delta\sigma/n\sigma$, where n is the surface density of molecular scatterers), for the $\nu=0$ to $\nu=1$ transition for CO as a function of the assumed orientation of the molecule on the surface, for three different tunneling energies. In these polar plots the point at 0° (vertically upward) corresponds to CO orientation normal to the oxide surface with the carbon atom closest to the oxide. The orientation dependence of the cross section does not depend strongly on the tunneling electron energy. However, there is a strong resonance at -1.8 eV, corresponding to a bound-state energy level 1.8 eV below the vacuum level for the $X\alpha$ potential used.

Figure 7 shows the predicted CO inelastic cross section of Kirtley and Soven for transitions from the vibrational ground state [$n=0$ for the harmonic oscillator model $E_n=(n+1/2)\hbar\omega_0$] to the first ($n=1$) and second ($n=2$) excited states, as a function of the energy of the tunneling electron with respect to the vacuum level of the molecule, assuming CO orientation normal to the interface with the carbon closest to the oxide. When the tunneling electron energy matches that of the molecular bound state, the predicted first-harmonic ($n=0$ to $n=1$) cross section can become quite large. The crucial point, however, is that the second-harmonic ($n=0$

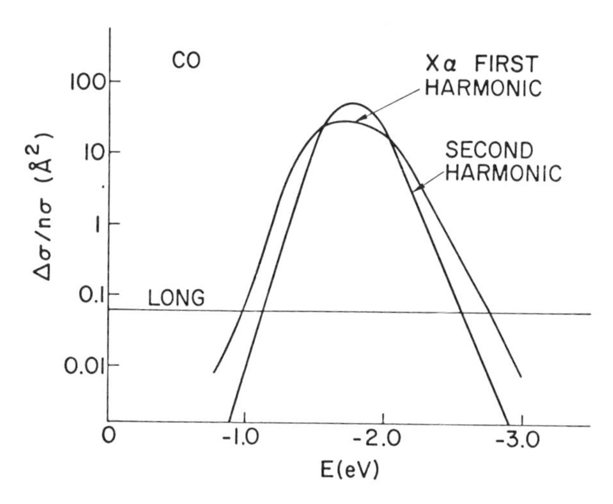

Figure 7. Inelastic tunneling cross sections for the short-range theory of Kirtley and Soven, for CO oriented normal to the interface, as a function of tunneling energy relative to the molecular vacuum level, for the first- and second-harmonic transitions. Included for comparison is the long-range theoretical prediction of Kirtley, Scalapino, and Hansma, for the same case. The short-range cross sections can become very large when the tunneling energy approaches a bound-state resonance. But at resonance, the higher-harmonic transitions should appear as well. Higher-harmonic transitions do not appear strongly in tunneling spectroscopy, indicating that the short-range interactions are probably not important. Reprinted, with permission, from reference 11.

to $n = 2$) cross section increases also, so that at resonance the second-harmonic term is as large as the first-harmonic term. Also included in Figure 7 is the predicted cross section for the long-range (KSH) theory, assuming the same orientation, with the oxygen 1.0 Å from the counterelectrode, an oxide dielectric constant of 3, an rms C–O bond displacement of 0.034 Å, and a dipole derivative of $1.32e$.

Kroeker, Kaska, and Hansma[38] have reported tunneling spectroscopy spectra for CO on Rh-doped alumina. They found values of 0.1% for $\Delta\sigma/\sigma$. This value is difficult to compare with theory since the absolute surface densities were not known. Assuming that the surface density was that of a close-packed array of hard spheres, this would imply a cross section of 0.03 $Å^2$, somewhat smaller than the KSH prediction.

In any case comparison of the long-range with the short-range cross sections shows that the long-range interaction dominates when the tunneling electron energy is off a molecular bound-state resonance: when the tunneling electron energy is at resonance, where the short-range interactions might be expected to dominate, higher-harmonic excitations should be seen. Experimentally the second-harmonic intensities are very small (~ 200 times smaller than the first-harmonic intensity for benzoic acid on alumina). Let me emphasize the argument, which should also apply to other electron vibrational spectroscopies: Short-range interactions should excite higher-harmonic transitions. The absence of these transitions in the vibrational spectra indicates that the interaction mechanism is probably long range.

One possible explanation for the relatively weak role played by the short-range interactions is that the molecules are in intimate contact with a metal electrode. The molecule–metal interaction should broaden the bound-state molecular energy levels, weakening the resonances. As is shown in Figure 7, the short-range contribution to the total cross section is relatively weak off resonance.

A second argument for the long-range nature of the interaction mechanism comes from an analysis of opposite bias intensity asymmetries in tunneling spectroscopy. The change in conductance due to the opening of a given inelastic tunneling channel in an Al–Al_2O_3-doped impurity–Pb junction is larger if the junction is biased with the Al negative than if the Al is biased positive with respect to the Pb. Figure 8 shows typical ratios of the conductance steps for the two opposite bias polarities of a junction doped with benzoic acid.[9] Similar results have been obtained for a large number of other molecular layers.[30] Other characteristics of the spectra—relative mode intensities, linewidths, and vibrational mode energies—appear to be the same for opposite bias polarities. The reason for the asymmetry in intensities is simple. More energetic electrons are more likely to penetrate

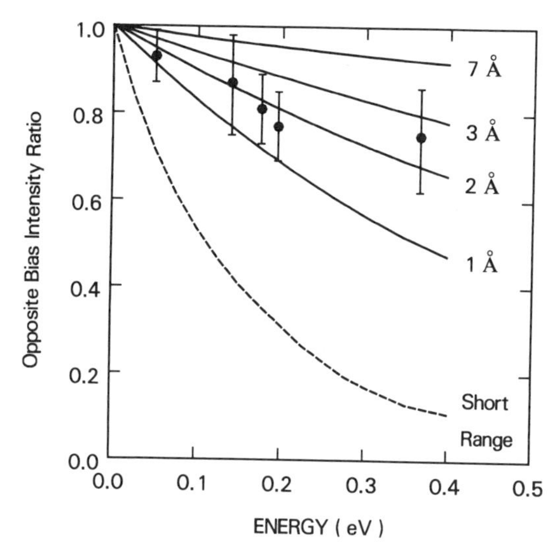

Figure 8. Comparison of the ratio of inelastic conductance steps for opposite bias voltages for the long-range theory of Kirtley, Scalapino, and Hansma (solid curves), an infinitely short-range theory (dashed curve), and experiment (solid curves with error bars), as a function of the vibrational energy loss. The long-range curves were generated for a randomly oriented dipole scatterer for various distances from one metal electrode. In all cases the barrier potential was assumed to be 2 eV and the barrier thickness was taken to be 15 Å. The long-range theory predicts relatively small asymmetries that agree with experiment for dipole–image-plane distances of a few angstroms. The short-range theory can be made to agree with experiment only if the dipole is placed very near the center of the barrier—much closer than is reasonable from other considerations. This tends to support the view that the interaction in tunneling spectroscopy is long range.

the barrier. The molecular impurities are located close to one edge of the tunneling barrier potential. Therefore, electrons tunneling inelastically in one direction lose energy before penetrating the bulk of the insulator. Electrons tunneling in the opposite direction lose energy after penetrating the bulk of the barrier. If we assume that the interaction between the tunneling electrons and the molecular layer is local, very simple calculations[39] show that the asymmetry should be large (dashed curve of Figure 8 with $a = 1$ Å, $\phi = 2$ eV, and $s = 15$ Å), much larger than is observed experimentally.

If the potential is long-range, however, the location of the scattering dipole within the barrier is not as important and the asymmetry should be reduced. Figure 8 shows the predictions for the opposite bias conductance step ratio of the theory of KSH, as a function of energy loss, for point dipoles located various distances from one metal surface, again assuming $s = 15$ Å and $\phi = 2$ eV. These curves were generated by taking the sums of the squares of the conductance steps for the dipoles oriented normal and parallel to the interface: it was assumed that the dipoles were randomly oriented in the barrier. Good agreement with experiment can be obtained if it is assumed that the scattering dipoles are located a few angstroms into the barrier region. To get comparable agreement for a short-range potential, one has to assume that the dipoles are within a few angstroms of the center of the insulating region. Both the manner in which inelastic electron tunneling junctions are made, and modeling of the current–voltage characteristics of the junctions,[28–33] indicate that this is not the case.

4. Conclusions

(1) Very simple potentials can be used to model intensities in inelastic electron tunneling spectroscopy with success: The partial charge model of Kirtley, Scalapino, and Hansma predicts the correct order of magnitude tunneling spectroscopy intensities; can be used in conjunction with infrared derived dipole derivatives to infer the orientations of formic acid, and methyl-sulfonic acid on alumina; and predicts the correct opposite bias asymmetry ratios of intensities.

(2) In addition, theoretical modeling of tunneling spectroscopy intensities indicates that Raman active as well as infrared active vibrational modes should be observable in tunneling spectroscopy, and that the orientational selection rules that are, for example, quite simple in infrared reflection spectroscopy from metals, are more complex in tunneling spectroscopy.

(3) It appears that the short-range portions of the total interaction potential need not be considered in modeling tunneling spectroscopy intensities. This is fortunate, since the calculations of the short-range effects are extremely cumbersome, while long-range potential calculations can use intensity data derived from optical spectroscopies to predict tunneling spectroscopy intensities.

However, there is a great deal that remains to be done before tunneling spectroscopy intensities can be used as a reliable analytical tool. (1) The relative contributions of the induced vibrational potential should be worked out in a fashion consistent with KSH. (2) Detailed comparison of theory with experiment should be done for a wider range of molecular layers. It would be especially helpful to have a comparison of the infrared and tunneling spectroscopy spectra of the same or very similar samples so that the effects of the top electrode can be judged.

References

1. I. Giaever, in *Tunneling Phenomena in Solids* (E. Burstein and S. Lundquist, eds.), pp. 255–271, Plenum Press, New York (1969).
2. W. L. McMillan and J. M. Rowell, Lead phonon spectrum calculated from superconducting density of states, *Phys. Rev. Lett.* **14**, 108–112 (1965).
3. J. Klein, A. Leger, M. Belin, D. Defourneau, and M. J. L. Sangster, Inelastic electron tunneling spectroscopy of metal–insulator–metal junctions, *Phys. Rev. B* **7**, 2336–2348 (1973).
4. G. Herzberg, *Infrared and Raman Spectra of Polyatomic Molecules*, Van Nostrand, New York (1945).
5. E. Evans and D. L. Mills, Theory of inelastic scattering of slow electrons by long-wavelength surface optical phonons, *Phys. Rev. B* **5**, 4126–4139 (1972).
6. H. Ibach, Comparisons of cross sections in high-resolution electron-energy-loss spectroscopy and infrared reflection spectroscopy, *Surf. Sci.* **66**, 56–66 (1972).
7. D. J. Scalapino and S. M. Marcus, Theory of inelastic electron–molecule interactions in tunnel junctions, *Phys. Rev. Lett.* **18**, 459–461 (1967).
8. J. Lambe and R. C. Jaklevic, Molecular vibration spectra by inelastic electron tunneling, *Phys. Rev.* **165**, 821–832 (1968).
9. John Kirtley, D. J. Scalapino, and P. K. Hansma, Theory of vibrational mode intensities in inelastic electron tunneling spectroscopy, *Phys. Rev. B* **14**, 3177–3184 (1976).
10. John Kirtley and James T. Hall, Theory of intensities in inelastic electron tunneling spectroscopy: Orientation of absorbed molecules, *Phys. Rev. B* **22**, 848–856 (1980).
11. John Kirtley and Paul Soven, Multiple-scattering theory of intensities in inelastic electron tunneling spectroscopy, *Phys. Rev. B* **19**, 1812–1817 (1979).
12. Elias Burstein and Stig Lundquist, eds. *Tunneling Phenomena in Solids*, Plenum Press, New York (1966).
13. A. D. Brailsford and L. C. Davis, Impurity assisted inelastic tunneling: One-electron theory, *Phys. Rev. B* **2**, 1708–1713 (1970).
14. L. C. Davis, Impurity-assisted inelastic tunneling: Many electron theory, *Phys. Rev. B* **2**, 1714–1732 (1970).
15. C. Caroli, R. Combescot, P. Nozieres, and D. Saint-James, A direct calculation of the tunneling current: IV. Electron–phonon interaction effects, *Solid State Phys.* **5**, 21–42 (1972).
16. T. E. Feuchtwang, Tunneling theory without the transfer Hamiltonian formalism: V. A theory of inelastic electron tunneling spectroscopy, *Phys. Rev. B* **20**, 430–455 (1979).
17. J. Bardeen, Tunneling from a many particle point of view, *Phys. Rev. Lett.* **6**, 57–59 (1961).
18. L. I. Schiff, *Quantum Mechanics*, p. 269, McGraw-Hill, New York (1968).
19. M. H. Cohen, L. M. Falicov, and J. C. Phillips, Superconductive tunneling, *Phys. Rev. Lett.* **8**, 316–318 (1962).
20. B. D. Josephson, Possible new effects in superconductive tunneling, *Phys. Lett.* **1**, 251–253 (1962).
21. J. D. Langen and P. K. Hansma, Can the concentration of surface species be measured with inelastic tunneling?, *Surf. Sci.* **52**, 211–216 (1975).
22. S. L. Cunningham, W. H. Weinberg, and J. R. Hardy, in *Inelastic Electron Tunneling Spectroscopy* (T. Wolfram, ed.), pp. 129–143, Springer-Verlag, New York (1977).
23. A. A. Cederberg, Inelastic electron tunneling spectroscopy: Intensity as a function of surface coverage, *Surf. Sci.* **103**, 148–176 (1981).

24. N. I. Bogatina, I. K. Yanson, B. I. Verkin, and A. G. Batrak, Tunneling spectra of organic solvents, *Sov. Phys.-JETP* **38**, 1162–1165 (1974).
25. M. G. Simonsen, R. V. Coleman, and P. K. Hansma, High-resolution inelastic tunneling spectroscopy of macromolecules and adsorbed species with liquid phase doping, *J. Chem. Phys.* **61**, 3789–3799 (1974).
26. K. W. Hipps, Ursula Mazur, and M. S. Pearce, A tunneling spectroscopy study of the adsorption of ferrocyanide from water solution by Al_2O_3, *Chem. Phys. Lett.* **68**, 433–437 (1979).
27. W. F. Brinkman, R. C. Dynes, and J. M. Rowell, Tunneling conductance of asymmetrical barriers, *J. Appl. Phys.* **41**, 1915–1921 (1970).
28. R. B. Floyd and D. G. Walmsley, Tunneling conductance of clean and doped Al–AlI–Pb junctions, *J. Phys. C* **11**, 4601–4614 (1978).
29. D. G. Walmsley, R. B. Floyd, and W. E. Timms, Conductance of clean and doped tunnel junctions, *Solid State Commun.* **22**, 497–499 (1977).
30. M. F. Muldoon, R. A. Dragoset, and R. V. Coleman, Tunneling asymmetries in doped Al–AlO_x–Pb junctions, *Phys. Rev. B* **20**, 416–429 (1979).
31. C. S. Korman, J. C. Lau, A. M. Johnson, and R. V. Coleman, Studies of aromatic-ring compounds absorbed on alumina and magnesia using inelastic electron tunneling, *Phys. Rev. B* **19**, 994–1014 (1979).
32. J. C. Lau and R. V. Coleman, Ag-vs-Pb electrodes in inelastic electron tunneling spectroscopy, *Phys. Rev. B.* **24**, 2985 (1981).
33. K. W. Hipps and Ursula Mazur, An inelastic tunneling spectroscopy study of some iron cyanide complexes, *J. Phys. Chem.* **84**, 3162–3172 (1980).
33. J. Rath and T. Wolfram, in *Inelastic Electron Tunneling Spectroscopy* (T. Wolfram, ed.), pp. 92–102, Springer-Verlag, New York (1977).
34. L. M. Godwin, H. W. White, and R. Elliatioglu, Comparison of experimental and theoretical inelastic electron tunneling spectra for formic acid, *Phys. Rev. B* **23**, 5688 (1981).
35. James T. Hall and Paul K. Hansma, Adsorption and orientation of sulfonic acids on aluminum oxide: A tunneling spectroscopic study, *Surf. Sci.* **71**, 1–14 (1978).
36. A. Hartstein, J. R. Kirtley, and J. C. Tsang, Enhancement of infrared absorption form molecular monolayers with thin metal overlayers, *Phys. Rev. Lett.* **45**, 201–204 (1980).
37. J. W. Davenport, W. Ho, and J. R. Schrieffer, Theory of vibrational inelastic electron scattering from oriented molecules, *Phys. Rev. B* **17**, 3115–3127 (1978).
38. R. Kroeker, W. C. Kaska, and P. K. Hansma, How carbon monoxide bonds to alumina supported rhodium particles, *J. Catal.* **57** 72–79 (1979).
39. I. K. Yanson, N. I. Bogatina, B. I. Verkin, and O. I. Shklyarevski, Asymmetry of tunnel spectrum intensities of impurity organic molecules, *Sov. Phys.-JETP* **35**, 540–543 (1972).

3

Tunneling Spectroscopies of Metal and Semiconductor Phonons

E. L. Wolf

1. Introduction

Tunneling spectroscopy of metal and semiconductor phonons began with the prize winning research of Esaki, Giaever, McMillan, and Rowell in the late 1950s to mid 1960s. Since that time there has been rapid and fruitful growth of the field. In fact, there have been at least as many (if not more) publications in this field as on the subject of most of this book, tunneling spectroscopy of molecular vibrations. Thus, it would be impossible in one chapter to treat the entire field in depth.

Fortunately, there have been a number of reviews written.[1-4] Thus, rather than skim over the surface of the entire field, I choose in this chapter to emphasize the experiments that have a special bearing on the rest of this book and the recent experiments on proximity effect tunneling that are at the frontier of research at present. Therefore, with due apology to early workers, the references and figures here have not been chosen reflecting historical priority of work but for convenience and clarity in illustrating important points relevant to inelastic electron tunneling spectroscopy (IETS).

In broad terms, there are two basic mechanisms by which the spectrum of quantized lattice vibrations, or phonons, of a metal (or metallic semiconductor) electrode, S, can become observable in the current–voltage ($I-V$)

E. L. Wolf • Ames Laboratory—USDOE and Department of Physics, Iowa State University, Ames, Iowa 50011. The Ames Laboratory is operated for the U.S. Department of Energy by Iowa State University under contract No. W-7405-Eng-82. This research was supported by the Director for Energy Research, Office of Basic Energy Sciences, WPAS-KC-02-02-02.

characteristic of a tunnel junction sandwich of the form $C-I-S$. The first mechanism, which makes possible an inelastic *threshold spectroscopy*, is in its simplest forms similar to the inelastic threshold spectroscopy of molecular vibrations described in the preceding chapter of this volume. Thus, measurements of the second derivative spectrum, d^2I/dV^2, as a function of junction bias V, yield a series of peaks appearing at bias voltages simply related to the phonon frequencies by the threshold relation $eV = \hbar\omega_{\text{ph}}$. In its simplest form, sketched in Figures 1a, 1b the threshold condition allows a qualitatively different alternative final state for the tunneling transition: rather than a single "hot" electron at energy eV above the Fermi energy of the electrode S the new possibility is for the electron to appear at or near the Fermi energy, $E = eV - \hbar\omega_{\text{ph}}$, giving up energy $\hbar\omega_p$ to the simultaneous creation of an elementary excitation. Because the two-particle final state is an added possibility (the elastic process of Fig. 1a, $eV > \hbar\omega_p$ still can occur) one may speak of an additional channel above threshold and expect a steplike increase in the conductance at the threshold bias. Since the required energy $\hbar\omega_p$ is available in either sign of bias $eV = \pm\hbar\omega_p$ one expects in the simplest case increases in the conductance symmetrically located about $V = 0$, as sketched in Figure 1c. This, of course, leads to an antisymmetric structure in the second derivative d^2I/dV^2, Figure 1d. A clear example of this behavior, shown in Figure 2a, is the measurement at low temperature of d^2I/dV^2 of a Ge tunneling *pn* junction[5] (Esaki diode). Here the electrodes C, S, are degenerately doped (metallic) regions of *p*- and *n*-type germanium and the insulator is the narrow depleted region of the semiconductor between the metallic regions. In fact, as we shall see later, further selection rules arising from conservation of crystal momentum restrict the observed phonons in this case to those near the *boundary* of the Brillouin zone. In other cases the spectrum of phonons observed in d^2I/dV^2 is closer to the total phonon density as a function of energy.

The second major spectroscopy of phonons in electron tunneling occurs when the electrode S is a *superconductor* (metals or degenerate semiconductors can quality) whose electron–phonon coupling is strong. In this case a more complicated but very complete and detailed image of the total phonon spectrum, $\alpha^2F(\omega)$, effective in producing the superconducting Cooper pairs, is present in the *tunneling density of states*, $N_T(E)$. Experimentally, this quantity is determined as the ratio $(dI/dV)_S/(dI/dV)_N$, where the subscripts S, N indicate the S electrode to be in the superconducting or normal state, respectively. Theoretically, $N_T(E)$ is directly related to the "density of states" or excitation spectrum of the superconductor:

$$N_T(eV) = \int_{-\infty}^{\infty} \text{Re}\left\{\frac{|E|}{[E^2 - \Delta^2(E)]^{1/2}}\right\} \frac{\partial}{\partial E} f(E + eV)\, dE = \frac{(dI/dV)_S}{(dI/dV)_N} \quad (1)$$

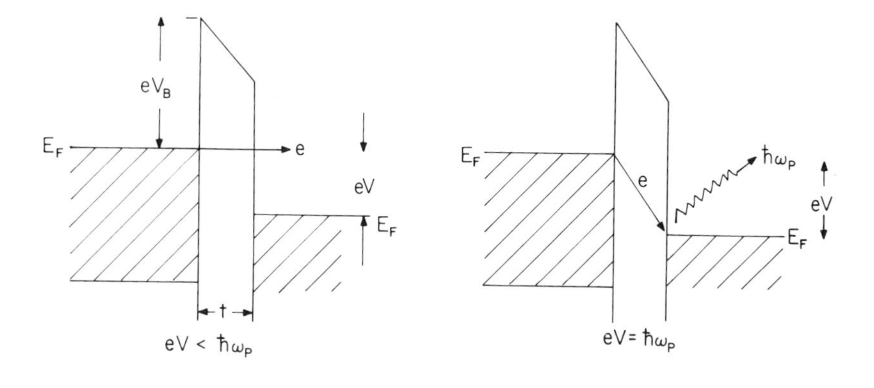

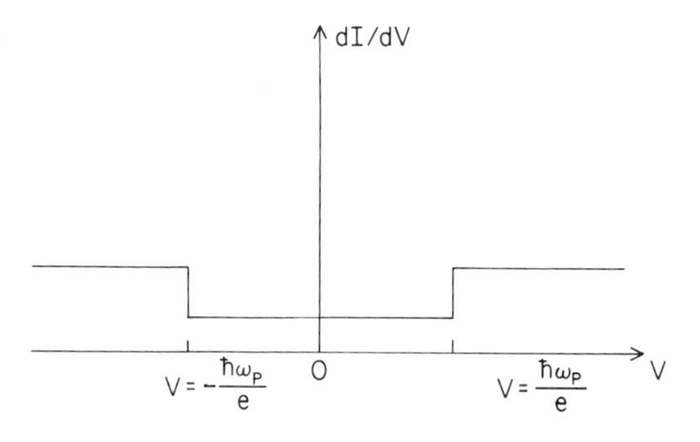

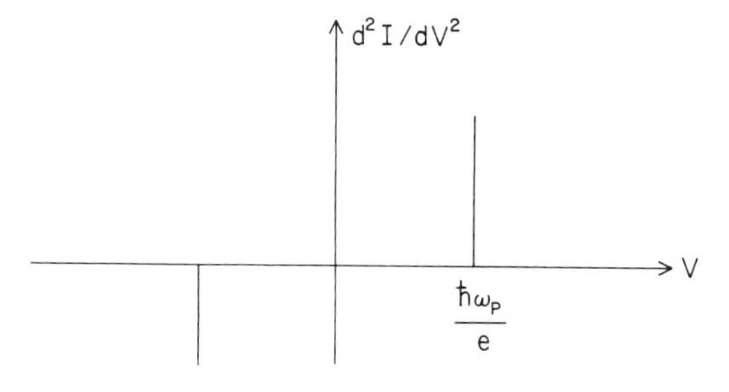

Figure 1. (a) A junction biased below threshold, $V < \hbar\omega/e$, injects electrons at energy up to eV into the positively biased electrode (shown on the right). They tunnel into vacant states above the Fermi level. (b) At threshold, the most energetic electrons, of energy $eV = \hbar\omega_p$, can create one vibrational quantum and still tunnel into a vacant state just above the Fermi level of the positively biased electrode. (c) Idealized conductance expected for simple threshold process has symmetric step increases at $eV = \pm\hbar\omega_p$. (d) The second derivative d^2I/dV^2 provides peaks at the threshold energies, yielding a spectrum of the excitation energies. This structure is antisymmetric about $V = 0$.

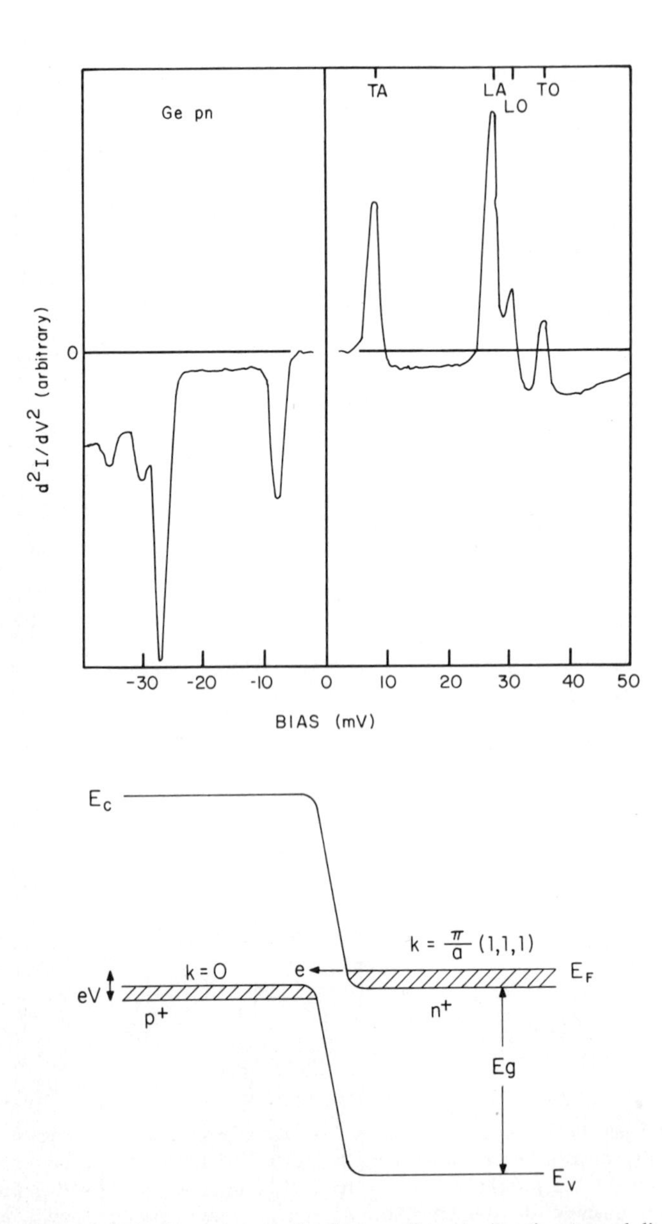

Figure 2. (a) Second derivative spectrum of a Ge tunneling $p-n$ junction (tunnel diode) at low temperature reveals zone boundary phonon energies (redrawn from work reported by R. T. Payne, reference 9). (b) Schematic diagram of tunneling $p-n$ junction.

Here f is the usual Fermi function

$$f=\left[1+\exp\left(\frac{E-E_F}{kT}\right)\right]^{-1} \tag{2}$$

with E_F the Fermi energy, so that the derivative $\partial f/\partial E$ in the integrand is, for small kT, a sharply peaked function whose full width at half-maximum is 3.5 kT, or about 1.3 meV at 4.2 K. Thus in the limit of $T=0$

$$N_T(eV)=\mathrm{Re}\left\{\frac{|eV|}{\left[(eV)^2-\Delta^2(eV)\right]^{1/2}}\right\} \tag{3}$$

This function, taking Δ as a constant Δ_0, is the famous BCS excitation spectrum,[6] with peaks at the gap edges, $E=\pm\Delta_0$. Verification of this function in early work by Giaever[7] (see Figure 3) helped establish the BCS theory and provided direct measurements of the gap parameters of various superconductors. Small deviations from the $\Delta=$ const prediction were early seen, however, in certain higher-T_c metals such as Pb. These deviations are most evident in the bias energy range corresponding to the phonon energies of the superconductor. It follows from what we have presented that the

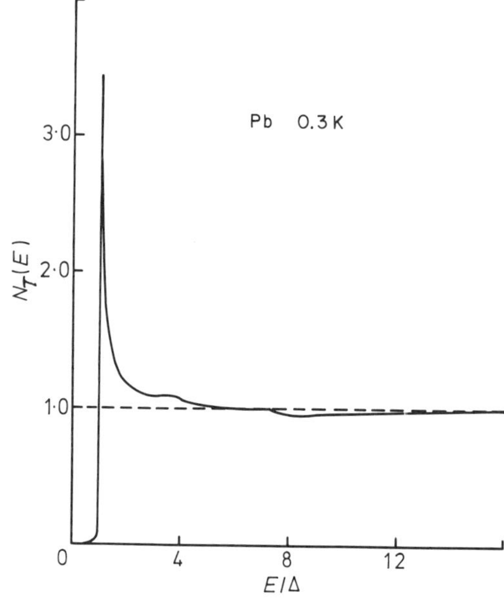

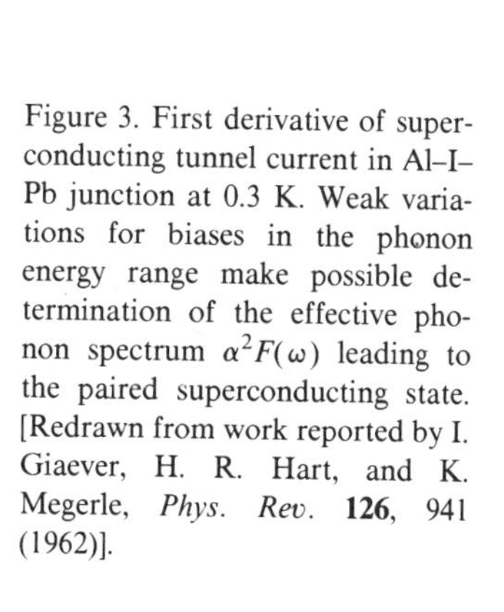

Figure 3. First derivative of superconducting tunnel current in Al–I–Pb junction at 0.3 K. Weak variations for biases in the phonon energy range make possible determination of the effective phonon spectrum $\alpha^2F(\omega)$ leading to the paired superconducting state. [Redrawn from work reported by I. Giaever, H. R. Hart, and K. Megerle, *Phys. Rev.* **126**, 941 (1962)].

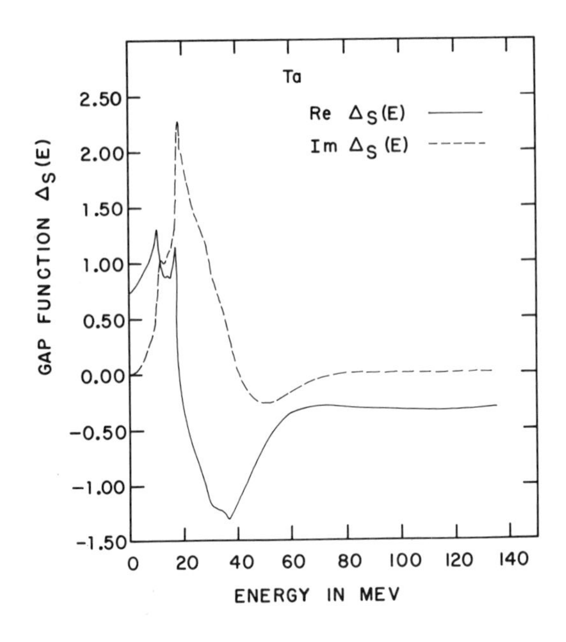

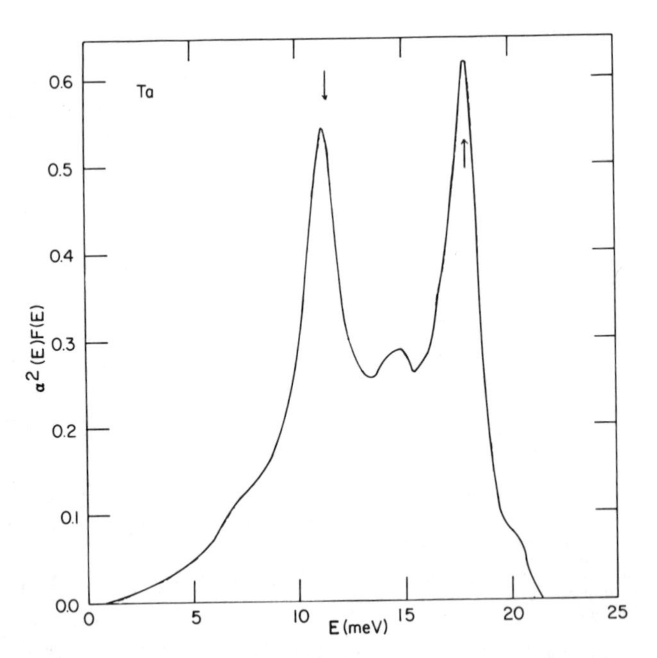

Figure 4. (a) Pair potential function $\Delta(E)$ inferred for Ta from tunneling measurements. The solid and dashed lines represent, respectively, the real and imaginary parts of the gap function. (Reference 8.) (b) Effective phonon spectrum $\alpha^2F(\omega)$ for Ta, determined by proximity electron tunneling spectroscopy (PETS). [E. L. Wolf, R. J. Noer, D. Burnell, Z. G. Khim, and G. B. Arnold, *J. Phys. F* **11**, L23 (1980).]

promised image of the phonon spectrum of the superconductor must reside in the dependence of the generalized gap parameter or *pair potential*, $\Delta(E)$, upon energy, $E = eV$. It is this dependence $\Delta(E)$ which leads to small deviations of $N_T(E)$ from its BCS form ($\Delta = \text{const}$).

Thus, the desired information about the phonon spectrum enters the measured $N_T(E)$, [Eq. (1)] via $\Delta(E)$ which occurs in the inverse radical. Fortunately, it is possible, making use of the full theory of strong coupled superconductivity[1] (the Eliashberg equations) to very accurately deduce from measurements of $N_T(E)$ not only $\Delta(E)$ [and the renormalization function $Z(E)$] but also the underlying spectrum of phonons, $\alpha^2F(\omega)$. It has been stated[1] that this approach provides the most detailed available probe of the superconducting state. As an example, Figure 4b shows the $\alpha^2F(\omega)$ function of Ta,[4] corresponding to the pair potential shown in Figure 4a.[8]

2. Threshold Spectroscopy of Normal-State Phonons

2.1. Semiconductors

The strongest known examples of phonon thresholds in tunneling occur in cases similar to the Esaki diode (see Figures 2 and 5), where *phonon assistance* is *required* for a tunneling transition to occur. In Figure 5a,[9] note that the conductance below 8 mV bias is negligibly small. In this case, at very low bias no elastic process (Figure 1a $eV < \hbar\omega_p$) can occur because it would violate an important selection role, the conservation of crystal momentum, **k**. Referring to Figures 2b and 5b, the electrons which tunnel from the n-type region have large k values lying near the boundaries of the Brillouin zone in the six equivalent conduction band "valleys" in the [111] directions. The final states, however, lie at the zone center, the location of the highest-energy levels in the valence band of Ge. (The usually filled states near $k = 0$ are made available for extra electron occupation in the degenerate p-type material by a heavy doping of an acceptor impurity such as boron.) Since the whole tunnel junction, $C-I-S$, occurs within a perfect Ge crystal lattice, with only the doping concentration changing with position, the transition of the electron from, e.g., $k = \pi/a$ (111) to $k = 0$ is forbidden, unless accompanied by another transition such that the initial total wave vector equals the total final wave vector. This can occur if a phonon of wave vector $k = \pi/a$ (111) is created to preserve the initial wave vector, leaving the electron free to make the transition to $k = 0$. [Of course, at high enough temperature a $k = -\pi/a$ (111) *thermally generated* phonon might exist and be absorbed, permitting the electron transition at $eV = 0$; our discussion is based on the assumption that thermal energy is essentially unavailable. This

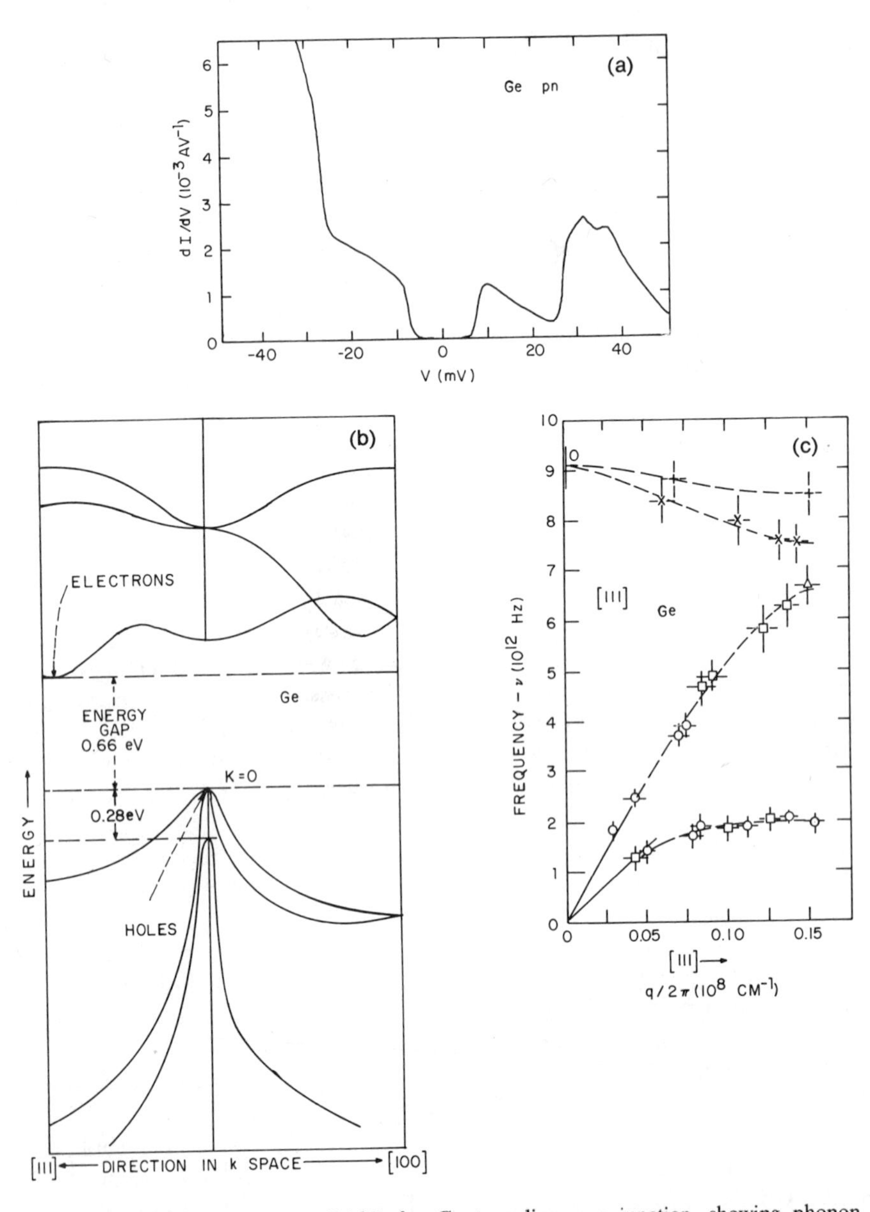

Figure 5. (a) First derivative, dI/dV, for Ge tunneling $p-n$ junction, showing phonon thresholds. The conductance is almost entirely phonon assisted as indicated by extremely small values for $|eV| < 8$ meV. (Redrawn from work reported by R. T. Payne, reference 9.) (b) Band structure of Ge, showing indirect gap with (111) valleys at the zone boundary. [Redrawn from work reported by H. Brooks, *Advances in Electronics and Electron Physics*, Vol. VII (L. Marton, ed.), Academic Press, New York (1955)]. (c) Phonon dispersion curves in the (111) directions for Ge, from neutron scattering measurements. [Redrawn from work reported by B. N. Brockhouse and P. K. Iyengar, *Phys. Rev.* **111**, 747 (1958).]

absorption most closely illustrates the concept of a *phonon-assisted transition*; the same term is applied when the phonon is generated rather than absorbed.] The entire $T=0$ current in the Esaki diode is thus phonon-assisted. It turns out that there are four different phonons (the four energies occur in the graph) of wave vector $k=\pi/a$ (111), able to provide the assistance. These phonon energies, clearly revealed in the d^2I/dV^2 spectrum of Figure 2a, are known also from inelastic neutron scattering measurements. The ω-k dispersion relation provided by such neutron scattering measurements in Ge are shown in Figure 5c.

A second illustration of threshold determinations in Ge is given in Figure 6, in the case of a metal–semiconductor or Schottky-barrier contact of a metal on degenerate n-type material.[10] In this case the conductance changes at threshold are relatively small, and a much larger elastic or nonassisted current flows even at $T=0$ and $V\simeq 0$. In this case, as illustrated in Figure 6c, direct electronic transitions can occur from the large Fermi surface of the metal to the various conduction band valleys of the Ge, conserving, as required, the component of k parallel to the interface. The large observed nonassisted current may indicate that the wave-vector values corresponding to the conduction band valleys are available in the metal, or that the selection rule is weakened by the fact that the two electrodes have separate crystal structures and may well be joined in a rather disordered interface region which may change the wave vector of the traversing electrons.

In such measurements on p-type Ge or Si, where the final electronic states are at $k=0$, the optical phonon branch at $k=0$ is observed more prominently. An example, Figure 7, is the prominent structure at 64.9 mV bias in a Pb Schottky barrier contact to boron-doped (p-type) silicon. An interesting detail of this measurement[11] is the satellite peak at approximately 78 mV. This has been identified as a local mode vibration of the lighter boron impurity atom in a cage of its four nearest-neighbor silicon atoms. The strength of this impurity vibration is relatively large here because of the high doping, 2.3×10^{20} atom/cm^3 of boron.

Finally, as an example of the many diverse phonon observations which have been made by this rather simple technique, Figure 8 shows a d^2V/dI^2 spectrum obtained[12] in contact to metallic $KTaO_3$. Here are located four separate longitudinal optical vibrations, the highest energy of which lies at 102.5 mV, suggesting the local motion of a strongly bonded object of small mass in this complicated structure.

It is necessary in some cases to look beyond the simple concept of opening an additional inelastic channel to explain fully the observed spectra. Comparison, indeed, of Figure 7 with Figure 1b and 1c reveals in the case of the p-type Si Schottky barrier contact, *failure* of the antisymmetry in d^2I/dV^2 expected for the purely inelastic mechanism. Other clear-cut cases

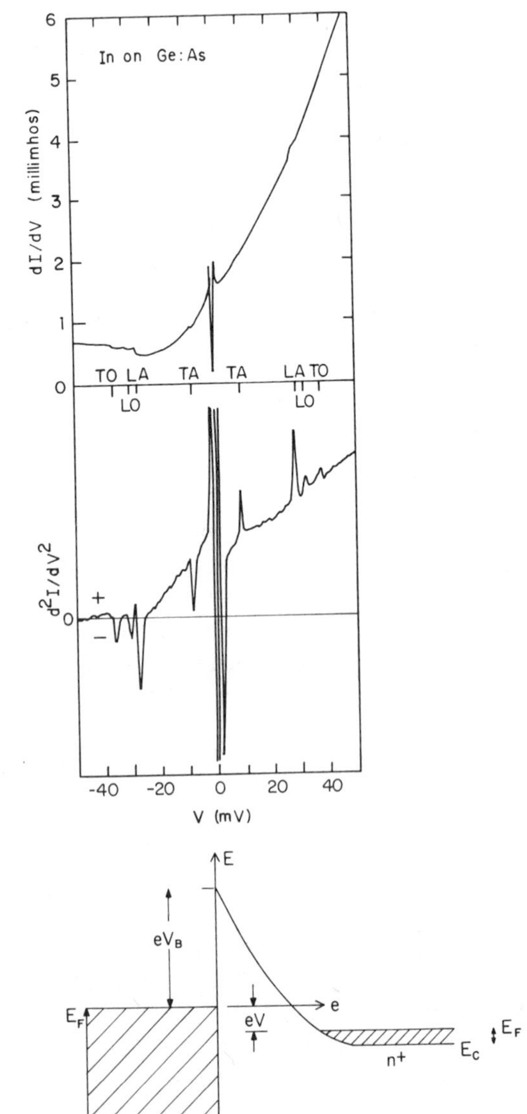

Figure 6. (a) First derivative (conductance) for Schottky barrier contact on heavily doped (degenerate) *n*-type Ge. (b) Second derivative of the same data as shown for (a). (Redrawn from work reported by F. Steinrisser, L. C. Davis and C. B. Duke, reference 10.) (c) Configuration of Schottky barrier tunneling contact to an *n*-type material.

in which the d^2I/dV^2 spectra are basically *symmetric* about $V=0$ are known—an added example is shown in Figure 9, in measurements on metallic CdS. Other cases[3] include GaAs, GaSb, InAs, and CdTe. In many of these materials a lattice vibration can be accompanied by strong local electric fields because of the different charges of the separate constituent

Figure 7. Second derivative d^2I/dV^2 spectrum for In Schottky barrier to heavily doped p-type Si. Note approximate symmetry of main phonon peaks at ± 64.9 mV, indicating a self-energy mechanism. Local mode vibration of B impurity is present at 78 mV bias. (Reference 11.)

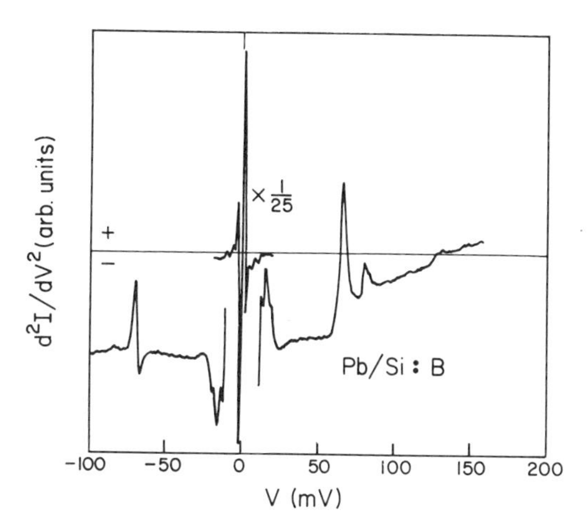

ions: these materials are said to be polar in nature. This can lead to especially strong electron–phonon coupling. The origin of the effects of Figures 7 and 9 certainly is in such coupling, which can become evident in the tunneling conductance in several different ways. Note that d^2I/dV^2, especially in Figure 9, deviates from its background even at bias energies *below* the corresponding threshold energy. How can this occur?

Consider an electrode in which electrons interact with a single Einstein phonon mode $\hbar\omega_0$. While electrons of energy $E < E_F + \hbar\omega_0$ cannot emit a real phonon, their behavior near this threshold can be modified by *virtual phonon emission*. The latter effect can be described by the real part of the

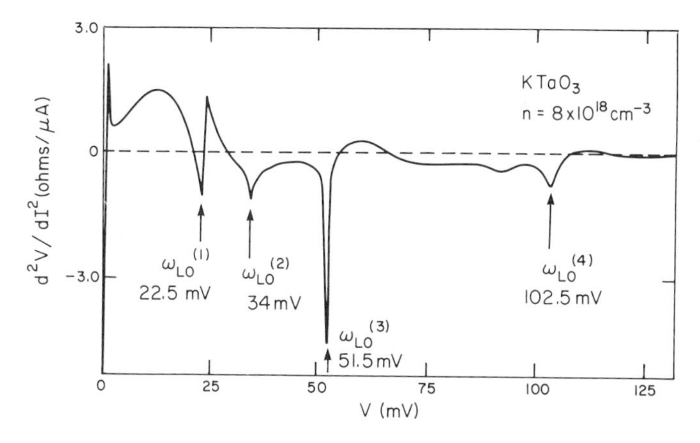

Figure 8. Phonons in $KTaO_3$, revealed by electron tunneling, (Redrawn from work reported by K. W. Johnson and D. L. Olson, reference 12.)

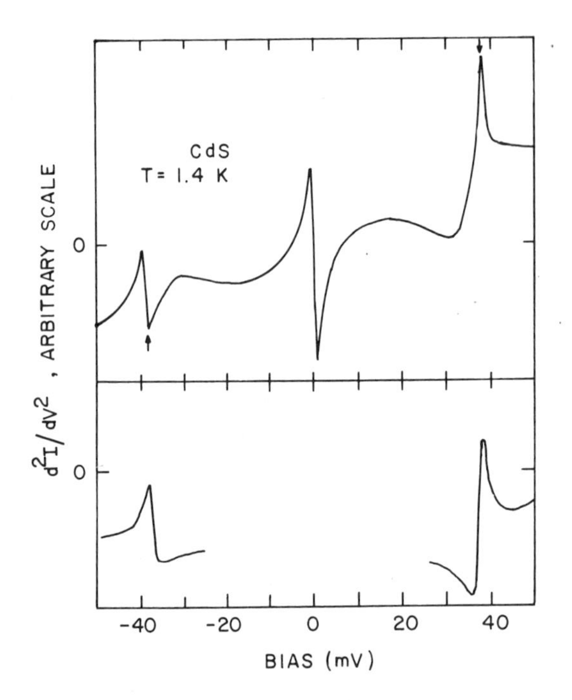

Figure 9. Self-energy effects in the derivative tunneling spectrum of a Au-degenerate CdS Schottky barrier contact, compared with a theoretical model calculation. [Redrawn from work reported by D. L. Losee and E. L. Wolf, *Phys. Rev.* **187**, 925 (1969).]

self-energy,[12] which enters the E-k relation of the electron,

$$E(k) = \frac{\hbar^2 k^2}{2m^*} + g^2 \ln\left|\frac{E - E_F - \hbar\omega_0}{E - E_F + \hbar\omega_0}\right| \tag{4}$$

Here g^2 is constant and we assume $E_F > \hbar\omega_0$. This expression is also valid for $|E - E_F| > \hbar\omega_0$; the lifetime against emission of a real phonon above threshold is described by an imaginary part of the self-energy, which we have neglected here. In this expression E represents the energy of the electron relative to the bottom of the band, rather than relative to the Fermi energy, as previously.

Physically Eq. (4) reflects the tendency of the moving electron (or hole) to cause a comoving polarization or distortion of the nearby ions from their usual lattice positions. (This polarization cloud can be expressed quantum mechanically as a cloud of virtual phonons.) As a result of this coupling to the ion motion, the $E(k)$ relation of the electron, and its consequences, notably the electron's group velocity, $\hbar^{-1}dE/dk$, may be strongly perturbed even below the crossing (at threshold) of the parabolic band $E=\hbar^2k^2/2m$ with the Einstein modes $E=E_F\pm\hbar\omega_0$. In order to deduce the consequences of such a perturbation, it is necessary to develop a more detailed mathematical picture of the tunneling current between two electrodes C, S. The conductance calculation, to which we now turn, is also needed in order to correctly describe the results with a superconductor electrode.

In order to calculate the conductance, dI/dV, let us first assume that an electron, represented by an incident wave $\psi=e^{ikx}$ of positive wave vector k, and energy $E=\hbar^2k^2/2m$, strikes a barrier of constant height $V_B>E$, which occurs in the interval $0<x<t$. The solutions of the Schrödinger equation

$$\frac{-\hbar^2}{2m}\frac{\partial^2}{\partial x^2}\psi+V(x)\psi=E\psi \tag{5}$$

within the classically forbidden barrier region $0<x<t$ are of the form $\psi=Be^{-\kappa x}+B'e^{\kappa x}$, where the real decay constant κ is

$$\kappa=\left[\frac{2m}{\hbar^2}(V_B-E)\right]^{1/2} \tag{6}$$

The exponentially decaying barrier function must be matched in magnitude $\psi(x)$ and derivative $m^{-1}\partial\psi/\partial x$ at $x=0$ to the incident plus reflected wave $e^{ikx}+Re^{-ikx}$, and at $x=t$ to the transmitted wave, Te^{ikx}. The probability current

$$j=\frac{i\hbar}{2m}\left(\psi\frac{\partial\psi^*}{\partial x}-\psi^*\frac{\partial\psi}{\partial x}\right) \tag{7}$$

is independent of x, and in particular is continuous at each boundary. The resulting exact barrier transmission factor D in the important case of small transmission, $\kappa t\gg 1$, is given by

$$D=|T|^2=\beta^2e^{-2\kappa t}, \qquad \beta\propto\kappa k/(\kappa^2+k^2) \tag{8}$$

The transmission factor D represents the fraction of the probability current $\hbar k/m$ carried by the incident wave e^{ikx} that is transmitted by the barrier.

It is often convenient to adopt the quasiclassical or WKB approximation,[13] applicable to a general barrier profile $V(x)$. In the quasiclassical case of a thick, gently sloping barrier $V(x)$, the WKB result is

$$D=\exp(-2K) \tag{9}$$

with

$$K=\int_{x_1(E_x)}^{x_2(E_x)}\kappa(x,E_x)\,dx \tag{10}$$

and

$$\kappa(x,E_x)=\{2m^*[V(x)-E_x]/\hbar^2\}^{1/2} \tag{11}$$

Here x_1 and x_2 are the classical turning points for kinetic energy E_x.

If the WKB formula is applied outside its range of validity, as to the square barrier, an error in the prefactor β^2 is introduced, evident in comparing Eqs. (8) and (9), but the exponent is still given correctly. The behavior of parameters in the exponent, Eq. (10), is of course of primary importance in determining the magnitude of the tunneling current. The relatively weak energy dependence introduced by the prefactor β^2 becomes important in the limit of small E, corresponding to a band edge, where we see from Eq. (8) that the prefactor, in this particular case, goes to zero linearly with wave vector k. We have written the energy as E_x, in extension to noninteracting electrons in two or three dimensions, where $E=E_x+E_{\parallel}$. The component of momentum $k_{\parallel}$ parallel to the junction is assumed to be conserved in the tunneling process.

If we now view the metal–insulator–metal tunnel junction in Figure 1, the top of the barrier corresponds to the conduction band edge of the insulator and κ is the decay constant in the forbidden gap at an energy $V(x)-E_x$ below the conduction band edge. In reality, the applied bias voltage V changes the potential barrier, which we sometimes write as $V(x,V)$. The barrier penetration factor thus strictly varies with the three independent parameters, E, $k_{\parallel}$, and V. Following common usage, however, we usually leave the V dependence understood and write $D(E,k_{\parallel})$. The diagram in Figure 1 indicates the conventions that we have adopted: energy is measured from the conduction band edge on the left-hand side; positive bias is applied to the right-hand electrode, causing electron flow from left to right.

We turn to evaluation of the current density J in a solid-state structure such as that of Figure 1. The current per unit area is obtained by integrating over all allowed k vectors, multiplying by $\hbar^{-1}\partial E/\partial k_x$ to give the probability current normal to the barrier, and multiplying by $-e$:

$$J=\frac{2e}{(2\pi)^3\hbar}\int dk_x\,d^2k_{\parallel}\,\frac{\partial E}{\partial k_x}[f(E)-f(E+eV)]\,D(E,k_{\parallel}) \tag{12}$$

The factor 2 comes from spin degeneracy, $(2\pi)^3$ from the k-space volume per state; and the Fermi functions,

$$f(E) \equiv f_R(E) = \{1 + \exp[(E - E_F)/k_B T]\}^{-1} \tag{13}$$

with E_F the right-hand Fermi energy, guarantee flow from occupied to unoccupied states. A more common form for J is obtained in the free-electron cases, where

$$E = E_x + E_{\parallel} \tag{14}$$

and $\partial E/\partial k_x dk_x$ can be replaced by dE. In the limit $T = 0$, the expression becomes, integrating over energies on the right,

$$J = \frac{e}{2\pi^2 h} \int_{E_F - eV}^{E_F} dE \int D(E, k_{\parallel})\, d^2 k_{\parallel} \tag{15}$$

For some purposes it is convenient to rewrite the $k_{\parallel}$ integral in Eq. (15) in terms of $E_{\parallel} = (\hbar^2/2m^*)k_{\parallel}^2$. Thus $(2\pi)^{-2} d^2 k_{\parallel}$ becomes $\rho_{\parallel}(E_{\parallel})\, dE_{\parallel}$, leading to

$$J = \frac{2e}{h} \int_{E_F - eV}^{E_F} dE \int_0^E \rho_{\parallel}(E_{\parallel}) D(E, E_{\parallel})\, dE_{\parallel} \tag{16}$$

with easy generalization to finite temperature, simply by inserting Fermi occupation factors in the integrand.

Having obtained correct expressions for the tunneling current density J, from which the conductance dJ/dV can be readily obtained, we return to incorporation of the electron–phonon coupling effects such as described by Eq. (4), in the case of the normal-state electrode. Following Davis and Duke,[14] this can be done by inserting a density of states factor $\text{Im}G(E', E)/\pi$ into the current expression, yielding, for arbitrary T,

$$J = \frac{2e}{h} \int_{-\infty}^{\infty} dE [f(E) - f(E + eV)]$$

$$\times \int_{-\infty}^{\infty} dE' \frac{\text{Im}\, G(E', E)}{\pi} \int_{k_{\parallel}(0)}^{k_{\parallel}(E')} D(E', k_{\parallel})\, d^2 k_{\parallel} \tag{17}$$

The distortion of the electron E, k relation by electron–phonon coupling, as for example represented by Eq. (4), leads to an altered density of states $\text{Im}\, G(E', E)/\pi$, which can be thought of crudely as a change in $\partial k/\partial E$ (the

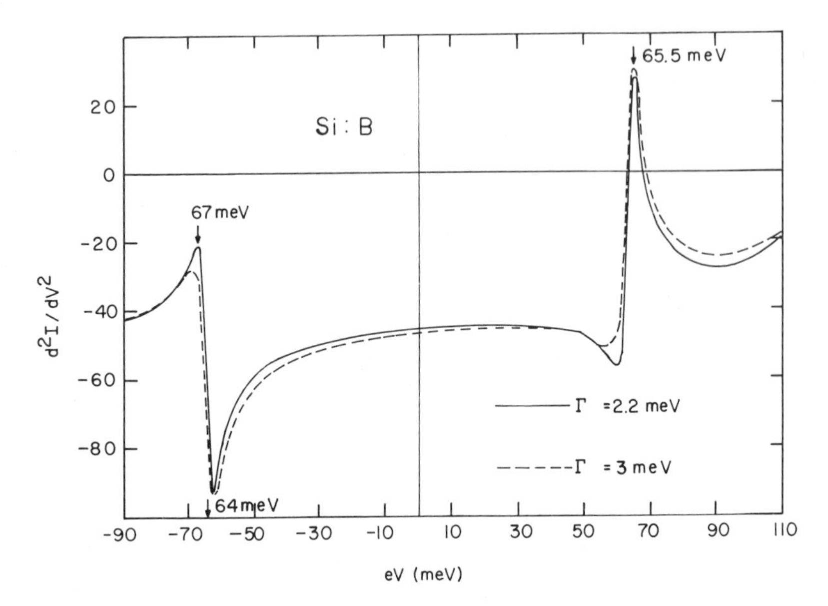

Figure 10. Calculated effect of strong electron–phonon coupling on Schottky barrier tunneling d^2I/dV^2 spectrum. (Redrawn from work reported by L. C. Davis and C. B. Duke, reference 14.) (For comparison, see Figure 7.) The two curves represent slight variations in model parameters.

density of states in one dimension) from a relation such as (4). A further change in the tunneling current may arise from the altered behavior of the penetration probability $D(E', k_{\|})$ when perturbations like Eq. (4) are present. Incorporation of these effects by Davis and Duke is illustrated in Figure 10 to explain tunneling data on metal p–Si contact junctions. The results are in surprisingly good agreement with experiment, see Figures 7 and 9.

2.2. Metals

While observation of very strong effects from normal-state phonons is limited to semiconductor–electrode junctions, weaker normal-state phonon structure has been often observed in metal–insulator–metal junctions. Examples of such observations[(15)] are given in Figures 11a and 11b, all representing normal-state electrodes. In Figure 11a, the Al phonon peak at 36 mV is clearly seen, with very weak but identifiable phonon peaks from Ag and Au in the two lower traces, as pointed out by the arrows. Ag and Au are metals with very weak electron–phonon couplings consistent, for example, with their inability to form a superconducting state at presently attainable temperatures. The metals on the right panel of Figure 11b, Sn, Pb, and In, respectively, are all strong-coupling superconductors (here, however, observed in the normal state) and correspondingly show stronger phonon peaks. The curves in this figure are the so-called even-conductance—if $dI/dV = \sigma(V)$, $\sigma_e = [\sigma(V) + \sigma(-V)]/2$—and the peaks are therefore interpreted as representing phonon emission. It has been demonstrated by Rowell *et al.*[16] that self-energy effects, as distinguished from simple phonon emission, are observable in the odd conductance $\sigma_0 = [\sigma(V) - \sigma(-V)]/2$, and that the phonon emission peaks disappear from this quantity in good junctions.

An example of the use of normal-state metal phonon spectra is illustrated in Figure 12, in which the change in shape of the Al phonon spectrum (taken directly form the normal-state d^2I/dV^2 spectra by subtracting a background) with change in film morphology is demonstrated.[17] The top curve here represents a clean bulklike Al film, while in the two lower panels the Al has been made "granular" by evaporation in the presence of O_2 partial pressure on the order of 10^{-5}–10^{-4} Torr. The grain size in the lower case is estimated to be only 20–30 Å, and is about twice that in the center panel. The change in shape of the phonon distribution

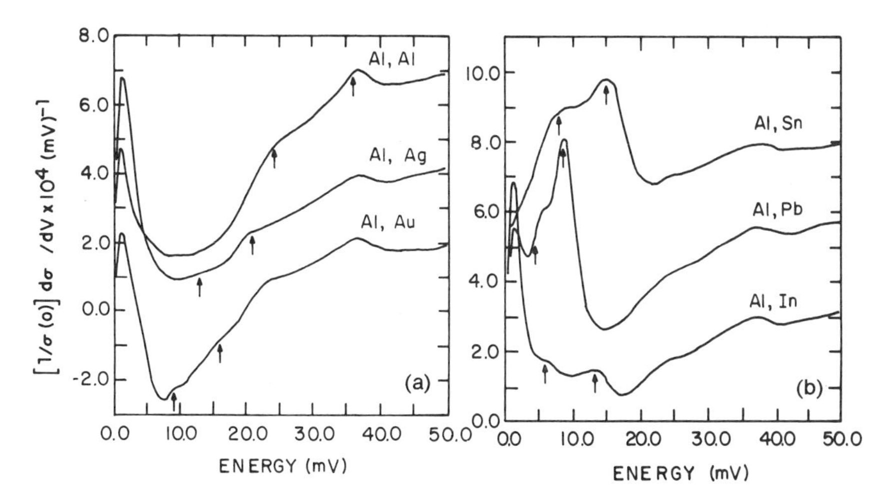

Figure 11. (a) Weak phonon effects observed in thin-normal-metal–film tunnel junctions. The metals Ag and Au have very weak electron–phonon coupling as compared with the phonons of Al. (b) The metals Pb and Sn reveal phonon peaks stronger than those of Al, while In appears to be comparable in coupling strength. (Redrawn from work reported by T. T. Chen and J. G. Adler, reference 15.)

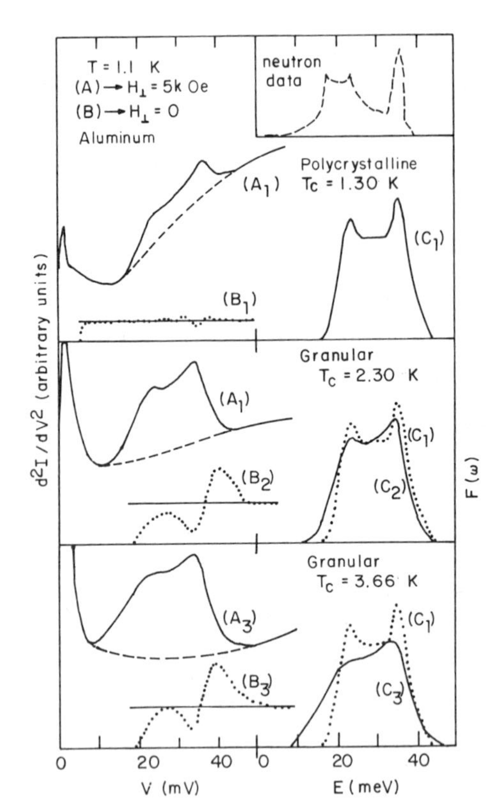

Figure 12. Approximate Al $F(\omega)$ functions extracted from normal-state Al films of varying morphology. In lower panels, the Al is "granular" (see text) which correlates with an increased superconducting T_c and evidently a broadened phonon spectrum. (Redrawn from work reported by A. Leger and J. Klein, reference 17.)

that occurs in the granular cases, particularly the appearance of a low-frequency "tail," correlates with an enhanced superconducting transition temperature. The curves labeled A in the figure, from which the phonon spectra $F(\omega)$ (curves C) are obtained, are taken with the Al films at 1.1 K made normal by the application of a 5-kG magnetic field. The dotted curves labeled B show the spectra when the Al is superconducting.

3. Superconductive Tunneling: The Effective Phonon Spectrum $\alpha^2F(\omega)$

The use of a superconducting electrode in the tunneling sandwich permits the most detailed study of the phonon spectrum $\alpha^2F(\omega)$ of strong coupled superconductors.[18] This study yields other information on the superconductor, including the energy-dependent gap function $\Delta(E)$ and renormalization function $Z(E)$ in both the superconducting and the normal states.[1]

3.1. Superconductivity

As we have seen above in Eqs. (1)–(3), the gap function $\Delta(E)$ enters the measured quantity, $N_T(E)$, in an inverse radical, $E/[E^2-\Delta^2(E)]^{1/2}$. Variation of $\Delta(E)$ in the range of phonon energies, $E \gg \Delta(E)$, is thus only weakly present in $N_T(E)$, and must be extracted by a rather involved "inversion procedure," as we will see.

The pair potential $\Delta(E)$ has as its limiting value at low energy the usual gap parameter $\Delta_0 = \Delta(\Delta_0)$. (Here energy is measured from E_F, midway across the superconducting energy gap of width $2\Delta_0$.) The $\Delta(\Delta_0)$ can be thought of as the binding energy (per quasiparticle) that two electrons at $E = \pm\Delta_0$ (and also opposite spin projections and $\mathbf{k}_1 + \mathbf{k}_2 = 0$) can gain by condensing to a pair. $\Delta(E)$ is the same binding energy for arbitrary $E > \Delta_0$; $\Delta(E)$ is in fact a strong function of E, typically showing a reversal of sign from attractive ($\Delta > 0$) to repulsive for E just above the range of phonon energies of the metal. This perhaps surprising behavior has its origin in the dynamics of the retarded (in time) attractive electron–ion–electron interaction which leads to the paired state. Stated too simply, passage of one "free" electron between a given set of positively charged metal ions will start these ions moving toward each other to create a local positive charge fluctuation. However, the relatively massive ions move slowly (the period $T = 2\pi/\omega$ is typically on the order of 3×10^{-13} sec) such than an electron at the Fermi velocity $\sim 10^8$ cm/sec may thus have moved a distance on the order of 3000 Å or $\sim$1000 lattice spacings before, after a fraction of the period T, the ions reach their point of closest approach, and peak local positive charge density. This latter condition is clearly most *favorable* for the passage of the *second* electron in the pair; subsequently the local charge deviation will change its sign. The idea that the timing or phase relation of the electron and ion motions strongly affects the interaction leads, in a better treatment, to $\Delta(E)$ variations very close to those deduced from the experiments.

3.2. The Tunneling Density of States in $C-I-S$ Junctions

Having said this much about $\Delta(E)$, can we understand the origin of Eq. (1)? Basically, $N_T(E)$ measures a final density of states for tunneling transitions from the counterelectrode C, and these represent quasiparticle excitations of the superconductor. Evidently for $E < \Delta_0$ there is a gap in the density of states. This is easily understood, as a consequence of the attractive electron–electron interaction which leads to the pairing of all electrons in the metal in its $T = 0$ ground state. Quasiparticle excitations above this ground state require a minimum energy Δ_0 closely related to the condensation energy per particle which must be supplied by the injected electron, thus requiring a minimum bias $V = \Delta_0/e$. The dispersion relation for the quasiparticle excitations E_k in terms of the normal metal excitations, $\varepsilon_{k>} = \hbar^2 k^2 > /2m - E_F$ (electrons, $k > k_F$) and $\varepsilon_{k<} = (E_F - \hbar^2 k <^2/2m)$ (holes, $k < k_F$) is $E_k = (\varepsilon_k^2 + \Delta^2)^{1/2}$. The tunneling current given, e.g., by Eq. (15), is expressed in terms of the band energy ε_k, but must be rewritten in terms of the real excitation energies of the superconducting system, E_k. The change of variable required is the same one that transforms the normal density of states $dn/d\varepsilon_k \equiv N_0$ to the superconducting density of excitations $dn/dE_k = N_0\, d\varepsilon_k/dE_k$. That is

$$d\varepsilon_k = \frac{d\varepsilon_k}{dE_k} dE_k = \frac{E_k}{\left(E_k^2 - \Delta^2\right)^{1/2}} dE_k \cong N_T(E)\, dE$$

which effectively inserts the tunneling density of states expression (3) into the integral for the current density J. In this fashion Eq. (15), for the superconducting case, leads to Eq. (1), after differentiation to produce dI/dV. By the simple strategem of dividing the conductance $(dI/dV)_S$ (sample superconducting) by its value when the sample is normal, all factors except the desired density of states factor $N_T(E)$ (3) effectively cancel out. The resulting simplicity is one of the beauties of the superconducting tunneling experiment and has been important in permitting precise quantitative comparison of such *normalized* data to sophisticated theory.

In the superconducting state, the phonon peaks occur at points of maximum negative slope of dI/dV. Thus, the effect is of opposite sign to that of the normal phonon emission process, and negative peaks in d^2I/dV^2

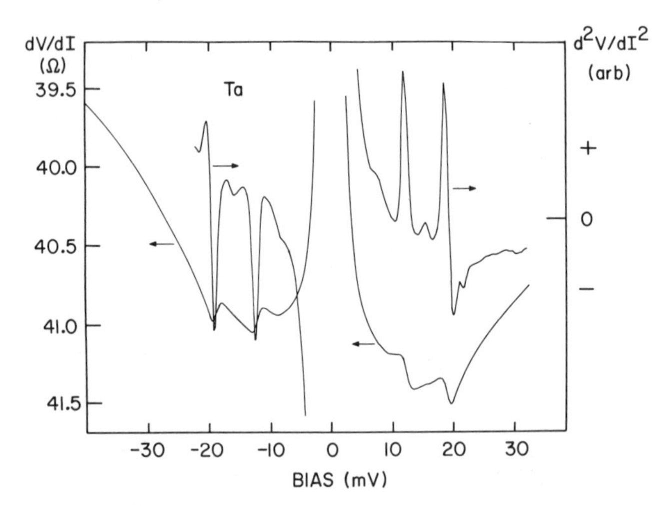

Figure 13. Directly traced tunneling derivative spectra on Ta metal in the superconducting state. From these spectra the superconducting pair potential and $\alpha^2F(\omega)$ functions shown in Figure 4 can be inferred. (Reference 8.)

and, hence, positive peaks in d^2V/dI^2, occur at phonon energies. This is illustrated in Figure 13, which also shows the symmetry of the phonon structure about $V=0$ in the conductance that one expects from Eqs. (1)–(3). These curves are directly traced experimental curves of the quantities $dV/dI=(dI/dV)^{-1}$ and

$$\frac{d^2V}{dI^2} = -\left(\frac{d^2I}{dV^2}\right) \Big/ \left(\frac{dI}{dV}\right)^3 \tag{18}$$

for a superconducting Ta electrode. Because of the slight variation of dV/dI in the phonon range of energies in Figure 13, the appearance of the d^2V/dI^2 spectrum is close to the negative of d^2I/dV^2. In numerical data reduction, of course, Eq. (18) is used with the measured dV/dI data to obtain d^2I/dV^2.

3.3. McMillan–Rowell Inversion for $\alpha^2F(\omega)$

How does one proceed[18] from data, such as shown in Figure 13, to the underlying pair potential $\Delta(E)$ and the effective phonon spectrum $\alpha^2(F(\omega)$?

First, we note that the desired phonon information is most directly contained in the departure of the normalized conductance (1) from its form in the BCS case where $\Delta(E)$ is just the constant Δ_0. Therefore the *BCS-reduced* conductance $\sigma/\sigma_{\mathrm{BCS}}-1$, where $\sigma=N_T(E)$ from Eq. (1) is constructed from the measured data, choosing an appropriate value for the BCS gap parameter Δ_0 in σ_{BCS}. An example of such a plot is shown in Figure 14, solid line.[19] The characteristic relative departures, here as great as 10%, indicate strong electron–phonon coupling.

At this point it is necessary to make contact with the underlying theory of strong-coupling superconductivity—the Eliashberg equations[20]—in order to obtain the effective phonon spectrum, $\alpha^2F(\omega)$, from normalized conductance data such as shown in Figure 14. These equations, written in their usual form for an isotropic superconductor,[1] are

$$\begin{aligned}
[1-Z(\omega)]\,\omega &= \int_{\Delta_0}^{\omega_c} d\omega' \,\mathrm{Re}\left[\frac{\omega'}{(\omega'^2-\Delta'^2)^{1/2}}\right] \\
&\quad\times \int \alpha^2(\omega_q)F(\omega_q)\left[D_q(\omega'+\omega)-D_q(\omega'-\omega)\right] d\omega_q \\
\Delta(\omega) &= \frac{1}{Z(\omega)} \int_{\Delta_0}^{\omega_c} d\omega' \,\mathrm{Re}\left[\frac{\Delta'}{(\omega'^2-\Delta'^2)^{1/2}}\right] \\
&\quad\times \left\{\int \alpha^2(\omega_q)F(\omega_q)\left[D_q(\omega'+\omega)+D_q(\omega'-\omega)\right] d\omega_q - \mu^*\right\}
\end{aligned} \tag{19}$$

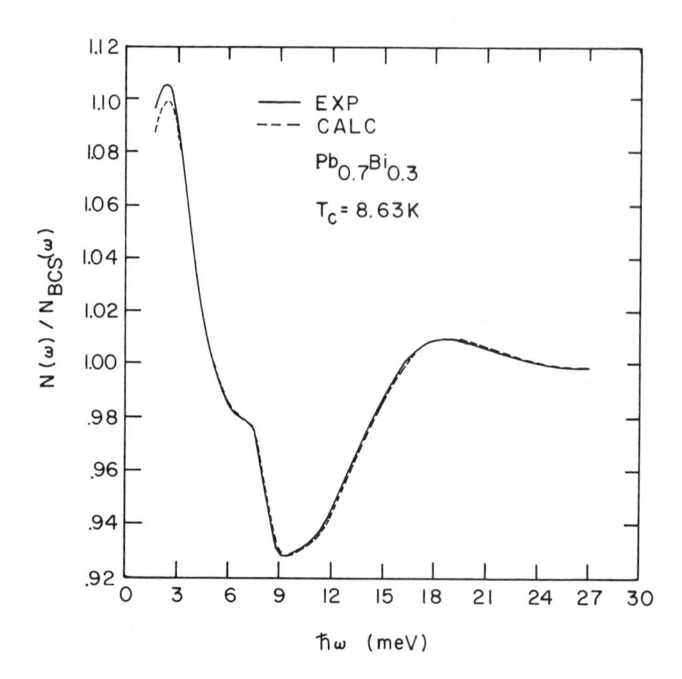

Figure 14. The reduced conductance σ/σ_{BCS} for a particular PbBi alloy junction of strong electron–phonon coupling. The solid curve is experimental, while the dashes represent the result of a variational calculation of $\alpha^2F(\omega)$ and $\Delta(E)$. (Redrawn from work reported by J. G. Adler and T. T. Chen, reference 19.)

Here $Z(\omega)$ is the renormalization function, whose limit at $\omega=0$ is $Z(0)=1+\lambda$, with λ the electron–phonon coupling constant. $Z(0)=1+\lambda=m^*/m$ is the enhancement, by the electron–phonon coupling, of the mass of the quasiparticle over that due simply to band structure effects, $F(\omega)$ is the phonon spectrum, while $\alpha^2(\omega_q)$ is an averaged squared matrix element. The symbol μ^* refers to a pseudopotential representing the repulsive Coulomb electron–electron interaction. The value of μ^* is found experimentally to be 0.1 ± 0.02 in a wide variety of superconductors.

The standard approach to inversion of the Eliashberg Eqs. (19) to find $\Delta(E)$, $Z(E)$, $\alpha^2F(\omega)$, and μ^* consistent with the reduced conductance data $\sigma/\sigma_{BCS}-1$ is a variational numerical procedure developed by W. McMillan and J. M. Rowell.[18] Trial functions for $\Delta(E)$ and $\alpha^2F(\omega)$ are assumed, a calculated $\sigma/\sigma_{BCS}-1$ is generated, and its differences from the experimental $\sigma/\sigma_{BCS}-1$ are used to generate improved $\Delta(E)$ and $\alpha^2F(\omega)$ functions. The process is repeated and typically converges to stable values after 8 or 10 iterations. The value of μ^* is chosen finally to given the experimental value of the energy gap $\Delta(\Delta_0)$.

The potential accuracy of this inversion method is evident in Figure 14, in that the calculated curve, shown dashed, is scarcely distinguishable from the experimental curve.[19] The full information thus made available is displayed in Figure 15, showing the $\alpha^2F(\omega)$ spectrum, and in Figure 16, which shows the derived functions describing the superconducting and normal states. The quantity $\phi(\omega)$ (upper left panel) is defined as $Z_s(\omega)\Delta_s(\omega)$. The renormalization function $Z_N(E)$ for the *normal* state (upper right panel) is obtained from the first equation of Eqs. (19) simply by setting $\Delta(E) = 0$ for the normal state.

As a further example of the high quality of $\alpha^2F(\omega)$ results obtainable from superconductive tunneling, and some insight into the role of the $\alpha^2(E)$ matrix element factor, Figure 17 shows the $\alpha^2F(\omega)$ function for trigonal crystalline mercury (Hg), which is a strong-coupling superconductor with $T_c = 4.5$ K and electron–phonon coupling parameter $\lambda = 1.6$. The dashed curve in Figure 17a is the tunneling $\alpha^2F(\omega)$ function obtained by Hubin and Ginsberg,[21] while the solid curve was obtained from careful neutron scattering measurements and subsequent fitting to an eight-neighbor Born–von Kármán lattice model, by Kamitakahara *et al.*[22] Inspection of this figure shows that the ratio $\alpha^2F(\omega)/F(\omega)$ is clearly a decreasing function of energy. Quantitative comparison of the observed ratio $[\alpha^2(\omega)]$ (dots, in the lower panel) to a pseudopotential theory calculation (solid curve, lower

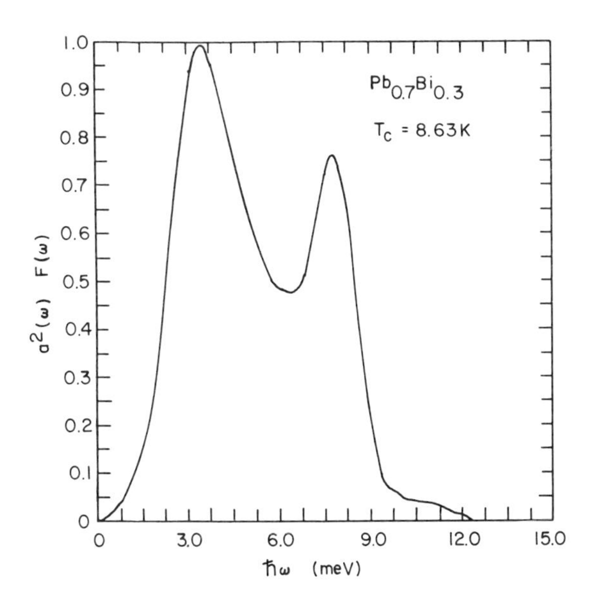

Figure 15. The $\alpha^2F(\omega)$ function (effective phonon spectrum) for the PbBi alloy whose reduced conductance σ/σ_{BCS} is shown in Figure 14.

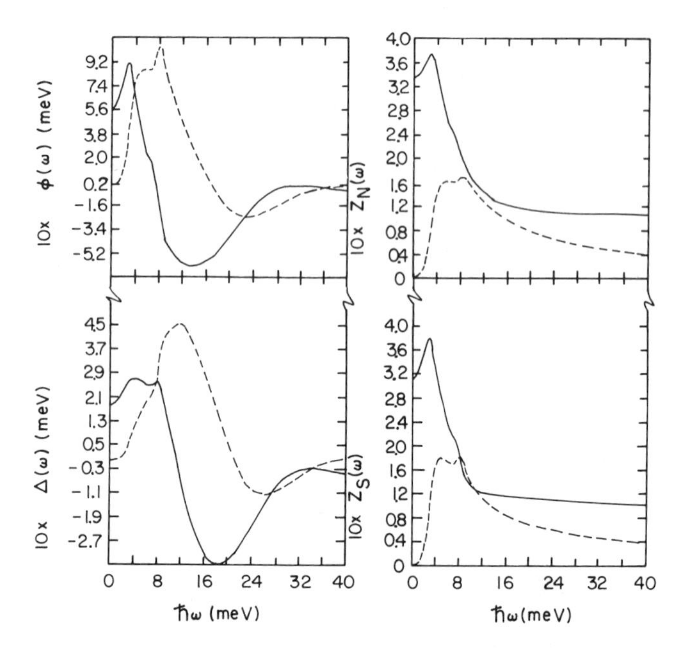

Figure 16. Derived superconducting functions for the PbBi alloy of the previous two figures (Redrawn from work reported by J. G. Adler and T. T. Chen, reference 19.) Upper left: the function $\phi(\omega)$ is defined as $Z_s(\omega)\Delta_s(\omega)$. Lower left: the superconducting pair potential $\Delta(\omega)$. Lower right: the renormalization function in the superconducting state, $Z_s(\omega)$. The $\omega = 0$ limit of this function is $1+\lambda$. Upper right: the renormalization function $Z_N(\omega)$ for the normal state. In all cases solid curves are real parts and dashed curves are imaginary parts.

panel) shows the observed $\alpha^2(\omega)$ (from tunneling and neutron measurements) to be surprisingly well predicted in both shape and absolute magnitude.

The interested reader is referred to more detailed summaries of such results for many (non-transition-metal) strong-coupling superconductors: an article by Allen and Dynes,[23] a more detailed review by the present author,[4] and numerical tabulations of the $\alpha^2F(\omega)$, $\Delta(E)$, $Z(E)$, and $N_T(E)$ functions obtained from tunneling.[24]

In summary, the work of McMillan and Rowell[18, 1] has been important, putting our present understanding of superconductivity in quantitative agreement with microscopic theory[20, 25] which forms an extension of the pairing theory of Bardeen, Cooper, and Schrieffer.[6] The practical limitation of this McMillan–Rowell tunneling spectroscopy using conventional $C-I-S$ junctions to non-transition-metal superconductors has spurred development of extensions of the method. One of these extensions, proximity electron tunneling spectroscopy (PETS) is the subject of the next section.

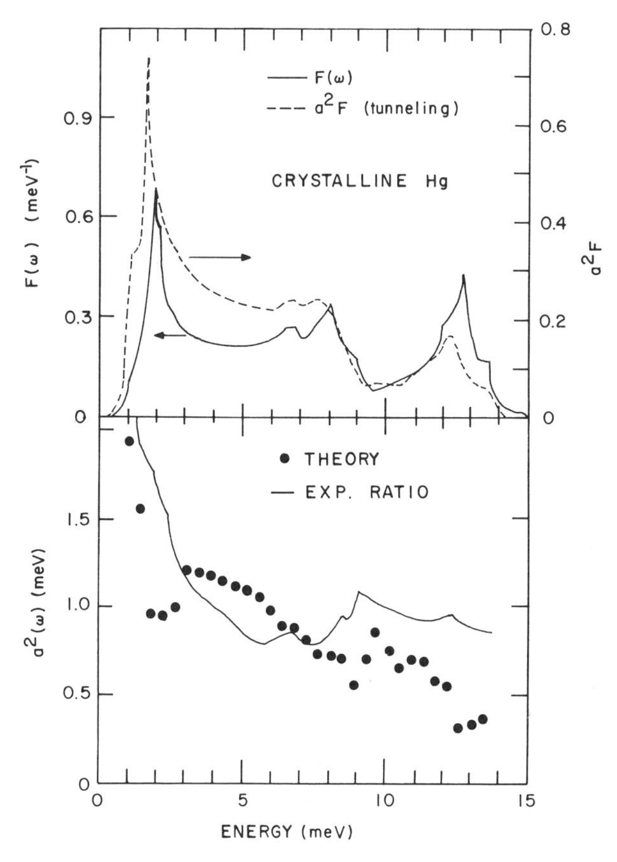

Figure 17. Comparison of $\alpha^2F(\omega)$ from tunneling with $F(\omega)$ from neutron scattering for trigonal crystalline Hg. (Redrawn from work reported by W. A. Kamitakahara, H. G. Smith, and H. Wakabayashi, reference 22.) The tunneling curve for $\alpha^2F(\omega)$ (dashed in upper panel) is from reference 21. Lower panel gives comparison of the matrix element factor $\alpha^2(\omega)$ as obtained from the ratio of $\alpha^2F(\omega)$ to $F(\omega)$ (solid line) and by pseudopotential calculation (dots).

4. Proximity Tunneling Methods

It is really suprising that the conventional tunneling spectroscopy based on $C-I-S$ junctions, where either the C or S layer is directly oxidized to produce the critical insulating barrier I, has been as widely successful as it has, in view of the fact that many metals form oxides of a noninsulating and even metallic nature. A good example of such a metal is vanadium, whose oxides, including V_2O_3, VO_2, and V_2O_5, exhibit a complex behavior including metallic and magnetic semiconducting phases. These properties preclude the fabrication of good tunnel junctions using oxidized vanadium as the barrier.[26] The alternative approach, depositing V onto a prepared counter-electrode–oxide combination, e.g., Al–Al_2O_3, does not work well either, for more subtle reasons. The only report of phonon structure in such junctions [necessary, of course, to carry out the McMillan–Rowell analysis for $\alpha^2F(\omega)$] is from workers who evaporated the V film onto an Al–Al_2O_3 structure held at cryogenic temperature, ~4.2 K, during the evaporation.[27] They found that unless the sample was maintained at low temperatures, <150 K, degradation of the junction characteristics would occur, apparently the result of migration of oxygen in the structure into the surface layers of V in contact with the Al_2O_3 barrier. It is not necessary to form the undesired vanadium oxide phases to degrade the tunneling characteristics, for even small (a few at. %) amounts of oxygen dissolved in V can destroy its superconductivity. This same sensitivity to O contamination is shared by Nb and other transition metals.

One method which may be used to circumvent this difficulty is to use a proximity junction of the form $C-I-NS$, where N is a thin layer of a metal, such as Al or Mg, which has weaker or nonexistent intrinsic superconducting properties but which has nearly ideal oxidation properties. The cleaned but reactive surface of the superconductor S, which is usually prepared as a foil in the proximity method,[28] is protected from contamination harmful to its superconductivity by the N layer, which acts as a diffusion barrier* for oxygen and other gases. This follows from our knowledge that oxidation, e.g., of Al, is a self-limiting process which preserves a sharply defined metal–oxide interface.

But how is it possible, from measurements on such $C-I-NS$ junctions, to learn about the phonon structure of the superconductor S? Would one expect to see phonon structure from the N layer metal, also, in this case? The answer to the first question lies in consideration of Andreev[29] "scattering of electrons into holes" at the NS interface, a process which depends on both pair potentials $\Delta_S(E)$, and $\Delta_N(E)$,[30] but which, in a limit of small d_N (N metal thickness), $d_N \to 0$, leads to a tunneling density of states which is very close to that of the S metal only, Eq. (3). In answer to the second question, concerning N-metal phonons: yes, in general, for larger $d_N \gtrsim 50$ Å these are evident in the tunneling conductance, much as if the N layer were a strong coupling superconductor.[31] This is a manifestation of the superconducting *proximity effect* in which an observable pair potential $\Delta_N(E)$ can be induced in even a nonsuperconductor such as Mg[32] by diffusion of Cooper pairs from an adjacent superconductor layer S. This provides a new opportunity, not present in conventional tunneling, for observing the phonon structure of inherently weak coupling metals, including Cu, Ag, and Au, as well as the Mg that we have mentioned.

*See note added in proof.

4.1. $C-I-NS$ Junctions in the Thin-N Limit

An idealized proximity model for discussing these effects[30] is illustrated in Figure 18, which shows a pair potential $\Delta(x)$ rising sharply at $x=d_N$ (the NS interface) from a low value Δ_N to a larger value Δ_S in the superconducting film. The sharpness of the jump is a reflection of the nature of the electron–phonon interaction, which can vary on the scale of the lattice constant.

The spatial independence of Δ_N for $0<x<d_N$ is a good approximation for $d_N \ll \xi_p$ where the proximity coherence length $\xi_p = \hbar v_{FN}/\pi\Delta_S$ is typically $\geqslant 500$ Å. In the same limit, $d_N \ll \xi_p$, the effect of the N layer in depressing the pair potential Δ_S near the NS interface, due to leakage of pairs into the N layer, is also a small effect.[30]

The magnitude of Δ_N, as we have mentioned, is enhanced by the proximity of S. This can be understood qualitatively from the following. To a useful approximation the pair potential $\Delta(x)$ can be regarded as the product of a pair density $F(x)$ multiplied by the local electron–phonon coupling $N(0)V(x)$.[33] The validity of this approximation follows from the second equation of Eqs. (19), realizing that the term involving $\Delta(\omega')$ is sharply peaked at $\hbar\omega'=\Delta(\omega')$. The pair density is a smoothly varying function on a distance scale of ξ, the coherence length, and hence will be substantially increased and remain essentially constant over the width, $d_N \ll \xi$, of the N film. Because of the assumed difference in the electron–phonon coupling $N(0)V$ of the N and S films, and the local nature of $V(x)$, there remains a sharp jump in $\Delta(x)$ at $x=d_N$.

The Andreev process may be most simply described if one imagines an electron quasiparticle in N at an energy $\Delta_N<E<\Delta_S$ propagating toward the NS interface. At the interface the propagation must stop, for the

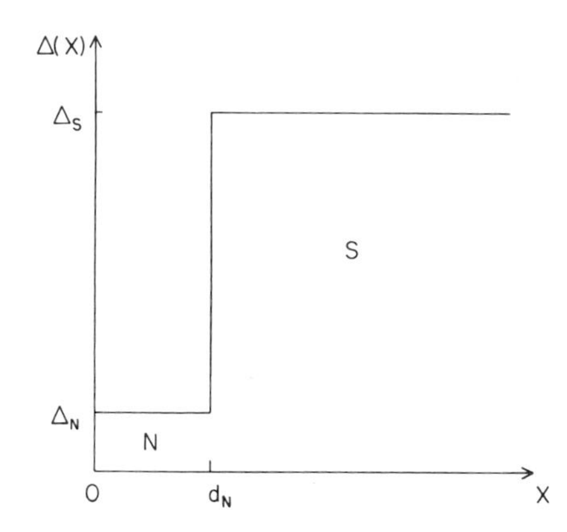

Figure 18. Schematic indication of pair potential variation with position in the NS junction.

available energy $E < \Delta_S$ is insufficient to create a propagating quasiparticle in S. The result predicted by Andreev is the joining of the initial quasielectron, energy E, with an electron below the Fermi surface at $-E$, to form a pair, moving to the right in S. The resulting hole propagates to the left in the N layer. The initial electron has been "scattered into a hole" by the pair potential discontinuity $\Delta_S - \Delta_N$.

The same scattering process also occurs for energies $E > \Delta_S$, with a scattering rate that depends upon both $\Delta_S(E)$ and $\Delta_N(E)$. The Andreev scattered quasiparticles return to the N–oxide interface and lead to a tunneling density of states at that position which also depends upon both Δ_S and Δ_N. Following the theory of Arnold,[30] an approximate expression for this proximity density of states, valid for energies in the phonon range $E \gg \Delta_N, \Delta_S$, is

$$N_T(E) = 1 + \tfrac{1}{2}\,\mathrm{Re}\,\frac{\Delta_N^2(E)}{E^2} + \tfrac{1}{2}e^{-2d/l}$$
$$\times \mathrm{Re}\left\{\frac{[\Delta_S(E) - \Delta_N(E)]^2}{E^2}\exp\left(2i\Delta_K^N d\right)\right\}$$
$$+ e^{-d/l}\mathrm{Re}\left\{\Delta_N(E)\frac{[\Delta_S(E) - \Delta_N(E)]}{E^2}\exp\left(i\Delta_K^N d\right)\right\}$$

where

$$\Delta_K^N = \frac{2Z_N(E)E}{\hbar v_{FN}} \tag{20}$$

Here $\Delta_N(E)$ is the induced pair potential, $Z_N(E)$ is the renormalization function, and v_{FN} is the Fermi velocity in the N layer. Scattering events in the N layer due to the N–S interface or to bulk impurities are characterized by the mean-free-path parameter, l. The oscillating exponential terms represent interferences arising from the small change in wave vector Δ_K^N due to Andreev reflection of electronlike $(k > k_F)$ into holelike $(k < k_F)$ quasiparticles at the NS interface. In the range of energies $E \leqslant 50$ mV and thickness $d \leqslant 100$ Å used in the study of S-metal phonons by proximity tunneling, $\Delta_K^N d_N$ is typically $\ll 1$, leaving the oscillatory exponentials near unity. In the same range of small $\Delta_K^N d$, substantial cancellation of terms in Δ_N occurs, making the strength of observed N-phonon structure a strong function of d_N. In the regime of ultrathin but nonzero values of d_N an approximate expression for the tunneling density of states is

$$N_T(E) = 1 + \tfrac{1}{2} e^{-2d/l} \mathrm{Re}\left[\frac{\Delta_S^2(E)}{E^2} \exp\left(\frac{i4\, dZ_N E}{\hbar v_{FN}} \right) \right] \tag{21}$$

for $E \gg \Delta_S$. Finally, substantiating our statement above, for $d \to 0$ and $d/l \to 0$, the result is

$$N_T(E) = 1 + \tfrac{1}{2} \mathrm{Re}\left(\frac{\Delta_S(E)}{E} \right)^2 \tag{22}$$

which is just the expansion of Eq. (3) for the conventional $C-I-S$ experiment, valid for the phonon range of energies, $E \gg \Delta_S$.

We now turn to summarize some results obtained in this fashion on "difficult" superconductors, including the V mentioned above. The study of phonons in weak-coupling N metals will then be summarized in turn.

4.2. Phonons in the Superconductor S

We begin by considering some recent results obtained using foil-based proximity junctions[28] whose pair potential profiles are believed to be similar to that sketched in Figure 18. (We have already shown, in the solid curve of Figure 4a, results from such a junction on tantalum.) The essential feature is that the superconductor of interest is present as a foil of thickness typically 50 μm, which has been cleaned and undergone recrystallization by heating to near its melting point in ultrahigh vacuum, typically 10^{-9} Torr.[28] The resulting atomically cleaned and locally flat surface is allowed to cool to near room temperature over a period of minutes and the Al deposition is made to thickness in the range 30–200 Å without delay, ensuring a nearly perfect $N-S$ interface. The coated foil is removed and allowed to oxidize, forming the Al_2O_3 tunnel barrier, while collodion is applied to mask off regions of the foil surface to insulate the In crossing counterelectrode strips which are subsequently deposited without delay; and the sample is cooled to 77 K as quickly as possible to preserve a sharply defined $N-S$ interface. The procedures have been used successfully with foils of Nb,[28] V,[34,35] Ta,[8] $Nb_{0.75}Zr_{0.25}$;[36] for proximity layers, Al and Mg have been found to behave in a similar fashion. Counterelectrodes have usually been In or Ag, but there is no inherent restriction to these metals.

The spectra in Figure 19 were obtained from an In–Al_2O_3–Al/V junction with about 20 Å of metallic Al, using a weak magnetic field to suppress the superconductivity of the In. Phonon structures from both V

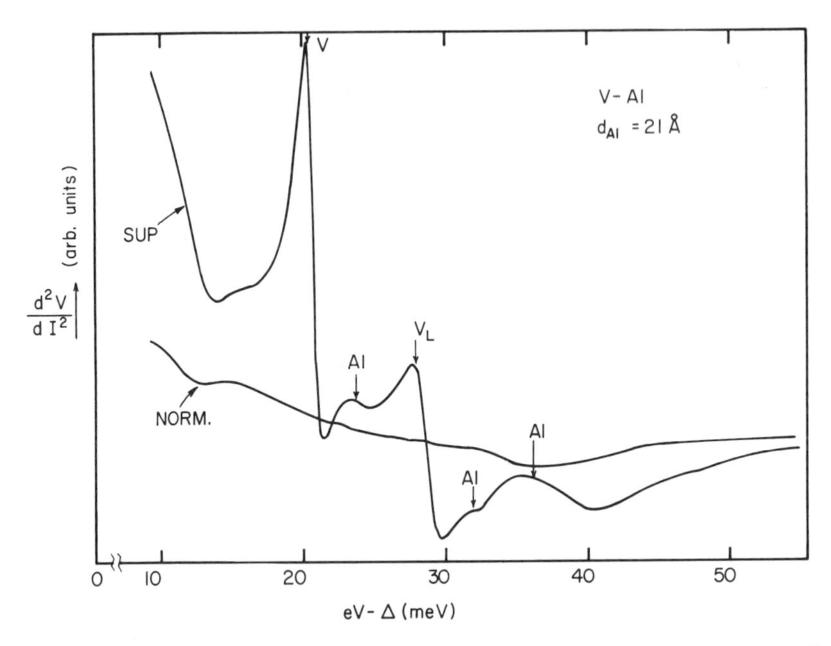

Figure 19. Tracing of second derivative spectra in superconducting and normal state for In–Al_2O_3–Al/V tunnel junction. Peaks from phonons of both metals are present, but V contributions dominate. (Reference 34.)

and Al are present (see arrows), so that the full expression for $N_T(E)$, Eq. (20), is really necessary to describe the contribution of the Al pair potential $\Delta_N(E)$ in this spectrum. That this is necessary can also be seen in Figure 20, where the dashed line, representing an inversion of the data neglecting $\Delta_N(E)$, leaves a noticeable discrepancy in the region of the main Al phonon peak, ~ 36 mV. Accordingly, analysis was continued using Eq. (20) with $\Delta_N(E)$ generated from an Al $\alpha^2F(\omega)$ function in the literature. Inversion of the data with $\Delta_N(E)$ input gave much improved agreement in $N_T(E)$ [the result is virtually indistinguishable from the measured $N_T(E)$ in Figure 20] and yielded the first $\alpha^2F(\omega)$ for V ever published,[34] as shown in Figure 21. The arrows in the figure locate peaks in the phonon spectrum $F(\omega)$ for V obtained from neutron scattering measurements.[37]

Fortunately, in many cases the simplified density of states, Eq. (21), is adequate to give an accurate description of proximity tunneling results. Such a case is shown in Figures 22 and 23, obtained from measurements on a foil of $NbZr_x$ alloy[36] at $x \sim 0.33$. This composition corresponds to the peak in the variation of T_c with x, whose value, 11 K, is substantially raised over the $T_c = 9.2$ K for pure Nb ($x = 0$). This alloy has been useful in

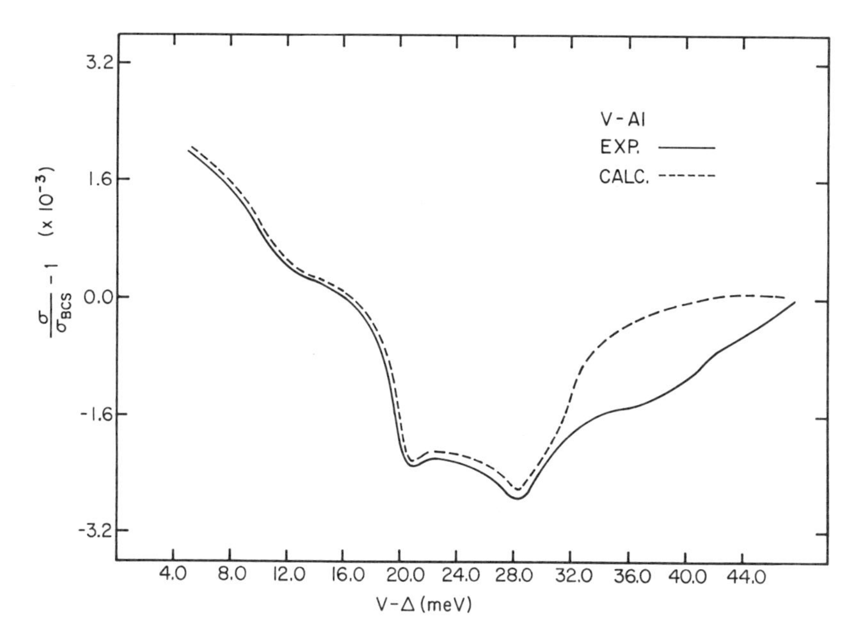

Figure 20. Reduced conductance $\sigma/\sigma_{BCS} - 1$ for Al/V junction of Figure 20. Dashed curve, calculated neglecting contributions of Al pair potential, is seriously in error, indicating importance of Al contributions. Reinversion of data (see text) including $\Delta_{Al}(E)$ produced calculated curve indistinguishable from experiment. (John Zasadzinski, Ph.D. thesis, Iowa State University, 1980, unpublished.)

superconducting magnets because of its high upper critical field $H_{c2} \simeq$ 11.5 T. Nevertheless, before proximity techniques were available, no tunneling phonon spectra had ever been obtained for this material, or for any transition-metal alloy, for that matter. With the proximity technique, however, strong $NbZr_x$ phonon spectra are easily obtained, as shown in Figure 22. The phonon-induced deviations in $N_T(E)$ here exceed 1%, even after some attenuation by scattering in the Al layer of thickness ~25 Å and scattering ratio $d/l \simeq 0.4$. Further, very little evidence for an energy-dependent Al pair potential is present in the experimental $\sigma/\sigma_{BCS} - 1$ function, and the inversion of the data using Eq. (21), ($\Delta_N = 0$) provides a good description (dashed line, Figure 22). The resulting $\alpha^2F(\omega)$ function for the $NbZr_x$ alloy is strongly broadened and has an increased weight at low energies, 5–10 mV, as shown in Figure 23. From these results a quantitative analysis of the changes in the superconductivity of $NbZr_x$ at $x = 0.33$ relative to $x = 0$ have been given.[36] The biggest effect is an approximate 40% increase in the parameter $\lambda = 2\int \alpha^2F(\omega)\,d\omega/\omega$ caused by the shift of spectral weight in $\alpha^2F(\omega)$ to low energies ("phonon softening"). This

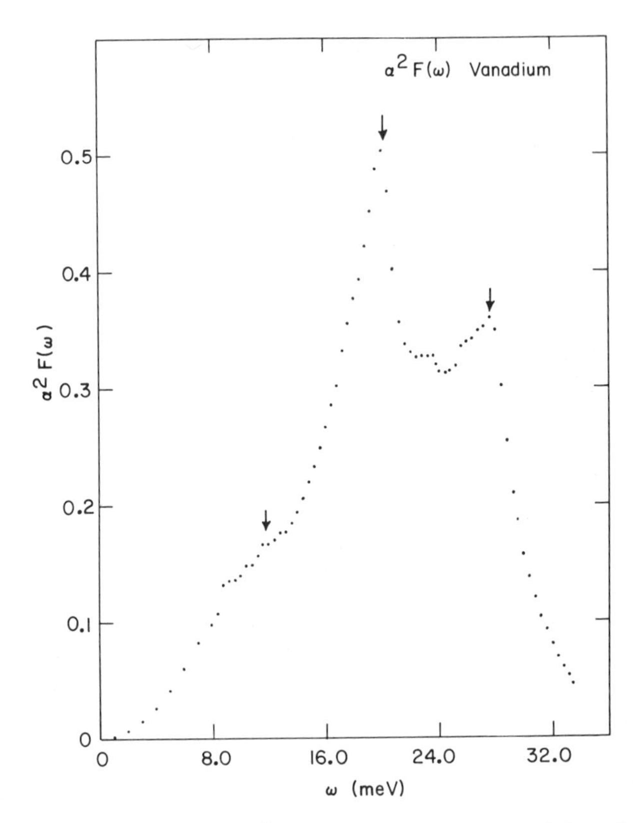

Figure 21. First published tunneling $\alpha^2F(\omega)$ function for V, obtained from data shown in Figures 20 and 21. Arrows locate phonon peaks obtained by neutron scattering. (Reference 34.)

central conclusion from proximity tunneling has recently been verified by neutron scattering measurements on $NbZr_x$ at nearly the same concentration x.[38]

Finally, there has been application of Eq. (21) to correct conventional tunneling results, some already in the literature, for the presence of unintentional "N layers" which in various ways correspond to surface degradation of the S electrode in nominal $C-I-S$ junctions. A classic example of this application of the proximity theory is to tunneling data on Nb, for which an early conventional study has provided unphysical results.[39] Sensible corrections for a probable surface contaminant layer of NbO and/or NbO_2 allowed physically acceptable results to be inferred.[40] A further recent example of this role of proximity analysis is given in the case of Nb_3Sn

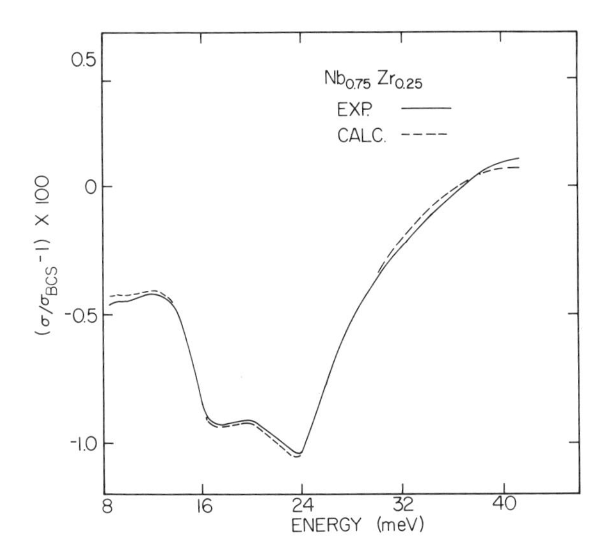

Figure 22. Reduced conductance for junction on $Nb_{0.75}Zr_{0.25}$ alloy, showing strong phonon-induced deviations. The calculated curve is obtained neglecting Al pair potential contributions. (Reference 36.)

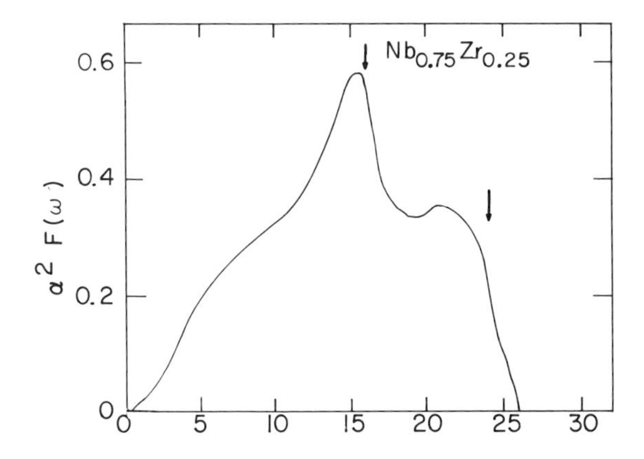

Figure 23. $\alpha^2F(\omega)$ function for NbZr alloy, corresponding to calculated curve in Figure 22. This spectrum is broadened and shifted toward low energies, relative to the spectrum of pure Nb, substantially increasing λ and T_c for the alloy. (Reference 36.)

tunneling data,[41] for which a reasonable analysis in terms of electron–phonon coupling, with $\lambda = 1.8$, can be obtained with sensible proximity corrections.

4.3. Phonons in the Proximity Layer N

As we have emphasized earlier, the superconducting proximity effect enhances the pair potential $\Delta_N(E)$ in the N layer. Referring to Eq. (20), one sees that the first two terms in the $E \gg \Delta_S, \Delta_N$ expansion of the density of states

$$N_T(E) \cong 1 + \tfrac{1}{2}\mathrm{Re}\left(\frac{\Delta_N(E)}{E}\right)^2 + \cdots$$

represent simply the first two terms in the corresponding expansion of the tunneling density of states for the N metal alone, namely, $N_T(E) = \mathrm{Re}\{(E)/[E^2 - \Delta_N(E)^2]^{1/2}\}$. Even though $\Delta_N(E)$ is enhanced in magnitude by proximity, its energy-dependent features are retained, and the corresponding $\alpha^2F(E)_N$ should provide a linearly scaled effective phonon spectrum of N.

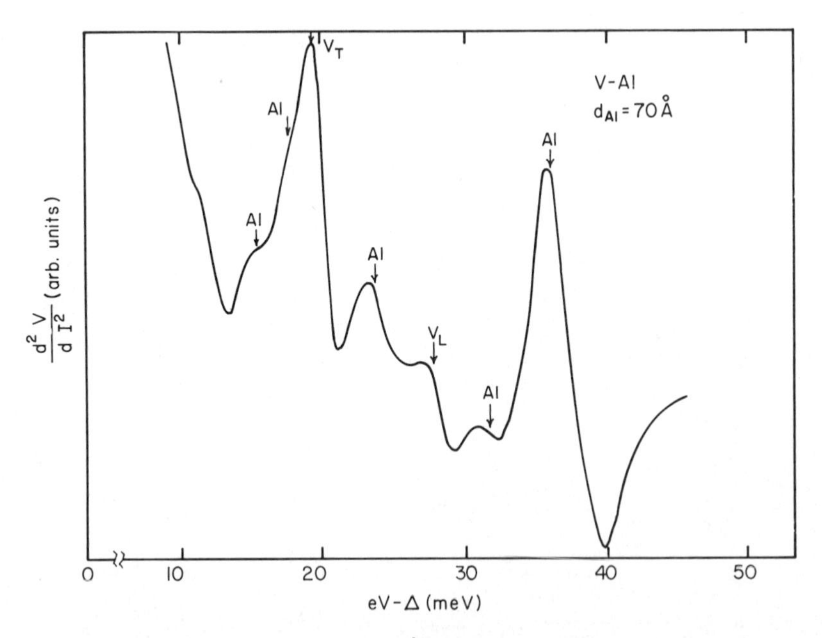

Figure 24. Second derivative spectrum of 70 Å Al layer on V showing prominent Al phonon structure. (Reference 35.)

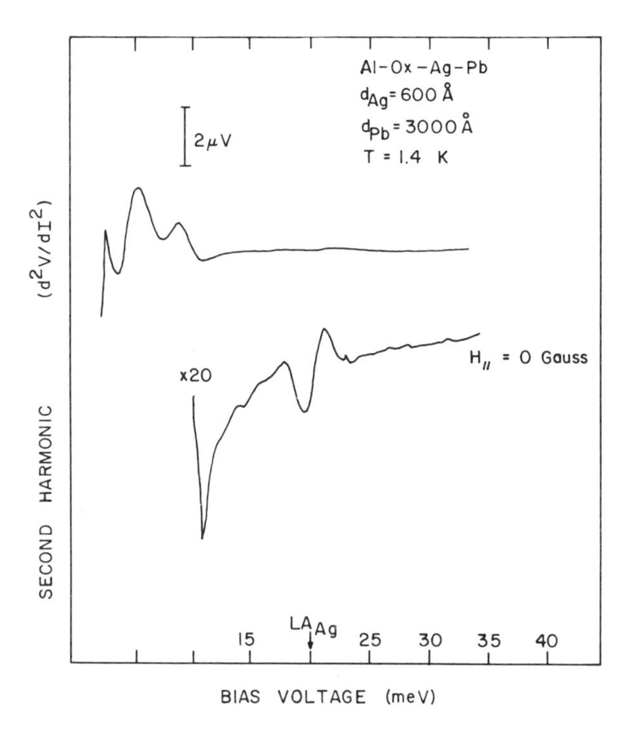

Figure 25. d^2V/dI^2 trace of Al–Al_2O_3–Ag(600 Å)/Pb(3000 Å) junction, showing Ag LA phonon peak at 20 mV. Note change of scale indicating weakness of Ag phonon peak relative to Pb phonon peaks. (Reference 42.)

To obtain $\Delta_N(E)$, Eq. (20) tells us that it will be helpful to increase d_N and d_N/l in order to attenuate the contributions from the third and fourth terms, which decay as $e^{-2d/l}$ and $e^{-d/l}$, respectively, and which contain undesired $\Delta_S(E)$ contributions. As we have noted earlier, the $\Delta_N(E)$ contributions from the third and fourth terms in Eq. (20), for small d_N, tend to cancel the Δ_N in the second term. Hence, small d_N is undesirable on this account, also, if we wish to maximize contributions from the N metal.

Data with strong Δ_N contributions, consistent with the analysis above are shown in Figure 24, from a junction with about 70 Å of Al metal on a V foil.[35] (Similar results had earlier been reported[31] for Al on Pb.) Here the 36-meV Al longitudinal phonon is greatly enhanced relative to the data of Figure 19, and the V phonons, somewhat reduced, are still present to a noticeable degree, indicated by the arrows in the figure. One would expect to be able to extract the Al $\alpha^2F(\omega)$ function from such data, by taking the V $\alpha^2F(\omega)$ as a known quantity, which has been achieved for Al in reference 35. This is an area of current activity in proximity tunneling research.

Although no other $\alpha^2F(\omega)$ function from a proximity-enhanced N-metal layer has ever been published, to our knowledge, considerable information about phonon energies in many metals, including several very weak coupling metals, e.g., Ag and Cu, has been obtained directly from d^2V/dI^2 traces. Figure 25 shows such a trace for Al–Al_2O_3–Ag/Pb junction,[(42)] in which the Ag LA phonon at 20 mV is weakly but clearly observed. It is a reasonable expectation that progress in this area will eventually permit quantitative $\alpha^2F(\omega)$ functions to be obtained in such cases.

5. Conclusions

(1) Interactions between electrons and phonons in both normal and superconducting systems allow several means of observing phonons via their effects on the tunneling current.

(2) The largest effects of phonons are present in Esaki tunnel diodes in indirect bandgap (degenerate) semiconductors, where phonon assistance is required to conserve crystal momentum in the tunnel transition. Inelastic threshold spectroscopy of the relevant phonons is possible in d^2I/dV^2 measurements.

(3) Weaker inelastic phonon excitations also occur in Schottky-barrier contacts and in normal metal evaporated film tunnel junctions.

(4) In the superconductors where electron–phonon coupling is the driving force in forming the superconducting state, an image of the phonon spectrum occurs in the tunneling density of states $N_T(E)$, measured as the ratio $(dI/dV)_S/(dI/dV)_N$ of conductances in a $C-I-S$ tunnel junction. Phonon peaks correspond to negative peaks in d^2I/dV^2, in this case.

(5) Some of the limitations of the conventional superconductive tunneling are being removed by the proximity electron tunneling spectroscopy, PETS. This permits superconductors S of unfavorable barrier-forming characteristics to be studied.

(6) The opportunity to study phonons in very weak-coupling, inherently nonsuperconducting, metals such as Mg, Cu, and Ag, is present in the PETS technique.

Note Added in Proof

The role of thin Al layers (N) as continuous diffusion barriers in NS junctions has been confirmed and elucidated in a recent paper: J. Kwo, G. K. Wertheim, M. Gurvitch, and D. N. E. Buchanan, XPS study of surface oxidation of Nb/Al overlayer structures, *Appl. Phys. Lett.* **40**, 675 (1982).

References

1. W. L. McMillan and J. M. Rowell, in *Superconductivity* (R. D. Parks, ed.), Vol. 1, pp. 561–613, Dekker, New York (1969).
2. C. B. Duke, *Tunneling in Solids*, Academic Press, New York (1969).
3. E. L. Wolf, in *Solid State Physics* (D. Turnbull and H. Ehrenreich, eds.), Vol. 30, pp. 1–91, Academic Press, New York (1975).
4. E. L. Wolf, Electron tunneling spectroscopy, *Rep. Prog. Phys.* **41**, 1439–1508 (1978).
5. L. Esaki, New phenomenon in narrow germanium $p-n$ junctions, *Phys. Rev.* **109**, 603–604 (1958).
6. J. Bardeen, L. N. Cooper, and J. R. Schrieffer, Theory of superconductivity, *Phys. Rev.* **108**, 1175–1204 (1957).
7. I. Giaever, Energy gap in superconductors measured by electron tunneling, *Phys. Rev. Lett.* **5**, 147–148 (1960).
8. E. L. Wolf, D. M. Burnell, Z. G. Khim, and R. J. Noer, Proximity electron tunneling spectroscopy. IV. Electron–phonon coupling and superconductivity of tantalum, *J. Low Temp. Phys.* **44**, 89–118 (1981).
9. R. T. Payne, Phonon energies in germanium from phonon-assisted tunneling, *Phys. Rev.* **139**, A570–A582 (1965).
10. F. Steinrisser, L. C. Davis, and C. B. Duke, Electron and phonon tunneling spectroscopy in metal germanium contacts, *Phys. Rev.* **176**, 912–914 (1968).
11. D. E. Cullen, E. L. Wolf, and W. D. Compton, Tunneling spectroscopy in degenerate p-type silicon, *Phys. Rev. B* **2**, 3157–3169 (1970).
12. K. W. Johnson and D. H. Olson, Electron tunneling into $KTaO_3$ Schottky barrier junctions, *Phys. Rev. B* **3**, 1244–1248 (1971).
13. G. D. Mahan, in *Tunneling Phenomena in Solids* (E. Burstein and S. Lundquist, eds.), pp. 305–313, Plenum Press, New York, 1969.
14. L. C. Davis and C. B. Duke, Interaction of tunneling electrons with collective modes of the electrodes, *Solid State Commun.* **6**, 193–197 (1968).
15. T. T. Chen and J. G. Alder, Electron tunneling in clean Al–insulator–normal-metal junction, *Solid State Commun.* **8**, 1965–1968 (1970).
16. J. M. Rowell, W. L. McMillan, and W. L. Feldman, Phonon emission and self-energy effects in normal-metal tunneling, *Phys. Rev.* **180**, 658–668 (1969).
17. A. Leger and J. Klein, Experimental evidence for superconductivity enhancement mechanism, *Phys. Lett.* **28A**, 751–752 (1969).
18. W. L. McMillan and J. M. Rowell, Lead phonon spectrum calculated from superconducting density of states, *Phys. Rev. Lett.* **14**, 108–112 (1965).
19. J. G. Adler and T. T. Chen, Strong coupling superconductivity of $Pb_{0.7}Bi_{0.3}$ alloys, *Solid State Commun.* **9**, 1961–1964 (1971).
20. G. M. Eliashberg, Interactions between electrons and lattice vibrations in a superconductor, *Sov. Phys.-JETP* **11**, 696–702 (1960).
21. W. N. Hubin and D. M. Ginsberg, Electron tunneling into superconducting mercury films at temperatures below 0.4 K, *Phys. Rev.* **188**, 716–722 (1969).
22. W. A. Kamitakahara, H. G. Smith, and N. Wakabayashi, Neutron spectroscopy of low frequency phonons in solid Hg, *Ferroelectrics* **16**, 111–114 (1977).
23. P. B. Allen and R. C. Dynes, Transition temperatures of strong-coupled superconductors reanalyzed, *Phys. Rev. B* **12**, 905–922 (1975).
24. J. M. Rowell, W. L. McMillan, and R. C. Dynes, The electron–phonon interaction in superconducting metals and alloys II, unpublished (to be submitted to *J. Phys. Chem. Ref. Data*).

25. D. J. Scalapino, J. R. Schrieffer, and J. W. Wilkins, Strong-coupling superconductivity I, *Phys. Rev.* **148**, 263–279 (1966).
26. R. J. Noer, Superconductive tunneling in vanadium with gaseous impurities, *Phys. Rev. B* **12**, 4882–4885 (1975).
27. B. Robinson and J. M. Rowell, A superconducting tunneling measurement of the annealing characteristics of the transition metals Nb and V, *Inst. Phys. Conf. Ser.* **39**, 666–670 (1978).
28. E. L. Wolf, J. Zasadzinski, J. W. Osmun, and G. B. Arnold, Proximity electron tunneling spectroscopy. I. Experiments on Nb, *J. Low Temp. Phys.* **40**, 19–50 (1980).
29. A. F. Andreev, The thermal conductivity of the intermediate state in superconductors, *Zh. Eksp. Teor. Fiz.* **46**, 1823–1828 (1964). [English trans. *Sov. Phys.-JETP* **19**, 1228–1281 (1964)].
30. G. B. Arnold, Theory of thin proximity-effect sandwiches, *Phys. Rev. B* **18**, 1076–1100 (1978).
31. P. M. Chaikin and P. K. Hansma, Observation of normal-metal phonons with proximity-effect tunneling, *Phys. Rev. Lett.* **36**, 1552–1555 (1976).
32. D. M. Burnell and E. L. Wolf, Tunneling observation of an induced pair potential in Mg, *Bull. Am. Phys. Soc.* **26**, 211 (1981).
33. G. Deutscher and P. G. de Gennes, in *Superconductivity* (R. D. Parks, ed.), Vol. 2, pp. 1005–1034, Dekker, New York (1969).
34. J. Zasadzinski, W. K. Schubert, E. L. Wolf, and G. B. Arnold, A proximity electron tunneling study of V and V_3Ga, in *Superconductivity in d- and f-Band Metals* (Suhl and Maple, eds.), pp. 159–164, Academic Press, New York (1980).
35. J. Zasadzinski, D. N. Burnell, E. L. Wolf, and G. B. Arnold, A superconducting tunneling study of vanadium, *Phys. Rev. B* **25**, 1622–1632 (1982).
36. E. L. Wolf, R. J. Noer, and G. B. Arnold, Proximity electron tunneling spectroscopy. III. Electron–phonon coupling in the Nb-Zr system, *J. Low Temp. Phys.* **40**, 419–440 (1980).
37. B. P. Schweiss, Phonon density of states of vanadium, Karlsruhe Research Report KFK 2054, pp. 11–12 (1974).
38. F. Gompf and B. Scheerer, The phonon density of states of $Nb_{.75}Zr_{.25}$, Karlsruhe Research Report KFK 3051, pp. 8–10 (1980).
39. J. Bostock, V. Diadiuk, W. N. Cheung, K. H. Lo, R. M. Rose, and M. L. A. MacVicar, Does strong-coupling theory describe superconducting Nb?, *Phys. Rev. Lett.* **36**, 603–606 (1976).
40. G. B. Arnold, J. Zasadzinski, and E. L. Wolf, A resolution of the controversy on tunneling in Nb, *Phys. Lett.* **69A**, 136–138 (1978).
41. E. L. Wolf, J. Zasadzinski, G. B. Arnold, D. F. Moore, J. M. Rowell, and M. R. Beasley, Tunneling and the electron–phonon-coupled superconductivity of Nb_3Sn, *Phys. Rev. B* **22**, 1214–1217 (1980).
42. B. F. Donovan-Vojtovic and P. M. Chaikin, Tunneling characteristics of trilayer tunnel junctions, *Solid State Commun.* **31**, 563–565 (1979).

4

Electronic Transitions Studied by Tunneling Spectroscopy

S. de Cheveigné, J. Klein, and A. Léger

1. Introduction

Over the past few years, vibrational tunneling spectroscopy has become an important surface and interface spectroscopy with applications to the study of adsorption phenomena and heterogeneous catalysis. But in these fields, the understanding of the *electronic* states of molecules is also of fundamental importance. What information can inelastic electron tunneling spectroscopy (IETS) provide?

Let us point out, first, that we would expect to obtain very intense spectra of electronic transitions. This is because tunneling intensities are—at least in a first approximation—thought to be proportional to optical oscillator strengths[1] (see Chapter 2). For vibrational modes these oscillator strengths are of the order of 10^{-5}, whereas for allowed electronic transitions they can exceed 10^{-1}. The study of these transitions, therefore, is an interesting test of the theory: Is tunneling spectroscopy really analogous to optical spectroscopy?

Electronic transitions have been seen by tunneling spectroscopy in a number of molecules[2-4] and in rare earth oxides.[5] After briefly describing the experimental techniques particular to the study of electronic transitions, we shall discuss the results obtained—and especially the intensities observed. The outlines of a theory appropriate for electronic transitions will be

S. de Cheveigné, J. Klein, and A. Léger • Groupe de Physique des Solides de L'Ecole Normale Supérieure, Université Paris VII, Tour 23, 2 place Jussieu, 75251 Paris Cedex 05, France.

sketched, and the practical application of this branch of tunneling spectroscopy will be evaluated.

2. Experimental

We shall only consider the aspects in which experimental techniques for the study of electronic transitions differ from those already described in previous chapters. The main reasons for these differences are as follows:

(a) electronic transitions take place at much higher energies than vibrational ones: several volts instead of several tenths of a volt;
(b) tunnel junctions tend to break down under biases higher than 2–3 V. One must therefore look for particularly low-energy electronic transitions, but not necessarily optically allowed ones, as we shall see.

Two types of systems have been studied. One can find large organic molecules with low-energy electronic transitions—dyes that are colored precisely because they have optically allowed transitions in the visible range. These molecules tend to be unstable and careful handling is required to avoid decomposition during the doping process. One possibility is to sublimate the material from alumina or graphite crucibles by very gentle heating—this method presents the advantage of allowing the whole fabrication process to take place in the vacuum system, but is only possible for the more stable molecules. The other possibility is to use liquid phase doping of the junctions (described in Chapter 1). Both Al–Al_2O_3–Pb and Mg–MgO–Pb junctions have been used for studying electronic transitions in large organic molecules.

Rare earth oxides also present very low energy electronic transitions. In this case junctions are made with a rare earth electrode which is oxidized by an O_2 glow discharge to form an oxide barrier whose transitions will be observed. A lead counterelectrode completes the junction.

Following junction fabrication, one needs to study the second derivative of the $I(V)$ characteristic to obtain a spectrum—as for vibrational transitions. But new problems appear because at biases beyond 0.5 V the $I(V)$ characteristic ceases to be even approximately linear in V, and eventually becomes exponential (as the Fowler–Nordheim regime is reached). This is because the bias applied to the junction is an appreciable fraction of the barrier height. In the high-voltage range the logarithmic derivative $(1/\sigma)\, d\sigma/dV$ (where $\sigma = dI/dV$) is better adapted to the observation of high-energy transitions than the more usual d^2V/dI^2.[6] Consider for

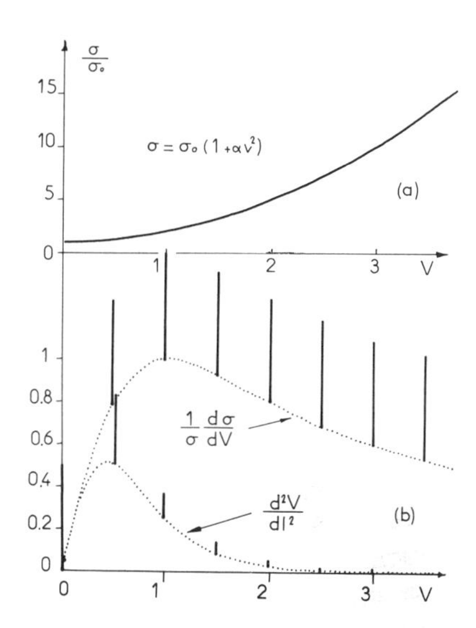

Figure 1. (a) Model of the junction conductance function of applied bias. (b) Peak heights in the second derivatives $(1/\sigma)(d\sigma/dV)$ and d^2V/dI^2 for identical jumps in the conductance situated at 0, 0.5, 1,... V.

example a small inelastic step at V_0 in a parabolic elastic conductance:

$$\sigma(V)=\sigma_0(1+\alpha V^2)[1+\beta Y(V-V_0)]$$

where Y is the Heavyside function. Then

$$\frac{1}{\sigma}\frac{d\sigma}{dV}\simeq\frac{2\alpha V}{1+\alpha V^2}+\beta\delta(V-V_0)$$

$$\frac{d^2V}{dI^2}\simeq\frac{2\alpha V}{\sigma_0^2(1+\alpha V^2)^3}+\frac{\beta\delta(V-V_0)}{\sigma_0^2(1+\alpha V^2)^2}$$

The first term in each case is the elastic background and the second is the inelastic contribution. This inelastic term remains constant in $(1/\sigma)\,d\sigma/dV$ whereas it decreases rapidly in d^2V/dI^2 as V increases (Figure 1).

3. Results

3.1. Rare Earth Oxides

Electronic transitions have been observed by tunneling spectroscopy in Ho–Ho_2O_3–Pb and Er–Er_2O_3–Pb junctions. The nature of the oxide—in both cases a sesquioxide—was determined by observation of its phonon

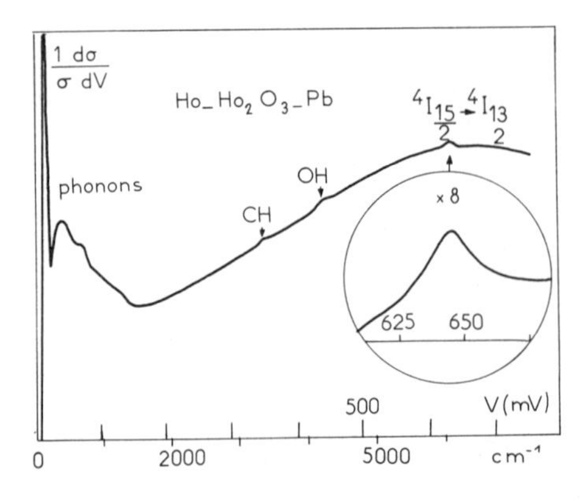

Figure 2. Tunnel spectrum of a Ho–Ho_2O_3–Pb junction ($T = 4.2$ K, $V_{mod} = 3$ mV rms). The $^5I_8 \rightarrow {}^5I_7$ transition of the HO^{3+} ion can be observed at 640 meV.

spectrum in the 20–90 meV range. Ho_2O_3 is known to present a transition at 625 meV ($^5I_8 \rightarrow {}^5I_7$ in the HO^{3+} ion) and Er_2O_3 presents a transition at 800 meV ($^4I_{15/2} \rightarrow {}^4I_{13/2}$ in the Er^{3+} ion). These are optically forbidden transitions, with very small oscillator strengths ($\simeq 10^{-6}$) for reasons of parity conservation: the initial and final states have the same quantum number l and the dipole moment matrix element between the two is practically zero. The first optically allowed transitions are at much higher energies (> 5 eV), too high to be seen by tunneling spectroscopy.

Both transitions are observed (Figures 2 and 3). It can be noted that the peaks are wider than the experimental resolution, even though f-electron levels are known to very sharp. This may be because of a less well ordered environment than in an oxide crystal or because of unresolved crystal field splitting (which is of the order of tens of millivolts).[7] The relative conductance variation is of the order of 10^{-3}. If we take into account the fact that only the ion layer against the counterelectrode will be seen by tunneling spectroscopy,[8, 9] we can estimate the oscillator strength observed as about

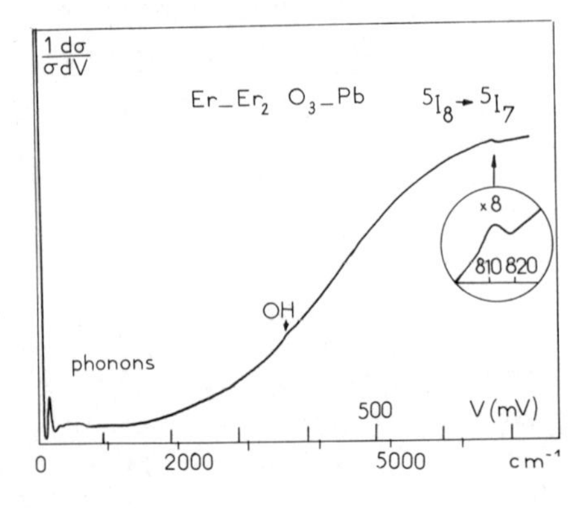

Figure 3. Tunnel spectrum of an Er–Er_2O_3–Pb junction ($T = 4.2$ K, $V_{mod} = 3$ mV rms). The $^4I_{15/2} \rightarrow {}^4I_{13/2}$ transition of the Er^{3+} ion is observed at 810 meV.

10^{-4}–10^{-5}, about an order of magnitude stronger than expected. In the absence of a proper theory for the excitation of electronic transitions by tunnel electrons, we can only suggest a line of thought. Since the matrix element of the dipole moment is nearly zero, it may be necessary to go to the next term in the development, the quadrupole term, which could give a larger contribution. Another possibility would be that the strong electric field E_x in the barrier mixes the $4f$ wave functions with functions of different parity (somewhat like the crystal field but without imposing the same symmetry conditions):

$$|\psi_{4f}\rangle \rightarrow |\psi_{4f}\rangle + \sum_i \frac{\langle \psi_i | E_x | \psi_{4f} \rangle}{E_i - E} |\psi_i\rangle$$

The mixing coefficient can be estimated at about 3×10^{-3}, taking the energy separation $E_i - E \simeq 5$ eV, a polarization of 0.5 V over 30 Å and a spatial extension for the $4f$ functions of 1 Å. In this case also, the matrix element would be enhanced. But whatever the mechanism, it is clear that tunneling selection rules are less strict than optical ones.

3.2. Large Molecules

Electronic transitions have been observed in a number of large organic molecules. Singlet–singlet transitions have been seen in

tetracyanin (1.3 eV) (Figure 4)
xenocyanin (1.3 eV) (Figure 5)

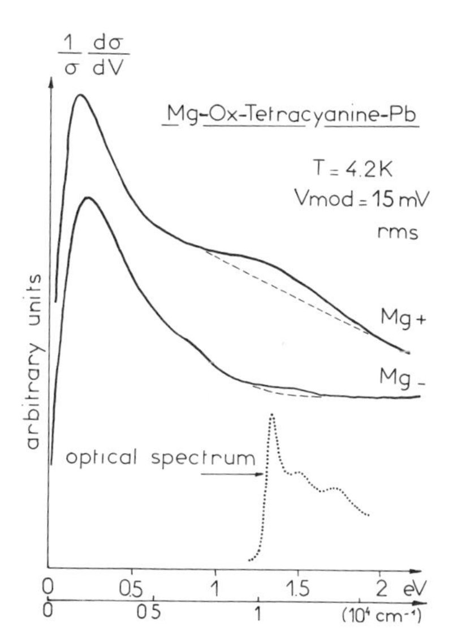

Figure 4. Tunnel spectra of tetracyanine, and the optical spectrum of the molecule in solution.

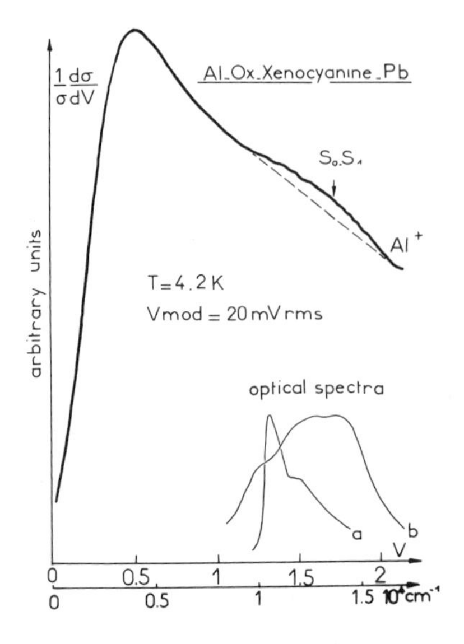

Figure 5. Tunnel spectrum of xenocyanine and the optical spectra of the molecule (a) in solution and (b) in a thin film.

bis(4-dimethylamino-dithiobenzil)nickel (1.25 eV)
pentacene (1.8 eV).

Singlet–triplet transitions have been seen in

β-carotene (not observed optically, expected at about 1.2 eV)
pentacene (0.8 eV) (Figure 6)

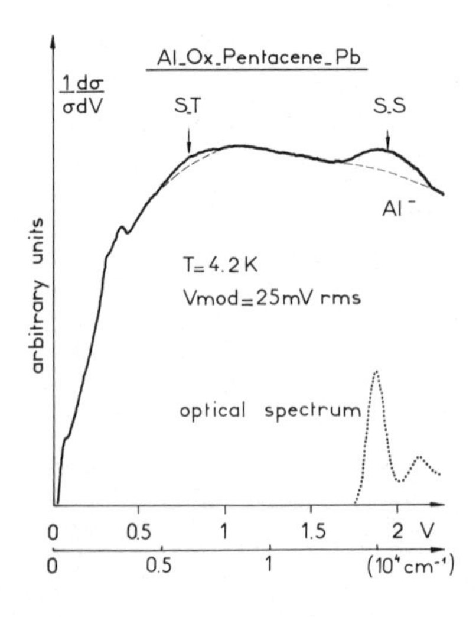

Figure 6. Tunnel spectrum of pentacene and the optical spectrum of the molecule in solution.

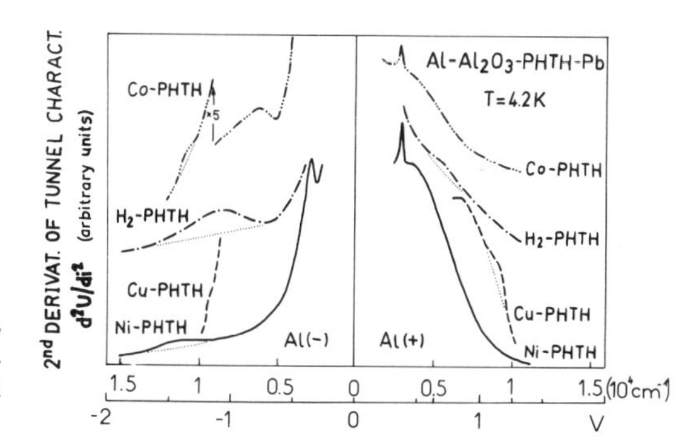

Figure 7. Tunnel spectra of Co-, H_2-, Cu-, and Ni-phthalocyanine. Reproduced with permission from reference 4.

H_2-phthalocyanine (not observed optically) (Figure 7)
Cu-phthalocyanine (1.15 eV) (Figure 7)
Co-phthalocyanine (1.2 eV) (Figure 7)
Ni-phthalocyanine (1.3 eV) (Figure 7)
Fe-phthalocyanine (not observed optically) (Figure 7)
NH-rhodanine-merocyanine (not observed optically) (Figure 8).

(the values in parentheses are the energies of the transitions as given by other measurements, generally optical).

A first remark that can be made is that the transitions are much wider than those observed in the optical spectra of free molecules. But, as illustrated in Figure 5, the width is comparable to that observed in the optical spectra of thin molecular films: The broadening is apparently due to a strong interaction of the molecules with a disordered environment.

A most remarkable point is that singlet–singlet and singlet–triplet transitions are observed with roughly the same intensities, although the first are optically "allowed," with oscillator strengths of about 10^{-1}, and the second are optically forbidden with oscillator strengths $\lesssim 10^{-3}$. Clearly optical selection rules, based on spin conservation, do not apply.

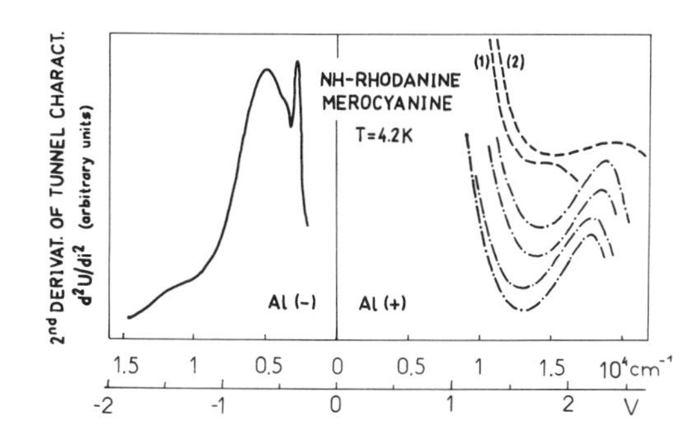

Figure 8. Tunnel spectra of NH–rhodaine–merocyanine. Curves plotted in different lines belong to measurements on junctions prepared in different experiments, and numbers indicate subsequent runs on the same junction. Reproduced with permission from reference 4.

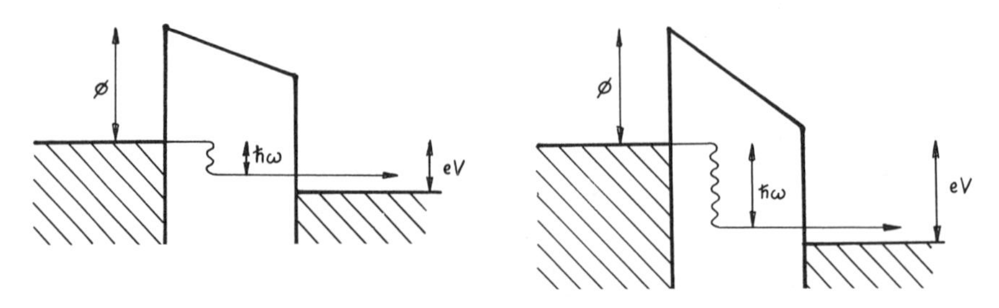

Figure 9. A very schematic image of tunneling electron: If it excites a high-energy transition, it is then faced with a higher barrier than if it only excites a low-energy transition.

Closer observation shows us that the observed peaks are much weaker than expected by comparison with vibrational transitions that have oscillator strengths $\simeq 10^{-5}$. A rough estimation shows that the conductance variation is about 50 times lower than expected for a singlet–singlet transition. This can be accounted for when one recalls that a tunneling electron that loses a lot of energy "sees" a much higher barrier than one that has only lost a little: It therefore has a much smaller tunneling probability (Figure 9). For example, this effect can be shown to attenuate by a factor of the order of 30 the conductance variation associated with a given transition when it takes place at 1.4 eV instead of 0.4 eV.[3]

So, to summarize, molecular electronic transitions are seen with the same intensity whether they are optically allowed or forbidden. And, if one corrects for the attenuation of the inelastic current due to barrier height effects, the intensity is roughly that expected for allowed transitions. Can this be explained theoretically?

We must begin by saying that a specific theory does not exist so far. We can only qualitatively explain the absence of the selection rule concerning the spin state of the molecule. The origin of the optical selection rule lies in the necessity of conserving the total spin of the system composed of the molecule plus the photon. Since the photon transfers no spin, it cannot change that of the molecule. If the fundamental has a spin state $S=0$ (singlet) (which is usually the case), the molecule cannot be excited to a spin state $S=1$ (triplet state) by a photon.

On the other hand, the rule can be lifted in the case of the excitation of a molecule by an electron—as is well known in electron impact spectroscopy.[10] If the incident electron is slow enough ($\simeq 30$ eV), it can be exchanged with a molecular electron, allowing the spin state of the molecule to change.

However, if the incident electrons are rapid, this exchange cannot take place and optical selection rules hold: only transitions between equivalent spin states occur. This appears in the theory of electron spectroscopy in a manner that is easily illustrated in the case of a hydrogen atom.[11] There are two terms in the scattering amplitude:

(a) There is a direct term, written in the Born approximation as

$$f_{n0} \propto \iint e^{i(k_0-k_n)r_2}\left(\frac{1}{|\mathbf{r}_1-\mathbf{r}_2|}-\frac{1}{r_2}\right)\psi_n^*(r_1)\psi_0(r_1)\,dr_1\,dr_2$$

here 1 represents the electron of the hydrogen atom; 2 represents the incident electron; ψ_0 and ψ_n are the initial and final states of the atom; k_0 and k_n are the initial and final wave numbers of the free electron.

This term reduces to zero if ψ_0 and ψ_n have different spin states since the Coulombian potential only acts on spatial variables. It can be written as follows:

$$f_{n0} \propto \frac{1}{K^2}\int \psi_n^*(r)e^{i\mathbf{K}\cdot\mathbf{r}}\psi_0(r)\,dr$$

where $\mathbf{K}=\mathbf{k}_0-\mathbf{k}_n$.

(b) There is also an exchange term, where the electrons 1 and 2 exchange in the collision:

$$g_{n0} \propto \iint e^{i(\mathbf{k}_0\mathbf{r}_2-\mathbf{k}_n\mathbf{r}_1)}\left(\frac{1}{|\mathbf{r}_1-\mathbf{r}_2|}-\frac{1}{r_1}\right)\psi_n^*(r_2)\psi_0(r_1)\,dr_2\,dr_1$$

This exchange now allows a change in the spin state of the molecule, while still conserving the total spin of the system.

The term can be written as follows:

$$g_{n0} \propto \frac{1}{k_0^2}\int \psi_n^*(r)e^{i\mathbf{K}\cdot\mathbf{r}}\psi_0(r)\,dr$$

The result can be generalized to molecules.[12] The point of interest to us is that the direct term—which reduces to zero if ψ_0 and ψ_n have different spin states—is strong for small momentum transfer **K**. But the exchange term, which allows transitions between different spin states becomes strong for small $\mathbf{k}_0$, that is for slow incident electrons. It seems to us that the same type of exchange can take place in a tunnel junction.

The effect has been explicitly calculated for free electrons—but not for tunneling electrons. The result is probably similar: singlet–triplet transitions would be stronger than in optical spectroscopy, but a precise estimation would be welcome!

4. What Are Not Electronic Transitions?

Particular care must be taken not to confuse electronic transition peaks with other noninelastic phenomena. One of these is a peak observed with Pb counterelectrodes due to quantum size-effect oscillations, that can be seen at 0.83 eV.[13] It can be distinguished from an inelastic peak because it only appears in the Al^- bias, it is temperature dependent, and above all it appears as a *dip*, not a peak since it is due to a conductance drop not an increase.

Another effect reported by Lüth *et al.*[4] is the appearance of peaks in the Al^+ bias that move up in energy from one bias sweep to the next (Figure 8). They attribute this effect to injection into the conduction band of the dopant layer, which then traps electrons. The energy level of the conduction band is therefore higher, due to the trapped electrons, hence the higher peak position at the next sweep.

5. Conclusions

(1) Electronic transitions can be observed by tunneling spectroscopy.

(2) Both optically allowed singlet–singlet transitions and optically forbidden singlet–triplet transitions are observed with roughly the same intensity. A theoretical analogy with optical spectroscopy is therefore inappropriate. One should look to the theory of electron spectroscopies, but this remains to be done.

(3) Parity conservation selection rules are slightly lifted, as shown in the case of rare earth oxides.

(4) The observed peaks are broad, partially owing to the ill-defined environment of the molecules in the tunnel junctions.

(5) Transitions beyond about 2 eV have not been observed because of junction breakdown, and because of the intensity of the elastic background.

(6) These last two points show that practical applications of IETS to the study of the electronic configuration of adsorbed molecules are rather limited.

Acknowledgment

We wish to thank Dr H. Lüth for allowing us to reproduce his curves and for useful comments on the electronic transitions observed.

References

1. D. J. Scalapino and S. M. Marcus, Theory of inelastic electron–molecule interactions in tunnel junctions, *Phys. Rev. Lett.* **18**, 459–461 (1967).
2. A. Léger, J. Klein, M. Belin, and D. Defourneau, Electronic transitions observed by inelastic electron tunneling, *Solid State Commun.* **11**, 1331–1335 (1972).
3. S. de Cheveigné, J. Klein, A. Léger, M. Belin, and D. Defourneau, Molecular electronic transitions observed by inelastic tunneling spectroscopy, *Phys. Rev. B* **15**, 750–754 (1977).
4. H. Lüth, U. Roll, and S. Ewert, Electronic transitions in some phthalocyanine and N–H rhodanine–merocyanine films studied by inelastic electron tunneling spectroscopy, *Phys. Rev. B* **18**, 4241–4249 (1978).
5. A. Adane, A. Fauconnet, J. Klein, A. Léger, M. Belin, and D. Defourneau, Observation of electronic transitions in rare earth oxides by inelastic tunneling, *Solid State Commun.* **16**, 1071–1074 (1975).
6. J. Klein, A. Léger, B. Delmas, and S. de Cheveigné, Second and third derivative of a tunnel junction characteristic. Application to the observation of electronic transitions, *Rev. Phys. Appl.* **11**, 321–325 (1976).
7. G. H. Dieke, *Spectra and Energy Levels of Rare Earth Ions in Crystals*, J. Wiley-Interscience, New York (1968).
8. S. Gauthier, J. Klein, A. Léger, S. de Cheveigné, and C. Guinet, Experimental study of relative intensities in inelastic electron tunneling spectra, *Phys. Rev. B* (in press).
9. J. Kirtley and J. T. Hall, Theory of intensities in inelastic electron tunneling spectroscopy. Orientation of adsorbed molecules, *Phys. Rev. B* **22**, 848–856 (1980).
10. J. P. Doering and A. J. Williams III, Low-energy, large angle electron-impact spectra: Helium, nitrogen, ethylene and benzene, *J. Chem. Phys.* **47**, 4180–4185 (1967).
11. B. L. Moiseiwitsch and S. I. Smith, Electron impact excitation of atoms, *Rev. Mod. Phys.* **40**, 238–353 (1968).
12. M. Matsuzawa, Electron-impact excitation cross sections of the lowest-lying triplet states of benzene, *J. Chem. Phys.* **51**, 4705–4709 (1968).
13. R. C. Jaklevic, J. Lambe, J. Kirtley, and P. K. Hansma, Structure at 0.8 eV in metal–insulator–metal tunneling junctions, *Phys. Rev. B* **15**, 4103–4104 (1977).

5

Light Emission from Tunnel Junctions

D. L. Mills, M. Weber, and Bernardo Laks

1. Introduction

A quick glance at the list of chapter titles in the present volume shows that electron tunneling spectroscopy has evolved from an esoteric curiosity in a few solid-state physics laboratories into a tool of considerable practical importance. For example, one may study the vibrational spectra of molecules bound to supported metal catalysts with this method, for geometries that closely mimic those of interest to catalytic chemists. While surface vibrational spectroscopy is a field in a stage of rapid development at the time of this writing, with several new techniques under exploration, tunneling spectroscopy is unique in its ability to explore such geometries.

The basic notion exploited in this form of spectroscopy is that as the electron tunnels through an oxide barrier between two metals, it may scatter inelastically from a molecule bound to the near vicinity of the barrier. As it does so, it excites a vibrational normal mode of the molecule, to lose the energy $\hbar\omega_0$, where ω_0 is the frequency of the mode excited. By scanning d^2G/dV_0^2, with G the junction conductance and V_0 the applied dc voltage, one obtains a spectrum with linelike features at the voltages $V_0 = \hbar\omega_0/e$ where excitation of the vibrational quanta first occurs.

As the electron tunnels through the junction, it may transfer a fraction of its energy not only to vibrational modes of molecules or phonons, but to

D. L. Mills and M. Weber • Department of Physics, University of California, Irvine, California 92717. **Bernardo Laks** • Instituto de Fisica, DFA, Universidade Estadual de Campinas, Campinas, San Paulo, Brazil.

any elementary excitation to which it may couple. From Chapter 4 of the present volume, for example, one sees that electronic transitions may be studied through tunneling spectroscopy.

The present chapter is devoted to the discussion of one other distinctly different tunneling process. This is one in which a *photon* is emitted in the tunneling process. Not very long ago, Lambe and McCarthy[1] reported the emission of light from tunnel junctions subjected to a dc bias voltage V_0. The light emitted from the junction is visible to the naked eye in a darkened room.* The phenomenon is most fascinating; one obtains emission of light from a device through which a dc current is flowing. Measurements of the frequency spectrum[1] of the light show a continuous spectral distribution with a cutoff frequency $\omega_c = eV_0/\hbar$. Thus, as the voltage is increased above two volts or so, the color of the light changes from red, to orange, to yellow and so on, until the junction burns out with increasing voltage. The conditions required to observe the effect are not exotic. The original experiments of Lambe and McCarthy were carried out at room temperature, for example. These authors did find it necessary to roughen the junction surface to see the emission. We shall see shortly that this provides a strong clue to the mechanism that controls the emission in a number of the devices constructed so far.

From the fact that one observes dim light from the junction in a darkened room, it is possible to estimate the probability that an electron emits a photon as it tunnels through the junction. The devices typically carry a current of 10 mA, so roughly 10^{17} electrons/sec traverse the junction. A dim light corresponds to roughly 10^6 photons/sec at the eye, and if the observer stands one meter away from the device with a pupil diameter of 3 mm, the solid angle subtended by the eye is roughly 3×10^{-5} sr. If the photons are distributed uniformly over a solid angle, then a total of 4×10^{11} photons/sec are emitted. The probability that an electron emits a photon as it tunnels is then in the range of 10^{-5}–10^{-6}. This is to be compared with 10^{-2} as the probability the electron will excite a molecular vibration upon tunneling[2] in junctions doped with molecules in the tunnel junction region. Thus, the tunnel junctions explored so far are very inefficient light emitters, and they have the possibility of becoming a useful new light source with continuously tunable color if this efficiency can be increased by a few orders of magnitude. We shall present a pessimistic view of the possibility of achieving such an improvement in performance with devices similar to those studied so far, but there is still the possibility that junction geometries quite different from those fabricated at this time may function far better.

*We wish to thank Professor P. K. Hansma for providing us with a laboratory demonstration of light emission from a tunnel junction.

Before we turn to a discussion of the particular geometries that have been explored to date, it will be useful to comment on general features of the emission mechanism. For this purpose, we need to recall some aspects of the optical response of electrons in metals. At visible frequencies, one is concerned with electromagnetic waves that have frequency ω so large that $\omega\tau \gg 1$, where τ is the electron relaxation time. Then the frequency-dependent dielectric constant of the material may be written $\varepsilon(\omega) = \varepsilon_\infty - \omega_p^2/\omega^2$, where ω_p is the plasma frequency of the electrons and ε_∞ the (frequency-dependent) contribution to the dielectric constant from interband transitions. Here we assume ε_∞ is real, but in general this quantity is complex. For most metals, with gold as an exception, throughout the visible the condition $\omega < \omega_p/(\varepsilon_\infty)^{1/2}$ is satisfied, so the dielectric constant $\varepsilon(\omega)$ is *negative*. If an interface is formed between a metal with negative dielectric constant and an oxide for which the dielectric constant is positive, then a special set of electromagnetic modes of the structure exist, as we shall see in the next section. They are called surface polaritons, and they are solutions of Maxwell's equations that propagate along the interface with electromagnetic fields that decay exponentially as one moves away from the interface in either direction.

Electrons can couple quite strongly to surface polaritons. One may see this by examining data on the inelastic scattering of electrons incident on a metal surface from the outside.* Such an electron beam, if its kinetic energy is sufficient, may excite surface polaritons with wave vector $k_\parallel$ parallel to the surface large compared to ω/c. In this regime, the frequency of the surface polariton is independent of $k_\parallel$, and is given by $\omega_p/(1+\varepsilon_\infty)^{1/2}$, for the metal–vacuum interface.† Here as an electron is reflected off the surface, the probability p that a surface mode is excited is given by‡

$$p = \Gamma \exp(-\Gamma) \tag{1}$$

where

$$\Gamma = \frac{\pi}{a_0 k_I \cos\theta_I} \frac{1}{(1+\varepsilon_\infty)^{1/2}} \tag{2}$$

with a_0 the Bohr radius, k_I the wave vector of the electron, and θ_I the angle

*See the experimental data by C. J. Powell, reference 3.

†A review of the properties of these modes is given in reference 4. See also reference 5.

‡This formula, strictly speaking, is valid only when the kinetic energy of the electron is large compared to $\omega_p/(1+\varepsilon_\infty)^{1/2}$, but it remains valid in a semiquantitative sense at lower kinetic energies. See the discussion in Chapter 3 of reference 6.

of incidence of the electron beam. One readily sees that Γ is the order of unity under typical conditions. For an electron with a kinetic energy of 50 eV, for example, the factor $a_0 k_I$ is roughly equal to 2.

The above remarks refer to an electron beam incident on the crystal from the outside. However, the electric fields of the surface polaritons have roughly the same strength *inside* the metal as they do outside,* so electrons inside the metal can be expected to couple efficiently to the surface modes. Thus, we expect the probability that an electron which tunnels through an oxide barrier excites a surface polariton to be substantial. Unfortunately, since the electromagnetic fields of the surface wave are bound to the interface, even if the excitation process proceeds with high efficiency, the junction will not radiate electromagnetic waves. It is necessary to decouple the energy stored in the surface polariton from the junction structure for the junction to emit radiation. If the junction surface is roughened, or the interfaces are perturbed from a perfect planar structure in any way, the surface polariton may be scattered off the junction into the vacuum to emerge as a photon. We remarked earlier that in their initial experiments, Lambe and McCarthy found it necessary to roughen the surface of the junction to observe radiation emitted from it. This led them to propose the two-step process just described as responsible for the emission, and the theoretical analyses that have appeared subsequently have focused on this picture, or modified forms of it. The present authors have also explored the possibility that the tunneling electrons may emit a photon directly, to find such emission is comparable in intensity to that from the two-step process, in the frequency regime where interband transitions are allowed. We comment further on this later in the present chapter.

There are three basic experimental configurations that have been explored in the experiments to date, and for each geometry the emission mechanism differs, though for each the basic two-step picture may be invoked to account for the principal features found in the data.† The first of these is the geometry employed by Lambe and McCarthy and later by others.[7, 8] The junctions may be regarded as planar structures with a film of aluminum first laid down on a substrate. This is then allowed to oxidize, and a film of silver or gold is then evaporated over the oxide layer. As we shall see, these latter materials endow the junction structure with "conve-

*In the limit $k_{\parallel} \gg \omega/c$, for the metal–vacuum interface, the fields inside are simply the mirror image of those outside. For more general values of $k_{\parallel}$, conservation of tangential components of **E** across an interface ensures that the field outside and the field inside are comparable in strength.

†This is true except possibly for junctions which incorporate gold as one of the constituents. Here we believe that in certain spectral regimes, direct emission of photons by electrons may occur. This will be discussed in more detail below.

nient" surface polaritons. Then in these structures, roughness introduced deliberately, or present as a consequence of the sample preparation procedure, plays an essential role in the two-step process. This roughness has transverse scale that ranges from a few tens of angstroms to a few hundred angstroms; an analysis of the frequency spectrum of the emitted radiation provides very direct evidence for the presence of roughness with a transverse scale of 50 Å or so.

A second geometry has been employed by Kirtley, Theis, and Tsang.[9] They begin with a holographic grating of sinusoidal profile, and then fabricate an Al–Al_2O_3–Ag junction on top of the grating. Thus they have a junction structure with the property that each interface has a profile that replicates that of the substrate. As the surface polariton propagates along such a structure, its fields are Bragg diffracted from it, to produce radiation that emerges in a well-defined direction for each choice of surface polariton wave vector. The gratings have periods the order of a micron, so the characteristic transverse scale of the modulations in the profile of the junction interfaces is two to three orders of magnitude larger than in junctions that radiate by virtue of microscopic roughness. The reader will require some understanding of the nature of the dispersion curves in the planar junction structures to appreciate that rather different surface polariton modes enter as intermediate states in these two cases.

Hansma and his colleagues have explored a third configuration, quite different from either of the two just described. They deposit blobs of noble metal on top of the oxide layer, with the thought that a given blob will have one or more electromagnetic resonances. An electron can then tunnel into one of these inelastically, to excite an electromagnetic resonance of the object in the process. The energy stored in the resonance can then be dissipated partly as radiation into the space above the junction, hopefully with efficiency larger than that realized on the junctions with roughened surfaces. In a certain formal sense, this configuration can be regarded as an extreme limiting case of the junction with slightly roughened surface. If one begins with a junction with small amplitude roughness present, to first approximation the electromagnetic normal modes of the structure are those of the junction with smooth interfaces. As the amplitude of the roughness is increased, at some point one will encounter new modes of excitation with fields localized to the near vicinity of bumps or other prominent features of the profile. Many years ago, Berreman[10] discussed the influence of such resonances on the reflectivity of ionic crystals in the infrared, and more recently Ruppin[11] has explored the interaction of optical radiation with bumps of hemispherical shape on a surface of Ag metal, to find a complex sequence of resonances in the visible. The geometry used by Hansma and his colleagues is one in which, in essence, such localized resonances play the primary role in the emission process, rather than extended modes of a roughly planar structure.

The remainder of this article is as follows. Section 2 presents a summary of the properties of the surface polaritons of the tunnel junction structure with planar interfaces. From this, we may appreciate certain qualitative features of the modes that control the emission in the first three geometries. Section 3 reviews the content of theories of the emission process for various geometries, along with some features of the data. Then in Section 4 we give concluding remarks and some speculations for the future.

2. Planar Tunnel Junctions and Surface Polaritons

All textbooks on electromagnetic theory discuss the reflection and refraction of electromagnetic waves from the planar boundary between two dielectric media. At the same time, all modern texts known to us with the exception of one* fail to note that the same interface can support the propagation of an electromagnetic wave along it. This wave has fields localized to the near vicinity of the interface, in the sense that they fall off exponentially with distance from it, as one moves into either medium. These modes are not a recent discovery, but they have been known since the turn of the present century, and have played a prominent role in various subareas of physics and engineering since that time.† Perhaps these fascinating normal modes of the radiation field have failed to attract general attention because, as we shall see shortly, a certain special condition must be realized before an interface can support them. The condition is in no sense highly restrictive, when viewed from the perspective of solid-state physics. For example, the modes may propagate on the surface of aluminum at all frequencies from the microwave, through the infrared and visible, and into the ultraviolet up to the energy of 10.6 eV. We begin with a discussion of these modes at a simple planar interface, then we summarize their properties for junction structures of interest to the present paper. We shall refer to this collection of modes as surface polaritons; the reader may wish to note that a volume that explores many basic properties of these modes will appear shortly.[5]

In Figure 1, we show the basic geometry to be explored initially. We have the interface between two dielectric media. The interface lies in the xy plane, with ε_0 the dielectric constant of the material in the upper half-space $z > 0$, and $\varepsilon(\omega)$ that in the lower half-space $z < 0$. We shall suppose that for

*A discussion of surface polaritons does appear in reference 12.

†Discussion of the early work on this mode is given in reference 13.

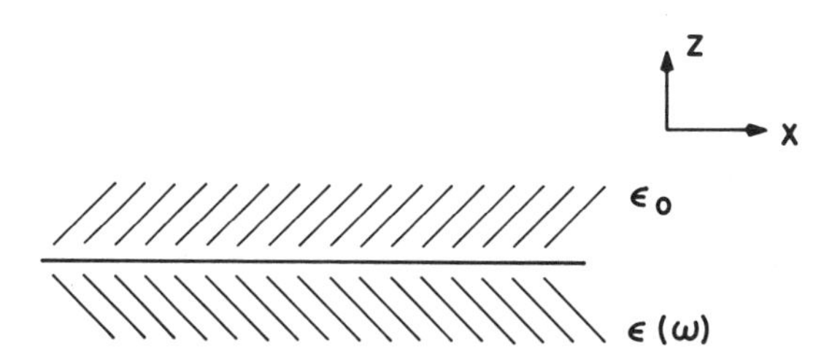

Figure 1. The geometry used to illustrate the basic properties of surface polaritons. In the upper half-space $z > 0$, we have a dielectric with dielectric constant ε_0, while in the lower half-space $z < 0$, we have a metal with dielectric constant $\varepsilon(\omega)$. The calculations are carried out for the choice $\varepsilon_0 = 3$, and $\varepsilon(\omega) = 1 - \omega_p^2/\omega^2$ with $\omega_p = 15$ eV, characteristic of aluminum.

$z > 0$, we have an insulator, and in the lower half-space a metal. We shall thus take ε_0 to be real and positive, and for $\varepsilon(\omega)$ we have $\varepsilon(\omega) = \varepsilon_\infty - \omega_p^2/\omega^2$, with ε_∞ the contribution to the dielectric constant of the metal from interband transitions.

Surface polaritons are electromagnetic waves with electric field in the xz plane, and with amplitude that decays exponentially as one moves away from the interface. Thus, we seek solutions to Maxwell's equations with the form

$$\mathbf{E}^{>}(x, z) = E^{>}\left(\hat{x} + i\frac{k_\parallel}{\alpha_>}\hat{z}\right) e^{ik_\parallel x} e^{-\alpha_> z} e^{-i\omega t}, \qquad z > 0 \tag{3a}$$

and also

$$\mathbf{E}^{<}(x, z) = E^{<}\left(\hat{x} - i\frac{k_\parallel}{\alpha_<}\hat{z}\right) e^{ik_\parallel x} e^{+\alpha_< z} e^{-i\omega t}, \qquad z < 0 \tag{3b}$$

In these expressions, the ratio between the $\hat{x}$ and $\hat{z}$ components in each medium is fixed by the requirement that $\nabla \cdot \mathbf{E} = 0$. For each Cartesian component of the electric field, the wave equation must be satisfied in each medium. For $z > 0$, this reads

$$\left[\frac{\partial^2}{\partial x^2} + \frac{\partial^2}{\partial z^2} + \frac{\omega^2}{c^2}\varepsilon(\omega)\right] E_{x,z}(x, z) = 0 \tag{4}$$

with a similar form for $z < 0$. For given values of ω and $k_\parallel$, the wave equation fixes both $\alpha_>$ and $\alpha_<$:

$$\alpha_>^2 = k_\parallel^2 - \frac{\omega^2}{c^2}\varepsilon_0 \tag{5a}$$

$$\alpha_<^2 = k_\parallel^2 - \frac{\omega^2}{c^2}\varepsilon(\omega) \tag{5b}$$

Note from Eq. (5a) that for $\alpha_>$ to be real, with the consequence that $\alpha_>^2$ is positive, we must have $k_\parallel > \omega(\varepsilon_0)^{1/2}/c$, and a similar constraint on the possible choice of $k_\parallel$ and ω follows from Eq. (5b).

For the pair of fields displayed in Eq. (3) to be a proper solution of Maxwell's equations, they must also satisfy suitable boundary conditions across the interface. One of these is that tangential components of **E** be conserved, and this requires that $E^> = E^<$. In addition to this, we require normal components of **D** be equal on each side of the interface, and this leads to a further constraint that may be written

$$\varepsilon(\omega) = -\varepsilon_0 \frac{\alpha_<}{\alpha_>} \tag{6}$$

The result in Eq. (6) constitutes an implicit dispersion relation of the wave; for a given choice of $k_\parallel$, the condition will be satisfied for one and only one frequency.

We may pause in the development to explore the consequences of Eq. (6). We said earlier that we shall assume ε_0 is real and positive, as is appropriate to the situation where the medium in the upper half-space is an insulating oxide. Also, $\alpha_<$ and $\alpha_>$ are necessarily positive, if Eqs. (3a, 3b) are to describe waves "bound" to the interface. Thus, Eq. (6) requires the dielectric constant $\varepsilon(\omega)$ to be a *negative* number. We assume that this requirement is the reason why many standard texts overlook the surface polariton solutions to Maxwell's equations. It is usual to suppose, either explicitly or implicitly, that the dielectric constant of matter is positive. For insulating materials, considerations of thermodynamic stability require the *static* (zero-frequency) dielectric constant to be positive, but even for insulators at finite frequencies there is no reason for such a constraint. For our nearly free electron metal with $\varepsilon(\omega) = \varepsilon_\infty - \omega_p^2/\omega^2$, we have $\varepsilon(\omega) < 0$ for *all* frequencies below the bulk plasma frequency $\omega_p/(\varepsilon_\infty)^{1/2}$. At this point the reader may appreciate that satisfaction of the condition in Eq. (6) does not lead to very unusual requirements on the materials used to form the interface.

When $\varepsilon(\omega) < 0$, it follows from Eqs. (5a, 5b) that the condition $\alpha_< > \alpha_>$ is always met. Thus, not only must we have $\varepsilon(\omega) < 0$, but the stronger condition $\varepsilon(\omega) \leqslant -\varepsilon_0$ results. For our model, the surface polaritons thus propagate on the interface for frequencies that satisfy

$$\omega \leqslant \frac{\omega_p}{(\varepsilon_\infty + \varepsilon_0)^{1/2}} \tag{7}$$

For the interface between aluminum and its oxide, we have $\omega_p \cong 15$ eV,

$\varepsilon_\infty \cong 1$ and it is reasonable to suppose $\varepsilon_0 = 3$, a value typical of wide gap insulators. With these numbers, the interface supports surface polariton propagation at all frequencies below 7.5 eV.

The implicit dispersion relation in Eq. (6) may be rearranged to read

$$\frac{c^2 k_\parallel^2}{\omega^2} = \frac{\varepsilon_0 \varepsilon(\omega)}{\varepsilon_0 + \varepsilon(\omega)} \tag{8}$$

and for the numbers just quoted, the dispersion curve is plotted in Figure 2a. The key points to note are that when $ck_\parallel \gg \omega$, the frequency of the wave is given by $\omega_p/(\varepsilon_\infty + \varepsilon_0)^{1/2}$ independent of $k_\parallel$ to good approximation, while for $\omega \ll \omega_p/(\varepsilon_\infty + \varepsilon_0)^{1/2}$ the dispersion curve hugs close to the "light line" $ck_\parallel/(\varepsilon_0)^{1/2}$, the dispersion relation of a photon in the oxide which propagates parallel to the interface. For the purposes of our later discussion, it will be important to note that in the visible ($\omega \cong 2.5$ eV), the wave vector of the wave is $k_\parallel = 2.5 \times 10^5$ cm^{-1}, corresponding to a wavelength of 2500 Å.

Next consider the nature of the electric fields stored in the wave. When $ck_\parallel \gg \omega$, one has $\alpha_> = \alpha_< = k_\parallel$. Thus, the fields are symmetrical on each

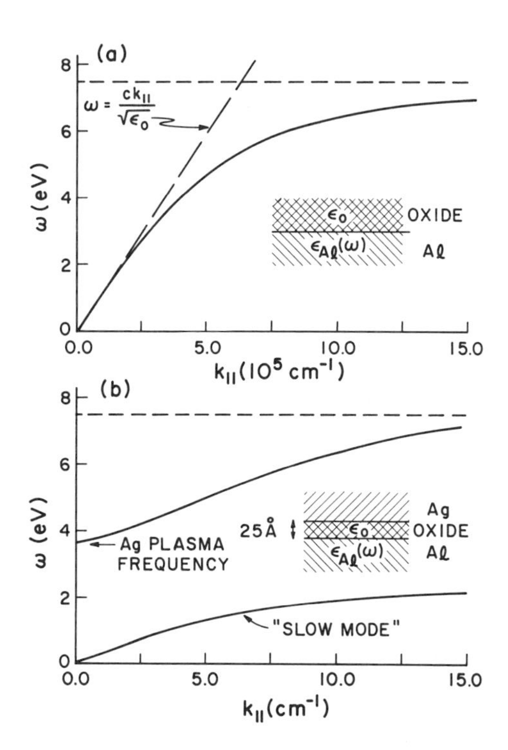

Figure 2. Surface polariton dispersion curves for propagation on (a) the interface between aluminum and its oxide, each assumed semiinfinite in extent, and (b) an interface between aluminum and Ag, with a thin oxide layer in between.

side of the interface, in the sense that on each side they decay to zero in the distance $l_>(k_\parallel)=l_<(k_\parallel)=k_\parallel^{-1}$. In each medium, the $\hat{x}$ and $\hat{z}$ components are equal in magnitude, and the magnitude of the fields inside the metal are the same as those in the oxide. This was the basis for the argument presented in Section 1, which used information on the interaction with a beam of electrons *outside* the metal to infer that electrons *inside* the metal couple strongly to the surface polariton. Even for more complex structures (discussed below) which support surface polariton propagation, similar statements may be made in the regime $ck_\parallel \gg \omega$.

The picture of the fields associated with the wave is very different when $\omega \ll \omega_p/(\varepsilon_0+\varepsilon_\infty)^{1/2}$ and one has $\omega \cong ck_\parallel/(\varepsilon_0)^{1/2}$. We now have $\alpha_< \cong \omega_p/c$, which is simply the inverse of the optical skin depth of the metal, which is typically 200 Å or so in the visible and near infrared. Hence $l_<(k_\parallel)\cong 200$ Å, and when ω is small compared to ω_p, $l_>(k_\parallel) \gg l_<(k_\parallel)$ so most of the energy density in the wave is stored in the *oxide* rather than in the metal. Furthermore, while the x components of the electric field on each side of the interface must be equal by the boundary condition on tangential components of **E**, the normal component in the oxide is larger by the factor $\varepsilon(\omega)/\varepsilon_0$. Again, this increases the fraction of the energy density of the wave that is stored in the oxide. One may quantify this as follows. Let $(dE_>/dt)$ be the energy per unit time carried by the wave which is stored in the half-space $z>0$, and let $(dE_</dt)$ be the fraction stored in the metal. It is intriguing that while the direction of flow of the energy density stored in the oxide is *parallel* to the wave vector, that stored in the metal is *antiparallel* to it. We have the relation[4]

$$\frac{dE_<}{dt} = -\frac{\varepsilon_0^2}{\varepsilon(\omega)^2}\frac{dE_>}{dt} \tag{9}$$

so the net energy flow is always parallel to the direction of the wave vector.

What is important for the purposes of the present discussion is to note that when $\omega \ll \omega_p/(\varepsilon_0+\varepsilon_\infty)^{1/2}$ and $\omega ck_\parallel/(\varepsilon_0)^{1/2}$, one has

$$\frac{dE_<}{dt} = -\frac{\omega^4\varepsilon_0^2}{\omega_p^4}\frac{dE_>}{dt} \tag{10}$$

so only a small fraction of the total energy density carried by the wave is stored in the metal. This means that electrons will couple to such surface polaritons much more weakly than the estimate in Section 1 suggests. We have a general principle here: electrons couple very strongly to surface polaritons when they have a wave vector $k_\parallel$ large enough to satisfy $ck_\parallel \gg \omega$, while the coupling becomes far less efficient as the light line is approached because then only a small fraction of the energy density in the wave resides in the metal.

We now turn to the nature of surface polaritons that propagate on the more complex tunnel junctions of interest here. We shall present only a summary of numerical calculations of the dispersion curves and also the fields associated with the modes, since the dispersion relations may only be obtained in implicit form for these more intricate geometries. The principles which emerged in the preceding discussion of the simple plane interface will provide a useful framework, however.

We approach the description of the full tunnel junction in a sequence of steps, beginning with a thin oxide layer sandwiched between a semi-infinite piece of aluminum, and a noble metal such as Ag. The dielectric response of aluminum is well described by choosing $\varepsilon_{\mathrm{Al}}(\omega)=1-\omega_p^2/\omega^2$, with ω_p the bulk plasma frequency equal to 15 eV. In Ag or Au, interband contributions enter the dielectric function importantly, and the electron density is lower. The bulk plasma frequency, defined as that frequency for which the (real part) of the dielectric function vanishes, thus lies far below the plasma frequency of aluminum. For silver metal, $\varepsilon_{\mathrm{Ag}}(\omega)$ vanishes at 3.75 eV.

Consider first the interface between Ag and Al, with no oxide present. The frequency region between the plasma frequency of Ag and Al is a "window" within which surface polaritons may propagate, since there $\varepsilon_{\mathrm{Al}}(\omega)$ is negative while $\varepsilon_{\mathrm{Ag}}(\omega)$ is positive. Everywhere else, both dielectric constants have the same sign, and as we have seen it is not possible for surface modes to "bind" to the interface. The nature of surface polaritons at the interface between conducting media was examined explicitly first by Halevi.(14)

Now if a thin layer of oxide is inserted between the two conductors, we may see that *two* surface wave branches must exist. This is clear because if d is the thickness of the oxide layer, and $k_{\parallel}d \gg 1$, we see from our earlier discussion we have to have very good approximation between one mode with fields highly localized to the Al–oxide interface at the frequency where $\varepsilon_{\mathrm{Al}}(\omega)=-\varepsilon_0$, and a second mode localized on the Ag oxide interface at the frequency for which $\varepsilon_{\mathrm{Ag}}(\omega)=-\varepsilon_0$. In Figure 2b, we show calculations of the dispersion relation for the surface modes of this structure. We indeed see a two-branch dispersion curve, with two modes for each value of $k_{\parallel}$. The high-frequency branch extends from the bulk Ag plasma frequency, up to the frequency for which $\varepsilon_{\mathrm{Al}}(\omega)=-\varepsilon_0$. This mode is very similar in character to that expected for the Ag–Al interface in the absence of the oxide layer, as described in the previous paragraph.

The low-frequency branch has an asymptotic frequency determined by the condition $\varepsilon_{Ag}(\omega) = -\varepsilon_0$ as $k_{\parallel} d \to \infty$, and thus in this limit has fields localized to the near vicinity of the oxide–silver interface. The key feature of the lower branch of the dispersion curve in Figure 2b is that in the visible its wave vector is *very large*. At 2.5 eV, the wave vector of the mode is 3×10^6 cm^{-1}, corresponding to a wavelength of only 200 Å. This is more than one order of magnitude smaller than the wavelength of a visible photon. These very short wavelength surface polaritons interact strongly with roughness on the interface that has a transverse scale a_0 so that $k_{\parallel} a_0 \cong 1$, i.e., they interact strongly with features that have a transverse scale of 30–50 Å. We shall see in the next section that light-emitting junctions of the first class discussed in Section 1 (interfaces roughened artificially, or with radiation observed from junctions with roughness present from the deposition process) involve this very short wavelength surface polariton as the principal mode excited by the tunneling electrons. The frequency spectrum of light emitted from such junctions thus contains information on roughness with such a small transverse scale; the present authors know of no other such short-wavelength probe of features on this length scale. This lower branch has been referred to in the recent literature as the "slow surface polariton." It was first discussed many years ago in a different physical context by Swihart,[15] and is also widely known as the Swihart mode.

Now let us complicate the geometry one more step by adding one more interface. Figure 3 shows the surface polariton dispersion curves for a semi-infinite aluminum substrate overlaid with 25 Å of oxide, then with 200 Å of Ag deposited on the top. As $k_{\parallel} \to \infty$, we now have *three* modes of the structure, since there is one more mode localized to the Ag–vacuum interface in this limit. We see three modes in Figure 3, with the dispersion curve of the slow mode rather unaffected by the presence of the new interface. The upper branch also has an appearance very similar to that in

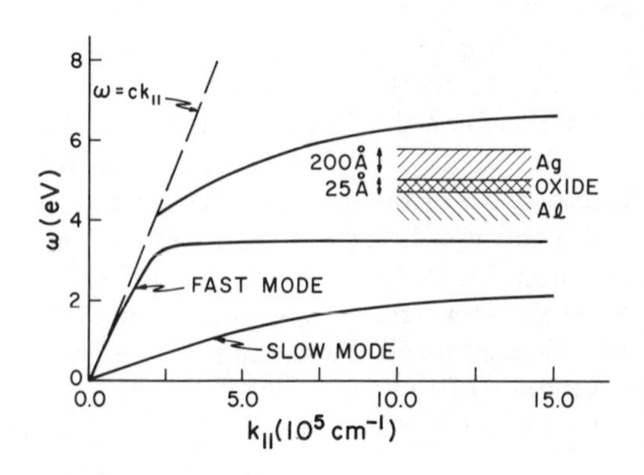

Figure 3. The dispersion curves for a structure with 25-Å layer of oxide and 200 Å of Ag deposited on the aluminum substrate.

Figure 2b, except the dispersion curve now ends when it strikes the "light line" $\omega = ck_{\parallel}$. As we saw earlier in the section, the structure cannot support propagation of a surface polariton to the left of the light line, where $ck_{\parallel} < \omega$, since the electromagnetic fields in the vacuum above the junction structure cannot be localized near it. That this is so can be seen by examining Eqs. (5a, 5b) with ε_0 set equal to unity. What happens in this case is that the upper branch merges with the light line, and as in the simple one-interface example explored explicitly earlier, as $k_{\parallel} \to \omega/c$ from the left, $\alpha_{>} \to 0$ and the wave looks more and more like a "true photon" which propagates parallel to the surface, with fields that extend down into the substrate structure one skin depth.

There is a new branch, labeled the "fast mode" in Figure 3. As $k_{\parallel} \to \infty$, its frequency approaches that for which $\varepsilon_{Ag}(\omega) = -\varepsilon_0$, but as $k_{\parallel}$ is decreased the dispersion curve of this mode approaches the light line $\omega = ck_{\parallel}$, then follows it down to zero frequency. Throughout the visible, we have $\omega \cong ck_{\parallel}$ to quite a good approximation, and the wavelength of the mode is roughly 5000 Å in the visible, a value very close to a vacuum photon of the same energy and more than twenty times longer than that of the "slow mode." The "fast mode" will not be affected greatly by roughness on the 30–50-Å scale, simply because its wavelength is very long. It will play an important role in emission from junctions grown on holographic gratings, however, since these structures have spatial periods quite comparable to the wavelength of the fast mode. The argument in the paragraph that contains Eq. (10) may readily be extended to these "fast modes," to show that when $\omega \cong ck_{\parallel}$, most of their energy is stored in the vacuum *above* the crystal, away from the electrons in the metal. Thus, electrons in the metal couple to the "fast mode" with much smaller efficiency than they do the "slow mode," to which a discussion very similar to that which surrounds Eqs. (1) and (2) may be applied.

In Figures 4–6, we present calculations of the magnitude and spatial variation of the electric fields associated with the structure used to generate the dispersion curves in Figure 3. In all these calculations, the value of the x component of electric field at the silver–air interface has been set equal to unity. On the whole, upon scanning the figures, one sees that for the range of wave vectors explored, the electromagnetic field associated with the modes extend throughout the tunnel junction structure. Figure 4 displays the spatial variation of the fields associated with the "slow mode," for selected frequencies. One sees that for large $k_{\parallel}$, the electromagnetic fields associated with the "slow mode" have their peak value near the oxide barrier; the fields of visible frequency are quite a bit more localized than suggested in the figure. Figure 5 shows the fields associated with the "fast mode." One may appreciate that for the "fast mode," a very large fraction

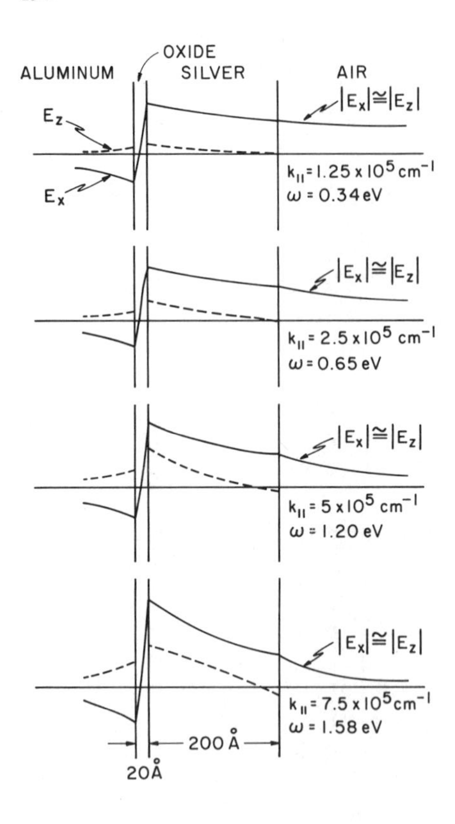

Figure 4. The x and z components of electric field associated with the "slow surface polariton" of Figure 3, calculated for selected points on the dispersion curve. The dielectric constants of all three media are assumed to be real, so E_z and E_x are 90° out of phase everywhere.

of its energy density resides in the vacuum outside the metal at visible frequencies, as the discussion which follows Eq. (10) suggests. As $k_{\|}$ increases, the fields associated with this mode become localized to the outermost interface, while Figure 6 shows that for large $k_{\|}$, the fields associated with the upper branch become localized to the $Al–Al_2O_3$ interface.

We still do not have the full picture of surface polariton dispersion curves of a tunnel junction structure, since in Figure 3 we have not included the influence of the interface between the aluminum film and the substrate upon which the entire structure rests. If this feature is added, then as $k_{\|} \to \infty$ one finds a *fourth* mode in which the fields are localized at the interface between the aluminum film and the substrate. The three low-lying dispersion curves of the whole structure have been studied by Kirtley and co-workers.[2] These authors find that in the visible, in addition to the "slow mode" and "fast mode" illustrated in Figure 3, there is a third branch for which the dispersion relation may be approximated by $\omega \cong ck_{\|}/(\varepsilon_s)^{1/2}$, where ε_s is the dielectric constant of the substrate. This mode has electro-

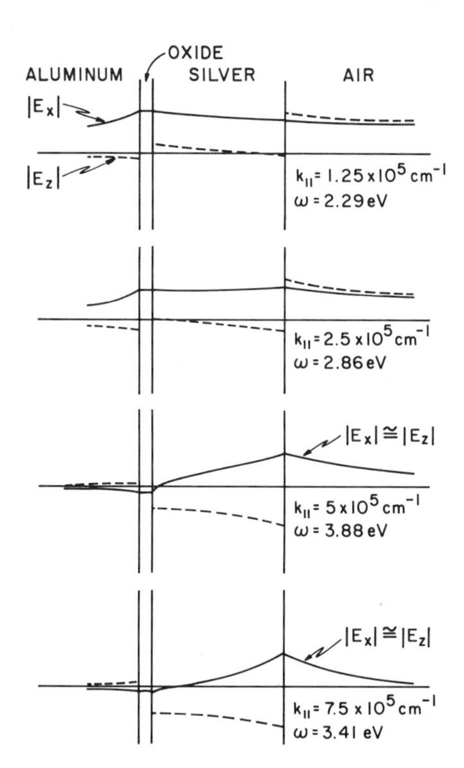

Figure 5. The x and z components of electric field associated with the "fast surface polariton" of Figure 3, calculated for the same points on the dispersion curve used to generate Figure 4.

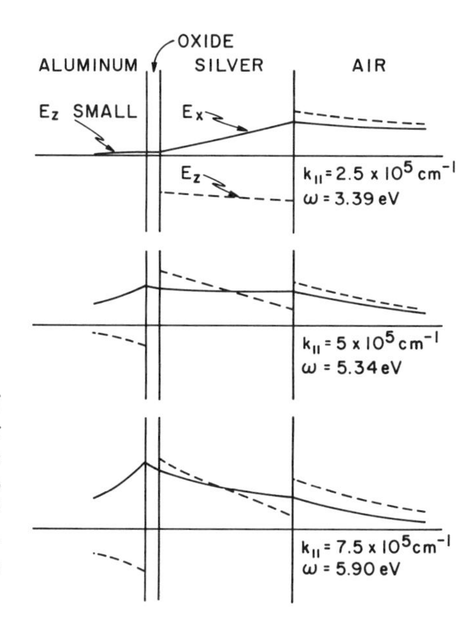

Figure 6. The x and z components of electric field associated with the upper branch of Figure 3, calculated for the same points on the dispersion curve used to generate Figure 4, save for $k_{\parallel} = 1.25 \times 10^5$ cm^{-1}, which for this branch would correspond to a frequency to the left of the "light line."

magnetic fields which peak in magnitude about the interface between the aluminum film and the substrate; as the frequency ω approaches $ck_{\parallel}/(\varepsilon_s)^{1/2}$, the fields penetrate deeply into the *substrate*, rather than extend far out into the vacuum as do those associated with the "fast mode" of Figure 3. As the frequency $\omega \to ck_{\parallel}/(\varepsilon_s)^{1/2}$, the attenuation constant analogous to the $\alpha_<$ of our earlier discussion approaches zero, to give a mode with electromagnetic fields confined to the substrate. In most of the tunnel junctions explored so far, the aluminum–substrate interface where the fields associated with the mode assume their maximum value lies two or three optical skin depths below the vacuum. As a consequence, this fourth mode plays little role in the emission process, so we shall concentrate on the dispersion curves illustrated in Figure 3 for the remainder of the present chapter.

The purpose of this section has been to introduce the reader to the basic properties of the surface polaritons that enter the interpretation of the data on light-emitting junctions, with emphasis on those features of the modes important in this particular phenomenon. We next turn to a discussion of the theory of the emission mechanism, and of experimental data taken on junction structures of each of the three types discussed in Section 1.

3. Light Emission from Tunnel Junctions: The Theoretical Picture and Examples

3.1. General Remarks

There have been two sets of theoretical papers which address the details of the light emission process. The present authors have developed a description which applies to planar structures,[16–18]* where deviations in the interface profiles from perfect flatness may be treated by perturbation theory. This theory has been applied to the analysis of the frequency spectrum of light emitted from junctions with small amplitude roughness of random character present,[7] and it can be extended to describe the saturation of the emission intensity as the roughness amplitude increases.[17] In addition, a number of features of the data on junctions grown on holographic gratings are accounted for by this theory[9, 18] though, as we shall see, a careful and quantitative comparison of theory and experiment[9] suggests the presence of a mechanism for coupling the tunneling electrons to photons not included in the theory's present form.

*Kirtley and co-workers have found some typographical errors in this paper, and have corrected one error. See the preprint cited in reference 9.

Scalapino and his colleagues have developed a theory of light emission valid in a very different limit. They consider tunneling from an aluminum substrate with oxide overlayer, both surfaces assumed planar, into a sphere of noble metal placed on top.[19] Once again, the tunneling electron excites electromagnetic resonances of the spherical particle–substrate combination, and these subsequently radiate into the vacuum above the substrate.

While the two sets of theories appear very different, in fact they are based on a very similar physical picture of the emission process. We shall not attempt to summarize the theoretical details of either theory here, since these are given fully in the literature. We shall be content to simply comment on the essential ideas and some results.

If a junction is subjected to a dc bias voltage V_0 then a dc current I will flow through the junction. In practice, this current is not perfectly steady, but there will be random noise necessarily present. Thus, the current is actually time dependent, and we write

$$I(t)=\bar{I}+\delta I(t) \tag{11}$$

where $\bar{I}$ is the average current, and $\delta I(t)$ the small-amplitude noise. It is the fluctuations in tunnel current which are responsible for exciting the radiation fields in the near vicinity of the tunnel junction structure. Thus, the theory contains two ingredients. One must first describe the frequency spectrum of the tunnel current fluctuations, and their spatial distribution. From this, one may then build a description of how the tunnel current fluctuations excite the electromagnetic field. Since, as we have seen, surface roughness or departures of the interfaces from planar form are required for the surface polaritons excited by the current fluctuations to radiate their energy into the vacuum, this feature must be included. We have proceeded by writing Maxwell's equations for the structure with the current fluctuations as a driving term, then we calculate the energy stored in the radiation field far from the junction within the framework of a description that treats deviations from perfect flatness as a perturbation. Scalapino *et al.* consider the normal modes of a sphere placed above the flat substrate, then treat the spherical particle as a point dipole radiator placed above a metallic substrate, again with tunnel current fluctuations as the excitation source.

Some years ago, Scalapino developed a description of the frequency spectrum of the fluctuations in tunnel current, for a junction subjected to a dc bias. If we let $\langle\delta I(t)\delta I(0)\rangle$ be the correlation function which describes these fluctuations, we may write

$$\langle\delta I(t)\delta I(0)\rangle=\int_{-\infty}^{+\infty}\frac{d\omega}{2\pi}|I(\omega)|^2e^{-i\omega t} \tag{12}$$

where one finds the very simple form

$$|I(\omega)|^2 = \begin{cases} e\bar{I}\left(1 - \dfrac{\hbar\omega}{eV_0}\right), & \hbar\omega < eV_0 \\ 0, & \hbar\omega > eV_0 \end{cases} \tag{13}$$

In both sets of theories, this expression forms the starting point of the analysis.

In addition to the frequency spectrum of the tunnel current fluctuations, one requires their spatial distribution. In our work, we assumed the fluctuations extended throughout the junction structure. Subsequent analyses[8, 9] of the variation of the intensity of the emission with thickness of the outermost film shows that in fact the fluctuations do not extend much beyond one "hot-electron" mean free path from the junction. Since the hot-electron mean free path does not differ greatly from the thickness of the outermost film in most of the junctions studied, for many purposes the difference between the predictions of our original model and the data are not great, though an account of variation of emission intensity with film thickness requires inclusion of the finite electron mean free path in the theory. Scalapino *et al.*, in their model of emission induced by tunneling into a small sphere, assume the tunnel current fluctuations are localized within the oxide barrier just below the point of contact with the sphere. Our view is that as in the planar structures, the fluctuations should be allowed to extend into the sphere, with a range controlled once again by the hot-electron mean free path. While inclusion of this feature will affect the intensity of the emission from a given spherical particle, many features of the theoretical predictions such as the frequency spectrum and angular distribution of the emitted radiation are insensitive to the details of the spatial distribution of the tunnel current fluctuations.

At this point, we hope the principal ingredients of the theories are clear. Rather than continue with a description of these, it may be best to turn to discussion of representative examples of the data, for junction structures of each class mentioned in Section 1. In the discussion that follows, we shall describe how features in the data illustrate aspects of the theory.

3.2. Light Emission from Slightly Roughened Junctions

As remarked earlier, the phenomenon of light emission from tunnel junctions was reported first by Lambe and McCarthy,[1] who measured the spectrum of the emission from Al–Al_2O_3–Ag and Al–Al_2O_3–Au junctions

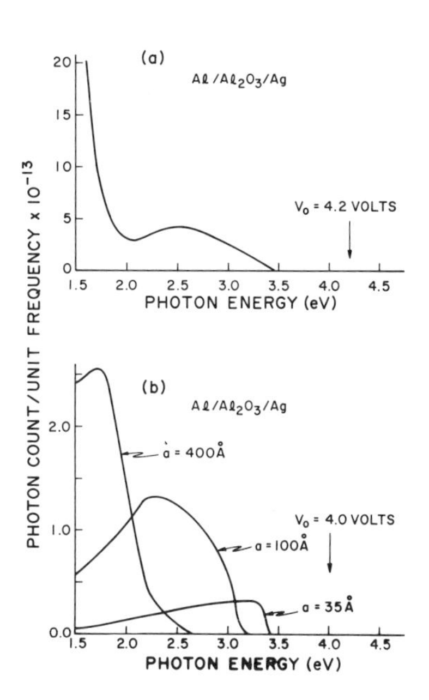

Figure 7. (a) The frequency spectrum of light emitted from an Al–Al_2O_3–Ag junction subject to a dc bias voltage of 4.2 V. The data are taken from the paper of Parvin and Parker. (b) The theoretical spectrum calculated by the present authors, for an emission from a tunnel junction similar to that used by Parvin and Parker (reference 7). It is assumed that the surface has roughness profile described by a Gaussian correlation function. The calculations are taken from reference (17).

which had been roughened by a chemical etching procedure. A more complete set of data, which explores a wide spectral range and also the angular distribution of the radiation, has been presented by Parvin and Parker.[7] They observed emission from similar structures with interfaces that one assumes are somewhat rough as a consequence of the preparation procedure.

In Figure 7a, we show data reported by Parker and Parvin, for a Al–Al_2O_3–Ag junction subjected to a bias voltage of 4.2 V. The thickness of the Ag film is 200 Å, so in fact the dispersion curves in Figure 3b describe the surface polariton modes that propagate on the structure. It is evident from a glance that the emission spectrum cuts off not at $eV_0/\hbar$, but rather at roughly 3.4 eV, which is close to the asymptotic frequency of both the "fast" and the "slow" mode dispersion curves in Figure 3a. This constitutes strong experimental evidence that, for this structure, it is indeed the surface polariton which is a necessary partner in the emission process.

Figure 7b presents a summary of our theoretical calculations of the frequency spectrum of the light emitted from the junction, on the assumption that it is the slow mode that plays the primary role. We have also calculated the emission mediated by the fast mode, to find this very much weaker, for any reasonable picture of the roughness on the junction. The crucial quantity which enters the analysis is a certain correlation function which characterizes the randomly rough surface.* Let $\zeta(\mathbf{x}_{\|})$ be the height of the surface above a plane that defines the average surface. We then require the quantity $\langle\zeta(\mathbf{x}_{\|})\zeta(\mathbf{x}_{\|}+\mathbf{r}_{\|})\rangle$, where the angle brackets denote an average over $\mathbf{x}_{\|}$. The resulting correlation function depends only on $\mathbf{r}_{\|}$ for a randomly rough surface, and in the interest of simplicity this has been taken to have the Gaussian form

$$\langle\zeta(\mathbf{x}_{\|})\zeta(\mathbf{x}_{\|}+\mathbf{r}_{\|})\rangle=\delta^2\exp\left(-r_{\|}^2/a^2\right) \tag{14}$$

where the length a, called the transverse correlation length, can be thought of crudely as the average distance between adjacent peaks on the rough surface.

Several qualitative conclusions may be drawn from Figure 7b. As the transverse correlation length increases through the range indicated, the peak in the spectrum of the emitted light shifts to the red, to eventually lie in the infrared. In essence, a surface polariton decouples most efficiently when the condition $k_{\|}a\cong 1$ is met, so as the transverse correlation length increases, it is the longer-wavelength modes of lower frequency that emit with greatest efficiency. Also, to obtain emission in the visible, the transverse correlation length must be quite short, to match the very short wavelength of the visible frequency slow modes. Our calculations suggest the transverse correlation length must be in the range of 30–50 Å to obtain this emission. As discussed below, studies of rough surfaces by optical reflectivity suggest that the transverse correlation length is much longer, of the order of 500 Å. This led us to suggest[16] that the surfaces are likely to contain roughness with two different length scales. There are "long-wavelength" features, with a transverse scale of 500–1000 Å, and these are responsible for the optical reflectivity anomalies. To explain the tunnel junction emission in the visible, one requires small scale structures with transverse scale of 30–50 Å in addition.

The data shown in Figure 7a lend strong support for this view. The dramatic tail which extends back into the near infrared is produced by the undulations on the 500–1000-Å scale, as one can see by comparison with the curve labeled $a=400$ Å in Figure 7b. The hump clearly present in the visible is then produced by roughness on the 50-Å scale.

As far as we can tell, emission of visible light by roughened tunnel junctions provides the first direct probe of roughness with such a small

*We have assumed throughout our work that only the outermost surface of the structure is rough. This is unlikely to be true in practice, though we believe no qualitative error is introduced by this assumption. It is likely that the absolute value of the calculated emission intensity could be off by as much as one order of magnitude or possibly more, to judge from an analysis of a related situation. See reference 20.

transverse scale. Actually, in the literature on the optical reflectivity of Ag surfaces, a picture nearly identical to that just described was proposed by Raether and his colleagues some time before the recent tunnel junction work.* When radiation is incident on the roughened interface between vacuum and the metal, the roughness couples the incident photon to surface polaritons, to produce a characteristic dip in the reflectivity a bit below the asymptotic frequency of the surface polaritons where $\varepsilon(\omega) = -1$. It is the analysis of the shape and position of this dip that leads one to deduce transverse correlation lengths on the 500–1000-Å scale. The Raether group found that as the amplitude of the roughness was increased, the dip shifted to lower frequencies. They interpreted this to mean that as the amplitude of the roughness was increased, the frequency of the surface polariton decreases. However, they were unable to obtain a quantitative account of the shift upon using values of the transverse correlation length in the range of 500–1000 Å. They could bring their theory of roughness-induced shifts in the surface polariton dispersion curve into agreement with the data, without affecting the other aspects of the analysis, if they assumed that small-scale features with transverse correlation length on the 50-Å scale were present also.

Thus, the interpretation of the tunnel junction data appears to offer support for the picture of roughness on silver surfaces very similar to that put forward earlier by the Hamburg group. We believe the very short wavelength "slow surface polaritons" may prove to be a most useful probe of features on the surfaces of films with a transverse length scale in the few tens of angstroms range.

Parvin and Parker have also measured the angular distribution of radiation emitted from their Al–Al_2O_3–Ag junctions, to find results in accord with the theoretical expectations, assuming the slow surface polariton enters as the "intermediate state." Thus for this system, the data compare favorably with the theory and the assumption that one particular surface polariton mode plays the dominant role.

In the original paper of Lambe and McCarthy, and also in the work of Parvin and Parker, the frequency spectrum of light emitted from Al–Al_2O_3–Au junctions was reported. For this structure, the asymptotic frequency of the slow surface polariton is a bit below 2.5 eV, so for voltages larger than this value the spectrum should "cut off" at this point if the same mechanism dominates. In fact, a continuous emission spectrum is observed that extends right up to $eV_0/\hbar$. We have found that for junctions with an outermost film made from Au, above 2.5 eV the electron may emit a photon

*See reference 21. This point is discussed in a chapter by Professor Raether which will appear in reference 5.

directly, without the surface polariton as an intermediary. Above 2.5 eV, in the spectral region where interband transitions contribute strongly to the dielectric constant of Au, we find the direct emission to be roughly as strong as that from the slow surface polariton mediated process.* Of course, an estimate of the relative strength of these two processes requires an assumption about the height of the roughness on the surface, and this is poorly known for the junctions employed in the experiments, but we do believe that above the onset of the interband transitions, the efficiency for direct emission of a photon may be quite comparable to that for the two-step process, and this may account for the observations in Au. Unfortunately, Parvin and Parker find the angular distribution of the emission above 2.5 eV to be quite different from that calculated for the direct emission process, but the theoretical angular distribution is, in this case, quite sensitive to the assumptions about the spatial distribution of the tunnel current fluctuations, and this is poorly known.

Finally, Lambe and McCarthy find that as the amplitude of the roughness is increased on their Al–Al_2O_3–Ag junctions, the emission intensity first increases, but then saturates. Such a result follows from the surface polariton picture, which shows the emission intensity to be proportional to the square of the roughness amplitude, and also to the mean free path of the surface polariton. As the amplitude of the roughness increases, the mean free path of the wave decreases, and one finds saturation of the emission intensity in the theory.[17]

In summary, the two-step process gives very good account of the data on Al–Al_2O_3–Ag junctions, and from the data one can learn about the nature of the roughness present on the samples. A second mechanism appears in the data on Al–Al_2O_3–Au, and our calculations suggest that *direct* emission of photons, without the surface polariton as an intermediate state, may possibly account for the photons emitted with energy above 2.5 eV.

3.3. Light Emission from Junctions Grown on Holographic Gratings

When theory and experiment are compared for the tunnel junction structures discussed in the previous section, the lack of precise information on the nature of the roughness present on various interfaces limits contact to the semiquantitative level. As we have seen, it is possible to extract information on the gross features of the roughness from the emission spectra, but a truly quantitative test of the theory and the basic emission mechanism cannot be obtained under controlled conditions with such samples.

*Please see Figure 6 of reference 15.

In a very beautiful set of measurements, Kirtley, Theis and Tsang[9] have explored light emission from a structure of known, controlled geometry. They begin with a grating of sinusoidal profile that is produced by a holographic process. The tunnel junction structure is then deposited on this grating as a substrate. If one assumes that each interface replicates the substrate profile, then each has a perfectly periodic profile with period a_0 that is the order of a micron in the structures fabricated so far.

While on the roughened tunnel junctions the "slow surface polaritons" evidently are responsible for the light emission, their wavelength (at visible frequencies) is very small compared to the periods of the holographic gratings. Thus, these waves will propagate so that they follow the grating profile in an adiabatic fashion, with only a small probability of radiating a photon. On the other hand, from our earlier discussion, the "fast surface polaritons" have wavelengths quite comparable to a_0 when their frequency lies in the visible, and these can be expected to radiate efficiently once they are excited.

For the junctions grown on a holographic grating, the frequency spectrum of the radiation emitted differs qualitatively from the data displayed in Figure 7. Consider a surface polariton of wave vector $\mathbf{k}_{\parallel}$ and frequency $\omega_s(\mathbf{k}_{\parallel})$ which propagates on a surface where a grating of small amplitude is imposed. Let the grooves of the grating be parallel to the $\hat{x}$ direction. The effect of the grating is to "mix" into the surface polariton solutions of Maxwell's equations with wave vector $\mathbf{k}_{\parallel} \pm \hat{x}2\pi/a_0$.* Recall from Section 2 that for any wave to have fields which decay exponentially with distance from the structure, as is the case for the fields associated with the surface polariton, one must have $k_{\parallel} > \omega/c$. If we have $\mathbf{k}_{\parallel} \pm \hat{x}2\pi/a < \omega/c$, then the effect of the grating is to mix fields of radiative character into the surface polariton; the surface polariton radiates a photon as it propagates across the grating. The frequency of the emitted photon equals that of the surface polariton itself, and the photon emerges in a well-defined direction which may be deduced by noting that its frequency is $\omega_s(\mathbf{k}_{\parallel})$ and its wave vector parallel to the surface is $\mathbf{k}_{\parallel} \pm \hat{x}(2\pi/a_0)$.

It follows from the above discussion that if one examines the frequency spectrum of radiation emitted in a given direction, one sees a *line* spectrum rather than the broad spectral distribution shown in Figure 7. This is easily appreciated from a simple special case. Let the z axis be normal to the tunnel junction structure, again with the x axis oriented normal to the grooves in the grating. Consider a photon emitted by the junction with wave vector in the xz plane, and which makes an angle θ with the z direction. Its

*A rather complete description of the influence of a periodic grating on the propagation of a surface polariton has been given by D. L. Mills, reference 22.

wave vector parallel to the surface is $k_{\parallel}^{(p)} = (\omega/c)\sin\theta$, so (if the grating amplitude is small), it must have its origin in radiation as a surface polariton with wave vector given either by $k_{\parallel}^{(\text{sp})} = (\omega/c\sin\theta) + 2\pi/a_0$, or one with wave vector given by $k_{\parallel}^{(\text{sp})} = (\omega/c\sin\theta) - 2\pi/a_0$. In Figure 8, for a particular choice of θ, in the $\omega - k_{\parallel}$ plane, we show a plot of the curves $\omega = (c/\sin\theta)(k_{\parallel} \pm 2\pi/a_0)$ superimposed upon the dispersion curve for the fast surface polariton. There are *two* frequencies which satisfy the constraint imposed by energy conservation. Thus, a scan of the frequency spectrum of the light emitted by the junction will show *two* lines, one at the frequency ω_+ and one at the frequency ω_- as illustrated in Figure 8.

The above discussion is appropriate to the limit where the amplitude of the grating is small. As the amplitude is increased, one will begin to see features in the spectrum that originate from surface polaritons with wave vector $k_{\parallel}^{(\text{sp})} = (\omega/c\sin\theta) \pm 2\pi n/a_0$, where n is an integer greater than unity. At the same time, one must recognize that the surface polariton dispersion curve is perturbed by the fact that the wave propagates on a grating of finite amplitude; in the literature, the nature of the dispersion relation has been explored in the limit of small amplitudes where the gaps that open up at the zone boundaries may be described by perturbation theory,[22] and also in the limit of finite grating amplitudes.[23]

The experiments of Kirtley, Theis, and Tsang indeed show line spectra, as expected from the above argument. The line labeled ω_- lies in the spectral range explored by them, and they see features for which $k_{\parallel}^{(\text{sp})} = (\omega/c\sin\theta) \pm 2\pi n/a_0$ with $n = 2$ and $n = 3$ in addition. From the data, these authors are able to measure the dispersion curve for the surface polaritons which propagate on the grating structure. These data amount to a *direct*

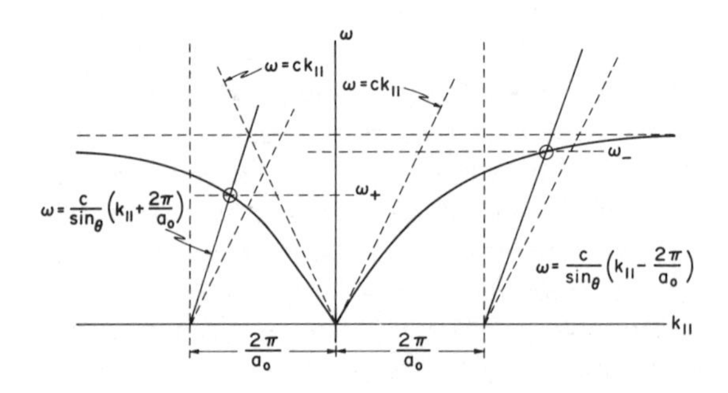

Figure 8. A geometrical construction which illustrates a graphical solution of the energy and wave vector conservation conditions, for emission of a photon by a surface polariton which propagates on a grating structure. See the text for a description of how the curves are constructed.

experimental confirmation that it is indeed the "two-step" mechanism described earlier that is responsible for the emission from the junctions used in this work.

We have also extended our theory to the case where the outermost surface of the tunnel junction structure has a sinusoidal profile,[18] to find good agreement with a number of principal features of the data. With this formalism, Kirtley and co-workers have carried out a careful and quantitative comparison between the theory and their data,[2] with the theory modified to incorporate the finite mean free path of the "hot electrons" driven through the junction by the dc voltage. The theory works remarkably well, and we refer the reader to their paper for a detailed description of the comparison.

There is one intriguing discrepancy between theory and experiment found by Kirtley *et al.* They find the absolute intensity of the emitted radiation is underestimated by the theory, by roughly a factor of 35. We do not regard this as a matter of any great concern by itself, since the theory as presently carried through assumes only that the profile of the outermost surface of the junction is not flat, while in the actual structures employed all interfaces have the sinusoidal profile which replicates that of the substrate. As we have seen earlier, the electromagnetic fields of the surface polariton extend throughout the entire structure, and earlier theoretical analyses of a similar situation* show that coupling between surface polaritons and photons may be enhanced substantially by the presence of roughness on more than one interface. Kirtley *et al.* have studied the variation of the emission intensity with thickness of the outer film, and argue that these data suggest that the coupling between the electron and the fast surface polariton is localized near the Ag–vacuum interface; such a coupling may be produced by the rapid spatial variation of the normal component of the vector potential near the surface, as one passes from the crystal to the interior of the metal. In studies of photoemission from aluminum, it has been argued earlier that precisely such coupling plays an important role.[24] This suggestion is most fascinating, and merits further theoretical attention. The data on light-emitting junctions in combination with the photoemission studies may lead to new and fundamental insights into the optical response of nearly free electron metals.

With the exception of the uncertainty in the nature of the coupling mechanism, the studies of light emission from tunnel junction structures fabricated on holographic gratings have verified very directly and quantitatively a number of predictions of the theory.

*See reference 20.

3.4. Light Emission from Small-Particle Junctions

In the two previous subsections, by and large the data may be interpreted within the framework of a theoretical description that treats the deviation from perfect flatness as a small perturbation. Hansma and his colleagues have studied emission of light from tunnel junctions that differ qualitatively from either of these two geometries.[25] They begin with a nominally flat oxidized aluminum surface, then deposit small gold particles on this substrate. The particle diameters are in the range of 100–300 Å. Finally, gold films were laid over this structure to form a tunnel junction. In a certain qualitative sense, one may view this as a very rough structure, where the influence of the roughness is not properly described by treating it as a modest perturbation on the properties of an idealized plane structure. Instead one has well-defined geometrical structures incorporated in the junction. There may then be *localized* modes of the electromagnetic field, as opposed to the surface polaritons considered above; these modes may be excited by a tunneling electron, to subsequently radiate a photon. The basic two-step process is still responsible for the emission, but this localized mode, rather than a surface polariton, plays the primary role as an intermediate state. The reader may recall that an isolated conducting sphere with radius $R \ll \omega_p/c$ has a dipole active localized plasma mode at the frequency $\omega_p/\sqrt{3}$, where again ω_p is the plasma frequency of the electrons contained within the sphere.

As discussed earlier, Scalapino and his colleagues[19] have worked out a detailed theory of light emission from a sphere placed above Al_2O_3–Al substrate, where the emission proceeds by virtue of electromagnetic resonances of the structure excited by tunnel current fluctuations localized near the point of contact between the sphere and the substrate. These authors present an elegant study of the normal modes of the structure and their excitation by the tunnel current fluctuations, to find an emission spectrum peak just below 2 eV, for spherical gold particles placed above the substrate. They predict the radiation should be p polarized, with no emission along the direction normal to the tunnel junction.

The experimental data show p-polarized emission with a strong, broad peak remarkably similar in shape to that obtained from the theory. The angular distribution is very similar to the theory also, with a prominent lobe directed some 60° from the normal. However, the emission intensity fails to vanish along the normal, and an s-polarized component of the radiation field is present also. Hansma *et al.* argue that there are two independent contributions to emission from the junctions. Upon supposing the s and p components of the "background emission" have identical angular distributions, a separation of the p-polarized portion of the emission into two components yields one with both angular and spectral distribution very close to that generated by the theory of Scalapino, Rendell, and Muhlschlegel. The background has an angular distribution very close to that we calculate for the two-step process mediated by the "slow surface polariton" on a roughened Au–Al_2O_3–Al tunnel junction structure. We have suggested that the background has its origin in this process;[15] there are surely regions of the surface that are relatively open with no gold particles in the near vicinity. These have roughness on the 50-Å length scale, and emit light in a manner very similar to that from the planar structures with roughened interfaces.

It should be noted that Hansma *et al.* fail to see emission above 2.3 eV in Au, so if we wish to interpret the contribution to the spectrum reported in this range by Lambe and McCarthy, and by Parvin and Parker as direct emission, an explanation of its absence here is required. As remarked earlier, reliable estimates of the relative intensities are unfortunately hard to generate.

3.5. Summary

The purpose of this section has been to review the principal qualitative features of the experimental data on light emission from tunnel junctions, with emphasis on how well these are described by current theories of the emission process. In each case the theory idealizes the actual physical situation quite considerably. In our work, for example, the deviations of the junction surface from perfect flatness is treated as a perturbation, and we consider such deviations to be present on the outermost film surface. In practice, the amplitude of the roughness present, or that of the diffraction grating, may be large enough that use of perturbation theory becomes dubious. Also roughness is present on all interfaces, not only the outer surface and this must influence the theoretical predictions substantially, as shown previously. Also, the numerical calculations for roughened junctions are all based on the use of the Gaussian form for the correlation function that describes the character of the roughness [Eq. (15)], and there is no guarantee this correctly represents the situation. The elegant analysis of the small-particle junctions carried out by Scalapino, Rendell, and Muhlschlegel assumes the small particle to be a perfect sphere, while the actual surface surely contains highly nonspherical globules as well. These authors emphasize, however, that the modes responsible for the emission have fields quite localized to the bottom of the sphere, near the oxide layer. Thus, the radius of curvature of the surface near the bottom is the primary parameter that enters, so the results may be extended to surfaces of more general form.

In view of these idealizations necessarily present in theories that may be used for detailed numerical results, we consider the agreement between theory and experiment to be very satisfactory. In our minds, there are two important issues that need further thought. The first is the presence of appreciable emission above the surface polariton cutoff in planar junctions which employ gold as the outer film. As discussed earlier, it is our view that this likely arises from direct emission of photons by the tunneling electrons in a "one-step" process, but more experiments should be carried out to see if this suggestion is reasonable. The analysis of the data taken on the junctions grown on holographic gratings carried out by Kirtley, Theis, and Tsang suggests that a new coupling mechanism yet to be incorporated into the theory must be added to the existing descriptions. This coupling mechanism, if it is indeed the same as that which enters the analysis of the photoemission data, may influence optical properties of nearly free-electron metals quite generally, so its elucidation is of very great interest. So, while considerable progress has been made toward an understanding of the emission mechanism, interesting issues remain.

4. Conclusions

In this chapter, we have attempted to outline the physics that controls light emission from tunnel junctions, and also to review the experiments performed to date. Once again, we regard this phenomenon as most fascinating. If it were possible to improve the efficiency of the junctions, then very likely usable devices would result. What is required is to improve the efficiency to the point where one may see the emitted light in a room with a normal level of illumination, as opposed to a darkened room required for present generation tunnel junctions. As remarked in Section 1, as the dc voltage is increased, the color of light emitted by the junction changes continuously. There would be many applications of such a source, if it were sufficiently intense. We can provide no simple prescription that may be implemented to greatly increase the efficiency of junctions; several interested experimentalists have already devoted thought to this issue, and we are aware of a number of attempts that are underway to fabricate new geometries aimed at increasing the emission efficiency. We have little new to add to the discussions that have taken place already, and we might do best by outlining briefly what limits emission efficiency in the present devices.

If we consider planar junction structures with roughened interfaces, then we have seen that the "slow surface polariton" excited by the tunneling electron is responsible for the emission. The wave vector $k_{\|}$ of a typical mode is very large compared to ω/c, so these modes are electrostatic in character, i.e., their properties are not at all influenced by retardation effects. In this sense, they are very similar to the surface polaritons of large wave vector (surface plasmons) that propagate on the interface between metal and vacuum. From the discussion in Section 1 [see Eq. (2)], we know electrons couple very strongly to these waves. However, while they are excited in a copious manner by injection of electrons through the junction, the probability that a photon is emitted by the slow surface polariton is very small. We have seen that the electromagnetic field energy of these modes is stored largely in the metallic portions of the junction structure when their frequency is in the visible; this means the mode has quite a short lifetime, since the oscillating electric fields associated with the mode drive currents in the metals, which are quite lossy at these frequencies. We have explicitly calculated the probability that the slow surface polariton emits a photon during its lifetime,(16) to find this is in the range of 10^{-5} to 10^{-6} for parameters we believe are relevant to typical tunnel junctions. The efficiency of the planar tunnel junctions thus seems limited by one's ability to decouple an appreciable fraction of the energy stored in the slow surface polariton from the surface, on a time scale set by its lifetime. A good part of the problem, incidentally, is that the electric fields associated with the slow surface polariton are localized about the wrong interface. As we have seen in Section 3, the fields of this mode are localized near the oxide barrier, which is typically a bit more than an optical skin depth below the outer junction surface. Roughness on the outer surface of the junction structure is thus much less effective in decoupling the wave than would be the case if the mode were localized on the outer surface; we have made a crude estimate of the probability of photon emission by a similar mode localized on the outer surface, using results obtained earlier by one of us,* to find emission probabilities roughly two orders of magnitude larger than those we obtain for the actual slow surface polariton on the tunnel junction structure. Adding roughness on the oxide–metal interfaces does not help much, because the electromagnetic fields associated with the emitted *photon* are quite small that far below the outer surface, so the matrix element is small for this form of coupling also.

One might think that by increasing the amplitude of the roughness the emission efficiency should increase, but in fact this not only increases the probability per unit time the mode will emit a photon, but at the same time it shortens the lifetime of the slow surface polariton. The end result is that as the roughness amplitude is increased, the emission intensity saturates.(17)

*The interaction of a surface polariton with surface roughness has been analyzed by D. L. Mills, reference 26. A factor of π was misplaced in some of the formulas, as pointed out in the Erratum to that reference.

It has been suggested that a *periodic* array of very large amplitude structures would support a new set of electromagnetic modes that are long lived, and which would radiate efficiently.[27] In essence, such a structure would consist of a periodic array of miniature antennas that would radiate coherently. Microfabrication technology in the development state may allow such a geometry to be realized in the near future. It is possible presently to create surfaces which are covered by hemispherical bumps a few thousand angstroms in diameter.[28] This is done by bombarding SiO_2 with charged particles of an appropriate energy. This produces hemispherical pits in the SiO_2 surface, and one then deposits a metal film on the SiO_2. These hemispheres are not arranged in a periodic array, but instead are scattered randomly over the surface. Nonetheless, it would be intriguing to see studies of tunnel junctions fabricated with such a surface. Quite recently, Rendell and Scalapino[29] have presented an analysis tunnel current emission from junction structures upon which regular microstructures have been impressed. These authors consider a variety of possible forms, with emphasis on effective circuit approximations that lead to rather simple formulas. We refer the reader to this paper for a detailed discussion of these geometries.

If we turn to the junctions grown on holographic gratings, the two factors which control the emission intensity are quite different. Here, as we saw in the last section, it is the fast surface polariton that is responsible for the emission. These modes can decouple from the surface quite efficiently, or at least much more efficiently than the slow modes just discussed.

We may illustrate this as follows. The "fast surface polariton" on the tunnel junction structure is not very different, at least in a qualitative sense, from the ordinary surface polariton on a metal–vacuum interface, as Kirtley and co-workers point out. The interaction of such a surface polariton with roughness has been analyzed in some detail by one of us,[26] in the limit that the roughness amplitude is small. From these results, we can easily form an expression for the probability P that the mode emits a photon in its lifetime. A rather simple formula may be obtained if we assume $k_{\|}a \ll 1$, where a is the transverse correlation length that enters Eq. (14), and also if we are in the regime where the frequency of the surface polariton is roughly $ck_{\|}$. The expression we find is simply

$$P = \tfrac{4}{3}(\delta a)^2 k_{\|}^4 \omega_p \tau \tag{15}$$

Here δ is the rms height of the roughness, ω_p the electron plasma frequency, and τ the electron relaxation time. [We have taken the dielectric constant of the metal to be given by $\varepsilon(\omega) = \varepsilon_\infty - \omega_p^2/\omega(\omega + i/\tau)$, then assumed $\omega\tau \gg 1$ in addition to the assumptions outlined above.] Typical parameters encountered for evaporated films are $\delta \cong 50$ Å and $a \cong 500$ Å (the small-scale roughness is ineffective in decoupling the fast mode), while for silver we have $\omega_p \cong 10^{16}$ rad/sec and $\tau = 10^{-14}$ sec. These numbers give

$$P \cong 0.06 \tag{16}$$

with $k_{\parallel} = 10^5$ cm^{-1}. Thus, simply when propagating on a rough surface, the "fast mode" is decoupled from the structure with efficiency far greater than the slow mode. The grating should decouple it very efficiently indeed.

However, in Section 2, we saw that most of the electromagnetic energy density associated with the fast mode lies in the vacuum outside the tunnel junction structure, at least in the visible. We see this from the figures displayed in Section 2, along with Eq. (10). Thus, while the fast mode decouples from the surface more efficiently than the slow mode, electrons necessarily couple to it far less strongly. We seem to be in a "no win" situation!

Note that as the wave vector of the fast mode increases (and hence its frequency, on the portion of the dispersion curve where $\omega \cong ck_{\parallel}$), the photon emission probability P increases, and at the same time a larger fraction of the energy density of the mode is stored within the metal. One sees this in Figure 5. The quantum efficiency of the junction should thus rise as the frequency of the emitted photon moves from the visible toward the ultraviolet. Such behavior is indeed found in the data of Kirtley, Theis, and Tsang.[2]

This concludes our discussion of a phenomenon we find most fascinating, and we look forward to further progress in our understanding of it.

References

1. John Lambe and S. J. McCarthy, Light emission from inelastic electron tunneling, *Phys. Rev. Lett.* **37**, 923–925 (1976); S. J. McCarthy and John Lambe, Enhancement of light emission from metal–insulator–metal tunnel junctions, *Appl. Phys. Lett.* **30**, 427–429 (1977).
2. J. R. Kirtley and James T. Hall, Theory of intensities of inelastic electron tunneling spectroscopy orientation of adsorbed molecules, *Phys. Rev. B* **22**, 848–856 (1980).
3. C. J. Powell, Characteristic energy losses of 8-keV electrons in liquid Al, Bi, In, Ga, Hg, and Au, *Phys. Rev.* **175**, 972–982 (1968).
4. D. L. Mills and E. Burstein, Polaritons: The electromagnetic modes of media, *Rep. Prog. Phys.* **37**, 817–926 (1974).
5. V. M. Agranovich and D. L. Mills, eds., *Electromagnetic Waves on Surfaces and Interfaces: Surface Polaritons and Their Interactions*, North Holland Publishing Company, Amsterdam (1982).
6. H. Ibach and D. L. Mills, *Electron Energy Loss Spectroscopy and Surface Vibrations*, Academic Press, San Francisco (1982).

7. K. Parvin and William Parker, Optical spectra and angular dependence of the visible light emitted by metal–insulator–metal tunnel junctions, *Solid State Commun.* **37**, 629–633 (1981).
8. N. Kroo, Zs. Szentirmay, and J. Felszerfalvi, Optical determination of mean free path of hot electrons in metals, *Phys. Status Solidi (b)* **102**, 227–234 (1980).
9. J. R. Kirtley, T. N. Theis, and J. C. Tsang, Diffraction-grating-enhanced light emission from tunnel junctions, *Appl. Phys. Lett.* **37**, 435 (1980); Light emission from tunnel junctions on gratings, *Phys. Rev. B* **24**, 5650 (1981).
10. D. W. Berreman, in *Localized Excitations in Solids*, p. 420 (R. F. Wallis, ed.) Plenum Press, New York, 1968.
11. R. Ruppin (to be published).
12. A. M. Portis, *Electromagnetic Fields; Sources and Media*, John Wiley & Sons, New York, 1978.
13. G. Joos, *Theoretical Physics*, Hafner, New York, 1934.
14. P. Halevi, Electromagnetic wave propagation at the interface between two conductors, *Phys. Rev. B 12*, 4032–4035 (1977).
15. J. C. Swihart, Field solution for a thin-film superconducting strip transmission line, *J. Appl. Phys.* **32**, 461–469 (1961).
16. Bernardo Laks and D. L. Mills, Photon emission from slightly roughened tunnel junctions, *Phys. Rev. B* **20**, 4962–4980 (1979).
17. Bernardo Laks and D. L. Mills, Roughness and the mean free path of surface polaritons in tunnel-junction structures, *Phys. Rev. B* **21**, 5175–5184 (1980).
18. Bernardo Laks and D. L. Mills, Light emission from tunnel junctions: The role of the fast surface polariton, *Phys. Rev. B* **22**, 5723–5729 (1980).
19. Daniel Hone, B. Muhlschlegel, and D. J. Scalapino, Theory of light emission from small–particle tunnel junctions, *Appl. Phys. Lett.* **33**, 203–204 (1978); R. W. Rendell, D. J. Scalapino, and B. Muhlschlegel, Role of local plasmon modes in light emission from small-particle tunnel junctions, *Phys. Rev. Lett.* **41**, 1746–1750 (1978).
20. D. L. Mills and A. A. Maradudin, Surface roughness and the optical properties of a semi-infinite material: The effect of a dielectric overlayer, *Phys. Rev. B* **12**, 2943–2958 (1975).
21. H. Raether, *Nuovo Cimento* (to be published).
22. D. L. Mills, Interaction of surface polaritons with periodic surface structure: Rayleigh waves and gratings, *Phys. Rev. B* **15**, 3097–3118 (1977).
23. Bernardo Laks, D. L. Mills, and A. A. Maradudin, Surface polaritons on large amplitude gratings, *Phys. Rev. B* **23**, 4965–4976 (1981).
24. H. J. Levinson, E. W. Plummer, and P. J. Feibelman, Effects on photoemission of the spatially varying photon field at a metal surface, *Phys. Rev. Lett.* **43**, 952–955 (1979).
25. P. K. Hansma and H. P. Broida, Light emission from gold particles excited by electron tunneling, *Appl. Phys. Lett.* **32**, 545–546 (1978). Arnold Adams, J. C. Wyss, and P. K. Hansma, Possible observation of local plasmon modes excited by electrons tunneling through junctions, *Phys. Rev. Lett.* **42**, 912–915 (1979).
26. D. L. MIlls, Attenuation of surface polaritons by surface roughness, *Phys. Rev. B* **12**, 4036–4046 (1975); Erratum: Attenuation of surface polaritons by surface roughness, *Phys. Rev. B* **14**, 5539 (1976).
27. P. K. Hansma (private communication).
28. E. Burstein (private communication).
29. R. W. Rendell and D. J. Scalapino, Surface plasmons confined by microstructures on tunnel junctions, *Phys. Rev. B* **24**, 3276–3294 (1981).

6

Comparisons of Tunneling Spectroscopy with Other Surface Analytical Techniques

L. H. Dubois

1. Introduction

When an organic chemist synthesizes a molecule for the first time, he employs a variety of techniques to characterize his new discovery. Probes ranging from x-ray diffraction to nuclear magnetic resonance are used to provide details on the structure and bonding in the sample. Since no single experiment can provide a complete spectrum of information, a variety of complementary studies must be performed. Similarly, to accurately characterize a surface, numerous techniques must be employed. Unfortunately, it is impossible to write a short chapter covering all surface sensitive probes in detail since volumes of material have been written on the subject. Even summarizing the advantages and disadvantages of the major techniques would require a multitude of pages and provide few new insights. Instead, I would like to take a different tack here and after introducing the basic areas of modern surface analysis, show how a variety of complementary surface analytical techniques can be brought to bear on a single chemisorption system. Hopefully this will prove to be an instructive approach.

Chapter 6 will be organized as follows: First, the most widely employed surface sensitive probes will be discussed as they relate to the determination

L. H. Dubois • Bell Laboratories, Murray Hill, New Jersey 07974.

of (1) surface chemical composition, (2) surface electronic structure, (3) surface geometry, and (4) surface vibrational properties. This latter point will be explored in more detail and comparisons between tunneling spectroscopy and other surface vibrational spectroscopies made. As mentioned previously, only a brief introduction to these four general topics will be presented here and the interested reader is referred to a variety of excellent books and review articles for more details on any particular subject. The chapter will conclude with an extensive discussion of the chemisorption of carbon monoxide on Rh/Al_2O_3 model catalysts. This is an adsorption system that has been studied by a wide variety of techniques, including many forms of surface vibrational spectroscopy. The complementary nature of the information obtained from many of these experiments will be emphasized.

2. Major Surface Analytical Techniques: A Brief Survey

Figure 1 is a schematic diagram of a generalized surface analysis apparatus. The three most common surface probes (shown at left) are photons, electrons, and ions. Incident beam energies range from a few millielectron volts to hundreds of kilovolts. The sample can be a metal, semiconductor, or insulator and may be studied in a variety of forms including single crystals, foils, evaporated films, and powders. The scattered

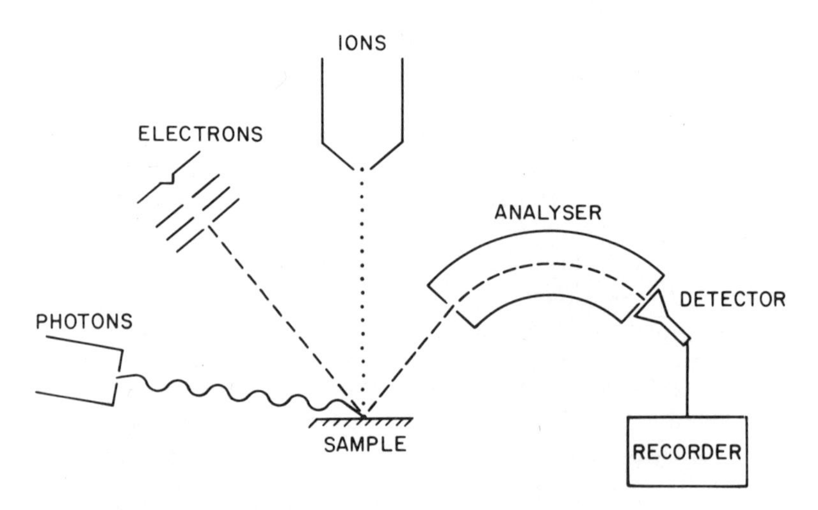

Figure 1. Schematic diagram of the major components of a generalized surface analysis apparatus including possible excitation sources, sample, analyzer, and detector.

(or transmitted) beam may be the same as the incident beam or may consist of secondary electrons (or ions) emitted by the surface. The energy, angular distribution, or mass of the outgoing beam is analyzed, detected, and finally recorded for further study. Surface sensitivity in all of these experiments is a consequence of the low penetration depth of the incident beam, the shallow escape depth of the outgoing beam or the very high surface area of the sample.

Although there are a wide variety of excellent techniques used to study surfaces, in the interest of brevity, only a few of these will be discussed here. Examples of the application of these techniques relevant to the materials employed in tunneling spectroscopy will be presented, where applicable.

2.1. Techniques for Studying Surface Chemical Composition

2.1.1. X-Ray Photoelectron Spectroscopy

X-ray photons (of energy $h\nu$, ~ 1 keV) incident on a surface can readily liberate core electrons (of binding energy E_b, $\sim 100-1000$ eV) localized on specific atoms. To a first approximation then, the kinetic energy ($E_k = h\nu - E_b$) distribution of the photoemitted electrons depends solely upon the atomic species present on or near the surface. In practice, however, the presence of neighboring atoms causes small shifts ($\sim 1-10$ eV) in the binding energies due to slight changes in the electrostatic potential at the core site. With these "chemical shifts" in the initial-state binding energies, x-ray photoelectron spectroscopy (XPS) allows us to determine not only the identity of surface species, but their chemical environment as well.

Because of its tremendous importance, x-ray photoelectron spectroscopy has been applied to studies on a variety of materials including single crystals, polycrystalline foils and films, and high-surface-area powders. Figure 2 presents typical XPS spectra for the O 1*s* and Al 2*p* levels of an evaporated aluminum film oxidized in a plasma discharge of O_2.[1] Contributions from both bulk and surface aluminum are clearly present. The binding energy difference between these two states (~ 2.5 eV) is characteristic of aluminum oxide formation.[2] Peaks are broadened on the low-energy side by core hole lifetimes, by the formation of electron-hole pairs, or by phonon creation.[3] The theory and application of XPS have been treated in several recent texts.[2-5]

2.1.2. Auger Electron Spectroscopy

Auger electron spectroscopy (AES) is one of the most useful tools for exploring the chemical composition of surfaces. Although the Auger process

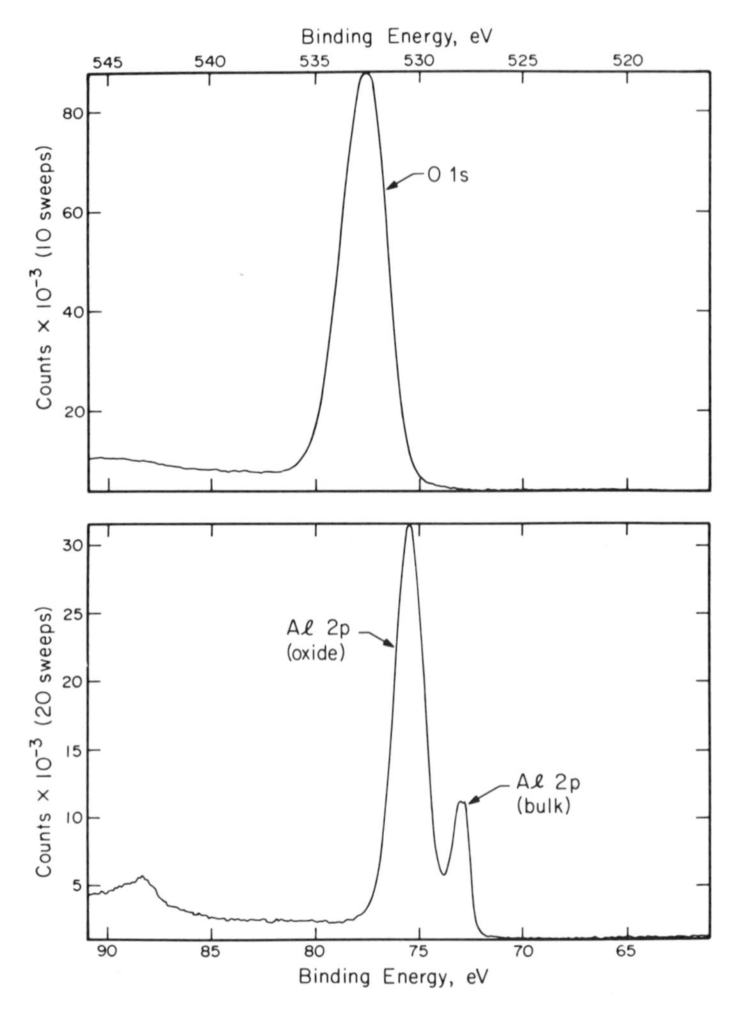

Figure 2. Representative spectra for the O 1*s* (top) and Al 2*p* (bottom) levels for an evaporated aluminum film oxidized in a plasma discharge of O_2 (reference 1).

was discovered in 1925[6] and applied to the study of surfaces in 1953,[7] it was not until the late 1960s that this technique became widely used for surface analysis. The first step in the Auger process is the formation of a core hole on or near the surface by an incident electron (2–5 keV) or high-energy photon. This vacancy is then filled by a second, outer electron and the released energy is transferred to a third electron in a radiationless process. The latter electron is ejected from the surface and its kinetic energy is measured by a suitable analyzer. The spectrum of the emitted electrons provides us with a fingerprint of the surface species since the energy levels of each atom are fixed (and well known). Thus, the identification of unknown surface species is made by reference to standard published spectra.

Like x-ray photoelectron spectroscopy, small chemical shifts in the measured electron distribution are observed in AES. An example of this is shown in Figure 3 for the case of aluminum, oxygen, and alumina. The Al_2O_3 Auger spectrum in the bottom trace is not merely the sum of the aluminum (top trace) and oxygen (middle trace) spectra, but contains new

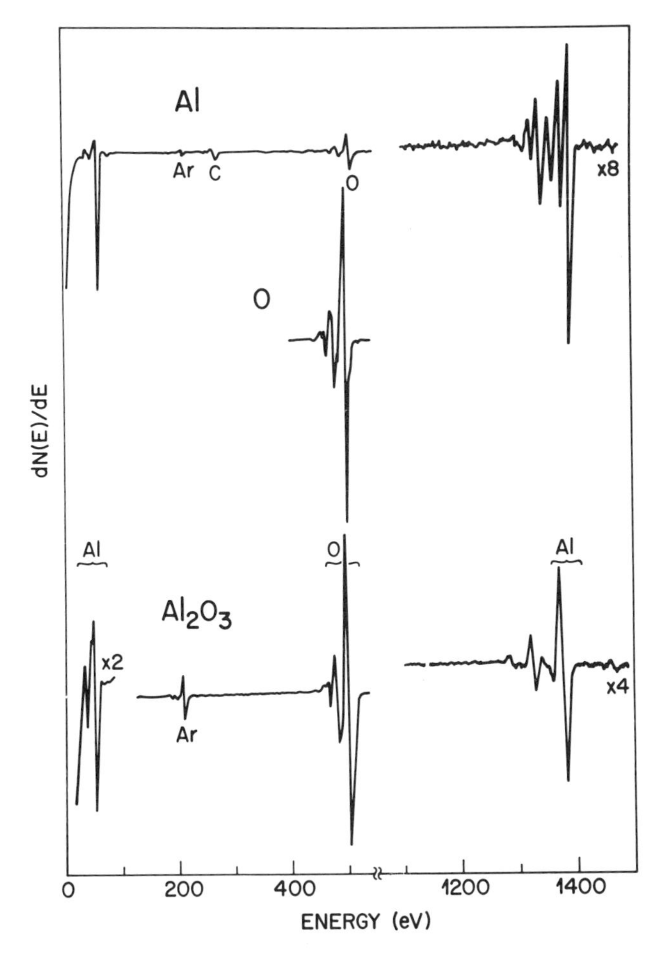

Figure 3. Auger electron spectra of aluminum (upper trace), oxygen (middle trace), and alumina (lower trace) recorded with an incident beam energy of 3 keV (reference 9).

structure characteristic of strong oxide formation (30–50 eV and 1300–1400 eV). Auger spectra are generally displayed in a derivative mode to suppress the broad, structureless secondary electron tail and to improve the signal-to-noise ratio.

Both Auger electron spectroscopy and x-ray photoelectron spectroscopy are routinely used to monitor surface cleanliness since contaminant concentrations of less than 0.1% of a monolayer are readily detectable. In both the upper and lower traces of Figure 3 small carbon, argon, and oxygen impurities can easily be seen. Again, AES can be used to study a wide variety of surfaces. A number of excellent review articles and compendia of spectra are available in the literature.[5,8–10]

2.1.3. Secondary Ion Mass Spectrometry

Kilovolt ions incident on a surface cannot only scatter (both elastically and inelastically) or diffract, but sputter away some surface species as well. A fraction of these particles are ionized as they leave the surface and can therefore be detected directly by a mass spectrometer. This forms the basis for secondary ion mass spectrometry (SIMS). SIMS is an extremely sensitive probe of surface impurities since both scattering cross sections and ion detection efficiencies are quite high. The limits of detection can approach 10^{-14} g of sputtered material or about 0.001% of a monolayer.[11] Because of this surface sensitivity, a wide variety of materials can be studied including metals, semiconductors, insulators, powders, foils, films, and even supported metal catalysts. The major drawback of this technique, however, is that the sample under study is damaged during analysis. A quantitative determination of the concentration of individual surface species is rather difficult to obtain at present owing to a lack of accurate information on absolute sputtering cross sections.

A very nice application of secondary ion mass spectrometry to the analysis of surface species present on a silver catalyst used for ethylene oxidation has been presented by Benninghoven.[11] A negative secondary ion spectrum of the first monolayer of the spent catalyst is displayed in Figure 4. The majority of the activation and contamination processes in a catalyst takes place in the top layer and this trace displays the wide variety of species present on the surface. The individual mass lines can be assigned using known isotopic ratios, by comparison with the secondary ion spectra of known compounds and by knowing something about the history of the sample. By comparing the absolute intensities of the individual mass lines with the results of measurements carried out on silver surfaces saturated with the given anion, the absolute coverage of the surface can be determined. Secondary ion mass spectrometry is unique in its ability to

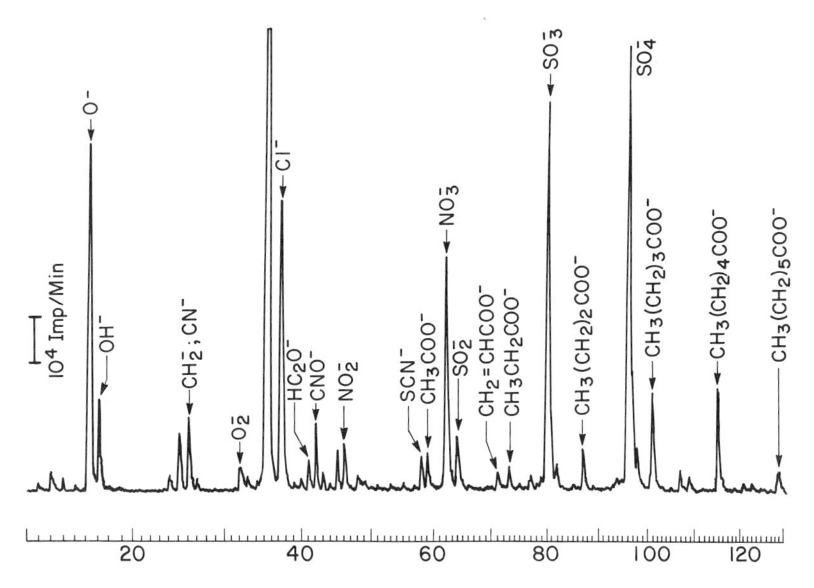

Figure 4. Negative secondary ion mass spectrum of the first monolayer of a spent silver catalyst used for ethylene oxidation (reference 11).

determine not only the chemical identity, but in many cases the structure of the adsorbed species—and this can be accomplished even when a mixture of species is present on the surface. Several excellent books and review articles have appeared in the literature in recent years.[11-13]

2.2. Determination of Surface Electronic Structure

2.2.1. Ultraviolet Photoelectron Spectroscopy

Incident photons and electrons are also useful probes for determining the electronic structure of surfaces. In ultraviolet photoelectron spectroscopy (UPS)[3-5] a photon beam between 10 and 100 eV incident on a surface can ionize valence electrons. The kinetic energy distribution of these photoemitted electrons is then a measure of the density of surface electron states. Because of the short escape depth of the outgoing low-energy electrons, ultraviolet photoelectron spectroscopy is an extremely surface sensitive probe. Like XPS, this technique can be used to distinguish between atoms weakly adsorbed on a surface (physisorbed) and those that are very tightly bound to the substrate (chemisorbed) simply by observing shifts in the initial-state levels.

Representative UPS spectra for the oxidation of a single-crystal surface of aluminum [(111) face] are shown in Figure 5.[14] The appearance of two chemically distinct oxygen-derived peaks are clearly seen in the Al $2p$ region of the spectrum. The 1.4-eV shifted peak arises from the interaction of the aluminum substrate with chemisorbed oxygen atoms. The other

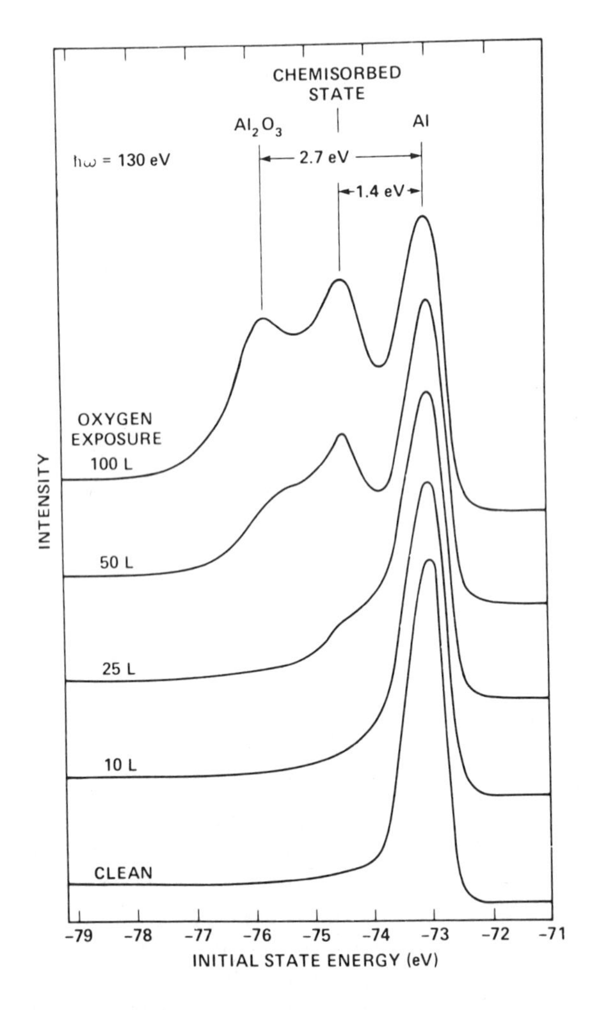

Figure 5. Representative ultraviolet photoelectron spectra ($h\nu = 130$ eV) for the progressive oxidation of an Al(111) single-crystal surface. The presence of both chemisorbed oxygen and bulk Al_2O_3 are unique to this surface (reference 14).

oxygen-derived peak, shifted 2.7 eV to higher binding energy is characteristic of the Al $2p$ level in aluminum oxide (Al_2O_3).[2] The oxygen exposure to the surface is measured in langmuirs (L), where $1\ L = 10^{-6}$ Torr sec of gas exposure and is approximately equal to monolayer coverage for molecules with unit sticking probability. For oxygen exposures above 1000 L only the second oxygen-derived peak is present in the spectrum. Thus, the changes in the surface electronic structure clearly demonstrate the transformation from chemisorbed oxygen to bulk oxide formation. The formation of a stable chemisorbed oxygen phase is unique to the (111) face of aluminum and is not observed on other single-crystal surfaces or on evaporated aluminum films.[14]

Analysis of the angular distribution of the photoemitted electrons from single-crystal surfaces can yield information on the geometry of adsorbed species. Angle-resolved ultraviolet photoelectron spectroscopy (ARUPS)[3,4,15] and the related technique of normal emission photoelectron diffraction (PhD)[16,17] have been successfully used to study the structure of both clean and adsorbate-covered metal surfaces. The interested reader is referred to these references for further details.

2.2.2. *Electron Energy Loss Spectroscopy*

In electron energy loss spectroscopy (ELS) an incident low-energy electron (10–1000 eV) undergoes an inelastic scattering event upon its reflection from a surface. The surface is left in an excited state and the reflected electron has lost a characteristic amount of energy. The primary beam can excite bound electrons to higher unfilled levels, ionize them, or excite surface plasmons (collective oscillations of conduction electrons). Each of these will yield a unique peak in the scattered electron distribution and therefore provide information about the surface electronic structure.[18,19]

Electron energy loss spectroscopy can also be used to study the progressive oxidation of aluminum. Figure 6 shows portions of energy loss spectra (incident energy 250 eV) from a polycrystalline aluminum sample in various stages of oxidation at room temperature.[20] The two peaks in the ELS spectrum of clean aluminum (upper trace) at 10.7 and 15.2 eV have been attributed to the excitation of surface and bulk plasmons, respectively. As the oxygen exposure is increased, two peaks at 7.5 and 21 eV grow in. The low-energy peak is probably a transition between the O $2p$ orbital and a final state just above the Fermi level while the high-energy loss has been attributed to the excitation of a bulk plasmon in amorphous alumina.

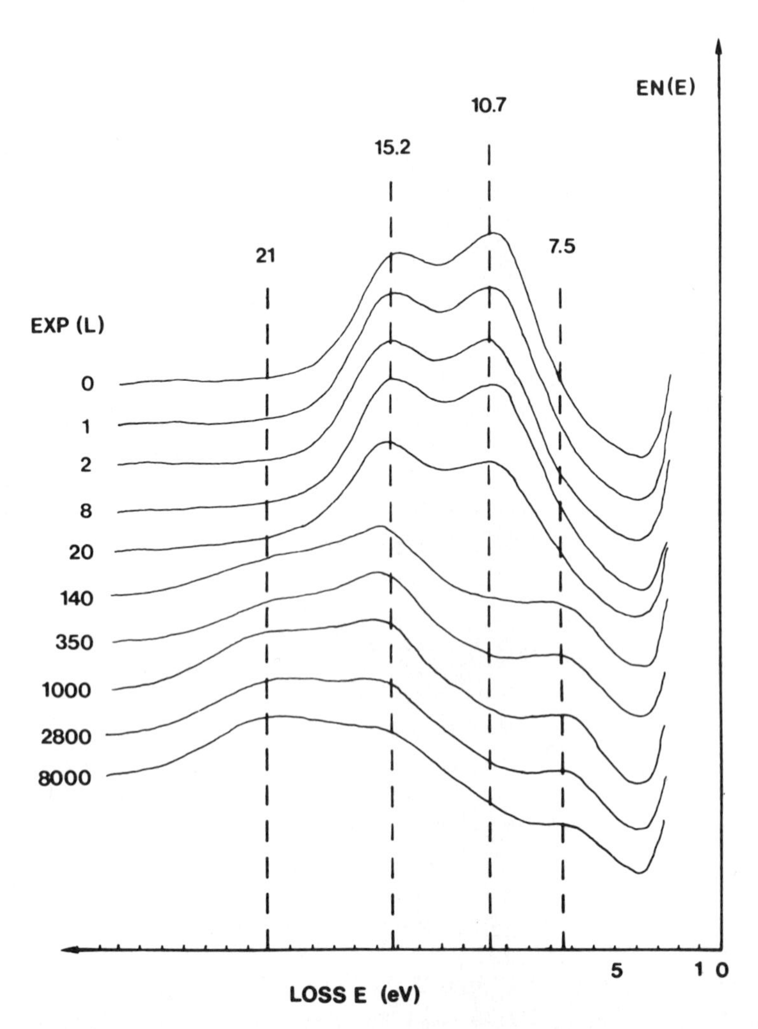

Figure 6. Electron energy loss spectra of a polycrystalline aluminum sample exposed to increasing doses of oxygen. The primary energy is 250 eV. The losses at 10.7 and 15.2 eV have been attributed to the excitation of surface and bulk aluminum plasmons, respectively. The peak of 21 eV is caused by the excitation of a bulk plasmon in amorphous alumina (reference 20).

2.3. Techniques for Surface Structural Analysis

Unlike most of the techniques discussed in the last two sections which can be used to study a wide variety of surfaces, the techniques for surface structural analysis are far more limited—electron or ion scattering experiments are useful for low-surface-area single-crystal experiments while transmission electron microscopy can only be used to study relatively thin samples.

2.3.1. Surface Extended X-Ray Absorption Fine Structure

An x-ray absorption spectrum of a solid exhibits an abrupt discontinuity at a photon energy equal to the binding energy of a core electron. This is followed by small (typically less than 5%) oscillations in the absorption coefficient. This fine structure extends to energies approximately 1 keV above the edge and provides structural information. Hence, the name of this technique—extended x-ray absorption fine structure (EXAFS). The yield of ejected photoelectrons is modulated as a function of incident photon energy due to interference between outgoing electrons and electrons backscattered from neighboring atoms. Thus, both the period and amplitude of these oscillations are dependent on the geometry and on the number of atoms surrounding the absorbing atom. Fourier analysis of this modulation yields interatomic distances with an accuracy of approximately ± 0.05 Å.[21,22]

EXAFS, like x-ray diffraction, is essentially a technique for studying the structure of bulk materials due to the long penetration depth of the incident x-rays. However, recent advances in this area have included studies on the structure of atoms adsorbed on single-crystal surfaces, hence, the new acronym SEXAFS—surface extended x-ray absorption fine structure.[23] Unfortunately, this technique has two major drawbacks. First, one of the physical inputs to the SEXAFS analysis is a set of phase shifts for electron scattering off of surface atoms. Experimental data from bulk materials can help in circumventing this problem, but uncertainty in these phase shifts is one of the limiting factors in the accuracy of this technique. Secondly, an intense, broad band source of x-rays is required for these studies, thus necessitating the use of synchrotron radiation.

The adsorption of iodine on silver has recently been studied by SEXAFS and representative results are shown in Figure 7.[24] In (a) the raw data for three different systems are shown: A, a 1/3 monolayer coverage of iodine on Ag(111); B, approximately 1–2 monolayers of iodine on Ag(111); and C, bulk AgI (a model compound). In (b) the normalized oscillations are displayed in momentum space after background subtraction. Fourier analysis of these data is shown in (c). Using the phase shifts and bond length

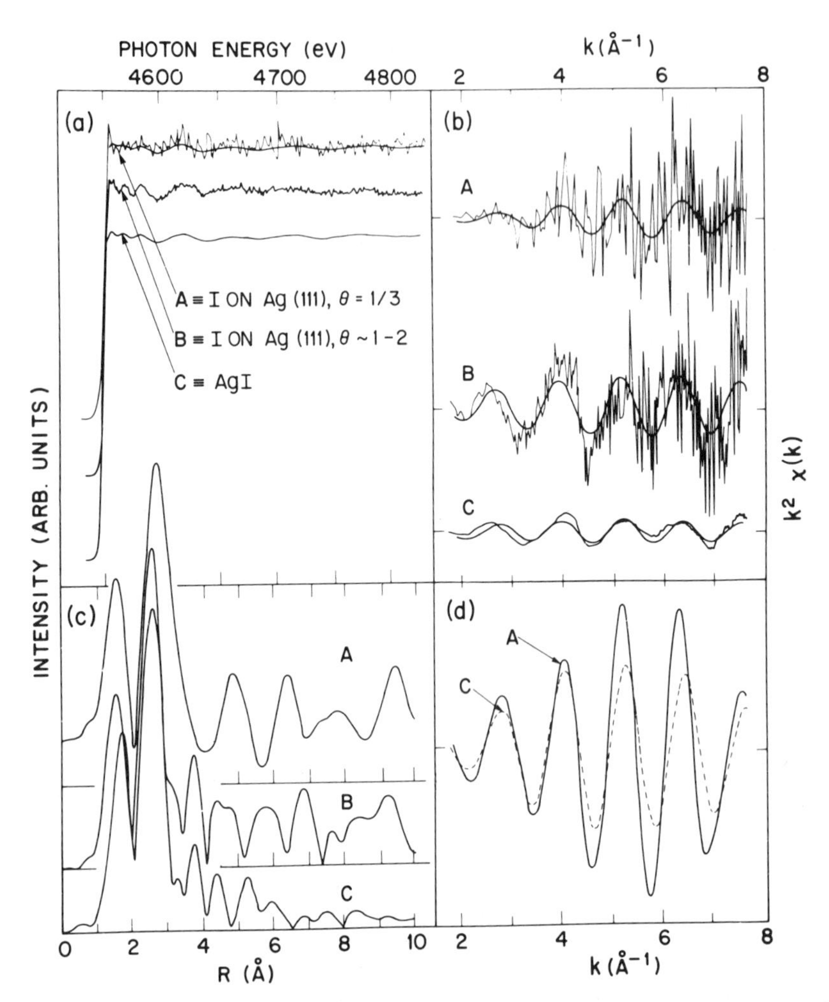

Figure 7. (a) Raw data in energy space for three different silver–iodine systems. Curves A and B are SEXAFS spectra for I_2 adsorption on Ag(111); curve C is an EXAFS spectrum of AgI. (b) Raw data in momentum space after background subtraction. (c) Fourier transforms in distance space of raw data from (b). The peaks at 2.6 Å have been arbitrarily set equal in height. (d) Normalized, retransformed data from (c) after filtering. Note the significant differences in both phase and amplitude (reference 24).

from silver iodide, a surface Ag–I bond length of 2.87 ± 0.03 Å is calculated and the iodine atoms sit in threefold hollow sites on the (111) surface. If the data of Figure 7c are filtered and retransformed, the curves shown in (d) are obtained. These traces simply clarify the structural difference between bulk AgI and iodine adsorption on Ag(111).

2.3.2. *Low-Energy Electron Diffraction*

In low-energy electron diffraction (LEED) studies, a collimated beam of electrons (20–200 eV) is incident on a single crystal surface. Between 1% and 10% of these electrons are elastically scattered by the atom cores of the surface (and an adsorbate, if present) and produce interferences that depend on the relative atomic positions of the surface under study. These diffracted electrons are detected by a fluorescent screen or Faraday cup. Analysis of the electron diffraction *pattern* yields information on the size and orientation of the surface unit cell. These diffraction patterns are relatively easy to interpret in many cases. Structural information (bond lengths, bond angles, adsorption sites) can only be obtained by analyzing the *intensity* of a given diffraction spot as a function of incident beam energy. These intensity verses incident energy curves ($I-V$ profiles) exhibit pronounced peaks and valleys which are indicative of constructive and destructive interference of the electron beam scattered from the surface (and near surface region) as the incident electron energy is varied. In addition to single scattering (Bragg) peaks, there are usually extra peaks due to the multiple scattering of electrons through the surface lattice. The analysis of LEED intensities requires a detailed theory of the diffraction process and this is not a trivial problem owing to these multiple scattering effects.[25,26] Nevertheless, a

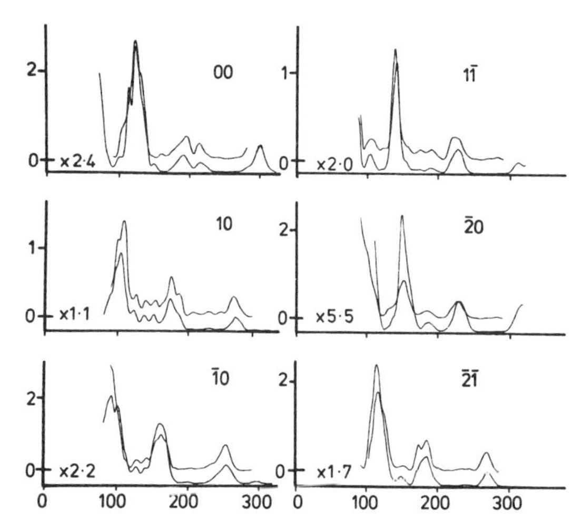

Figure 8. Low-energy electron diffraction structural analysis of the (100) surface of MgO. Comparisons between experimental (lower traces) and theoretical (upper traces) intensity vs incident energy curves for six different beams are displayed. The theoretical spectra were calculated with no relaxation of the surface and are displaced upwards by a constant amount (reference 28).

tremendous number of very accurate studies dealing with both clean and adsorbate-covered surfaces have been performed.[26,27] The results of one such study will be summarized below.

Magnesium oxide is an important material as an insulator in tunnel junctions and has also been studied by low-energy electron diffraction. Figure 8 shows LEED $I-V$ profiles for several diffraction spots from the (100) surface of MgO.[28] The theoretical spectra (upper traces) were calculated assuming that the surface was a simple termination of the bulk structure. The agreement with experiment in both peak positions and in their relative intensities is quite good. Slight expansion or contraction of the surface layer did not improve the fit significantly. The mean optimum surface layer relaxation was found to be a 0.3% (0.006 Å) contraction inwards toward the bulk with an rms deviation of 1.6%.

2.3.3. *Transmission Electron Microscopy*

A combination of high-resolution electron microscopy and selective chemisorption can give direct information on both the size and shape of small metal particles supported on an inert substrate. Transmission electron microscopy (TEM) provides a two-dimensional projection of a thin (<5000 Å) sample with very high resolution (on the order of angstroms).[29] The use of high incident beam energies (>100 keV) and novel imaging techniques permits the observation of small surface defects and of adatom diffusion along a surface.[30] An excellent example of the use of transmission electron

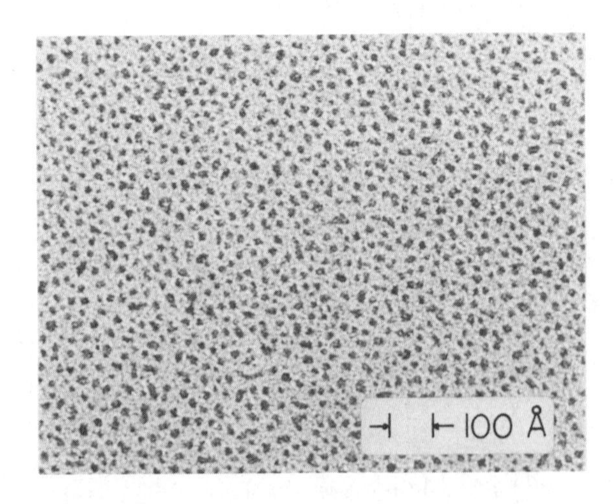

Figure 9. Transmission electron micrograph of platinum particles evaporated onto a carbon substrate (reference 31). Typical metal particle diameters are on the order of 20 Å.

microscopy to study small metal particles is shown in Figure 9.[31] These platinum particles were evaporated onto a carbon substrate and are on the order of 20 Å in diameter. Platinum was simultaneously deposited onto an oxidized aluminum strip in the presence of 10^{-4} Torr of carbon monoxide or other adsorbate and then covered with a lead top electrode for tunneling spectroscopy studies. In this way both the particle size and particle size distribution of the species present in the junction can be determined. Furthermore, the surface area of these small samples can be estimated.

2.3.4. Gas Adsorption

The exposed area of powders, polymers, or other high-surface-area materials can readily be determined using simple gas adsorption measurements. This method, developed by Brunaer, Emmett, and Teller[32] (BET), involves the measurement of an adsorption isotherm for nitrogen (or other inert gas) at relatively low temperatures where only physical adsorption can take place. The number of molecules required to give a monolayer can be calculated from the isotherm and multiplication of this quantity by the cross-sectional area of the adsorbed molecule yields the surface area of the sample.

Extension of these studies to high-surface-area supported metal catalysts where the exposed metal is only a small fraction of the total surface area is straightforward.[22] In this case selective hydrogen or carbon monoxide chemisorption at much higher temperatures is used. The assumption is that (1) hydrogen molecules will dissociate into atoms, (2) that each exposed metal site will bond only a single hydrogen atom, and (3) that hydrogen will not stick to the support. These assumptions have now been verified in several cases and this technique has gained widespread acceptance in the chemical industry.

2.3.5. Scanning Electron Microscopy

The topography of relatively thick samples can be studied using scanning electron microscopy (SEM).[33] In this case a highly focused electron beam (1–50 keV) is scanned over the surface of a sample and both the secondary and backscattered electrons are collected. Resolution is generally determined by the spot size of the focused electron beam and is on the order of 100 Å. In both scanning and transmission electron microscopies, intense beams of high-energy electrons are incident on a sample and can in principle severely damage the surface. Fortunately, electron–atom scattering cross sections are quite small at these energies.[33]

2.4. Observation of Surface Vibrational Modes

Recently there has been an increased interest in the vibrational properties of surfaces and adsorbates, in part due to the rapid growth in the number of available experimental techniques. Since inelastic electron tunneling falls into this category, more time will be spent discussing related methods of surface vibrational spectroscopy. A variety of approaches, including the scattering of photons, electrons, and neutrons, have proved successful in these studies, but as in the experimental methods described above, each has its own characteristic advantages and disadvantages. A summary of the techniques to be discussed in this section is presented in Table 1.

2.4.1. Infrared Spectroscopy

During the past fifty years, infrared (IR) spectroscopy has proven to be a very powerful technique for studying the molecular structure of matter in the solid, liquid and gaseous phases. The pioneering work of Eischens, Pliskin, and Francis in 1954 was the first application of vibrational spectroscopy to the study of molecules adsorbed on metal surfaces.[34,35] The results of these first experiments clearly point out several of the major advantages and disadvantages of transmission infrared spectroscopy.[36,37] The advantages are threefold: First, the experiments are exceedingly easy to set up and on the scale of other surface sensitive probes, are relatively inexpensive to perform. Sample cells are generally Pyrex tubes or small stainless-steel chambers with infrared transparent windows and fit directly into commercially available infrared spectrometers with little or no modification. Secondly, infrared spectroscopy has an inherently high resolution ($\ll 1$ cm^{-1}), although most studies are carried out with between 2 and 8 cm^{-1} resolution due to the spectral width of the observed vibrations.

Most researchers use conventional dispersive (either prism or grating) spectrometers, although Fourier transform instruments have become quite popular recently.[38] These spectrometers can provide an advantage over dispersive units in terms of higher signal-to-noise ratios and much lower data acquisition times (since the entire spectrum is scanned at once). Furthermore, time-resolved studies of surface species can readily be performed yielding information on the kinetics of adsorption and on surface reactions.[38] This leads to the final advantage of transmission infrared spectroscopy: the ability to perform experiments between ultrahigh vacuum (UHV, $<10^{-9}$ Torr) and several atmospheres of background gas pressure. This is probably the biggest advantage of using photons to study the structure of adsorbed species.

Table 1. Techniques of Surface Vibrational Spectroscopy

	Inelastic electron tunneling	Transmission infrared	Reflection-absorption infrared	Raman spectroscopy	High-resolution electron energy loss spectroscopy	Inelastic neutron scattering
Spectral range (cm^{-1})	240–8000	1500–3300	1500–3300	1–4000	300–5000	16–2000
Resolution (cm^{-1})	1–10	1–i0	1–10	1–10	40–80	4–200
Sample area (mm^2)	1	10^6	100	$10–10^6$	10	10^8
Sensitivity (% monolayer)	0.1	0.1	0.5	1	0.1	1
Substrate	Metal oxide	Metal oxide	Metal, metal oxide	High surface area or roughened Ag, Au, Cu	Metal, Semiconductor, insulator	Graphite, metal oxide, powdered metals
Adsorbate	Many	Many	Few, mainly CO	Few, pyridine and related compounds	Many	Hydrogen, hydrocarbons
Maximum pressure	Sample prep ~1 atm	>1 atm	10 Torr	>1 atm	10^{-5} Torr	1 atm
Theory	Electron–dipole, evolving	Photon–dipole, well established	Photon–dipole, well established	Surface-enhanced Raman, evolving	Electron–dipole, evolving	Neutron–nucleus, well established
Orientational selection rule	Yes, weak	Weak, if any	Yes	No	Yes, depending on mechanism	No

Unfortunately, transmission infrared spectroscopy has several major disadvantages[36,37]: First, due to the small absorption cross section for infrared photons, most studies have been carried out on high-surface-area powdered materials (generally metal oxides), materials that lack a well-defined surface structure and composition. Furthermore, owing to light adsorption by the oxide support below $\sim 1200\ cm^{-1}$, low-frequency vibrational modes, which are essential in structural determination, are completely masked. However, N–O ($1400–1800\ cm^{-1}$), C–O ($1800–2100\ cm^{-1}$), and C–H ($2800–3300\ cm^{-1}$) stretching vibrations can still be observed. Several research groups have attempted to solve some of these problems by studying adsorption on small metal particles of known structure (supported organometallic clusters),[39] on thin evaporated films,[40] or by a multiple reflection arrangement.[40] None of these solutions is particularly attractive because of the lack of detailed knowledge of the substrate.

Since metal oxides (including silica, alumina, and magnesia) are generally employed as substrate materials, comparisons between infrared and tunneling spectroscopy can readily be made. This is clearly shown in Figure 10 for the chemisorption of benzaldehyde on alumina.[41] Sample preparation for transmission infrared spectroscopy (upper trace) has been described extensively in the literature.[36,37] The preparation of tunnel junctions (lower trace) is discussed in Chapter 1 of this book. The infrared spectrum has reasonable resolution and sensitivity, but alumina absorption below approximately $1000\ cm^{-1}$ blocks the observation of all low-frequency modes. This is clearly not the case for the tunneling spectrum where the resolution and sensitivity are the same from 200 to $4000\ cm^{-1}$. Metal loading on the support can limit the IR spectral range even further. The Raman spectrum of this system (middle trace) will be discussed presently.

Extension of these infrared studies to well-characterized single-crystal surfaces has proven quite difficult until recently. The development of single reflection or reflection–absorption infrared (RAIR) spectroscopy can be traced to the work of Pritchard *et al.*[42] This technique allows studies on single-crystal surfaces and relies on either wavelength modulation[42,43] or polarization modulation[44] to improve surface sensitivity and increase the signal-to-noise ratio. Both of these schemes allow observation of vibrational spectra with approximately 0.01% absorption.

Despite this increased sensitivity, most studies to date have dealt with the adsorption of carbon monoxide on metal surfaces. CO has an extremely high dipole derivative and thus gives exceptionally strong vibrational spectra. An example of this is shown in Figure 11 for the adsorption of CO on Ru(001).[45] Clearly the sensitivity is available (since less than 0.3% of a monolayer of CO can be observed), but the frequency range is still somewhat limited. This problem may be solved shortly with the advent of new, more sensitive detectors in the $200–2000\ cm^{-1}$ region.

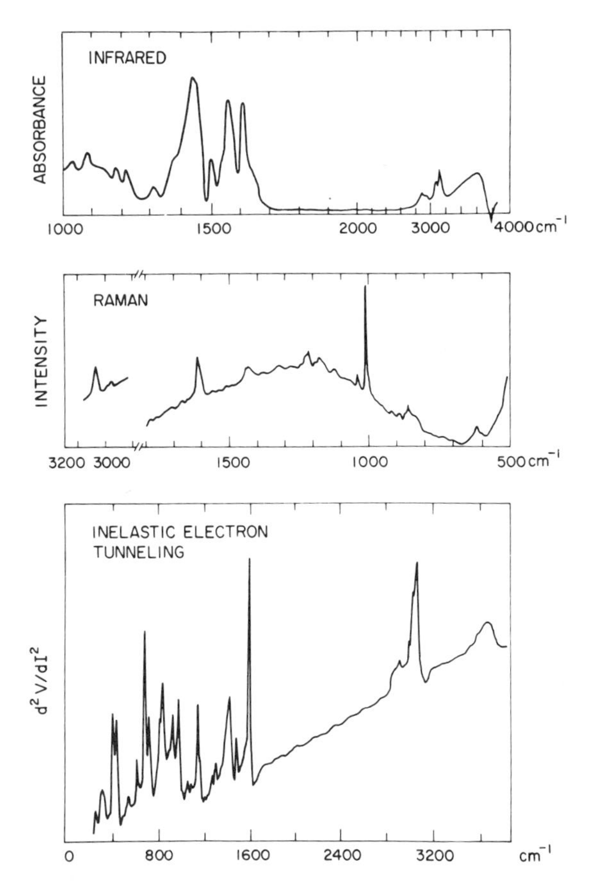

Figure 10. A comparison of infrared, Raman, and tunneling results for the same system: benzaldehyde adsorbed on alumina (reference 41). The tunneling spectrum has good resolution from 300 to 3800 cm^{-1}, while both the infrared and Raman spectra are quite limited. The agreement between peak positions determined by the three spectroscopies is excellent in the regions where they overlap.

Since we are now dealing with single-crystal substrates under ultrahigh vacuum conditions, a variety of complementary techniques can be used to study the surface–adsorbate interaction. However, we do encounter a new problem not necessarily present on supported metal particles: the normal dipole selection rule. This orientational selection rule states that only vibrations with an oscillating dipole moment perpendicular to the surface can be excited by an incident infrared photon.[43] This is a consequence of the dielectric response of the metal in which the screening effect of the conduction electrons prevents a tangential field from being established on the metal surface. Thus many important vibrational modes may never be

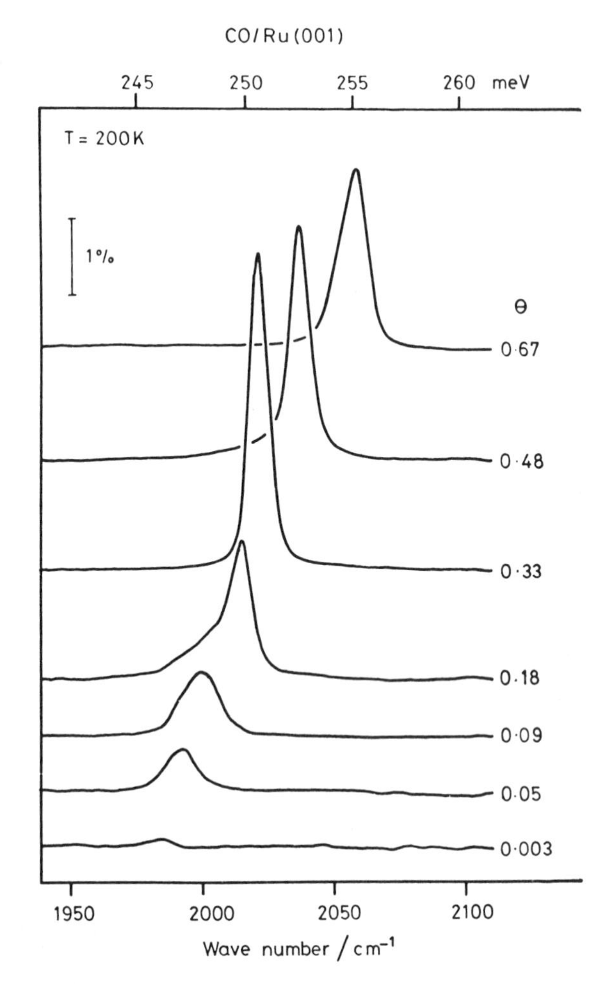

Figure 11. Reflection–absorption infrared spectrum of carbon monoxide adsorbed on Ru(001) as a function of surface coverage at 200 K (reference 45). Possible causes of the shift in the C–O stretching vibration are discussed in references 47–50.

observed. Pearce and Sheppard have proposed a similar selection rule for the transmission infrared spectra of supported metals.[46] Here they assume that the bands due to molecular vibrations parallel to the metal crystallites are reduced in intensity, but not entirely eliminated.

A variety of mechanisms have been proposed to explain the extensive shift in the C–O stretching vibration shown in Figure 11. These will not be discussed here and the interested reader is referred to references 47–50 for details.

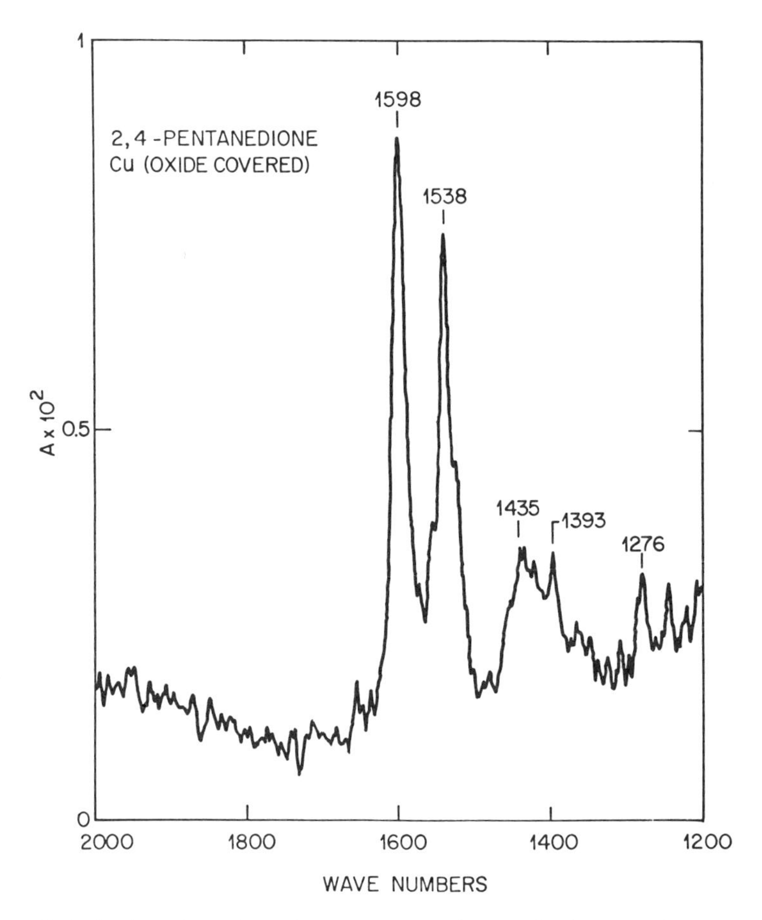

Figure 12. Reflection–absorption infrared spectrum of an oxidized copper film exposed to 0.010 *M* 2,4 pentanedione in ethanol (reference 51). The resolution here is 2 cm^{-1}

Extension of these studies to the adsorption of hydrocarbons on metal oxide surfaces (a system close to the hearts of the tunneling spectroscopist) has met with limited success. Figure 12 is a single reflection infrared spectrum of 2,4 pentanedione adsorbed on an evaporated and subsequently oxidized copper film.[51] Like tunneling spectra, the surface is not extremely well characterized; however, unlike tunneling spectra, there is no top metal electrode to complicate the analysis. The spectral range of these studies is still limited, although the resolution is quite good. As is the case with most alchohols, aldehydes, and acids, 2,4 pentanedione loses its hydrogens and adsorbs as the anion.

2.4.2. Surface Raman Spectroscopy

A casual look at Raman spectroscopy might lead the uninformed reader to believe that this is the ultimate technique for exploring the vibrations of adsorbed molecules. The resolution is quite high (only limited by the quality of the spectrometer); the complete vibrational spectrum can be easily scanned; low-frequency vibrational modes within several wave numbers of the exciting laser line can be seen; and the laser, optics, and monochromator are external to the vacuum system so that alignment and maintenance are much easier. Since all of the light is in the visible portion of the spectrum, no special infrared transparent windows are required. Furthermore, studies can be carried out from UHV to atmospheric pressure (even experiments at the solid–liquid interface can be performed) and the spectrometer does not need to be purged to remove background water vapor or carbon dioxide as in the case of infrared spectroscopy. Indeed it appears that Raman spectroscopy is the ideal method for studying the vibrational spectrum of adsorbed molecules. There is one major problem, however: sensitivity. Since typical Raman cross sections are on the order of 10^{-30} cm^2/sr molecule, observation of submonolayer quantities of materials on single-crystal surfaces is exceedingly difficult. This has led to studies in essentially two different areas: experiments on high-surface-area supported materials[52] and studies on the giant Raman effect (or surface enhanced Raman spectroscopy—SERS),[53,54] found to take place on roughened silver, copper, gold, and mercury surfaces.

Conventional Raman spectroscopy has been applied to the study of molecules adsorbed on high-surface-area oxides and zeolites as well as to supported metal catalysts.[52] Pyridine has historically been the system of choice for the development of surface Raman spectroscopy due to its large polarizability during a molecular vibration. Whereas the C–O stretching mode of adsorbed carbon monoxide (the most well-studied adsorbate in the infrared) yields an intense band near 2000 cm^{-1}, the in-plane symmetric ring deformation modes of pyridine near 1000 cm^{-1} have large Raman scattering cross sections. Also, like the frequency of the C–O stretching band, the position of the ν_1 ring breathing mode can be used to predict the structure of adsorbed pyridine since this mode has been shown to shift in a predictable way when pyridine interacts with various surface sites.[52] A representative Raman spectrum for the adsorption of pyridine on porous silica glass is shown in Figure 13.[55] Despite the high surface area of the sample and high polarizability of the adsorbate, the signal is not exceedingly intense. The same can be said about the Raman spectrum in Figure 10 (middle trace), which adds little new information to that obtained from either infrared or inelastic electron tunneling.

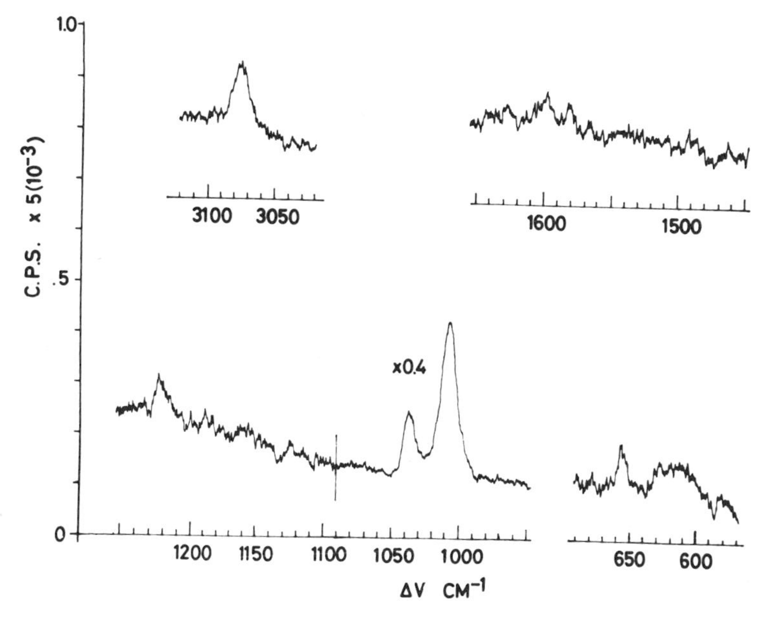

Figure 13. Raman spectrum of pyridine adsorbed on a porous silica-glass disk at 293 K (reference 55). The signal-to-noise ratio in the spectrum is quite poor despite the relatively high surface area of the substrate.

There are several problems with applying this technique to studies on high-surface-area materials, not the least of which is that the substrate material is not well characterized. Other problems include sample heating by the intense laser beam and background or impurity fluorescence.(52,56) Finally, even on these high-surface-area materials, molecules of catalytic interest such as CO and hydrogen yield extremely weak signals.(56)

Enhanced Raman spectra, spectra 10^4-10^6 times more intense than originally anticipated, were first observed for the adsorption of pyridine on silver electrodes in an electrochemical cell.(57) Subsequently, the enhanced Raman spectrum of a variety of other molecules adsorbed mostly on silver, but also on gold, copper, and now mercury surfaces have been reported.(53,54) Experiments take place not only in electrochemical cells, but in ultrahigh vacuum chambers as well, and of course in everything in between. Enhancement factors of as little as 100 to as high as 10^6 have been measured depending on the substrate, substrate preparation, and the molecule under study. In fact, each group studying this phenomenom has its own recipe for preparing "active" surfaces. Theories to try and explain this anomalously

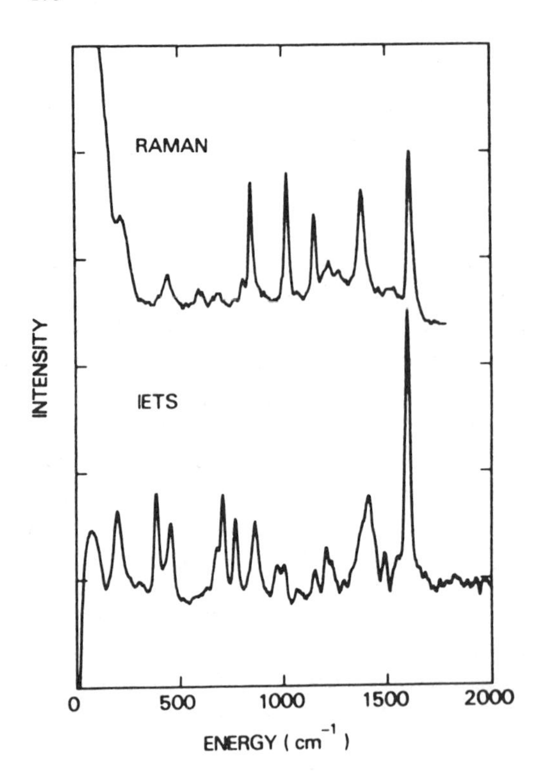

Figure 14. Comparison of the tunneling and Raman spectra of an Al–AlO_x–4-pyridine-COOH–Ag junction (reference 60). The Raman cross section for this monolayer coverage is enhanced by a factor of 10^4 over solution cross sections.

high signal abound, and they change rather quickly depending on their agreement with new experimental findings. Several of the major theoretical treatments have been summarized recently.[53,54,58]

An interesting study combining surface-enhanced Raman spectroscopy and inelastic electron tunneling spectroscopy has been undertaken by Tsang, Kirtley, and Bradley.[59] They studied the chemisorption of 4-pyridine carboxylic acid in the standard tunnel junction configuration with a top silver electrode. Measured enhancement factors were approximately 10^4. A representative spectrum is shown in Figure 14.[60] The positions and therefore the mode assignments determined by these two techniques are quite similar, although the relative peak intensities are very different. Hopefully a study of these intensity differences can provide more information on the mechanism of enhanced Raman spectroscopy.

Even though this technique allows us to observe the vibrational spectra of submonolayer quantities of certain molecules, the substrate morphology is generally not well characterized. Furthermore, these studies are restricted at present to only a few metals and to a limited class of adsorbates.

2.4.3. High-Resolution Electron Energy Loss Spectroscopy

High-resolution electron energy loss spectroscopy (ELS) is a technique which has recently become available for studies of the vibrational spectra of adsorbed atoms and molecules. In this technique a collimated beam of monochromatic electrons (~ 5 eV incident energy and 6–10 meV, full width at half-maximum) is inelastically scattered from a metal surface and the energy distribution of the specularly reflected beam is recorded. The incident electrons excite surface vibrational modes and will therefore lose energy corresponding to the frequency of the vibration(s) involved.[19] Both theory and experiment have shown that scattering in the specular direction occurs via a long-range electron–dipole interaction between the incident electrons and the adsorbed gas molecules. Since only those vibrations that give rise to a changing dipole moment perpendicular to the surface can be excited, high-resolution ELS and infrared spectroscopy provide similar information.[19,61]

Representative high-resolution ELS spectra are shown in Figure 15. These traces, taken from the work of Demuth *et al.*,[62] clearly show several of the major advantages of this technique. First, the entire infrared region of the spectrum (from 300 to 4000 cm^{-1}) can be scanned, and unlike optical measurements, all of these experiments can be accomplished without changing windows, prisms (gratings), or lenses. Secondly, for strong scatterers ELS is sensitive to less than 0.1% of a monolayer (approximately 10^{11} molecules), far more sensitive than most other techniques. Because of both this surface sensitivity and the nature of the inelastic scattering process, single-crystal metal surfaces make ideal targets. Thus, studies can be carried out on clean, well-characterized substrates under ultrahigh vacuum conditions. Also, because of this surface sensitivity ELS can be used to detect hydrogen both through its vibration against the substrate and against other

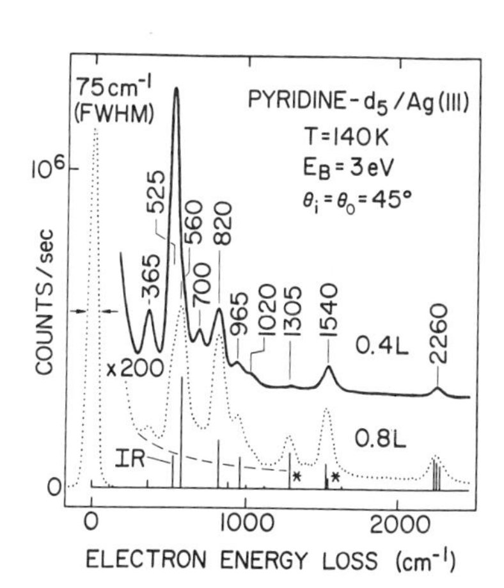

Figure 15. High-resolution electron energy loss spectra of 0.4L (solid line) and 0.8L (dotted line) exposures of deuterated pyridine to Ag(111) (reference 62). The dashed line under the 0.8L curve shows the spectrometer background. The infrared absorbances of gas phase pyridine are indicated at the bottom; the absorbances denoted by an asterisk have been reduced by one-third.

heavier adsorbed atoms. Finally, due to the low incident beam energies and beam currents, high-resolution electron energy loss spectroscopy is a nondestructive technique which can be used to probe the structure of weakly adsorbed molecules.

While discussing the advantages of high-resolution ELS, mention should be made of two of its major disadvantages. First, by optical standards the

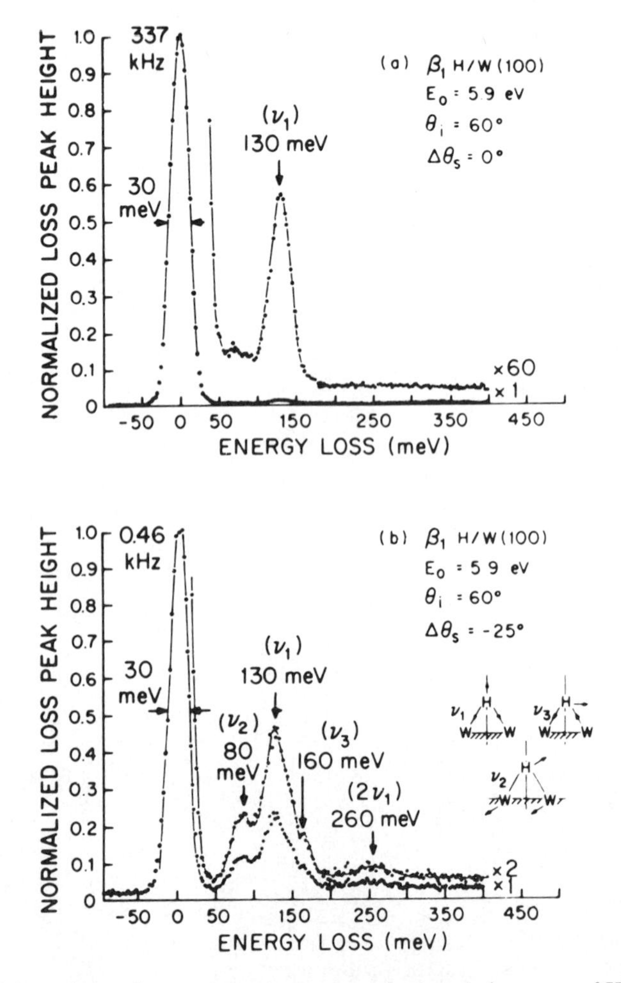

Figure 16. High-resolution electron energy loss spectra of a saturated coverage of H on W(100) (β_1 phase) at 300 K (reference 64). Spectra are normalized to unity at the elastic peak with the absolute elastic intensity shown beside the peak. Only dipole scattering is observed in the specular direction (a), while off-specular scattering in (b) (25° towards the surface normal) shows the presence of all modes. Assignment of the modes is indicated schematically in (b).

term "high resolution" is a misnomer since the resolution is limited, at present, to about 60 cm^{-1} (the full width at half-maximum of the elastic scattering peak). Peak assignments can be made much more accurately (± 5 cm^{-1}), however. This limits the use of isotopic substitution and the analysis of closely spaced vibrational modes. The second major drawback is that the maximum pressure under which experiments can be carried out is approximately 5×10^{-5} Torr due to electron–gas collisions inside the spectrometer. High-pressure catalytic reactions and chemisorption at the solid–liquid interface cannot be readily studied. Nevertheless, a tremendous number of studies on the adsorption of atoms, diatomic molecules, and large hydrocarbons on transition metal surfaces have been performed.

The normal dipole selection rule mentioned earlier can break down when we collect electrons out of the specular direction and therefore vibrations of all spatial orientations can be observed.[63] This is due to a new short-range scattering mechanism in which the incident electron and the adsorbate interact to form a short-lived negative ion state. This species decays and the ejected electron is no longer emitted in the specular direction. An example of this type of scattering is shown in Figure 16 for the chemisorption of hydrogen on tungsten.[64] In the specular direction (upper trace) only the vibration corresponding to the symmetric stretch (ν_1) of a bridge bonded species is observed. This mode must be perpendicular to the surface. As we move away from the specular direction and towards the surface normal (lower trace), two other vibrations corresponding to the wagging mode (ν_2) and asymmetric stretching mode (ν_3) both parallel to the surface are observed. Thus, in principle a complete vibrational spectrum of the adsorbed species can be obtained.

2.4.4. *Inelastic Neutron Scattering Spectroscopy*

The last technique to be discussed in this section is inelastic neutron scattering spectroscopy.[65,66] As shown in Table 1, the theory of this technique is well established. Experimentally, it resembles high-resolution electron energy loss spectroscopy in several respects. Schematically, a thermal beam of neutrons (2–200 meV) exits a reactor and passes through a monochromator or filter before impinging on a sample. The scattered neutrons are energy and momentum analyzed either by Bragg reflection or by time of flight and the resulting spectrum plotted out.

There are significant differences between inelastic electron scattering and inelastic neutron scattering, however. First of all, since the neutron is uncharged it interacts weakly with matter and therefore does not preferentially scatter from the surface of materials. This limits studies to high-surface-area substrates (>20 m^2/g). Materials such as charcoals, graphites, zeolites, metal oxides, Raney nickel, and transition-metal powders have all been studied. However, the spin-incoherent cross section of hydrogen is one or two orders of magnitude greater than that for most other atoms so that the scattering intensities are sufficient to study molecular vibrations in submonolayer hydrogenous films. This low scattering for most adsorbate (and substrate) atoms allows studies to be carried out at relatively high pressures. Furthermore, since the incident-neutron–adsorbate interaction is quite weak, multiple scattering effects can usually be neglected. Thus, calculations of both peak positions and their relative intensities are quite straightforward. Also, analysis of the angular distribution of the elastically scattered neutrons (diffraction pattern) can provide structural information.

If a mixture of species is present on a surface, then quantitative information on the relative population of each species can be obtained. Finally, since peak intensities are only proportional to the square of the vibrational amplitude there are no orientational selection rules and all vibrational modes can be observed. Thus, low-frequency modes which may have a small dipole derivative and will not be observed by either infrared or high-resolution electron energy loss spectroscopy may have very intense peaks in the inelastic neutron scattering spectrum.

There are several drawbacks to inelastic neutron scattering besides the high surface area of the sample required. At present the range of observed frequencies is limited to below 2000 cm^{-1}. This range does not allow the observation of C–H stretching vibrations. These modes are extremely helpful in determining the state of hybridization of adsorbed hydrocarbons. However, a greater limitation is the resolution—typically 50–100 cm^{-1}, depending on the incident neutron wavelength. Finally, as mentioned earlier, one is limited to studying the low-frequency vibrations of hydrogen or other very light atoms. However, one can think of this as an advantage since this makes inelastic neutron scattering and transmission infrared spectroscopy of high-surface-area materials complementary techniques.

An interesting example of the application of inelastic neutron scattering to the adsorption of benzene on Raney nickel is shown in Figure 17a.[67] The calculated spectrum in (b) [using a valence force field approximation and the model compound $CrC_6H_6(CO)_3$] agrees quite well with the observed spectra if the added peaks due to chemisorbed hydrogen are taken into account. These authors conclude that there is a significant perturbation of the adsorbed molecule because of the strong bonding to the nickel surface, which results in an approximately 20% weakening of both the C–C stretching and C–C interaction force constants. The benzene molecule is oriented parallel to the metal surface and bound to a single nickel atom. As you can see, despite the drawbacks of this technique, good quality vibrational spectra and therefore good quality structural determinations can be obtained under the proper conditions.

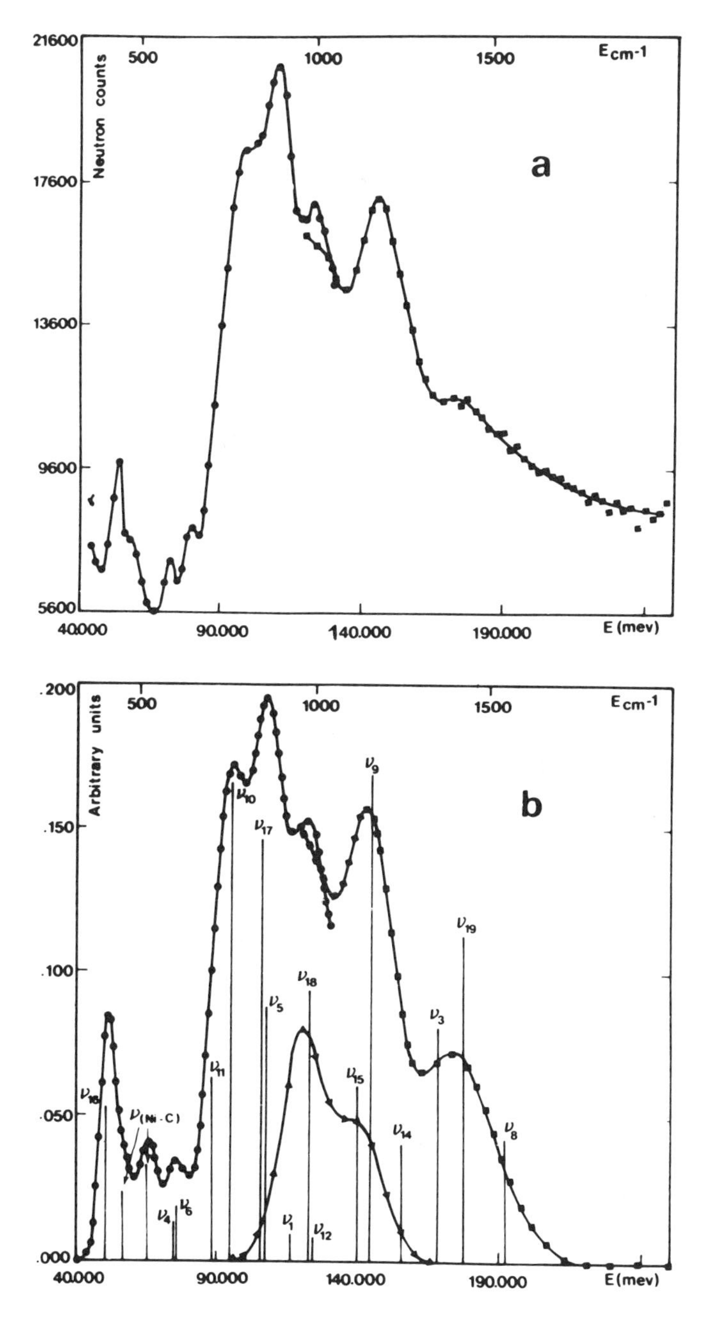

Figure 17. (a) Observed inelastic neutron scattering spectrum of benzene adsorbed on Raney nickel at 77 K (reference 67). (b) Calculated spectrum of adsorbed benzene (upper trace) with added experimental spectrum of adsorbed hydrogen (lower trace).

If one looks at Table 1 again, one can see a tremendous variety in the techniques discussed. Each method has its own strengths and weaknesses, but more importantly, each method is superior to the others under certain conditions. This then is the "message" of this section: that only after the problem is chosen and the nature of the information required decided upon, then and only then can the proper choice of technique be made.

3. The Application of Modern Surface Analytical Techniques to the Characterization of Carbon Monoxide Adsorbed on Alumina Supported Rhodium

Modern surface science has become an alphabet soup of acronyms—from LEED and SIMS to ARUPS and SERS; even EELS and SEXAFS have crept into our vocabulary. Beneath all of these letters (and techniques) lies one unifying goal—a desire to understand the detailed nature of solid surfaces.

Now that the major surface analytical techniques have been briefly introduced, we can turn to a discussion of the complementary nature of these experimental methods and show how they can be used to characterize a single chemisorption system: the adsorption of carbon monoxide on alumina supported rhodium particles. The choice of an example to discuss is rather difficult since all of the previously mentioned techniques cannot be used to study the same type of adsorption system. Furthermore, even though studies can be performed, in many cases they have not been. The choice of CO on rhodium/alumina was made because of the large variety of data available on this system from a number of the techniques described earlier, especially techniques involving surface vibrational spectroscopy. Hopefully the interested reader will not only see how inelastic electron tunneling fits into the scheme of modern surface science, but also (and more importantly) the complementary nature of the information that can be obtained by applying a variety of surface sensitive probes to the study of a single adsorption system.

3.1. Sample Preparation and Morphology

3.1.1. High-Surface-Area Samples

High-surface-area rhodium on alumina model catalysts are relatively easy to prepare and details are provided in numerous references.[68–73] Unfortunately they are rather difficult to characterize accurately. These samples are generally formed from an aqueous slurry of a rhodium salt ($RhCl_3$) and Al_2O_3. The resulting mixture is filtered, dried, and pressed into disks or pellets of the appropriate size. Samples are calcined (heated) in air and then reduced in flowing hydrogen (470–720 K) before carbon monoxide is adsorbed. This heat treatment is used to control the rhodium particle size and particle size distribution. The metal loading is controlled by varying the relative concentrations of $RhCl_3$ and Al_2O_3. Typical samples have metal loadings between 1 and 15% by weight and substrate surface areas around 100 m^2/g.

The details of this preparation process can be varied slightly depending on the type of experiment performed. High-quality infrared spectra can be obtained by spraying the aqueous $RhCl_3/Al_2O_3$ slurry onto an ir transparent window.[68,70] Small samples can also be pulverized and ultrasonically dispersed in alcohol and then deposited onto carbon films for transmission electron microscopic analysis.[69] Recent innovations have included studies of supported rhodium carbonyl compounds in an attempt to obtain a more uniform particle size distribution.[74]

The high-surface-area rhodium on alumina model catalyst system is rather unique in that a tremendous number of very accurate experiments have been performed in an attempt to obtain clean, reproducible, and well characterized samples. The surface area (BET) of the alumina substrate is generally supplied by the manufacturer and as stated earlier is approximately 100 m^2/g in most studies. The pore size (approximately 500 Å) and pore size distribution (200–1000 Å) of these substrates has also been measured.[70] Unfortunately details on a microscopic scale are difficult to determine. Al_2O_3 is amorphous, so that the geometry of the surface species is not known. The number of surface and near surface hydroxyl groups is also not accurately known, so even the chemical composition of the surface is uncertain. Furthermore, it has been reported that the catalytic activity of metal oxide substrates can vary with time as the nature of the surface slowly changes.[75] Several research groups have attempted to characterize metal oxide samples using x-ray photoelectron spectroscopy.[75,76] Although this technique is quite sensitive to surface structure and stoichiometry, no generally applicable correlations can be made in the case of commercial grade aluminas.[1,76]

Fortunately the structure of the rhodium crystallites is far more certain and, more importantly, is not extremely sensitive to the details of the substrate preparation. Yates *et al.* used transmission electron microscopy combined with hydrogen and carbon monoxide chemisorption data to give a very complete description of the surface of the rhodium particles.[69] Photomicrographs of 1% rhodium on alumina samples (reduced at 470 K) showed only two-dimensional metal rafts; no three-dimensional metal particles were observed. This is in agreement with hydrogen chemisorption data (number of hydrogen atoms/number of metal atoms ≅ 1). Particle sizes varied between 7 and 35 Å with an average raft diameter of 15.1 Å. Detailed examination of the electron micrographs showed that the most probable shape of the two-dimensional rafts was round. For a 10% sample (reduced at 750 K) two-dimensional rafts were observed with diameters less than 20 Å. In the size range between 20 and 40 Å, both rafts and three-dimensional particles were observed. Particles larger than this were all found to be more than one atom thick. The average particle size was 26 Å. Again this is in excellent agreement with hydrogen chemisorption data.

Thus, as mentioned earlier, metal particle sizes and particle size distributions can readily be varied. Although the structures of the high-surface-area rhodium on alumina model catalysts are rather complex, by combining these studies with experiments on low-surface-area samples, our understanding of this chemisorption system can be further increased.

3.1.2. Low-Surface-Area Samples

Low-surface-area rhodium on alumina model catalysts are also straightforward to prepare and allow us to apply several new techniques to the analysis of the surface structure. The details of the preparation of these samples are presented in Chapter 13 and the procedure will only be outlined briefly here.[77,78] In the tunnel junction configuration approximately 800–2000 Å of aluminum is first deposited in high vacuum onto a glass slide (or other suitable substrate). Transmission electron microscopy studies by Kroeker *et al.* indicate that the evaporated aluminum layers are polycrystalline with grain sizes on the order of microns.[77] These thin aluminum films are then oxidized in air prior to placement in a second, ultrahigh-vacuum evaporator for the rhodium deposition. This oxide layer now has a dual purpose; it serves not only as the insulating barrier necessary for tunneling spectroscopy, but also as the Al_2O_3 support for the rhodium particles.

These oxidized aluminum layers have been studied by a large number of techniques and are now quite well characterized. The layers are 10–30 Å thick, are amorphous, and contain no pin holes or large regions of unoxidized aluminum.[79,80] This latter point is clearly shown by the observation of a tunneling current in a completed junction. Both Auger electron spectroscopy (Chapter 13)[78] and x-ray photoelectron spectroscopy (Figure 2)[1] indicate that the chemical composition and the electronic structure of the oxidized aluminum film are similar to that of bulk Al_2O_3. These techniques are sensitive to both the aluminum oxidation state and to the stoichiometry of the oxide layer. The chemistry of the oxide layer has also been explored extensively.[81,82] Hansma *et al.* used electron microscopy,

BET surface area measurements, thermal gravimetric analysis, and studies on butene isomerization to clearly show that aluminum oxidized in this manner resembles γ-alumina in both its physical *and* catalytic properties.[81] Finally, by raising the sample temperature above 720 K in oxygen, aluminum metal oxidizes thickly enough so that the resultant oxide layer can be directly analyzed by electron diffraction. The oxide layer is found to be γ-alumina.[83]

The small rhodium particles are then deposited onto the alumina substrate by evaporating rhodium either in the presence of carbon monoxide (1×10^{-5} Torr) or in high vacuum followed by exposure to CO. The sample loading can be controlled by varying the extent of metal deposition. Rhodium evaporated in this manner has been shown to agglomerate into small, highly dispersed particles on the alumina support. Kroeker *et al.* used transmission electron microscopy to show that typical rhodium particle diameters were 20–30 Å for a 4 Å metal thickness.[77] These particles are similar in both size and distribution to those formed from the reduction of transition metal salts on alumina to prepare commercial catalysts. Auger electron spectroscopic studies on similarly prepared samples (Chapter 13) show the presence of "bulklike" rhodium metal and attenuation of the substrate aluminum oxide transitions.[78] To date these small rhodium particles have not been further characterized.

For tunneling spectroscopy the junctions are completed by evaporation of the top lead electrode. For high-resolution ELS and reflection infrared studies, the samples must remain in the ultrahigh-vacuum chamber, free from contamination.

3.2. Vibrational Spectroscopic Analysis

Much effort has gone into understanding the vibrational spectrum of carbon monoxide adsorbed on alumina supported rhodium model catalysts. In 1957 Yang and Garland published the first transmission infrared study on highly dispersed rhodium particles.[68] These authors provided convincing evidence for three distinct types of adsorbed CO: linear (RhCO), bridged (Rh_2CO), and twin [$Rh(CO)_2$]. Since that time at least eight other research groups have successfully repeated their experiments with little change in the original interpretation. This work has been reviewed recently.[71,84]

Figure 18, which shows the chemisorption of carbon monoxide on a 2.2% by weight Rh/Al_2O_3 sample as a function of adsorbate pressure, is typical of the observed spectra.[70] The modes labeled 2070 and 1870 cm^{-1} are probably the easiest to understand. By comparison with the infrared spectra of model organometallic compounds of known molecular struc-

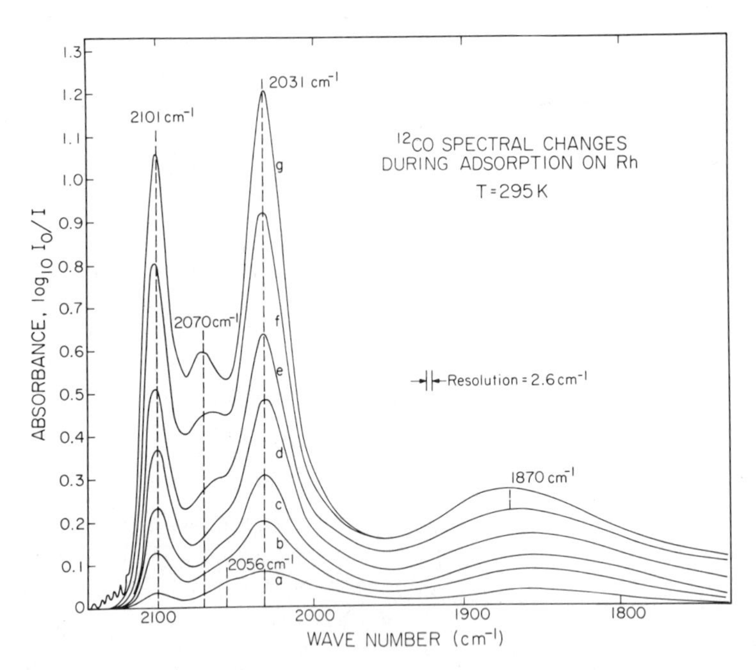

Figure 18. Transmission infrared spectra of carbon monoxide absorbed on a 2.2% by weight rhodium on alumina model catalyst at 295 K (reference 70). The CO pressure in the sample cell is (a) 2.9×10^{-3} Torr, (b) 4.3×10^{-3} Torr, (c) 5×10^{-3} Torr, (d) 8.3×10^{-3} Torr, (e) 0.76 Torr, (f) 9.4 Torr, and (g) approximately 50 Torr.

ture,[85,86] these peaks can be assigned to the carbon–oxygen stretching vibrations of linearly (2070 cm^{-1}) and bridged (1870 cm^{-1}) CO species. Sheppard and Nguyen have shown this to be a very reasonable method for analyzing the vibrational spectra of adsorbed CO.[71] The frequency shifts observed in these two peaks with increasing coverage (2056 → 2070 cm^{-1}; 1855 → 1870 cm^{-1}) can be explained by a variety of mechanisms including local field effects,[47,48] vibrational coupling,[48] and dipole–dipole interactions.[49,50] The interested reader is referred to one of these references for further information. The relative abundance of these two species increases with increasing metal loading.

The two modes observed at 2101 and 2031 cm^{-1} are assigned to the symmetric and asymmetric carbon–oxygen stretching vibrations of a twin CO species [$Rh(CO)_2$] by analogy with the infrared spectra of $Rh_2(CO)_4Cl_2$ (2095, 2043 cm^{-1}) and $Rh_2(CO)_4Br_2$ (2092, 2042 cm^{-1}).[87] By analyzing the relative intensity of these two modes Yates *et al.* concluded that the OC–Rh–CO bond angle must be near 90° for this species.[70] Figure 18 clearly shows that the doublet increases in intensity with increasing CO exposure without a change either in frequency or in relative intensity. The frequencies are invariant to within 1 cm^{-1} over the entire coverage range studied (a 30-fold range of infrared intensity). Thus, Yates *et al.* concluded that the twin CO species only occur on isolated rhodium atoms.[70] For rhodium atoms within a cluster, CO adsorption on neighboring metal atoms should lead to interactions producing an increasing C–O frequency with increasing gas exposure. Such a shift is observed for the linear and bridged species as well as for carbon monoxide adsorbed on bulk rhodium.[84,88] This view is confirmed by Yao and Rothschild, who suggest that on their RH/AL_2O_3 samples, Rh $\cdots$ Rh distances of order 8 Å are necessary for $Rh(CO)_2$ formation.[72] This assignment is also in agreement with carbon monoxide BET surface area measurements.[70]

Analysis of adsorbed ^{13}CO vibrational spectra is in agreement with the above assignments.[70]

No studies on the electronic structure of carbon monoxide adsorbed on high-surface-area alumina supported rhodium model catalysts have been reported to date. However, by analogy with infrared, XPS, and uv reflectance studies on zeolite supported rhodium particles, Primet concluded that the dicarbonyl species (two CO molecules per metal atom) can only occur on oxidized (+1) rhodium surface sites.[73]

As mentioned previously, because of light adsorption by the alumina support below ~ 1500 cm^{-1}, low-frequency vibrational modes (metal–carbon stretching and bending vibrations) are completely masked. Nevertheless infrared spectroscopists have gained a tremendous amount of detailed structural information on adsorbed CO from analyzing a rather limited data base. Some of these low-frequency modes have now been observed by tunneling spectroscopy.[79,89] A representative spectrum is shown in Figure 19.[77] Although the aluminum oxide phonon band is clearly observed (928 cm^{-1}), its intensity is quite weak and does not obscure the low-frequency modes. Thus we see the biggest advantage of inelastic tunneling spectroscopy over conventional transmission infrared spectroscopy to study model supported metal catalysts.

The character of all of the modes observed in the tunneling spectrum of Figure 19 have been well established.[77,89] The two high-frequency bands (1721 and 1942 cm^{-1}) are assigned to the carbon–oxygen stretching vibration of the adsorbed bridged (1721 cm^{-1}) and linear (twin) (1942 cm^{-1}) CO species. These modes are both broadened and down-shifted in frequency due to the presence of the top lead electrode. The explanation for this effect is discussed in Chapter 1 and in several references.[90,91] By the use of isotopic substitution ($^{13}C^{16}O$, $^{12}C^{18}O$ adsorption, Figure 20)[92] and of selective sulfur poisoning of specific sites on the rhodium surface,[89] conclusive vibrational assignments can be obtained. The character of these modes can

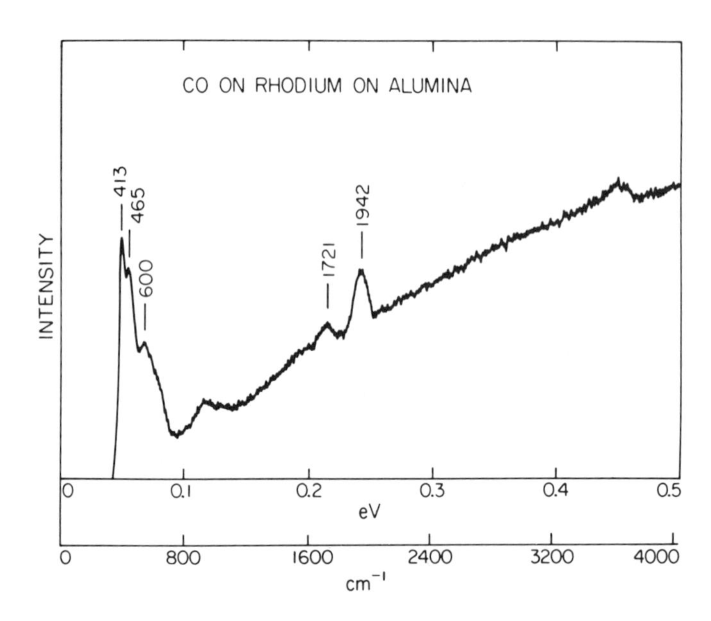

Figure 19. Differential tunneling spectrum of carbon monoxide chemisorbed on alumina supported rhodium particles (approximately 30 Å in diameter) (reference 77). The small peak at 928 cm^{-1} is due to the alumina support. These peak positions have not been corrected for shifts due to the superconducting lead electrode.

be confirmed by varying the metal loading and isolating the individual surface species.[89,92] The assignments for all of the observed modes (from both tunneling spectroscopy and transmission infrared spectroscopy) are summarized in Table 2.

Questions have been raised recently as to the applicability of using inelastic electron tunneling spectroscopy to study chemisorption on model supported metal catalysts. First, one wonders whether the species formed in the tunnel junctions at 4 K are the same as those found at elevated temperatures and pressures on alumina supported rhodium crystallites. And secondly one questions the effect of the top metal electrode on the nature of the adsorbed species.[91] Recent experiments can now lay these fears to rest. The vibrational spectra of similarly prepared samples *without* the upper electrode can be obtained using either high-resolution electron energy loss spectroscopy[78] or reflection–absorption infrared spectroscopy.[93] Representative high-resolution ELS spectra are shown in Figure 21.[78] The lower trace in this figure displays the vibrational spectrum of the oxidized aluminum substrate. The broad peak centered just below 900 cm^{-1} and asymmetric to lower energies is the aluminum oxide phonon band. The

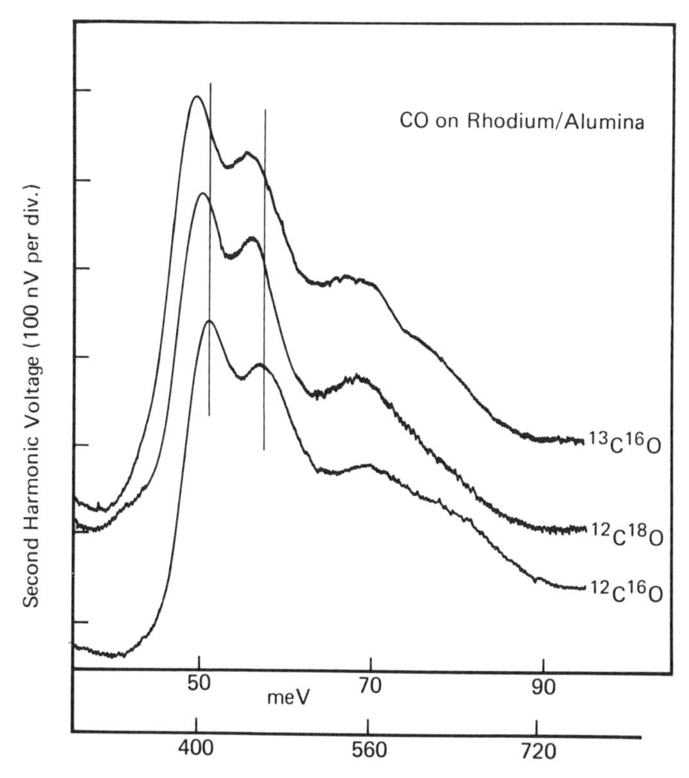

Figure 20. Differential tunneling spectrum of $^{13}C^{16}O$ (upper trace), $^{12}C^{18}O$ (middle trace), and $^{12}C^{16}O$ (lower trace) adsorbed on 30–40-Å diameter rhodium particles supported on alumina (reference 92). Junctions were formed at room temperature. Frequency shifts upon isotopic substitution indicate that the two lowest-energy peaks are due to bending modes (reference 89). The corresponding rhodium–carbon stretching bands overlap near 550 cm^{-1} and are not well resolved here. The presence of the multiply bonded species is seen in the broad shoulder near 640 cm^{-1}. These peak positions have not been corrected for shifts due to the superconducting lead electrode.

Table 2. Vibrational Modes Observed from the Chemisorption of $^{12}C^{16}O$ ($^{13}C^{16}O$) on Rh/Al_2O_3 Metal Catalysts

	Observed frequencies (cm^{-1})		
	Tunneling spectroscopy[77,89]		Infrared spectroscopy[70]
Species	δ_{Rh-CO}	ν_{Rh-CO}	ν_{C-O}
Linear	469 (454)	589	2070 (2024)
Twin	416 (402)	573	2101, 2031 (2056, 1987)
Bridged	—	605 (589)	1870 (1832)

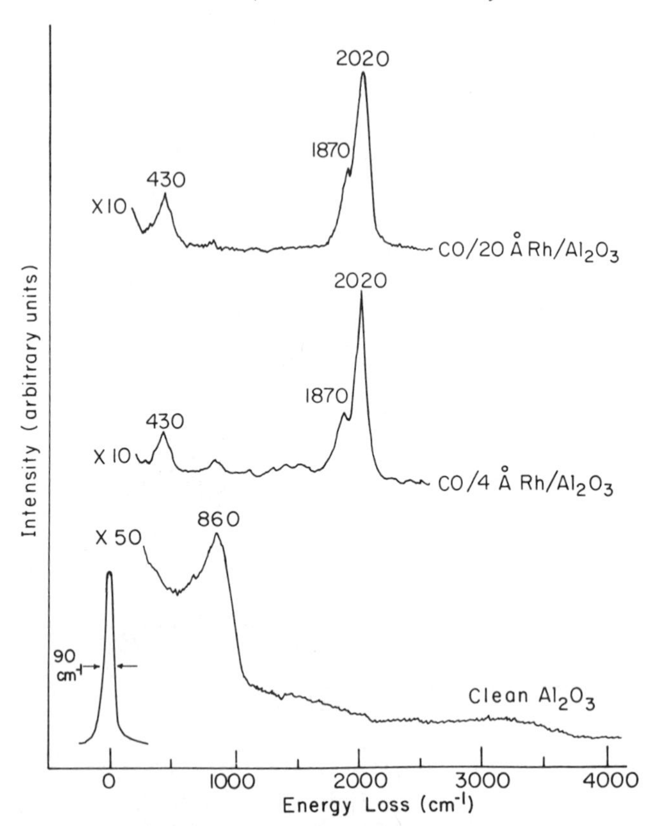

Figure 21. The high-resolution electron energy loss spectrum for the oxidized aluminum film used as the support is shown in the lowest trace. The broad band, asymmetric to lower wave numbers and centered around 860 cm^{-1}, is the aluminum oxide phonon band. The upper two traces show vibrational spectra for two different amounts of rhodium evaporated onto the alumina support in 1×10^{-5} Torr CO (4 Å and 20 Å average metal thickness). (Reference 78.)

middle and upper traces of Figure 21 show the vibrational spectra of CO chemisorbed on highly dispersed rhodium particles supported on the oxidized aluminum. The results are not particularly sensitive to the metal loading. Bands observed in the low-frequency region are similar to those found using tunneling spectroscopy (Figures 19 and 20). Unfortunately the ELS spectrum lacks sufficient resolution to clearly identify individual Rh–CO bending and stretching modes between 400 and 600 cm^{-1}. Both the high-resolution electron energy loss spectrum (shown here) and the reflection–absorption infrared spectrum (not shown) between 1600 and 2200 cm^{-1} are similar to that of the low-exposure transmission infrared spectrum (Figure 18). These studies along with experiments on the chemical nature and structure of the substrate clearly indicate that similar species are being observed in all cases.

3.3. ^{13}C Nuclear Magnetic Resonance Studies

Recently, ^{13}C nuclear magnetic resonance (NMR) spectroscopy and transmission infrared spectroscopy have been used to help understand the adsorption of carbon monoxide on alumina supported rhodium particles.[94,95] This represents the first example of the use of solid-state NMR to study chemisorption on dispersed metal surfaces. This technique also provides us with a method of calculating absolute site populations since the integrated intensity of the NMR spectrum is linearly proportional to the surface concentration. Furthermore, the average resonant frequency, or center of mass, of the line shape can be compared with chemical shifts of known compounds for identification of surface species in a fingerprinting fashion. Finally, differences in the spin-lattice relaxation time (T_1) of the adsorbed species may be used to help distinguish between adsorbed species.

Figure 22 shows a ^{13}C NMR spectra of ^{13}CO adsorbed on a 2.2% Rh on Al_2O_3 model catalyst.[95] The peak at -177 ppm ($T_1 = 5.6$ msec) corresponds to 58% of the adsorbed ^{13}CO. By selective exchange studies with gas phase ^{12}CO and by comparison with the ^{13}C NMR spectrum of suitable model compounds, this peak can be assigned to the twin CO species. The broad peak centered at -199 ppm ($T_1 = 64$ msec) contains contributions from linearly bonded (-184 ppm) and bridged bonded (-228 ppm) carbon monoxide. These assignments were based on similarities with relevant rhodium carbonyls and yield a ratio of linear to bridged species of approximately 2 to 1. The measured relaxation times and site

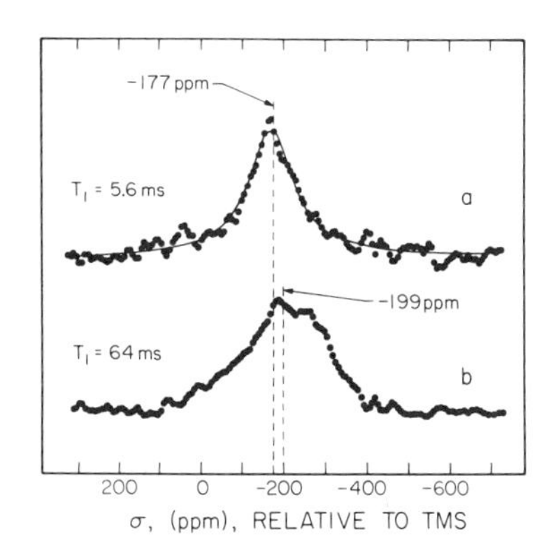

Figure 22. ^{13}C NMR spectrum of ^{13}CO adsorbed on a 2.2% by weight Rh on Al_2O_3 model catalyst (reference 95). The peak in *a* at -177 ppm ($T_1 = 5.6$ msec) is assigned to the twin CO species. The broad peak in *b* at -199 ppm ($T_1 = 64$ msec) contains contributions from both linearly bonded (-184 ppm) and bridge bonded (-228 ppm) carbon monoxide.

distributions are independent of the equilibrium ^{13}CO pressure from 10^{-3} to 50 Torr. One might expect the linear and bridge bonded CO species to have the same ^{13}C relaxation times because of rapid interconversion and diffusion on the rhodium crystallites.[72] One would also not anticipate exchange with the dicarbonyl species, formed on isolated atomically dispersed sites, and thus a substantially different relaxation time is expected.

By combining careful infrared and NMR studies on a given sample, the molar integrated intensities for the symmetric and asymmetric stretches of the dicarbonyl, for the linear and for the bridged species, can be determined. They are 74, 128, 26, and 85×10^6 cm/mole. These values are substantially higher than the value measured for gas phase CO (5.4×10^6),[96] but are in reasonable agreement with measurements on similar metal–carbonyl systems.[77,84] These molar integrated intensities are for only one metal loading and at one CO coverage and cannot be taken as universal values. More recent NMR and ir studies show the sensitivity of these measurements to the experimental conditions.[97]

3.4. Adsorbate Structure and Bonding from Studies of Model Systems

Studies on related model systems—rhodium cluster carbonyls and carbon monoxide adsorption on single-crystal rhodium surfaces—can provide new information on both the structure and bonding of CO on Rh/Al_2O_3 catalysts. As mentioned previously, assignments of the vibrational modes to individual CO species were made by comparison with the infrared spectra of model organometallic compounds of known molecular structure.[85–87] A list of several compounds and their vibrational frequencies between 1700 and 2200 cm^{-1} is presented in Table 3. This table clearly shows the close similarity between the carbonyl species observed to form on alumina supported rhodium model catalysts and on rhodium cluster carbonyls. It also shows that similar species are formed from the chemisorption of carbon monoxide on bulk rhodium[71,84,88]—both single crystals and evaporated films.

X-ray diffraction studies of several rhodium carbonyls yield Rh–C bond lengths of approximately 1.864 Å for the linear species[98] and 2.01 Å for the bridged species.[99] C–O bond distances are typically 1.155 Å.[98,99] The OC–Rh–CO bond angle measured in solid $Rh_2(CO)_4Cl_2$ is 91°,[100] in good agreement with the 90° angle determined from the infrared spectra of the dicarbonyl species observed on Rh/Al_2O_3. A low-energy electron diffraction intensity analysis for the linear CO species formed on Rh(111) (the lowest free energy face of rhodium) yields Rh–C and C–O bond lengths of 1.96 and 1.06 Å, respectively.[101] Because of the similarity

Table 3. Selected C–O Vibrational Frequencies for Carbon Monoxide Bonded to Rhodium (All Frequencies in cm^{-1})

		Site			
System	Technique	3-fold	Bridge	Atop	Twin
$CO/Rh/Al_2O_3$[a]	Tunneling	—	1721[b]	1942[b]	1942[b]
$CO/Rh/Al_2O_3$[c]	Transmission infrared	—	1870	2070	2031, 2101
CO/Rh film[d]	Transmission infrared	—	1852, 1905	2055	(2111)
CO/Rh(111)[e]	High-resolution ELS	—	1870	2070	—
$Rh_2(CO)_4Cl_2$[f]	Solid, infrared	—	—	2043, 2095	—
$Rh_2(CO_8)$[g]	Solution, infrared	—	1845, 1861	—	2061, 2086
$Rh_4(CO)_{12}$[h]	Solid, infrared	—	1848	—	2028–2105
$Rh_6(CO)_{16}$[h]	Solid, infrared	1770	—	—	2016–2077

[a]Reference 77.
[b]Peaks shifted due to the presence of the top lead electrode.
[c]Reference 70.
[d]Reference 88.
[e]Reference 84.
[f]Reference 87.
[g]Reference 85.
[h]Reference 86.

between the vibrational spectra, these values can be taken as reasonable estimates of the structural parameters for the species observed to form on alumina supported rhodium crystallites.

Analysis of the low-frequency (<1000 cm^{-1}) vibrational modes by comparison with the infrared and Raman spectra of model rhodium cluster carbonyls of known molecular structure is rather difficult at present. No normal coordinate calculations have been carried out on these compounds to date. Although numerous vibrational modes between 390 and 520 cm^{-1} have been observed in $Rh_4(CO)_{12}$ and $Rh_6(CO)_{16}$,[86] no simple comparisons can be made with the data presented here.[102]

The accepted picture of carbon monoxide bonding to metals is by electron transfer from the 5σ orbital of CO to the metallic *d* orbitals and by backbonding of the metallic electrons into the empty $2\pi^*$ orbital of the adsorbate. This scheme has been used both by surface scientists and by inorganic chemists to explain the infrared spectra of chemisorbed carbon monoxide and of metal carbonyls. This scheme has also been used to explain the valence band photoelectron spectrum of these species. The UPS spectra of CO on Rh(111)[103] and of $Rh_6(CO)_{16}$[104] are shown in Figure

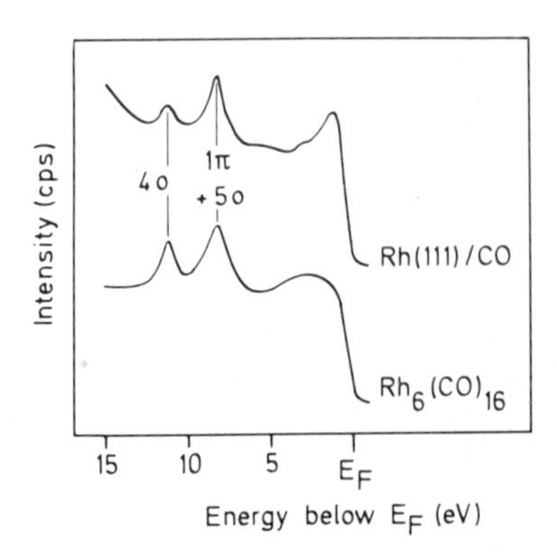

Figure 23. Valence band photoelectron spectra of CO chemisorbed on Rh(111) ($h\nu = 27.5$ eV) (reference 102) and of hexarhodium hexadecacarbonyl ($h\nu = 40.8$ eV) (reference 103). The Fermi level of the metallic system is lined up with the onset of emission from the *d* band of the cluster (reference 104).

23.[105] The maximum in the emission between 0 and 6 eV is interpreted as arising from the Rh *d* states. The onset of the emission of the lower spectrum is aligned with the Fermi level of the metallic rhodium (upper spectrum). The two CO derived levels (centered at about 8 and 11 eV) are qualitatively similar in both spectra. Conrad *et al.* [who originally made the comparison between $Rh_6(CO)_{16}$ and CO adsorbed on Pd(111)] concluded that the electronic properties of both systems as probed by UPS are nearly identical.[103] Judging by the vast differences in particle size here and the similarities in vibrational spectra and bond lengths, one might expect the UPS spectra of CO adsorbed on alumina supported rhodium to also be quite similar.

4. Conclusions

The results of all of these studies provide us with a clear picture of the structure and bonding of carbon monoxide adsorbed on model rhodium–alumina catalysts. This latter section clearly emphasizes several of the major points raised in this chapter:

(1) Only through a combination of complementary surface sensitive probes can we attempt to understand the subtle complexities of solid surfaces.

(2) The proper choice of techniques can only be made after the problem is carefully chosen and the nature of the information required decided upon.

(3) Although a tremendous number of sophisticated techniques used to determine surface composition, bonding, and structure exist, significant improvements in both sensitivity and resolution will be useful in the future.

References

1. H. E. Evans, W. M. Bowser, and W. H. Weinberg, An XPS investigation of alumina thin films utilized in inelastic electron tunneling spectroscopy, *Appl. Surf. Sci.* **5**, 258–274 (1980).
2. C. D. Wagner, W. M. Riggs, L. E. Davis, and J. F. Moulder, *Handbook of X-Ray Photoelectron Spectroscopy* (G. E. Muilenberg, ed.), Perkin-Elmer Corporation, Eden Prairie, Minnesota (1978).
3. T. N. Rhodin and J. W. Gadzuk, in *The Nature of the Surface Chemical Bond* (T. N. Rhodin and G. Ertl, eds.), pp. 113–274, North-Holland Publishing Company, Amsterdam (1979).
4. *Photoemission and the Electronic Properties of Surfaces* (B. Feuerbacher, B. Fitton, and R. F. Willis, eds.), John Wiley and Sons, Chichester (1978).
5. T. A. Carlson, *Photoelectron and Auger Spectroscopy*, Plenum Press, New York (1975).
6. P. Auger, On the Compound Photoelectric Effect, *J. Phys. Radium* **6**, 205–208 (1925).
7. J. J. Lander, Auger peaks in the energy spectra of secondary electrons from various materials, *Phys. Rev.* **91**, 1382–1387 (1953).
8. G. Ertl and J. Küppers, *Low Energy Electrons and Surface Chemistry*, pp. 17–52, Verlag Chemie, Germany (1974).
9. L. E. Davis, N. C. MacDonald, P. W. Palmberg, G. E. Riach, and R. E. Weber, *Handbook of Auger Electron Spectroscopy*, 2nd Ed., Physical Electronics Industries, Inc., Eden Prairie, Minnesota (1976).
10. D. T. Hawkins, *Auger Electron Spectroscopy, A Bibliography: 1925–1975*, Plenum Press, New York (1977).
11. A. Benninghoven, Surface investigation of solids by the statical method of secondary ion mass spectroscopy (SIMS), *Surf. Sci.* **35**, 427–457 (1973).
12. *Proceedings of the Second International Conference on Secondary Ion Mass Spectroscopy (SIMS II)*, Stanford University, August, 1979 (A. Benninghaven, C. A. Evans, Jr., R. A. Powell, R. Shimitu, and H. A. Storms, eds.), Springer-Verlag, New York (1979).
13. W. L. Braun, *A Review and Bibliography of Secondary Ion Mass Spectrometry (SIMS)*, Air Force Materials Laboratory Technical Report AFML-TR-79-4123 (1980).
14. S. A. Flodström, C. W. B. Martinsson, R. Z. Bachrach, S. B. M. Hagström, and R. S. Bauer, Ordered oxygen overlayer associated with chemisorption state on Al(111), *Phys. Rev. Lett.* **40**, 907–910 (1978).
15. N. V. Smith, in *Photoemission in Solids I* (M. Cardona and L. Ley, eds.) *Topics in Applied Physics*, Vol. 26, pp. 237–264, Springer-Verlag, New York (1978).
16. N. V. Smith, P. K. Larsen, and S. Chiang, Anisotropy of core-level photoemission from InSe, GaSe and cesiated W(001), *Phys. Rev. B* **16**, 2699–2706 (1977).
17. S. D. Kevan, D. H. Rosenblatt, D. Denley, B.-C. Lu, and D. A. Shirley Photoelectron-diffraction measurements of sulfur and selenium adsorbed on Ni(001), *Phys. Rev. B* **20**, 4133–4139 (1979).
18. G. Ertl and J. Küppers, *Low Energy Electrons and Surface Chemistry*, pp. 53–66, Verlag Chemie, Germany (1974).
19. H. Froitzheim, in *Electron Spectroscopy for Surface Analysis* (H. Ibach, ed.), pp. 205–250, Springer-Verlag, New York (1977).
20. F. Pellerin, C. LeGressus, and D. Massignon, A secondary electron spectroscopy and electron loss spectroscopy study of the interaction of oxygen with a polycrystalline aluminum surface, *Surf. Sci.* **103**, 510–523 (1981).
21. P. A. Lee and J. B. Pendry, Theory of extended x-ray absorption fine structure, *Phys. Rev. B* **11**, 2795–2811 (1975).

22. J. H. Sinfelt, Structure of metal catalysts, *Rev. Mod. Phys.* **51**, 569–589 (1979).
23. P. H. Citrin, P. Eisenberger, and R. C. Hewitt, SEXAFS studies of iodine adsorbed on single crystal substrates, *Surf. Sci.* **89**, 28–40 (1979).
24. P. H. Citrin, P. Eisenberger, and R. C. Hewitt, Extended x-ray-absorption fine structure of surface atoms on single-crystal substrates: Iodine adsorbed on Ag(111), *Phys. Rev. Lett.* **41**, 309–312 (1978).
25. J. Pendry, *Low Energy Electron Diffraction*, Academic Press, New York (1974).
26. M. A. Van Hove and S. Y. Tong, *Surface Crystallography by LEED*, Springer-Verlag, New York (1979).
27. D. G. Castner and G. A. Samorjai, Surface structures of adsorbed gases on solid surfaces. A tabulation of data reported by low-energy electron diffraction studies, *Chem. Rev.* **79**, 233–252 (1979).
28. C. G. Kinniburgh, A LEED study of MgO(100): III. Theory at off-normal incidence, *J. Phys. C* **9**, 2695–2708 (1976).
29. A. V. Crewe, J. Wall, and J. Langmore, Visibility of single atoms, *Science* **168**, 1338–1340 (1970).
30. M. S. Isaacson, J. Langmore, N. W. Parker, D. Kapf, and M. Utlaut, The study of the adsorption and diffusion of heavy atoms on light element substrates by means of the atomic resolution STEM, *Ultramicroscopy* **1**, 359–376 (1976).
31. P. K. Hansma, unpublished observations.
32. S. Brunauer, P. H. Emmett, and E. Teller, Adsorption of gases in multimolecular layers. *J. Am. Chem. Soc.* **60**, 309–319 (1938).
33. O. C. Wells, *Scanning Electron Microscopy*, McGraw Hill Book Company, New York (1974).
34. R. P. Eischens, W. A. Pliskin, and S. A. Francis, Infrared spectra of chemisorbed carbon monoxide, *J. Chem. Phys.* **22**, 1786–1787 (1954).
35. R. P. Eischens and W. A. Pliskin, The infrared spectra of adsorbed molecules, *Adv. Catal.* **10**, 2–56 (1958).
36. L. H. Little, *Infrared Spectra of Adsorbed Species*, Academic Press, London (1966).
37. M. L. Hair, *Infrared Spectroscopy of Surface Chemistry*, Dekker, New York (1967).
38. A. T. Bell, in *Vibrational Spectroscopies of Adsorbed Species* (A. T. Bell and M. L. Hair, eds.), ACS Symposium Series, Vol. 137, pp. 13–36, American Chemical Society, Washington, D.C. (1980).
39. B. Besson, B. Moraweck, A. K. Smith, J. M. Basset, R. Psaro, A. Fusi, and R. Ugo, IR and EXAFS characterization of a supported osmium cluster carbonyl, *J. Chem. Soc. Chem. Commun.* 569–571 (1980).
40. L. H. Little, in *Chemisorption and Reactions on Metal Films* (J. R. Anderson, ed.), Vol. 1, pp. 490–513, Academic Press, New York (1980).
41. P. K. Hansma, Inelastic electron tunneling, *Phys. Rep.* **30**, 145–206 (1977).
42. J. Pritchard, T. Catterick, and R. K. Gupta, Infrared spectroscopy of chemisorbed carbon monoxide on copper, *Surf. Sci.* **53**, 1–20 (1975).
43. J. Pritchard, in *Chemical Physics of Solids and their Surfaces* (M. W. Roberts and J. M. Thomas, eds.), Vol. 7, pp. 157–179, The Chemical Society, London (1978).
44. A. M. Bradshaw and F. M. Hoffmann, The chemisorption of carbon monoxide on palladium single-crystal surfaces: IR spectroscopic evidence for localized site adsorption, *Surf. Sci.* **72**, 513–535 (1978).
45. A. Bradshaw, Vibrational spectrum of $Ru_3(CO)_{12}$; Analogy with the adsorption system CO/Ru (001), *J. Chem. Soc. Chem. Commun.* 365–366 (1980).
46. H. A. Pearse and N. Sheppard, Possible importance of a "metal-surface selection rule" in the interpretation of the infrared spectra of molecules adsorbed on particulate metals;

Infrared spectra from ethylene chemisorbed on silica-supported metal catalysts, *Surf. Sci.* **59**, 205–217 (1976).
47. J. G. Roth and M. J. Dignam, Concerning the influence of inert gas adsorption on the infrared spectrum of OH groups on powdered silica, *Can. J. Chem.* **54**, 1388–1393 (1976).
48. M. Moskovits and J. W. Hulse, Frequency shifts in the spectra of molecules adsorbed on metals, with emphasis on the infrared spectrum of adsorbed CO, *Surf. Sci.* **78**, 397–418 (1978).
49. R. M. Hammaker, S. A. Francis, and R. P. Eischens, Infrared study of intermolecular interactions for carbon monoxide chemisorbed on platinum, *Spectrochim. Acta* **21**, 1295–1309 (1965).
50. M. Scheffler, The influence of lateral interactions on the vibrational spectrum of adsorbed CO, *Surf. Sci.* **81**, 562–570 (1979).
51. D. L. Allara, in *Vibrational Spectroscopies of Adsorbed Species* (A. T. Bell and M. L. Hair, eds.), ACS Symposium Series, Vol. 137, pp. 37–50, American Chemical Society, Washington, D.C. (1980).
52. B. A. Marrow, in *Vibrational Spectroscopies of Adsorbed Species* (A. T. Bell and M. L. Hair, eds.), ACS Symposium Series, Vol. 137, pp. 119–140, American Chemical Society, Washington, D.C. (1980).
53. R. P. Van Duyne, in *Chemical and Biochemical Applications of Lasers* (C. B. Moore, ed.), Vol. 4, pp. 101–186, Academic Press, New York (1979).
54. T. E. Furtak and J. Reyes, A critical analysis of theoretical models for the giant Raman effect from adsorbed molecules, *Surf. Sci.* **93**, 351–382 (1980).
55. T. A. Egerton, A. H. Hardin, Y. Kozirovski, and N. Sheppard, Reduction of fluorescence from high-area oxides of the silica, γ-alumina, silica-alumina and Y-zeolite types and Raman spectra for a series of molecules adsorbed on these surfaces, *J. Catal.* **32**, 343–361 (1974).
56. W. Krasser and A. Renouprez, in *Proceedings of the International Conference on Vibrations in Adsorbed Layers*, Jülich (H. Ibach and S. Lehwald, eds.), pp. 175–180, Kernforschungsanlange, Jülich, Federal Republic of Germany (1978).
57. D. L. Jeanmaire and R. P. Van Duyne, Surface Raman spectroelectrochemistry, *J. Electroanal. Chem.* **84**, 1–20 (1977).
58. E. Burstein, C. Y. Chen, and S. Lundquist, in *Proceedings of Joint US–USSR Symposium on the Theory of Light Scattering in Condensed Matter* (J. L. Birman, H. Z. Cummings, and H. K. Reband, eds.), pp. 479–498, Plenum Press, New York (1980).
59. J. C. Tsang, J. R. Kirtley, and J. A. Bradley, Surface-enhanced Raman spectroscopy and surface plasmons, *Phys. Rev. Lett.* **43**, 772–775 (1979).
60. J. R. Kirtley, in *Vibrational Spectroscopies of Adsorbed Species* (A. T. Bell and M. L. Hair, eds.), ACS Symposium Series, Vol. 137, pp. 217–245, American Chemical Society, Washington, D.C. (1980).
61. H. Ibach, Comparison of cross-sections in high resolution electron energy loss spectroscopy and infrared reflection spectroscopy, *Surf. Sci.* **66**, 56–66 (1977).
62. J. E. Demuth, K. Christmann, and P. N. Sanda, The vibrations and structure of pyridine chemisorbed of Ag(111): The occurrence of a compressional phase transformation, *Chem. Phys. Lett.* **76**, 201–206 (1980).
63. J. W. Davenport, W. Ho, and J. R. Schrieffer, Theory of vibrationally inelastic electron scattering from orientated molecules, *Phys. Rev. B* **17**, 3115–3127 (1978).
64. W. Ho, N. J. Dinardo, and E. W. Plummer, Angle-resolved and variable impact energy electron vibrational excitation spectroscopy of molecules adsorbed on surfaces, *J. Vac. Sci. Technol.* **17**, 134–140 (1980).
65. H. Taub, in *Vibrational Spectroscopies of Adsorbed Species* (A. T. Bell and M. L. Hair,

eds.), ACS Symposium Series, Vol. 137, pp. 247–280, American Chemical Society, Washington, D.C. (1980).
66. J. Howard and T. C. Waddington, in *Advances in Infrared and Raman Spectroscopy* (R. J. H. Clarke and R. E. Hester, eds.), Vol. 7, pp. 86–222, Heyden and Sons, Ltd., London (1980).
67. H. Jobic, J. Tomkinson, J. P. Candy, P. Fouilloux, and A. J. Renouprez, The structure of benzene chemisorbed on Raney nickel; a neutron inelastic spectroscopy determination, *Surf. Sci.* **95**, 496–510 (1980).
68. A. C. Yang and C. W. Garland, Infrared studies of carbon monoxide chemisorbed on rhodium, *J. Phys. Chem.* **61**, 1504–1512 (1957).
69. D. J. C. Yates, L. L. Murrell, and E. B. Prestridge, Undispersed rhodium rafts: Their existence and topology, *J. Catal.* **57**, 41–63 (1979).
70. J. T. Yates, Jr., T. M. Duncan, S. D. Worley, and R. W. Vaughan, Infrared spectra of CO on Rh, *J. Chem. Phys.* **70**, 1219–1224 (1979).
71. N. Sheppard and T. T. Nguyen, in *Advances in Infrared and Raman Spectroscopy* (R. J. H. Clark and R. E. Hester, eds.) Vol. 5, pp. 67–148, Heyden and Sons, Ltd., London (1978); and reference therein.
72. H. C. Yao and W. G. Rothschild, Infrared spectra of chemisorbed CO on $Rh/\gamma\text{-}Al_2O_3$: Site distribution and molecular mobility, *J. Chem. Phys.* **68**, 4774–4780 (1978).
73. M. Primet, Infrared study of CO chemisorption of zeolite and alumina supported rhodium, *J. Chem. Soc. Faraday Trans. 1* **74**, 2570–2580 (1978).
74. G. C. Smith, T. P. Chojnacki, S. R. Dasgupta, K. Iwatate, and K. L. Waters, Surface supported metal cluster carbonyls. I. Decarbonylation and aggregation reactions of rhodium clusters on alumina, *Inorg. Chem.* **14**, 1419–1421 (1975).
75. H. Vinel, J. Latzel, H. Noller, and M. Ebel, Correlation between results of x-ray photoelectron spectroscopic studies and catalytic behavior of MgO, *J. Chem. Soc. Faraday Trans. 1* **74**, 2092–2100 (1978).
76. J. L. Ogilvie and A. Wolberg, An internal standard for electron spectroscopy for chemical analysis studies of supported catalysts, *Appl. Spectrosc.* **26**, 401–403 (1972).
77. R. M. Kroeker, W. C. Kaska, and P. K. Hansma, How carbon monoxide bonds to alumina-supported rhodium particles; Tunneling spectroscopy measurements with isotopes, *J. Catal.* **57**, 72–79 (1979).
78. L. H. Dubois, P. K. Hansma, and G. A. Somorjai, The application of high-resolution electron energy loss spectroscopy to the study of model supported metal catalysts, *Appl. Surf. Sci.* **6**, 173–184 (1980).
79. M. F. Muldoon, R. A. Dragoset, and R. V. Coleman, Tunneling asymmetries in doped $Al\text{-}AlO_x\text{-}Pb$ junctions, *Phys. Rev. B* **20**, 416–429 (1979).
80. D. G. Walmsely, R. B. Floyd, and W. E. Timms, Conductance of clean and doped tunnel junctions, *Solid State Commun.* **22**, 497–499 (1977).
81. P. K. Hansma, D. A. Hickson, and J. A. Schwarz, Chemisorption and catalysis on oxidized aluminum metal, *J. Catal.* **48**, 237–242 (1977).
82. W. M. Bowser and W. H. Weinberg, The nature of the oxide barrier in inelastic electron tunneling spectroscopy, *Surf. Sci.* **64**, 377–392 (1977).
83. A. F. Beck, M. A. Heine, E. J. Caule, and M. J. Pryor, The kinetics of the oxidation of Al in oxygen at high temperature. *Corrosi. Sci.* **7**, 1–22 (1967).
84. L. H. Dubois and G. A. Somorjai, The chemisorption of CO and CO_2 on Rh(111) studied by high resolution electron energy loss spectroscopy, *Surf. Sci.* **91**, 514–532 (1980).
85. R. Whyman, Dirhodium octacarbonyl, *J. Chem. Soc. Chem. Commun.* 1194–1195 (1970).
86. W. P. Griffith and A. J. Wickham, Vibrational spectra of metal–metal bonded complexes of group VIII, *J. Chem. Soc.* **1969**, 834–839.

87. C. W. Garland and J. R. Wilt, Infrared spectra and dipole moments of $Rh_2(CO)_4Cl_2$ and $Rh_2(CO)_4BR_2$, *J. Chem. Phys.* **36**, 1094–1095 (1962).
88. C. W. Garland, R. C. Lord, and P. F. Troiano, An infrared study of high-area metal films evaporated in carbon monoxide, *J. Phys. Chem.* **69**, 1188–1195 (1965).
89. R. M. Kroeker, W. C. Kaska, and P. K. Hansma, Sulfur modifies the chemisorption of carbon monoxide on rhodium/alumina model catalysts, *J. Catal.* **63**, 487–490 (1980).
90. J. R. Kirtley and P. K. Hansma, Effect of the second metal electrode on vibrational spectra in inelastic-electron tunneling spectroscopy, *Phys. Rev. B* **12**, 531–536 (1975).
91. A. Bayman, P. K. Hansma, and W. C. Kaska, Shifts and dips in inelastic electron tunneling spectra due to the tunnel junction environment, *Phys. Rev. B*, **25**, 2449–2455 (1981).
92. R. M. Kroeker and P. K. Hansma, unpublished observations.
93. L. H. Dubois and D. L. Allara, unpublished observations.
94. T. M. Duncan, J. T. Yates, Jr., and R. W. Vaughan, ^{13}C NMR of CO chemisorbed onto dispersed rhodium, *J. Chem. Phys.* **71**, 3129–3130 (1979).
95. T. M. Duncan, J. T. Yates, Jr., and R. W. Vaughan, A ^{13}C NMR study of the adsorbed states of CO on Rh dispersed on Al_2O_3, *J. Chem. Phys.* **73**, 975–985 (1980).
96. R. A. Toth, R. H. Hunt, and E. K. Plyler, Line intensities in the 3-0 band of CO and dipole moment matrix elements for the CO molecule, *J. Mol. Spectrosc.* **32**, 85–96 (1969).
97. T. M. Duncan, The nature of molecules adsorbed on catalytic surfaces: Pulsed nuclear magnetic resonance and infrared absorbance studies, Ph.D. thesis, California Institute of Technology, 1980.
98. E. R. Corey, L. F. Dahl, and W. Beck, $Rh_6(CO)_{16}$ and its identity with previously reported $Rh_4(CO)_{11}$, *J. Am. Chem. Soc.* **85**, 1202–1203 (1963).
99. O. S. Mills and E. F. Paulus, Trimeric Pi-cyclopentadienyl-carbonylrhodium, *J. Chem. Soc. Chem. Commun.* 815–816 (1966).
100. L. F. Dahl, C. Martell, and D. L. Wampler, Structure and metal–metal bonding in $Rh_2(CO)_4Cl_2$, *J. Am. Chem. Soc.* **83**, 1762–1763 (1961).
101. M. A. Van Hove, L. H. Dubois, R. J. Koestner, and G. A. Samorjai, The structure of CO, CO_2 and small hydrocarbons (acetylene, ethylene, propylene, methylacetylene) adsorbed on Rh(111) and Pt(111) studied by LEED and HREELS, Supplement Le Vide, *Les Couches Minces* **201**, 287–290 (1980).
102. R. M. Kroeker, P. K. Hansma, and W. C. Kaska, Low energy vibrational modes of carbon monoxide on iron, *J. Chem. Phys.* **72**, 4845–4852 (1980).
103. W. Braun, M. Neumann, M. Iwan, and E. E. Koch, Energy level shifts of CO chemisorbed and condensed on Rh(111), *Solid State Commun.* **27**, 155–158 (1978).
104. H. Conrad, G. Ertl, H. Knözinger, J. Küppers, and E. E. Latta, Polynuclear metal carbonyl compounds and chemisorption of CO on transition metal surfaces, *Chem. Phys. Lett.* **42**, 115–118 (1976).
105. Y. Takasu and A. M. Bradshaw, in *Chemical Physics of Solids and Their Surfaces* (M. W. Roberts and J. M. Thomas, eds.), Vol. 7, pp. 59–88, The Chemical Society, London (1978).

7

The Detection and Identification of Biochemicals

Robert V. Coleman

1. Introduction

In this chapter, we review the application of tunneling spectroscopy to the study of biological molecules. These studies have included molecules with a wide range of molecular weights and include amino acids, bases, nucleotides, nucleic acids, and proteins. These compounds represent the important building blocks of biological materials, and detailed descriptions of them can be found in standard texts on biochemistry.[1,2] Using inelastic electron tunneling spectroscopy (IETS) techniques to examine these molecules involves the same problems as encountered in many other IETS studies, although resolution problems associated with molecules of large molecular weight play an increasingly important role.

The research up to the present time has involved the accumulation of data and comparison of the IET spectra to known infrared and Raman data. Initial applications to studies such as radiation damage have been carried out, but major applications to biological experiments are still in the exploratory development stage.

This chapter will present the IET spectra obtained from biological molecules and will review the agreement with data obtained from infrared and Raman spectra of the same molecules. The adsorption of the molecule on the oxide substrate is an important factor in determining the intensity

---

Robert V. Coleman • Department of Physics, University of Virginia, Charlottesville, Virginia 22901. Experimental work supported in part by the National Science Foundation and the Department of Energy.

and resolution which can be obtained from a given molecule. Chemical reactions between the molecular side groups and the oxide surface affect both the frequency and intensity of the IETS modes, while surface orientation and molecular configuration also influence IETS mode intensity. These factors are difficult to analyze as the molecule becomes larger and more complex, but IETS applied to biological compounds offers an interesting potential for learning more about such factors.

The spectra presented here show that highly satisfactory IET spectra can be obtained from many biological molecules while the general techniques can be applied to almost any biological compound. The use of the techniques as a tool in biological research is still being evaluated, and this review is designed to aid the reader in his own assessment and evaluation of possible applications.

Table 1. Selected Amino Acids

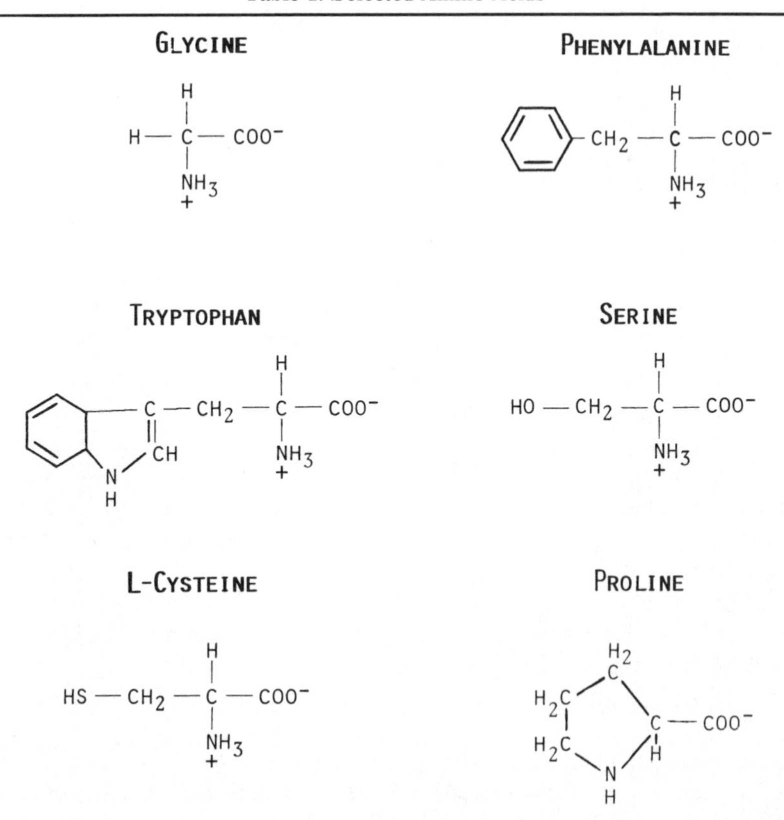

2. IET Spectra of Biological Compounds

2.1. Amino Acids

Some of the earliest IETS work on biological molecules was done with amino acids.[3–5] These compounds represent a range of molecular weights and, as far as IETS is concerned, represent a bridge between the problems encountered with small molecules and those experienced for much larger molecules, such as polyamino acids, proteins, and nucleic acids. Table 1 shows the composition of selected amino acids for reference.

The IET spectra exhibited by small α-amino acids such as glycine can readily be compared with the corresponding infrared or Raman spectra and precise correspondence between modes observed in all three spectroscopies can be established. The IET spectrum of glycine obtained by Simonsen

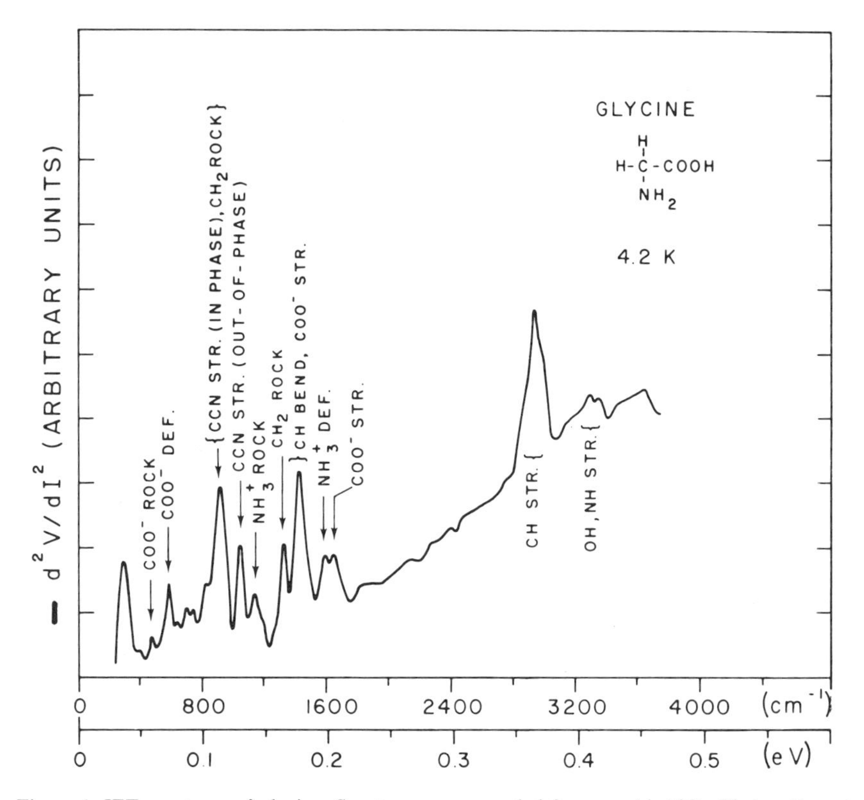

Figure 1. IET spectrum of glycine. Spectrum was recorded from an Al–AlO_x–Pb junction at 4.2 K. Junction was doped using an H_2O solution. (From reference 5.)

et al.[5] is shown in Figure 1 with mode identifications taken from a detailed infrared study by Tsuboi *et al.*[6]

The α-amino acids, either in the solid state or at their isoelectric point in aqueous solution, exist almost entirely as dipolar ions[7] and various modes are identified in the infrared analysis with the COO^- (carboxyl ion) and the NH_3^+ ion. The IET spectra for glycine and other α-amino acids show corresponding modes at the same frequencies and appear to exist as dipolar ions when adsorbed on the aluminum oxide barrier of the tunnel junction.[5] This is consistent with the fact that almost all molecules containing a COOH group adsorb on alumina by donating a proton to the surface and forming a CO_2^- ion. The surface adsorbed state of the α-amino acids

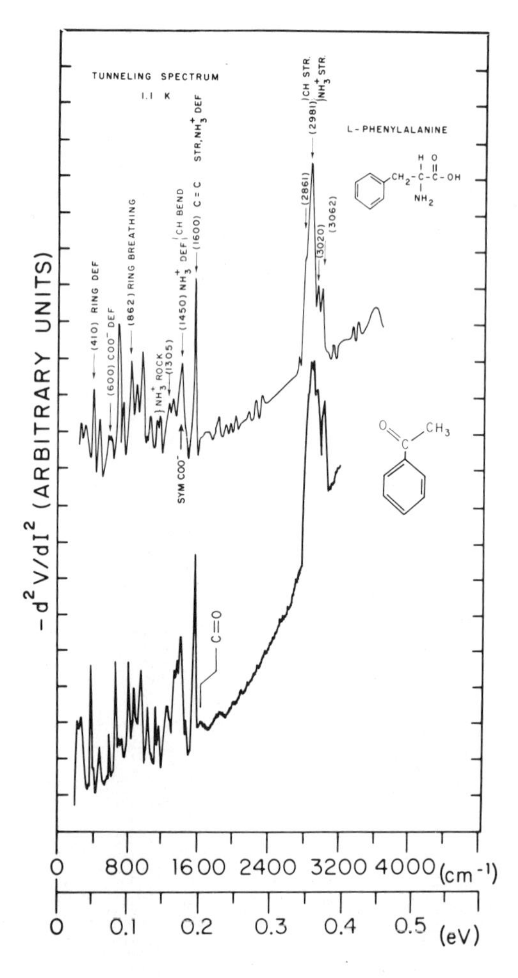

Figure 2. IET spectra of L-phenylalanine (200 Ω) (H_2O solution) (upper curve) and aceotphenone (800 Ω) (ethyl alcohol solution) (lower curve). The two spectra show a strong similarity in both mode energy and relative intensity. Reproduced from *Inelastic Electron Tunneling Spectroscopy*, p. 36 (T. Wolfram, ed.), Springer-Verlag, New York (1978).

studied by IETS therefore corresponds closely to the solid state or solution forms which have been previously studied by infrared and Raman spectroscopy.

The strong IETS modes have broad natural linewidths and closely spaced vibrational modes of markedly different intensity are often difficult to resolve. Therefore, broad, intense IETS peaks can have intensity contributions from a number of different vibrations.

In the case of amino acids containing aromatic rings, the IET spectrum is often dominated by the vibrational modes of the ring. An example of the IET spectrum obtained for L-phenylalanine is shown in Figure 2. This is compared to the IET spectrum of acetophenone, also shown in Figure 2. The majority of the modes observed for each molecule show identical frequencies and relative mode intensities. Therefore, the strongest modes can reasonably be identified with the fundamental aromatic ring modes. The side groups of each molecule interact with the surface and contribute modified modes such as the COO^- stretching modes in L-phenylalanine and the strongly depressed C=O stretch mode in acetophenone.

In a number of experiments, IETS has been extended to polyamino acids containing from two to five separate amino acids. An IET spectrum of L-tryptophyl–L-phenylalanine is shown in Figure 3. The number of strong modes observed is approximately the same as the number present in the

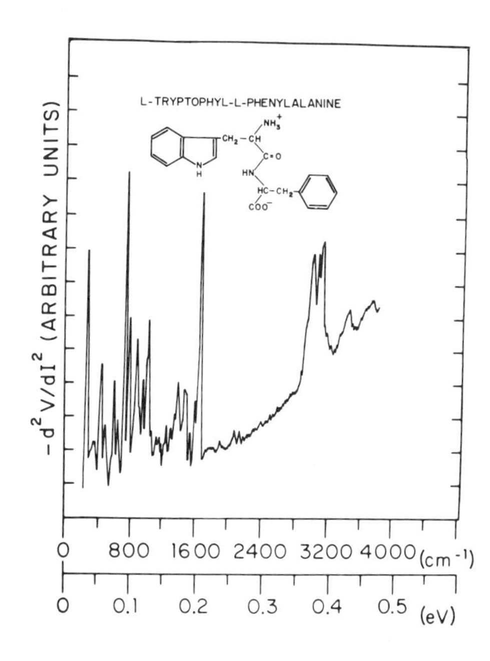

Figure 3. IET spectrum of the dipeptide L-tryptophyl–L-phenylalanine. Spectrum obtained at 4.2 K from an Al–AlO_x–Pb junction doped from a solution of peptide, NH_4OH, and H_2O (810 Ω).

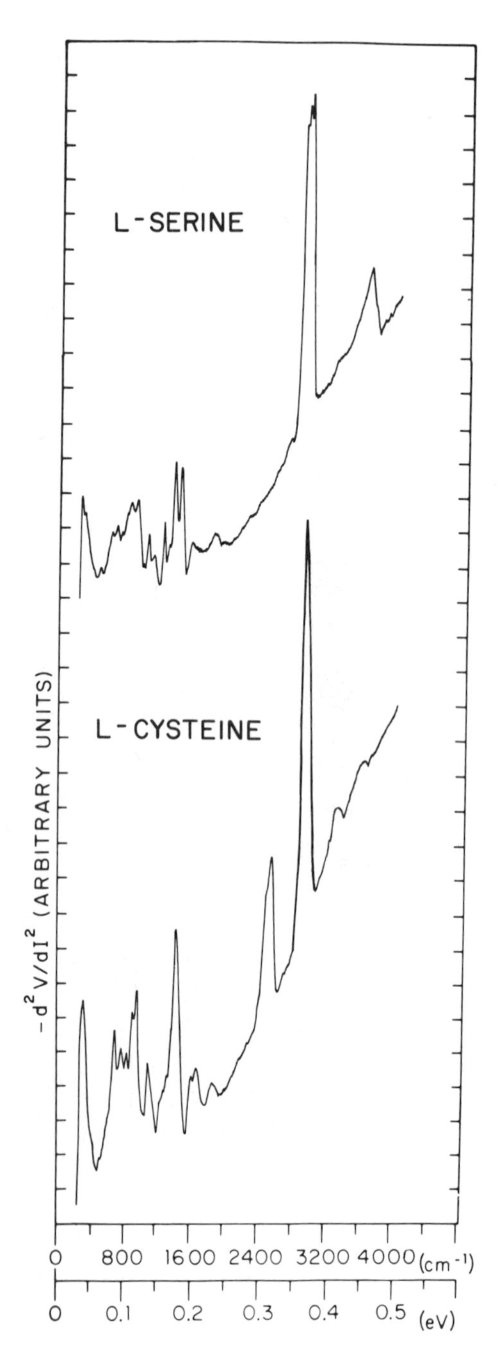

Figure 4. IET spectra of L-serine (158 Ω) (upper curve) and L-cysteine (153 Ω) (lower curve). Both spectra were recorded at 4.2 K from Al–AlO_x–Pb junctions doped from H_2O solutions. Strong S–H stretching mode is observed near 2600 cm^{-1} in L-cysteine.

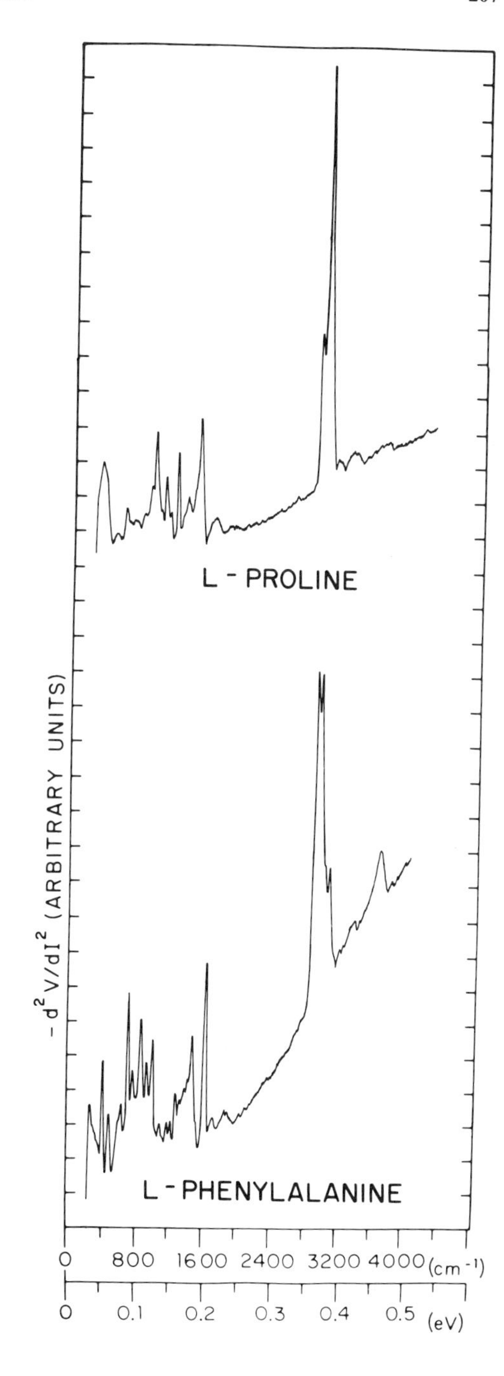

Figure 5. IET spectra of L-proline (320 Ω) (upper curve) and L-phenylalanine (320 Ω) (lower curve). Both spectra were recorded at 4.2 K from Al–AlO_x–Pb junctions doped from H_2O solutions. CH modes dominate in the spectrum of proline, while aromatic ring modes dominate in the spectrum of L-phenylalanine.

spectrum of L-phenylalanine. However, shifts in relative intensity and frequency occur between the two spectra and they are easily distinguishable. Analysis again suggests that the aromatic ring modes are dominant and that overlapping modes from the various constituents of the molecule are unresolved within the broad envelopes of the main peaks.

Several more examples of α-amino acid spectra are shown in Figures 4 and 5. The IETS spectra clearly differentiate between amino acids with different R groups but containing the common carboxyl group and the free, unsubstituted amino group on the α-carbon (see Table 1). Special bonds such as the S–H stretching mode in L-cysteine near 2600 cm^{-1} are easily detected. Proline (Figure 5), which does not contain the common amino

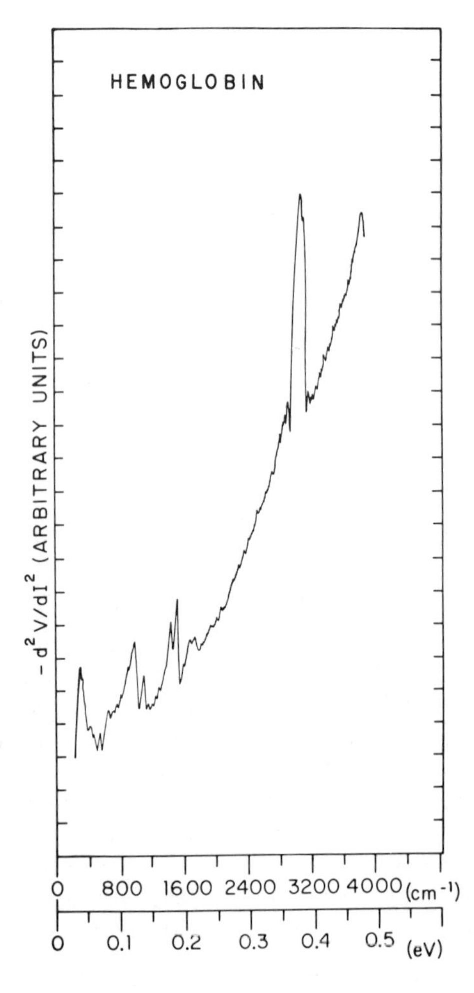

Figure 6. IET spectrum of hemoglobin (3520 Ω). Spectrum was recorded at 4.2 K from an Al–AlO$_x$–Pb junction doped from an H_2O solution. Strong CH modes dominate the spectrum.

group on the α-carbon, shows a substantial change in relative mode intensity, but exhibits modes very close in energy to those observed in L-serine (Figure 4). These spectra are both dominated by strong CH and CH_2 mode contributions.

The extension of IETS to peptides and proteins, for which amino acids are the building blocks, has not been explored systematically. A few preliminary experiments indicate that resolution of individual modes becomes very difficult and that only a broad envelope is recorded in the IET spectrum. This is particularly true of globular proteins such as myoglobin and hemoglobin, where IET spectra appear to show strong resolved modes only in the CH regions. This is evident in the IET spectrum of hemoglobin, as shown in Figure 6.

In globular proteins, the α-helical regions or domains are connected by nonhelical regions, and this permits the polypeptide to be wound into a three-dimensional structure 30 Å or more in diameter. This contributes substantially to the overall tunneling barrier resistance, but the tunneling electrons may not penetrate the molecule sufficiently to couple strongly to the individual molecular groups. In the absence of strong coupling to fundamental ring modes, etc., the modes due to small CH fragments in the tunneling electron path may show an enhanced relative intensity.

2.2. Pyrimidine and Purine Bases

Excellent IET spectra can be obtained from the pyrimidine bases adsorbed on the alumina barrier of Al–AlO_x–Pb tunnel junctions. Typical spectra are shown in Figures 7, 8, and 9 for uracil, thymine, and cytosine. Intense, well-resolved ring modes are observed which are typical of the IET spectra observed for single aromatic ring compounds where a specific site or side group can adsorb on the alumina substrate.

For the bases, the most probable mechanism for bonding is coordination at the aluminum ion site on the surface by the ring nitrogen or oxygen substituent. Studies of the bonding of heavy metal ions to pyrimidine bases in solution indicate that coordinate bonding occurs at the ring nitrogen (the most basic site). Steric considerations imposed by the surface can also make the oxygen substituent a possible candidate for the adsorption interaction.

The junctions were prepared by doping from an aqueous solution of the corresponding base. The poor solubility properties of the bases in water is the major limitation for application of the IETS method and only relatively low resistance junctions can be obtained for most bases. Nevertheless, the IET spectra show strong intensity, and systematic comparisons to Raman and infrared data can be made for most of the bases occurring in common nucleotides.

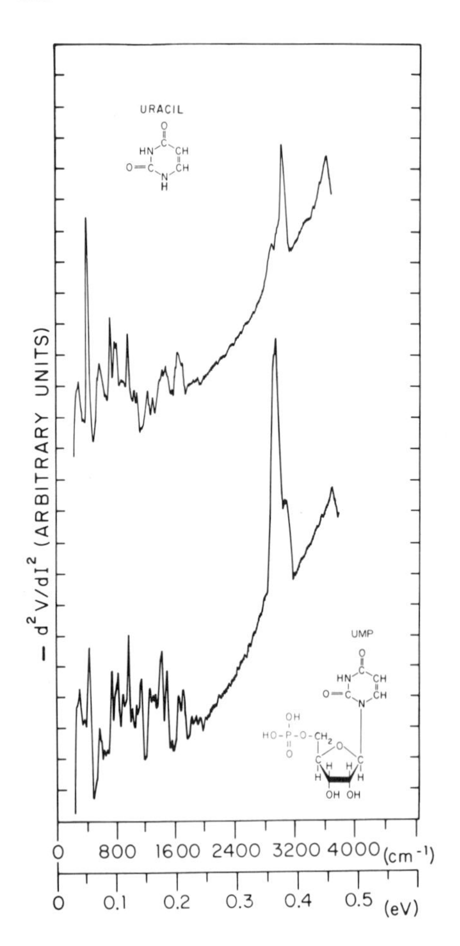

Figure 7. IET spectra of 5′-UMP (695 Ω) (lower spectrum) and uracil (56 Ω) (upper curve). Both spectra were recorded at 4.2 K from Al–AlO_x–Pb junctions doped from H_2O solutions. (From reference 8.)

Clark and Coleman[8] have made a detailed comparison of the IETS modes observed for the bases uracil, thymine, and cytosine to the infrared and Raman data, as reported by Susi and Ard,[9] Lord and Thomas,[10] and Angell.[11] The systematic comparison of mode energies shows that within $\pm 10\ cm^{-1}$, all of the major IETS modes below 1700 cm^{-1} correlate closely with similar modes observed in the Raman and infrared spectra. Intense modes in the IET spectra can be assigned to ring deformation modes and other ring vibrations, as well as to vibrations of specific C–H and N–H groups. The identifications and assignments used in the Raman and infrared analysis work out equally well in the IETS analysis, as summarized by Clark and Coleman.[8]

The number of C–H stretching modes above 2800 cm^{-1} observed in the IET spectra of the basis is comparable to those observed in Raman and

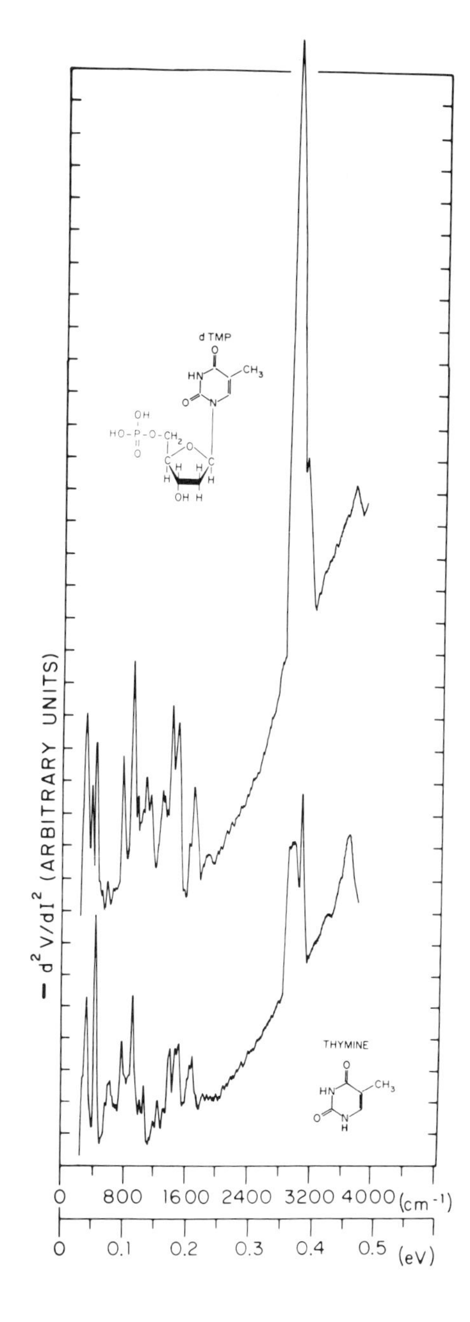

Figure 8. IET spectra of 5′-dTMP (539 Ω) (upper spectrum) and thymine (50 Ω) (lower spectrum). Both spectra were recorded using Al–AlO_x–Pb junctions doped from H_2O solutions. Spectra were recorded with the junction at 4.2 K. (From reference 8.)

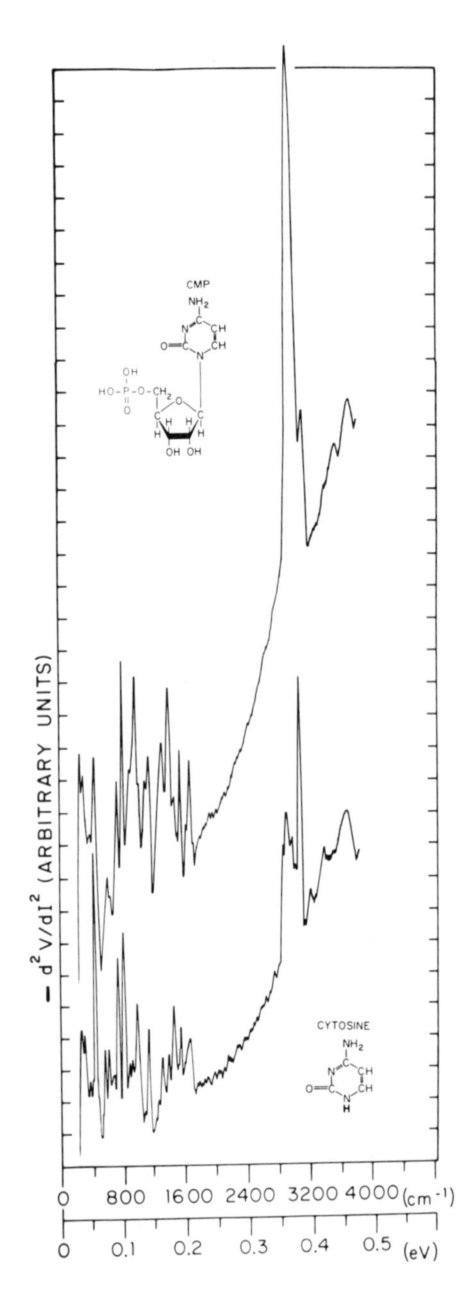

Figure 9. IET spectra of 5′-CMP (1274 Ω) (upper spectrum) and cytosine (932 Ω) (lower spectrum). Both spectra were recorded at 4.2 K from Al–AlO_x–Pb junctions doped from H_2O solutions. (From reference 8.)

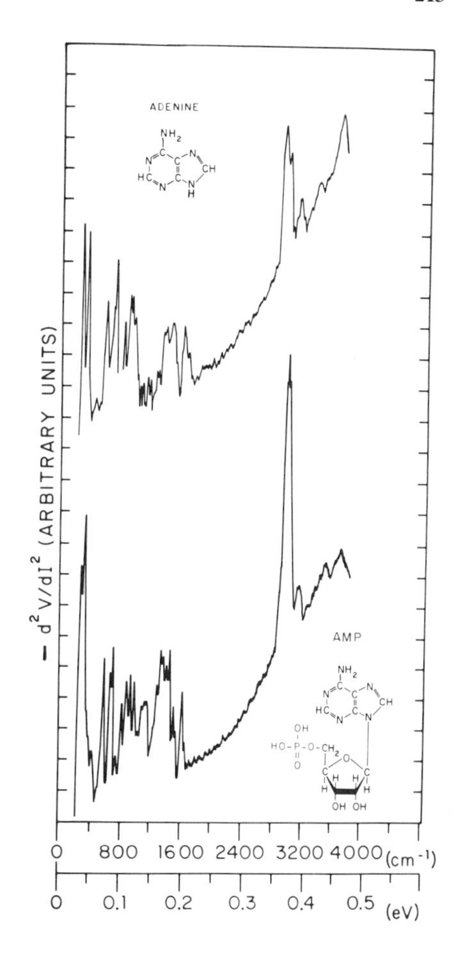

Figure 10. IET spectra of 5′-AMP (1818 Ω) (lower spectrum) and adenine (498 Ω) (upper curve). Both spectra were recorded at 4.2 K from Al–AlO_x–Pb junctions doped from H_2O solutions. (From reference 8.)

infrared spectra, but energy shifts on the order of 50 cm^{-1} are observed. These shifts may be associated with hydrogen bonding due to surface interactions or to cover electrode effects, but no definite conclusions can be made at this time.

The purine bases are somewhat more difficult to prepare for observation of IET spectra, and so far only adenine has been studied in detail. Figure 10 shows a reasonably well-resolved IET spectrum of adenine and a systematic comparison with Raman and infrared results has been carried out by Clark and Coleman.[8] The major IETS modes below 1700 cm^{-1} again correlate closely with the mode energies observed in Raman and infrared spectra. Corresponding modes occur within ± 10 cm^{-1} for all three spectroscopies. The mode assignments are consistent with the work of Lord and Thomas[10] and Angell[11] and are associated with coupled C=C and

C—N vibrations, ring and C—H bending vibrations, and C—H deformation vibrations. A strong NH_2 scissoring vibration is also resolved at 1645 cm^{-1} The C–H and N–H stretching modes above 2800 cm^{-1} again show substantial shifts on the order of 50 cm^{-1} when compared to the corresponding infrared modes.

2.3. Nucleotides and Nucleosides

The solubility and adsorption properties of the nucleotides are substantially better than the corresponding base residues, and IETS junctions are easily fabricated with resistances in the range 500–1000 Ω. The presence of the sugar and phosphate moieties stabilizes the surface adsorption properties and junction resistances are consistently reproduced for a given concentration of the dopant solution. The nucleosides which consist of only the base and sugar can be used to fabricate equally good IETS junctions, although less work has been done on them. The first IET spectra of nucleotides were obtained by Simonsen *et al.*,[5] while more extensive IETS studies were carried out by Clark and Coleman.[8]

The main vibrational modes observed in the nucleotide IET spectra show dominant contributions from the modes of the base ring. Alterations introduced by addition of the sugar–phosphate groups, aside from the specific mode contributions of these groups, are limited to small shifts in the vibrational energies and variations in the relative peak intensities. The shifts of the base ring modes as observed for the pure base versus the mononucleotide are on the order of 10 cm^{-1} and are similar to the shifts observed in the infrared and Raman spectra of these molecules.[9–11] IET spectra of 5′-UMP, 5′-dTMP, 5′-CMP, and 5′-AMP were also included in Figures 7–10, which showed spectra of the base residues. A spectrum of 5-GMP is shown in Figure 11.

The lower-lying modes of the nucleotide spectra are contributed by the base ring, while specific mode contributions of the sugar groups appear in well-defined regions of the spectrum. Coupled C–O and C–C stretch modes are observed in the range 1000–1125 cm^{-1}, while C–H and CH_2 bending vibrations contribute in the range 1360–1460 cm^{-1} and are superimposed on the base ring modes in the same range. Very strong aliphatic C–H stretching modes are present near 2900 cm^{-1} and the aromatic C–H stretching mode is observed above 3000 cm^{-1}. In addition to the modes summarized above, broad peaks associated with OH stretching vibrations are observed in the range 3400–3700 cm^{-1}. As shown in the IET spectrum of D-ribose in Figure 12, two main peaks are resolved in the OH stretching region. The maximum at 3476 cm^{-1} is identified with the OH groups on the

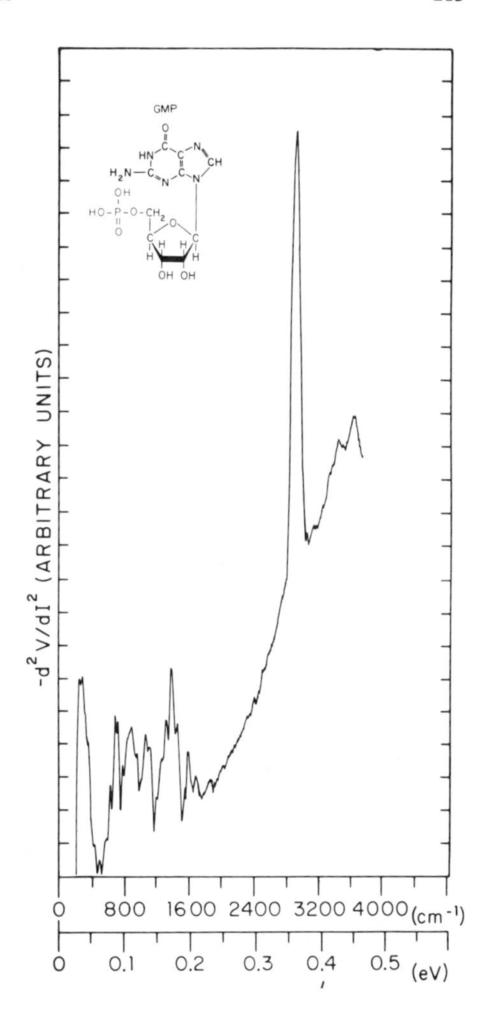

Figure 11. IET spectrum of 5′-GMP (1302 Ω). Spectrum was recorded at 4.2 K from an Al–AlO_x–Pb junction doped from an H_2O solution. (From reference 8.)

sugar, while the peak at 3641 cm^{-1} is identified with the OH groups on the alumina tunneling barrier.

In aqueous solutions above pH 7, the mononucleotides show vibrations attributed to the phosphomonoester group $ROPO_3^{2-}$ at 989 and 1100 cm^{-1}, as confirmed by both Raman and infrared spectra. The modes observed in this region for the IET spectra of the mononucleotides are also observed in the nucleosides and the IET spectra of both compounds are almost identical, as shown in studies by Clark and Coleman.[12] The nucleotides are generally doped from solutions at or above pH 7, so that the absence of the

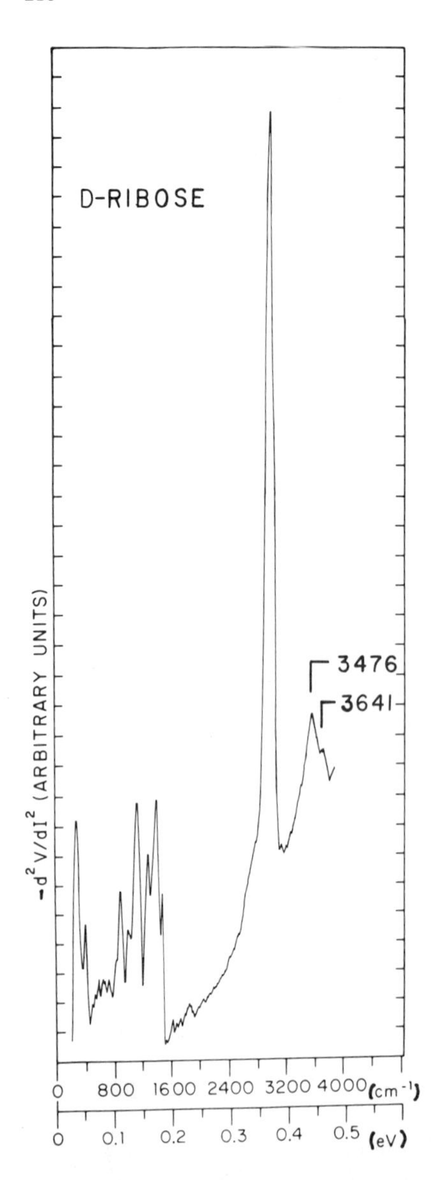

Figure 12. IET spectrum of D-ribose (390 Ω). Spectrum was recorded at 4.2 K from an Al–AlO_x–Pb junction doped from an H_2O solution. Strong peak at 3476 cm^{-1} is due to OH group of the sugar. Shoulder at 3641 cm^{-1} is associated with the surface OH groups on the aluminum oxide. (From reference 8.)

modes suggests that the $-PO_3^{2-}$ moiety is protonated on the surface or has reacted with the exposed Al^{3+} ions.

The ability of IETS to distinguish between various derivative molecules containing adenine has been tested by Clark and Coleman.[12] Eight different adenine derivatives, including adenosine, deoxyadenosine, 5′-AMP, 5′-ADP, 5′-ATP, 3′:5′-CAMP, and ApA were compared using IET spectra. In the range 550–1600 cm^{-1} there are 12 dominant vibrational modes which occur at the same energy to within approximately ± 5 cm^{-1} for all eight derivatives. The nine-membered purine ring, represented by adenine, has 15 fundamental in-plane ring vibrations as well as CH in-plane and out-of-plane vibrations in the range 550–1600 cm^{-1}. The 12 IETS modes common to all adenine derivatives are characteristic of the adenine base ring, while additional IETS modes exhibit energy shifts which can be used to distinguish between the various derivatives.

The most sensitive IETS modes lie below 550 cm^{-1} and high-sensitivity spectra for the eight adenine derivatives are shown in Figure 13. These low-lying modes clearly shift in energy according to the specific molecule and give an unambiguous identification of the various derivatives. These

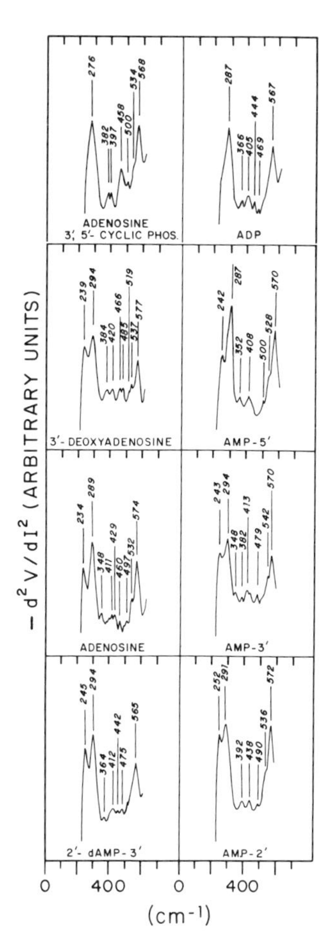

Figure 13. IET spectra below 600 cm^{-1} for selected nucleotides containing the base adenine. (From reference 12).

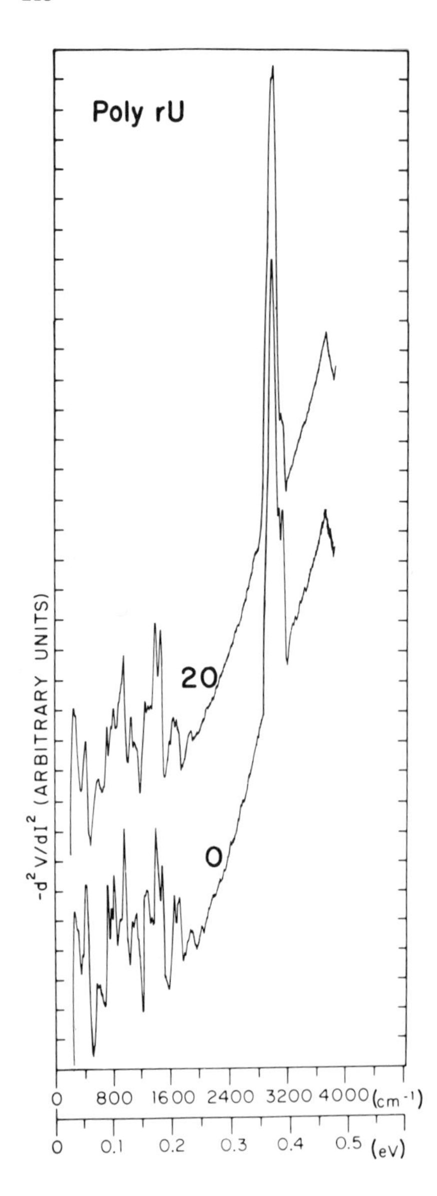

Figure 14. IET spectrum of polyuridylic acid, including uv radiation spectrum. (O) Control spectrum, 689 Ω; (20) 20-min exposure, 500 Ω. Damage is similar to that observed for the mononucleotide 5′-UMP. No dimer formation is observed. (From reference 8.)

low-lying ring modes are more strongly influenced by changes in the adsorption of side groups or changes in the surface configuration and molecular bond energies induced by changes in the attachment sites of the sugar and phosphate groups.

Preliminary results of IET spectroscopy on polynucleotides and polynucleosides indicate that reasonably well-resolved spectra can be obtained.

The number of intense modes resolved remains comparable to the number observed in the mononucleotides and suggests that the IET spectrum is unaffected by using the polymeric form of the molecule. An example of the IET spectrum obtained for polyuridine is shown in Figure 14 and can be compared to the spectrum of 5′-UMP shown in Figure 7. The two spectra show almost identical mode energies and relative intensities.

Attempts to obtain useful IET spectra from nucleic acids have been more successful than with proteins, although resolution of individual modes becomes more difficult. IET spectra of DNA and RNA were first published by Hansma and Coleman.(13) The early IET spectra demonstrated that the technique would work, but clearly indicated that improved resolution would be a problem. More recent work has made some progress on this problem, but more improvement will be necessary before significant application of the technique can be made. IET spectra obtained from DNA and RNA are shown in Figure 15 and represent amplified spectra in the range up to 2000 cm^{-1}. A detailed comparison of the weak IETS modes superimposed on the larger spectral envelope to the Raman modes observed for solid calf thymus DNA shows a surprisingly good correlation, as shown in Table 2. A comparable number of modes are observed in both spectroscopies and the majority of them agree within less than 10 cm^{-1}. The tendency of the IET spectrum to form an envelope dominated by strong CH modes is already evident, however, and this will limit resolution, as was also the case for the globular proteins.

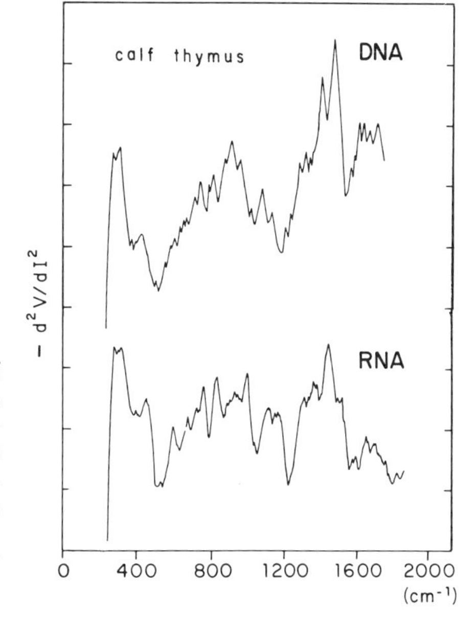

Figure 15. Amplified IET spectra of calf thymus DNA (2000 Ω) and yeast RNA (1113 Ω). Spectra were recorded from Al–AlO_x–Pb junctions at 4.2 K. Mode energies observed for DNA are compared to Raman mode energies in Table 2. Reproduced from *Inelastic Electron Tunneling Spectroscopy*, p. 49 (T. Wolfram, ed.), Springer-Verlag, New York, (1978).

Table 2. Wave Numbers for IETS and Raman Spectra of Calf Thymus DNA[a]

IETS (cm^{-1})	Raman solid (cm^{-1})	IETS (cm^{-1})	Raman solid (cm^{-1})
263		1000 m	1013 m
295		1058 s	1062 m
368		1113 m	
415 m	415 w		1144 vw
494 w	496 s	1189 m	1183 m
566 w	570 w	1215 m	1208 m
592 w	597 w	1258 m	1247 s
621 w	625 w	1294 m	1306 s
637 w		1315	
657 w	666 m		1335 s
	683 w	1371 s	1372 s
698 m		1447 s	1449 w
729 s	731 s	1537 m	1537 w
776 sh		1557 m	
795 s	786 vs	1581 ms	1586 s
858 m		1608 m	1612 w
889 s	873	1639 m	
940 m		1682 ms	1670 s
	963 w		

[a]Abbreviations: s = strong; m = medium; w = weak; sh = shoulder; underline indicates same mode in IETS and Raman.

3. Surface Adsorption and Orientation Effects on the IETS of Nucleotides

In all applications of IETS, the interaction of the molecule with the barrier substrate and the overlaid metal electrode can have either direct or indirect effects on the spectrum. In the case of nucleotides this problem has been examined, particularly with respect to the application of IETS to the study of uv-radiation damage, to be discussed in more detail in Section 4.

The base and sugar rings in nucleotides are connected by a glycosidic bond to form a compound which can constrain the orientation of the rings relative to the surface. Various studies[14] involving x-ray diffraction and other methods show that the sugar ring can rotate about the glycosidic bond and can assume either of two positions relative to the base approximately 150° apart. At either of these positions, the relative sugar position can change within a $\sim 45°$ range in response to interactions with the base ring. Further details are given in an article by Tso.[14]

The constraints on the geometry of the complex will, in general, orient both rings at an angle to the surface and stabilize the relative surface orientation of the molecule. Indirect information on the stability of the

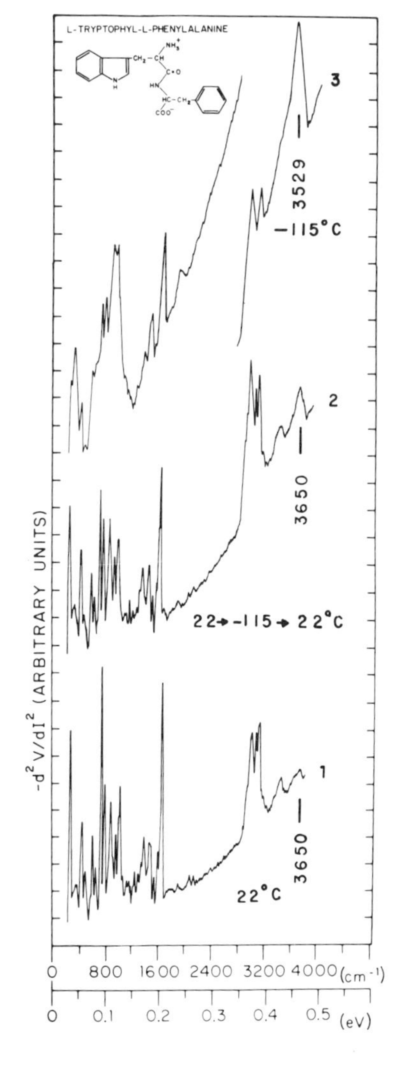

Figure 16. IET spectra of the dipeptide L-tryptophyl–L-phenylalanine. (1) Spectrum obtained after evaporation of the Pb electrode at 22°C, 810 Ω. (2) Doped substrate cooled to −115°C and then reheated to 22°C before evaporation of the Pb electrode, 258 Ω. (3) Doped substrate cooled to −115°C before evaporation of the Pb electrode, 270 Ω. This spectrum shows a reduced ring mode intensity and an enhanced and downshifted surface O–H stretching mode at 3529 cm^{-1}. All spectra were recorded at 4.2 K from Al–AlO_x–Pb junctions doped from a solution of peptide, NH_4OH, and H_2O. (From reference 8.)

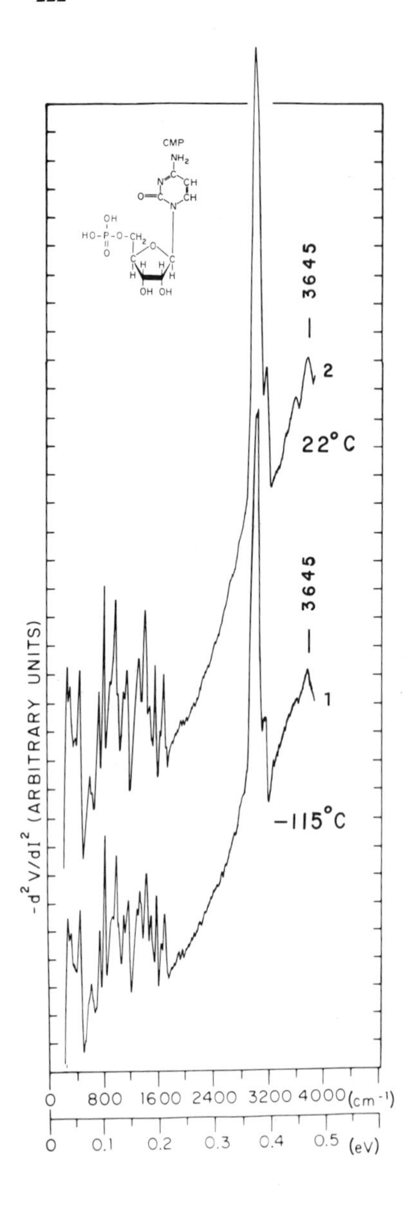

Figure 17. IET spectra of 5′-CMP obtained with Pb electrode evaporated at −115°C (1) and at 22°C (2). The two spectra are nearly identical, with no downshift of the OH mode and no intensity reduction for the cooled substrate. (From reference 8.)

compound has been obtained in the uv radiation experiments which require cooling of the substrate in order to avoid heating during radiation exposure.

The IETS mode intensity of single-ring compounds or unconstrained multiring compounds can be nearly quenched by cooling the substrate to −120°C before evaporation of the overlaid metal electrode. For example, amino acid complexes show very large substrate cooling effects, as shown for L-tryptophyl–L-phenylalanine in Figure 16.

The effects of substrate cooling on the IETS intensity are connected with the equilibrium OH structure on the alumina surface which changes as the substrate is cooled. This change in OH structure is accompanied by a downshift of the O–H stretching modes and an enhancement of the O–H stretching mode intensity, as indicated in Figure 16. Neither of these mode changes are observed for cooled substrates doped with nucleotides, as shown in Figure 17 for 5′-CMP prepared at room temperature and at $-115°C$. Both spectra show identical modes and relative intensities. More detailed studies of the effects of the OH structure on the barrier characteristics and IET spectra can be found in a paper by Dragoset *et al.*[15] The point is that the nucleotide IET spectra obtained from the uv radiation experiments discussed in the next section are completely unperturbed by the required substrate cooling.

4. uv Radiation Damage Studies with IETS

One of the applications of IETS which immediately suggests itself is the study of radiation damage to molecules, which will be reviewed in Chapter 9. In the case of biological molecules, the technique has been applied by Clark and Coleman[8] to the study of uv radiation damage in nucleotides. As outlined in Section 3, the unique surface-molecule-interface behavior observed for the nucleotides tends to ensure a straightforward observation of primary bond damage, and this section will briefly describe the results of such an IETS application.

Except for 5′-AMP, the lowest singlet state in the nucleotides is a $\pi\pi^*$ state, which appears as a strong band in the uv adsorption spectrum peaked at 2600–2800 Å. In AMP, the lowest singlet state is a $n\pi^*$ state which appears slightly to the red of a strong $\pi\pi^*$ singlet band, with a maximum at 2600 Å. In the sugar moiety, the adsorption band peaks at 1800 Å in the vacuum ultraviolet.

The junction substrate of Al–AlO_x doped with the nucleotide was placed on a liquid-nitrogen-cooled copper mount in a vacuum of 10^{-9} Torr. The uv source was a 1000-W xenon–mercury arc lamp with the slightly defocused output of the lamp directed full strength on the barrier and adsorbed molecules after passing through an infrared filter. The radiation fluences were calculated to be 10 J/cm^2 for a 1-min exposure and 600 J/cm^2 for a 60-min exposure. The average photon wavelength was estimated to be ~ 2500 Å, so that a 1-min exposure would correspond to a fluence of $\sim 1.26 \times 10^3$ photons/$Å^2$. Immediately after finishing the radiation exposure, the Pb electrode was evaporated onto the junction, which was then warmed to room temperature and removed from the vacuum system for mounting in the helium dewar and measurement of the IET spectrum.

The subsequent evaporation of the Pb electrode shields the irradiated molecules from further exposure to oxygen and water vapor. In this respect, the tunnel junction fabrication acts as a type of matrix isolation technique, allowing the results of primary radiation damage to be studied with reduced interference from secondary reaction of the radiation products with each other and with the surroundings.

Typical results are represented by spectra of 5′-CMP, as shown in Figure 18, which cover uv exposure times of 0, 10, 30, 45, and 60 min. The nucleotides containing pyrimidine bases damage more rapidly than those containing purine bases and all show the same type of major damage. A uv exposure time is increased, selected peaks show a substantial reduction in intensity with the strongest reductions occurring for peaks assigned to the base ring modes and to C–H out-of-plane bending modes. The strong ring mode below 500 cm^{-1} shows a rapid reduction in intensity and indicates a disruption of the pyrimidine ring in the initial stages of the damage sequence. The double bond modes are less damaged by uv radiation, and in the case of nucleotides containing purine bases, the growth of several double bond modes near 1600 cm^{-1} is observed during irradiation. At large radiation exposure all of the IET spectra are dominated by the bending and stretching vibrations of CH and CH_2 groups and by strong alumina phonon modes.

The sensitivity to uv radiation damage was measured for five different nucleotides and the relative order of damage rates are listed below:

$$\text{UMP} > \text{dTMP} > \text{CMP} > \text{GMP} > \text{AMP}$$

UMP is the most sensitive to uv radiation damage, while AMP is the least sensitive. This relative order of uv radiation resistance is directly correlated with the relative stability of the base residues as measured by the delocalization energy per π electron. A higher resonance energy implies a greater delocalization of the π electrons and a higher probability of converting the radiation excitation energy into nondissociative channels. The radiation resistance, as measured by IETS, follows exactly the same order as the resonance energy calculated for the ground and first excited states of the purine and pyrimidine base residues of the respective nucleotides.

Further details of the uv damage experiments on nucleotides and bases can be found in reference 8. The brief review given here shows that the IETS technique can yield quite reasonable results and that the various preparation and measurement steps carry through without complicating the interpretation.

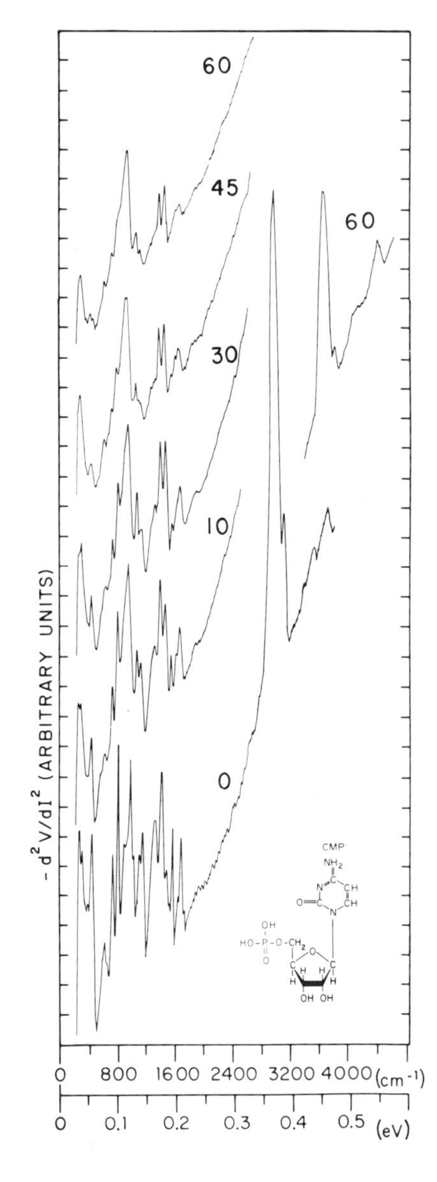

Figure 18. Ultraviolet radiation series for 5′-CMP. Junction substrates were held at ~ −100°C during irradiation and evaporation of the Pb electrode. Spectra were recorded at 4.2 K from junctions doped from H_2O solutions. (O) Control spectrum, 450 Ω; (10) 10-min exposure, 166 Ω; (30) 30-min exposure, 117 Ω; (45) 45-min exposure, 160 Ω; (60) 60-min exposure, 90 Ω. (From reference 8.)

The techniques outlined above would not detect radiation effects resulting from an aqueous environment, as is present for many biological systems. Such effects could be explored by irradiating the compounds before preparation of the tunnel junction, but the appropriate sequence of steps has not been worked out.

5. Conclusions

The IET spectra reviewed in this chapter are representative of the constituents of most biological systems and demonstrate that adequate resolution can be obtained for the majority of these compounds. The presentation of the spectra and the comparison to infrared and Raman data have established that systematic interpretation of the modes is possible and, except for the modes arising from specific surface-reacted groups, the IETS modes show a close correspondence to the vibrational modes observed in infrared and Raman spectra of solutions and solids. Applications of IETS to biological experiments are still in the development stage and considerably more exploratory work must be done. An application to uv radiation damage in nucleotides has been carried out and the results show that the experimental steps and interpretations work out very well. Important conclusions about the application of IETS to biological compounds are as follows:

(1) Tunneling spectra of the basic building blocks of biological systems, such as amino acids and nucleotides, give excellent IETS results.

(2) All α-amino acids can be easily distinguished by IETS.

(3) Nucleotides containing different bases are easily distinguished using IETS, and the spectra are dominated by modes from the base ring. Different nucleotide derivatives containing the same base can be distinguished by analyzing the low-lying ring modes.

(4) Polyamino acids, polynucleotides, and polypeptides can easily be examined with IETS techniques, but individual mode resolution and assignment is difficult.

(5) Extension of IETS to proteins and nucleic acids is straightforward. DNA and RNA show spectra quite comparable to infrared and Raman spectra, while for globular proteins individual modes are hard to resolve except for the dominant CH modes.

(6) An application of IETS to uv radiation damage in nucleotides shows that detailed information can be obtained about selective bond damage in different nucleotides.

Acknowledgments

The author wishes to acknowledge the contributions of Dr. J. M. Clark, Dr. M. G. Simonsen, and Estelle Phillips to many of the experiments reviewed in this chapter.

References

1. Albert L. Lehninger, *Biochemistry: The Molecular Basis of All Structure and Function*, 2nd Edition, Worth, New York (1975).
2. Lubert Stryer, *Biochemistry*, W. H. Freeman, San Francisco (1975).
3. Michael G. Simonsen and R. V. Coleman, Inelastic tunneling spectra of organic compounds, *Phys. Rev. B* **8**, 5875–5887 (1973).
4. Michael G. Simonsen and R. V. Coleman, Tunneling measurements of vibrational spectra of amino acids and related compounds, *Nature* **244**, 218–220 (1973).
5. Michael G. Simonsen, R. V. Coleman, and Paul K. Hansma, High resolution inelastic tunneling spectroscopy of macromolecules and adsorbed species with liquid-phase doping, *J. Chem. Phys.* **61**, 3789–3799 (1974).
6. Masamichi Tsuboi, Takaharu Onishi, Ichiro Nakagawa, Takehiko Shimanouchi, and San-Ichiro Mizushima, Assignments of the vibrational frequencies of glycine, *Spectrochim. Acta* **12**, 253–261 (1958).
7. John T. Edsall, John W. Otvos, and Alexander Rich, Raman spectra of amino acids and related compounds. VII. Glycylglycine, cysteine, cystine and other amino acids, *J. Am. Chem. Soc.* **72**, 474–477 (1950).
8. J. M. Clark and R. V. Coleman, Inelastic electron tunneling study of uv radiation damage in surface adsorbed nucleotides, *J. Chem. Phys.* **73**, 2156–2178 (1980).
9. H. Susi and J. S. Ard, Vibrational spectra of nucleic acid constituents—I: Planar vibrations of uracil, *Spectrochim. Acta, Part A* **27**, 1549–1562 (1971).
10. R. C. Lord and G. J. Thomas, Jr., Raman spectral studies of nucleic acids and related molecules—I: Ribonucleic acid derivatives, *Spectrochim. Acta, Part A* **23**, 2551–2591 (1967).
11. C. L. Angell, An infrared spectroscopic investigation of nucleic acid constituents, *J. Chem. Soc. Part I*, 504–515 (1961).
12. James M. Clark and R. V. Coleman, Inelastic electron tunneling spectroscopy of nucleic acid derivatives, *Proc. Nat. Acad. Sci. USA* **73**, 1598–1602 (1976).
13. Paul K. Hansma and R. V. Coleman, Spectroscopy of biological compounds with inelastic electron tunneling, *Science* **184**, 1369–1371 (1974).
14. Paul O.P. Ts'o, in *Basic Principles in Nucleic Acid Chemistry* (P.O.P. Ts'o, ed.), Chap. 6, pp. 481–517, Academic Press, New York (1974).
15. R. A. Dragoset, E. Phillips, and R. V. Coleman, to be published.

8

The Study of Inorganic Ions

K. W. Hipps and Ursula Mazur

1. Introduction

The majority of tunneling studies to date have focused on the adsorption of organic alcohols, acids, and amines on alumina and, less frequently, magnesia. This preoccupation with reactive organics is probably due to (a) the existence of a large body of ir, ESR, and chemical information concerning the adsorption of these materials on bulk oxides; (b) the desire of the experimenter to control the cleanliness of the surface by using vapor phase adsorption within the confines of the vacuum system whenever possible; (c) the high C–C and C–H bond energies of most of the systems studied preclude "backbone" reactions of the adsorbates and limit chemical modifications on adsorption to changes associated with the reactive group (OH, CO_2H, NH_2). This last factor, coupled with the rather large heats of formation of metal–oxygen and metal–amine bonds, limits the breadth of chemical and physical changes which can occur in adsorption and with deposition of the top metal electrode. For example, one does not usually observe a large ($\geqslant 1\%$) change in C–C or C–H motions of adsorbed organics once the image–dipole correction has been made. The chemical specificity of reaction evidenced by these organic systems tends, therefore, to provide a somewhat simplistic picture of the processes which occur in the production of doped tunnel junction.

Inorganic ions provide a much richer spectrum of possible surface, and/or top metal, reactions. In the case of isocyanate(OCN^-)-like ions,

---

K. W. Hipps and Ursula Mazur • Department of Chemistry and Chemical Physics Program, Washington State University, Pullman, Washington 99164. Tunneling studies were supported by the National Science Foundation under Grant DMR-7820251 and DMR-8115978.

two possible coordination sites are available and the oxide metal ion may bond with either the oxygen or nitrogen end of the molecule. Variations in the distribution of M–O- and M–N-type sites can occur with temperature, extent of hydration of the surface, and other factors. Further, the possibility of a bridging type structure which connects the oxide metal ion with a top electrode metal atom cannot be rejected *a priori*. In the case of transition metal complex ions, an even broader spectrum of final states of the doped junction exists. Consider for example the ferrocyanide ion $[Fe(CN)_6^{-4}]$. Surface water and/or oxide may displace one or more CN^- group from the coordination sphere of the iron. Charge transfer between Fe^{+2} and the top metal may lead, in the extreme case, to a top-metal–cyanide complex with iron atoms dispersed in the first few atomic layers of the top electrode. In fact, there is no *a priori* guarantee that the complex first exposed to the oxide surface will survive the adsorption and/or top metal deposition process. The experimenter must therefore be especially attentive to identifying the detailed chemical nature of the species seen in the tunneling spectra of inorganic ions. He cannot assume, as one does with reactive organics, that adsorption affects only a limited segment of the supported species and that the top metal plays only a minor role in the observed positions of bands.

Fortunately, experimental evidence obtained to data indicates that chemical and electrical processes associated with inorganic ion adsorption are usually more significant than effects due to top metal interaction. We may, therefore, often use tunneling spectroscopy to study processes associated with the adsorption of inorganic ions on oxide surfaces. The chemical complexity of possible adsorption processes discussed above becomes an exciting factor in the study of inorganic ions by tunneling spectroscopy. Further, there are no competing techniques which can provide the information available from tunneling spectroscopy.

Because of their nature, it is generally not possible to dope tunnel junctions with inorganic ions from any medium other than solution. As tunneling is currently practiced, solution phase doping occurs in the ambient atmosphere of the laboratory. This situation precludes the study of oxygen- and/or water-sensitive compounds. This restriction can be removed, and we will discuss the experimental modifications necessary for extending tunneling studies to include air-sensitive materials in a later section of this chapter.

2. Why Study Inorganic Ions by Tunneling Spectroscopy?

There are two viewpoints from which tunneling studies of inorganic ions may be viewed. In cases where the adsorption process occurs without significant chemical modification of the adsorbed ion, tunneling may be considered as an alternative *molecular* spectroscopy. In this case, incorporation of the species of interest in the tunnel junction is simply a preparative technique similar in concept to the matrix isolation method. The alternative viewpoint is that tunneling spectroscopy is a *surface* spectroscopy which is utilized to study the adsorption of inorganic ions on oxide surfaces.

A molecular spectroscopist is primarily interested in the location and symmetry of states associated with a known molecular configuration. His primary interest, therefore, is in *selection rules* and *usable spectral range*. The surface spectroscopist, on the other hand, is primarily interested in *identification* of, and *detection limits* for, species formed on a surface by the adsorption process. These are not mutually exclusive requirements inasmuch as knowledge of selection rules and a wide spectral range assists in the identification process. The emphasis is, however, decidedly different. One can conceivably identify a species of interest from one or two bands (the location of CO stretches in metal carbonyl systems for example) without significantly contributing to a knowledge of the molecular force field and other quantities of interest to a molecular spectroscopist. The justification for the use of tunneling to study inorganic ions will therefore depend on the viewpoint of the investigator. In Section 2.1 we will emphasize the molecular spectroscopist's view, while in Sections 2.2 and 2.3 identification and detection will be of primary interest. Throughout, we will use the term "molecule" to include ionic molecular species.

2.1. Direct Observation of Transitions Forbidden in Photon Spectroscopy

2.1.1. Vibrational Transitions

2.1.1.1. Theoretical Background. Within the context of the Born–Oppenheimer approximation, a molecule may be visualized as a collection of atoms or ions bound together by a well-defined potential into a configuration which, at reasonable temperatures, is nearly static. This picture is sufficiently classical to allow the concepts of equilibrium nuclear geometry and molecular force field to be transferred, more or less intact, from the domain of classical mechanics to that of quantum mechanics. These concepts, and that of electronic density maps, form the basis of how we conceptualize molecules and ions. Vibrational and rotational spectroscopy, being the primary tools with which molecular force fields are obtained, are fundamentally important to our understanding of molecular systems. Limitations inherent in the application of a given vibrational spectroscopy will generate a corresponding uncertainty in the potential function inferred from the bands observed by that method. In order to obtain the best possible picture of the vibrational potential for a given molecular or ionic system, it is necessary to orchestrate the use of several spectroscopic techniques in order to overcome the limitations of individual methods. The use of several methods is especially important for high-symmetry transition metal complex ions. This subsection will be devoted to demonstrating how tunneling spectroscopy can contribute to the determination of molecular vibrational potentials. The discussion will be presented for nonlinear molecules in the quadratic approximation.[1,2]

Consider an N-atom system bound by Hooke's-law-type forces. The kinetic energy of this system T is given by

$$T=(1/2)\sum_{j=1}^{N} m_j\left[(dx_j/dt)^2+(dy_j/dt)^2+(dz_j/dt)^2\right] \quad (1)$$

where x_j is the x Cartesian component of the displacement of atom j from its equilibrium position. This expression can be simplified by transforming to mass weighted Cartesian displacement coordinates, q_i, where

$$q_1=(m_1)^{1/2}x_1, \qquad q_2=(m_1)^{1/2}y_1,\ldots,q_{3N}=(m_N)^{1/2}z_N \quad (2)$$

Denoting the column vector those elements are $q_1, q_2, \ldots, q_{3N}$ by $\mathbf{q}$, the expression for the kinetic energy becomes

$$2T=\dot{\mathbf{q}}'\dot{\mathbf{q}} \quad (3)$$

where $\mathbf{q}'$ is the transpose of $\mathbf{q}$, and $\dot{\mathbf{q}}=d\mathbf{q}/dt$. Similarly, the potential energy, U, may be expressed by

$$2U=\mathbf{q}'\mathsf{U}\mathbf{q} \quad (4)$$

where U is the $3N\times 3N$ matrix whose ijth element is given by

$$U_{ij}=\left(\frac{\partial^2 U}{\partial q_i \partial q_j}\right)_{\mathbf{q}=0}=\frac{1}{(m_l m_n)^{1/2}}V_{ln}$$

where $\mathbf{V}$ is independent of the atomic masses, and l and n correspond to the Cartesian coordinates associated with mass weighted coordinates i and j.

The Lagrangian $\mathfrak{L}$ in the mass weighted system is, therefore,

$$2\mathfrak{L}=\dot{\mathbf{q}}'\dot{\mathbf{q}}-\mathbf{q}'\mathsf{U}\mathbf{q} \quad (5)$$

the application of Lagrange's equations of motion to (5) yields

$$\ddot{\mathbf{q}}=-\mathsf{U}\mathbf{q} \quad (6)$$

Equation (6) represents $3N$ coupled second-order differential equations with constant coefficients. These equations may be decoupled by transforming to a basis in which U is diagonal.

Let L be the matrix of orthonormal eigenvectors of U, such that

$$\mathsf{L}'\mathsf{U}\mathsf{L} = \Lambda \tag{7}$$

where $\Lambda_{ij} = \lambda_i \delta_{ij}$. Inserting Eq. (7) into Eq. (6) and rearranging yields

$$\ddot{\mathbf{Q}} = -\Lambda \mathbf{Q} \tag{8}$$

where $\mathbf{Q} = \mathsf{L}'\mathbf{q}$. Solutions of Eq. (8) are of the form

$$Q_i = A_i \sin(2\pi c \omega_i t + a_i) \tag{9}$$

$$\omega_i = (\lambda_i)^{1/2}/(2\pi c) \tag{10}$$

with ω_i in cm^{-1} and c is the speed of light. The coordinates Q_i are the *normal coordinates* of the vibrational problem in the quadratic approximation. They are the natural coordinates of the vibrating molecule in the sense that they diagonalize (decouple) Eq. (5). That is, the Lagrangian in the Q basis is

$$2\mathfrak{L} = \sum_{i=1}^{3N-6} \left(\dot{Q}_i^2 - \lambda_i Q_i^2\right) \tag{11}$$

For a nonlinear molecule, Eq. (11) contains only $3N-6$ terms because six of the λ_i will vanish. The associated Q_i represent rotational and translation motions of the molecule. Equivalent results are obtained by first removing the rotational and translational motions via an intermediate basis, $\boldsymbol{\eta}$ say, and then proceeding to find the eigenvectors of U on the $\boldsymbol{\eta}$ basis. Proceeding by this latter path demonstrates that complete specification of U requires, in general, $(3N-6)(3N-5)/2$ potential constants. Since there will be no more than $3N-6$ observable frequencies for a given isotopic composition molecule or ion, the difficulty of specifying U will increase sharply with N. If the molecule or ion in question has some symmetry, the number of unique elements of U will be reduced. In general, however, the number of unique frequencies of motion will also decrease due to the presence of symmetry-induced degeneracies. It is therefore important to establish a procedure for determining the number of independent elements of U and the number of unique frequencies. A convenient approach is based on group theory.[3,4]

If G is the point group associated with the ion or molecule in question, the normal coordinates will transform as the base of irreducible representations of G, Γ^k. The set of Q_i may therefore be equivalently expressed as $\{Q(\Gamma^k_{\alpha,l})\}$ where α labels the component of the Γ^k basis and l is used to identify the coordinates when a given irreducible representation appears more than once. Since the Hamiltonian of the vibrational problem must transform as the totally symmetric (a_1) irreducible representation

$$2H = \sum_{k,\alpha,l} \left[P^2_{k,\alpha,l} + \lambda_{k,l} Q^2(\Gamma^k_{\alpha,l}) \right] \tag{12}$$

where $\lambda_{k,l}$ is independent of α.

The number of vibrational fundamentals (n_f) is therefore equal to the number of $\lambda_{k,l}$ and is given by

$$n_f = \sum_k n_k \tag{13}$$

where n_k is the number of times the kth irreducible representation appears in the decomposition of the vibrational coordinate basis. The n_k may be found from the Cartesian displacement basis, allowing n_f to be determined without solving the vibrational problem.[1-4]

The number of independent elements of U may also be estimated without a knowledge of the detailed form of U. Consider a basis transformation of U such that the new basis is adapted to the symmetry group G, but not necessarily the normal coordinate basis. This transformation will block-diagonalize U. Each block will be associated with a particular irreducible representation basis component, and will therefore be $n_k \times n_k$ in size. The number of independent components of these small matrices will be $n_k(n_k + 1)/2$. The number of independent elements of U, n_u, will therefore be

$$n_u = \sum_k n_k(n_k + 1)/2 \tag{14}$$

Unless $n_k = 1$ for all k, the number of independent elements of U must exceed the number of frequencies. Symmetrical isotopic substitution cannot affect the results of Eq. (14) since the symmetry of U is maintained. It will, in general, modify the way in which the potential terms in the $n_k \times n_k$ submatrices of U are combined with mass terms. Symmetrical isotopic substitution will therefore, at most, assist in determining the elements of the $n_k \times n_k$ submatrices associated with vibrational modes of symmetry species Γ^k. The significance of this restriction will be made apparent in the following paragraphs. Because of the extreme difficulty in isolating and identifying isotopically asymmetrical transition metal complex ions, symmetrical isotopic species are usually the only isotopic species available to the spectroscopist.

The preceding discussion suggests that, provided a sufficient number of symmetrical isotopic species are available, all the elements of **V** (or **U**) may be determined. This is *not* the case. Although high structural symmetry induces a sharp reduction in the number of independent components of V, it also reduces the number of fundamentals which can be observed by ir and/or Raman spectroscopy. Perhaps the most familiar symmetry generated reduction is the so-called *exclusion rule*. Any molecule having inversion symmetry can have its fundamental vibrations divided into three disjoint classes; those which are ir active, those which are Raman active, and those which are neither ir or Raman active. The immediate consequence of the exclusion rule is that no more than half of a systems fundamentals may be observed by ir (or Raman) spectroscopy. ir and Raman spectroscopy combined cannot be used to observe fundamentals which are doubly (both ir and Raman) forbidden. Doubly forbidden fundamentals are "invisible" to photon spectroscopy.

The presence of "invisible" fundamentals in highly symmetrical complex ions means that there is one or more $n_k \times n_k$ block of V which cannot be determined by direct observation with photonic spectroscopy. While these doubly forbidden modes may sometimes be observed in combination with allowed modes in the ir, or as overtones in Raman, their assignment is often questionable. There are, in fact, very few metal complexes for which all of the elements of V are known.

Table 1 lists the ir active, Raman active, and doubly inactive mode symmetries for some representative point groups. It is clear from this table that the occurrence of "invisible" modes is common. Even the relatively low symmetry groups such as C_{3V} and C_{4V} can support inactive vibrations.

Let us now consider complexes of the form $M(XY)_6$ in some detail. We expect a total of $3(13)-6=33$ vibrations. Group theoretical analysis indicates that these motions will be of the following symmetry types: $2a_{1g}+2e_g$

Table 1. Infrared and Raman Active and Inactive Vibrations of Various Symmetry Groups According to the Conventions of Reference 3

Group	ir active	Raman active	Inactive
O_h	t_{1u}	a_{1g}, e_g, t_{2g}	a_{2g}, t_{1g}, a_{1u} a_{2u}, e_u, t_{2u}
O	t_1	a_1, e, t_2	a_2
D_{4h}	a_{2u}, e_u	$a_{1g}, b_{1g}, b_{2g}, e_g$	$a_{2g}, a_{1u}, b_{1u}, b_{2u}$
C_{4v}	a_1, e	a_1, b_1, b_2, e	a_2
C_{3v}	a_1, e	a_1, e	a_2
C_{2v}	a_1, b_1, b_2	a_1, a_2, b_1, b_2	None
T_d	t_2	a_1, e, t_2	a_2, t_1

$+t_{1g}+4t_{1u}+2t_{2g}+2t_{2u}$. There will therefore be only 13 distinct frequencies produced by the 33 modes, i.e., $n_f=13$. U can be block diagonalized to form a $(2\times2)a_{1g}$ block, two $(2\times2)e_g$ blocks, three $(1\times1)t_{1g}$ blocks, three $(4\times4)t_{1u}$ blocks, three $(2\times2)t_{2g}$ blocks, and three $(2\times2)t_{2u}$ blocks. Repeated matrices of a given type are equivalent. The number of independent elements of U is, therefore, $n_u=3+3+1+10+3+3=23$. Of the 13 unique frequencies, there are six Raman active (a_{1g}, e_g, t_{2g} types), four ir active (t_{1u}), and three inactive (t_{1g} and t_{2u}); only 10 of the fundamentals are observable by photon spectroscopy. The t_{1g} and t_{2u} blocks of V contain a total of four independent potential constants. If ir and Raman spectroscopy

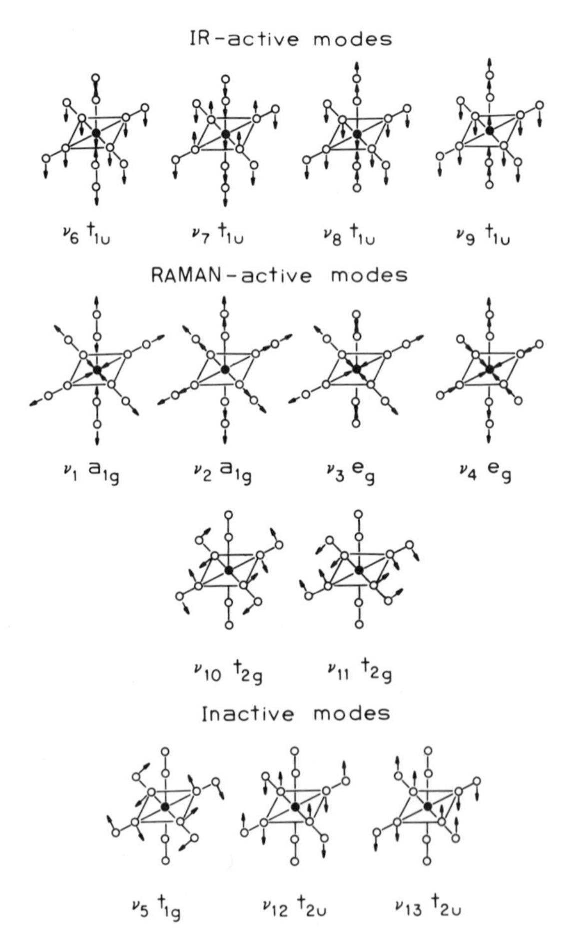

Figure 1. Schematic representation of the normal modes of an octahedral $M(XY)_6$ system.

are the only techniques used, about 20% of the potential matrix must be found from assignment of combination bands. A pictorial presentation of these modes is given in Figure 1.

In the case of $Fe(CN)_6^{-4}$, the t_{1g} mode has been estimated to occur at 350 cm^{-1} from a study of ir combination bands. There are at least three, and perhaps four, other fundamentals which occur between 410 and 300 cm^{-1} in energy. Sorting out the energy of $\nu(t_{1g})$ from the multiplicity of combination bands which result requires some initial assumptions about the form of V. These assumptions are then modified to produce, presumably improved, elements of V. The two t_{2u} frequencies probably lie in the region of 400 and 100 cm^{-1}, respectively, but have not been identified from combination bands. Clearly, a spectroscopic method which allows the direct observation of forbidden modes would greatly enhance our ability to extract molecular potentials from vibrational spectra.

2.2.1.2. Tunneling and Forbidden Transition. There are two separate mechanisms by which forbidden transitions can be observed in tunneling spectroscopy. We may classify them as *structural* and *dynamic* mechanisms.

The structure of a tunnel diode defines the maximum symmetry which an incorporated ion may possess. The maximum site symmetry depends upon the crystallographic structure of the oxide support, and the asymmetrical potential normal to the oxide surface due to the adsorbate–top-metal interface. In the case of $Al-AlO_x-M$ junctions, the maximal site symmetry is no greater than C_{3V}. For $Mg-MgO-M$ junctions, C_{4V} is the maximum possible site symmetry. Both groups have only one one-dimensional irreducible representation which transforms as a forbidden vibration. In principle, therefore, many of the transitions which are forbidden for the high-symmetry complex ion become allowed in the low-symmetry surface environment. For example, the t_{1g} motion of the $Fe(CN)_6^{-4}$ ion is split into two components by C_{4V} site symmetry. One of the components being both ir and Raman active.

The magnitude of the splitting induced by low-symmetry sites is often surprisingly small. $K_4M(CN)_6 \cdot 3H_2O$ ($M = Fe, Ru$) crystallize with polytypic unit cells having C_1 and C_2 site symmetries for the hexacyanide ion.[5] Thus, on the basis of symmetry we might expect the ir and Raman spectra of these crystals to be very complex. In fact there is very little band splitting and the Raman spectra of these crystals and of aqueous solutions are quite similar.[5] One must extensively distort the site symmetry of the hexacyanide ion, as in the case of $K_4Fe(CN)_6 \cdot \frac{1}{2}H_2O$, in order to see splittings of about 1%.[5,6] Band intensities are generally more sensitive to site symmetry than band energies. A 10% admixture of allowed mode to a forbidden one produces a readily observable band while modifying the energy of the band by only about 1%.

The dynamics of the electron scattering process in tunneling is decidedly different from the dynamics of photon absorption or scattering. While much of the theoretical work to date has focused on ir or Raman allowed vibrations interacting with the tunneling electron (see previous chapters), there is a growing body of experimental and theoretical evidence demonstrating that optically forbidden transitions can be observed by tunneling spectroscopy.[6–9] Thus, even in those cases where structural effects are insufficient to allow a band to be observed, the dynamics of the tunneling process may provide an allowing mechanism.

Tunneling spectroscopy may, therefore, supplement ir and Raman spectroscopy in those cases where the symmetry of the species of interest requires the presence of forbidden vibrations. The observation of these "invisible" modes by tunneling spectroscopy provides a two-fold benefit to the molecular potential analyst. The direct observation of forbidden fundamentals allows a more accurate and complete determination of the potential energy matrix. Further, in those cases where the forbidden mode may be seen in combination with ir allowed bands, it reduces the uncertainty in the assignment of potential constants for other modes by providing an increased number of potential constants at the outset of the analysis.

The relationship, therefore, of ir, Raman, and tunneling spectroscopy is one of interdependence. Polarized ir and Raman are used to locate allowed fundamentals while tunneling provides forbidden fundamentals. Combination bands observed in the ir are then analyzed and the V matrix determined by utilizing the results of all three methods. This partnership will be further strengthened by the development of empirical or theoretical selection rules for tunneling which apply to photon forbidden transitions.

2.1.2. Electronic Transitions

Chapter 4 presented a detailed review of the application of tunneling to the study of electronic transitions. In this subsection we will present our view of an important area of future application of tunneling to electronic transitions. We will again focus on transition metal complexes as the species of interest.

Transition metal complexes generally have a rich spectrum of electronic states which are due to metal localized transitions between components of the *d* orbitals. The energies of these transitions are determined by[3,4,10]

1. Interaction between the *d* electrons and the coordinated ligands, approximated by crystal or ligand field theory;
2. Interelectronic interaction between electrons occupying the *d* orbitals; this interaction is given in terms of the Racah parameters, *B*, and *C*;
3. Spin-orbit interactions between *d* orbital electrons.

Generally, one can distinguish three limiting cases. If the crystal (or ligand) field terms dominate over the electronic interaction (B and C) and spin-orbit coupling, one is said to have the *strong-field case*. If interelectronic interaction is the largest term in the Hamiltonian, while the spin-orbit term is the smallest, one is said to have a *weak-field* case. If neither of these conditions apply, this is the *intermediate-field case*.

In order to determine electronic state energies in the strong-field limit, one visualizes the d orbital energies as being defined by the crystal (or ligand) field. For octahedral or tetrahedral symmetry complexes, the five d orbitals split into a threefold (t_2) and twofold (e) set of levels separated by energy, Δ,[3,4,10] the value of Δ for the tetrahedral case being roughly half that for a similar octahedral complex. As the symmetry of the complex decreases, the number of nondegenerate components of the d orbitals increases. Table 2 gives the decomposition of d orbitals into their irreducible symmetry components for some common groups. The next step (in a perturbation sequence) is to add the number of electrons appropriate for the ion of interest. One then includes the $\Sigma e^2/r_{ij}$ interelectronic repulsion term in order to obtain *term energies* and *terms* ($^{2S+1}\Gamma$). The last step, in the strong field limit, is to include spin-orbit coupling in order to obtain levels (Γ).

Calculation of level energies in the weak field limit begins with the determination of free-ion terms (^{2S+1}L). These terms are then perturbed by the crystal field to yield terms ($^{2S+1}\Gamma$), and these are further split by spin-orbit coupling to yield levels (Γ).

Diagrams of term energies for all the d^n configurations in octahedral and tetrahedral symmetry are readily available in the form of Tanabe–Sugano diagrams.[3,4,10] The lower energy portions for d^4, d^5, and d^6 octahedral complexes are shown in Figure 2. Note that when $\Delta = 0$, the energies are just those of the free-ion terms (shown on left of diagram). $\Delta/C \rightarrow \infty$ is the strong-field limit. The diagrams are drawn so that the ground term of the ion is always at $E/B = 0$.

The parameter Δ depends on both the ligand and the central ion. Δ values generally increase as one moves down a column of the Periodic Table.[4,10] As a function of ligand, Δ decreases in the order $CN^- > NO_2^- > NH_3 > H_2O >$ oxalate $> Cl^- > I^-$.[3,4,10] $Fe(H_2O)_6^{+2}$, for example, is a weak-field ion having a ^5E to $^5T_2$ splitting of 1.3 eV.[94] $Fe(CN)_6^{-4}$, on the other hand, is a strong-field ion with a $^3T_1$ to $^1A_1$ splitting of 2.7 eV.[4] The

Table 2. Components of the *d* Orbitals in Various Symmetry Complexes

O	D_{4h}	D_3	D_{2d}
$e_g + t_{2g}$	$a_{1g} + b_{1g} + b_{2g} + e_g$	$a_1 + 2e$	$a_1 + b_1 + b_2 + e$

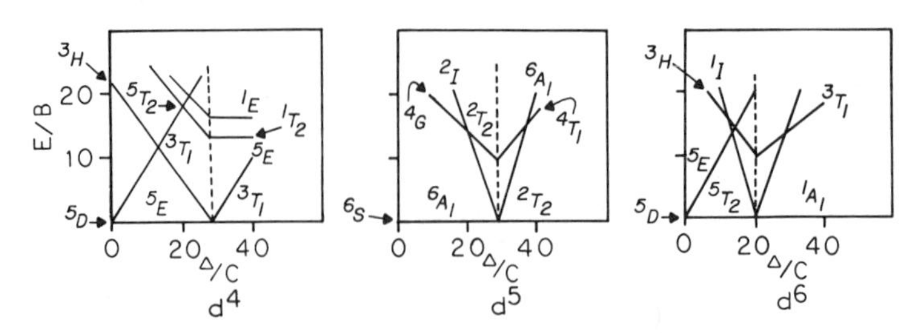

Figure 2. Low-energy electronic term diagrams for d^4, d^5, and d^6 octahedral metal complex ions.

d^5 Fe^{+3} and Ru^{+3} tris diimine complexes are strong field systems. The $^2T_2$ ground term is split by the trigonal ligand field into $^2A_2$ and ^2E terms which are *estimated* to have 0.1 to 0.6 eV separation.[11,12]

One further refinement to this simple theory is required. When spin-orbit coupling is taken into account, the highest electronic degeneracy supported by octahedral symmetry is three for even electron systems and four for odd. The 15-fold degenerate $^5T_2$ term, for example, must split into at least five separate levels with total energy spread of the order of the spin-orbit coupling parameter ζ. Figure 1 and Table 3 show that there will be many transition-metal complexes having a rich spectrum of excited states within about 1 eV of the ground state. The vast majority of these states are not observable by photon absorption spectroscopy because they are either spin forbidden ($\Delta S \neq 0$), or they are interconfigurational.

When all of the above interactions are included, it turns out that many transition metal complexes have electronic transitions in the vibrational region of the spectrum. Further, for those complexes having no inversion center, the intensities observed should exceed those observed for Ho_2O_3 (See Chapter 4) by at least one order of magnitude because of the increased

Table 3. Representative Values of the Racah Parameters, *B* and *C*, and the Spin-Orbit Coupling Parameters, ζ; All Values Are in cm^{-1}[a]

Ion	B	C	ζ
Cr^{+3}	1030	3850	273
Fe^{+2}	1058	3901	410
Cu^{+2}	1238	4659	829
Rh^{+2}	620	4002	1615

[a]Values taken from reference 4.

ligand–*d*-orbital mixing relative to ligand–*f*-orbital mixing. Tunneling is potentially the technique of choice for studying these systems. So why have none of these transitions been seen?

We are presently attempting to observe low-lying electronic transitions in transition-metal complexes. There are, however, difficult chemical problems which must first be analyzed and then surmounted. These problems are as follows:

1. Alumina and magnesia are active reducing agents.[13,14] Less active barriers are needed to allow the study of reducible complexes.
2. Alumina ions on the alumina surface can compete with the transition metal as binding sites for the ligands. Alternately, surface oxide ions can displace ligands from the coordination sphere of the complex. Chemically stable complexes must be studied, or less reactive barriers are needed.
3. Care must be taken to avoid complex–top-metal reactions.

2.2. Impregnation Catalysts

In terms of tonnage of catalyst, supported metals are the primary form of commercial catalyst. The most common supports are alumina, silica-alumina, charcoal, and zeolites. Typically, these catalysts are prepared by the following sequence of steps:

1. Adsorption (impregnation) of the metal of interest as a complex ion from solution. Although it is often convenient to distinguish between adsorption-type and impregnation-type catalysts,[15] we will use the terms interchangeably here.
2. A drying step to remove solvent.
3. Reduction, usually with H_2 near 300°C, of the adsorbed ions to metal.

During any or all of the above steps, processes leading to crystallite or aggregate formation can occur. The end result of these processes is the formation of crystalline or raftlike metal particles on the oxide surface. Since catalysis is primarily a surface process, the activity of the catalyst is found to be inversely related to the mean size of these rafts. Further, the cost/activity ratio will decrease as the raft size decreases. These considerations have led to considerable interest in the production of highly dispersed supported metal catalysts.

There have been relatively few spectroscopic studies of the early (adsorption) stage of catalyst formation. These have primarily concerned the adsorption of platinum and palladium complexes on silica.[15–17] A smaller number of papers have considered platinum, palladium, and ruthenium on alumina.[16,18] The primary tools in the above studies were uv–visible reflectance spectroscopy, XPS, and stoichiometric chemical analysis. Infrared spectroscopy, in most cases, cannot be used because of the intense substrate absorption below about 1100 cm^{-1}. In the case of the adsorption of the chloroplatinate ion, for example, all of the molecular fundamentals occur in the opaque region. While Raman spectroscopy has been applied to a few systems, problems associated with thermal and photodecomposition in the laser beam restrict its application.[19] Raman studies of impregnation catalysts also require high-surface-area substrates and high adsorbate coverages. High surface area samples exhibit intense diffuse reflectance and limit the sensitivity of the technique.

The dearth of vibrational spectroscopic studies of the impregnation process has led to a rather strange situation. There are well-developed empirical and thermodynamic rules describing the general requirement for the occurrence of adsorption, but little direct knowledge of the surface species formed or of their formation mechanism. Thus, it is known that the isoelectric point of the oxide, the pH of the impregnation solution, the dielectric constant of the impregnation solvent, and the number of labile ligands all play a role in the adsorption process.[17] The details of bonding of a complex to the oxide surface, however, are known in only a few cases. And in these cases, the knowledge is often based on the indirect evidence of elution studies. Clearly, there is a need for a spectroscopic method which will allow the chemical and structural nature of the adsorbate to be studied. Tunneling is such a method.

Tunneling provides access to the metal–ligand vibrational energy region below 1100 cm^{-1}, as shown by the representative spectra in Section 5. Further, metal–ligand bands appear with considerable intensity allowing very small surface concentrations to be observed. Adsorption on alumina from aqueous solutions in the 3–10 pH range can be easily studied, and most solvents appropriate for metal complex solution have no bands in the metal–ligand region of the spectrum. Tunneling spectroscopy can, and will, contribute significantly to our understanding of the impregnation process.

Tunneling may also be useful in analyzing later steps in the catalyst formation process. Consider, for example, high-activity Fischer–Tropsch catalysts prepared by impregnation with potassium salts of group VIII metal carbonyl ions followed by hydrogen reduction at 500°C.[20] The vibrational motions of the impregnated complex on the surface, the intermediates formed during the early stages of the reduction process, and finally, reactions of the reduced catalyst should be observable by tunneling spectroscopy. This suggestion is far from speculative. Tunneling spectra of metal carbonyls adsorbed on alumina from solution (the first step) and of

CO on metal particles (the last step) have already been obtained. We expect that tunneling studies of the genesis of catalysts will soon be appearing in the literature.

In summary, tunneling spectroscopy is a proven tool for studying the first step of catalyst generation—the impregnation step. Further, tunneling can, and will, be applied throughout the life cycle of many impregnation catalysts.

2.3. Speciation of Metal Ions in Natural Waters

Many metals are found in natural systems at concentrations approaching and occasionally exceeding toxic levels. For this reason it is desirable to understand the metal containing species in these systems and the factors affecting their transport. Numerous studies have concluded that adsorption onto suspended and bottom sediments is an important process in controlling dissolved metal concentrations and metal availability in the environment.[21–23] Adsorption is also important in soil–water interactions and in waste water treatment.[24,25]

Studies of trace metal adsorption phenomena have been carried out using naturally occurring solids and solutions in order to closely simulate environmental conditions. Model systems which are less realistic but better defined have also been developed. Among the important advances in "model" adsorption experiments in recent years has been an increased awareness of the importance of complexing ligands on metal ion adsorption.[26] Both inorganic and organic complexing agents are present in all natural aquatic systems, and have a dramatic effect on metal ion behavior and on the surface properties of potential adsorbents.

Most natural or synthetic complexing agents found in natural waters are present at higher concentrations than the trace metal ions with which they chelate. They include small inorganic ions (for example Cl^-, SO_4^{-2}), natural degradation products of plant and animal tissue, e.g., amino acids, and species derived from chemicals applied either deliberately or inadvertantly by man. Examples of the latter case are detergents and cleaning agents, pigment production by-products, nitrolotriacetate (NTA), ethylenediaminetetraacetic acid (EDTA), pesticides, ionic and nonionic surfactants, and a large group of synthetic macrocyclic compounds.

The types of metal species present in aqueous systems depends upon the stability of the hydrated metal ions and the tendency of the ions to form complexes with other organic and inorganic ligands. The influence of complexing agents on the adsorption behavior of trace metal ions on natural oxide materials is not well understood.

Generalized models for metal ion adsorption in the presence or absence of complexing ligands which consider various modes of interaction have been developed.[27] These models are based principally on studies involving pH and concentration parameters. Typically, cationic species adsorb in a narrow pH region ($6 \rightarrow 10$) in which partitioning of the species changes from nearly all in the solution phase to nearly all adsorbed as the pH increases. The reaction is often accompanied by the release of protons to the solution, either by ion exchange at the surface or by hydrolysis occurring simultaneously with adsorption (as in the case of uncomplexed metal ions).

The extent of adsorption of anionic species from water solution onto oxide surfaces varies from nearly complete coverage at low pH to no adsorption at higher pH. Protons are sometimes consumed when the ions adsorb.

The different types of complex adsorption behavior are most certainly related to the surface orientation and chemical composition of the adsorbate. Structural characterization of the surface species can be effectively accomplished through the utilization of spectroscopic techniques. Numerous studies analyzing the adsorption process of metal ions and complexes have not employed such techniques. The structural assignments proposed in these investigations were inferred from chemical and kinetic data and as such remain unsubstantiated. There are relatively few reports on the application of spectroscopic techniques to the study of adsoption at the water–oxide interface. These studies utilize ESR and optical spectroscopy (ir and uv–visible) as primary probes for structural analysis.

The high sensitivity of tunneling spectroscopy to ions adsorbed from aqueous solution makes it an attractive method for studying model systems. The broad pH range over which alumina based tunnel diodes may be doped includes the entire range of interest in most speciation studies. Further, as we demonstrate in Section 5, tunneling spectroscopy can be used to distinguish ligand-vs-metal complex adsorption on the oxide surface. On the negative side of the ledger, the tunneling spectroscopist does not have much control over the crystallographic form of the oxide and must dry the surface before it can be measured. In light of the near total lack of surface vibrational information on these systems, these restrictions do not seem overly severe for model studies. It is advisable, however, to supplement tunneling studies of these systems with ir spectroscopic studies of the same species adsorbed on bulk oxides under similar conditions. In the case of metal–amino-acid complexes, for example, the NH and CH vibrations occur within the ir "window" of alumina and reflect the bonding state to some extent. Comparison of ir and tunneling data, therefore, can provide justification for the application of the tunneling method to the particular cases.

3. Doping Techniques and Insulator Surfaces

3.1. Solution Phase Doping of Alumina Barriers

Solution pH, solvent dielectric constant, and solvent vapor pressure are critical factors in solution phase doping of inorganic ions. Solution pH is important for at least two reasons: (1) the surface charge of alumina in contact with an aqueous solution depends on pH,[17] and (2) the stability of a complex relative to an aquated metal ion or hydroxide may depend on pH.[28] Alumina in contact with an aqueous solution of $pH < 8$ is positively charged. Thus, negative ions tend to concentrate in the vicinity of the oxide surface. Above pH 8, alumina is negatively charged and positive ions preferentially associate with the surface. The extent of adsorption of a given ion, therefore, is dramatically affected by the pH of the solution, with positive ions adsorbing best at high pH.

This simple picture fails at very high and/or very low pH. In sufficiently basic solution, most metal complexes are converted to metal hydroxides, while at very low pH aquated metal ions form. Quantifying "very high" and "very low" requires a knowledge of the stability constants of the complex and the metal hydroxide. Some complexes are quite stable over the 3–10 pH range, while others are not.[28] An additional complication is the solubility of alumina and aluminum. As a matter of practical utility, the pH 3–10 range is available to the tunneling spectroscopist.

Figure 3 presents the tunneling spectra obtained from a series of Al–AlO_x–Pb junctions that were allowed to stand in various pH solutions for different lengths of time. Solutions were made basic with NaOH and acid with HCl. Clearly, the pH 3–10 range can be employed in doping experiments if exposure times of $\leqslant 15$ minutes are acceptable. Longer periods of exposure (up to 30 minutes) are practical for the pH 6–3 range. The high pH failure mode appears to be due to aluminum reacting with the base to form an excessively thick and irregular barrier. Visible striations appear on the aluminum film and the junction resistances increase dramatically.

The dielectric constant of the solvent is an important factor in determining the extent of adsorption of inorganic ions and the stability of metal complexes. Dissociation constants are exponentially dependent on the dielectric strength of the solvent in the absence of specific reactions.[29] The ability to adjust the solvent dielectric constant is therefore important to the molecular spectroscopist wishing to study low-stability complex ions. To the extent that solvation competes with adsorption, a decrease in dielectric constant of the solvent can increase the extent of adsorption of an ion. This is because the free energy of solvation generally becomes more

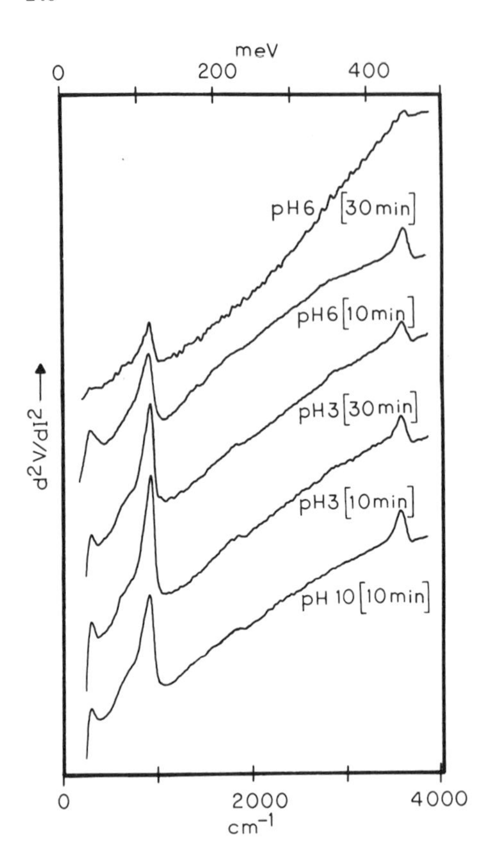

Figure 3. Tunneling spectra obtained from Al–AlO_x–Pb type diodes whose oxides were exposed to water solutions of the indicated pH for the indicated periods of time prior to lead deposition.

negative with increasing dielectric constant. In those cases where specific reactions occur with hydroxylic solvents, it is especially important to have a range of solvents available.

The vapor pressure at 25°C of the solvent also plays a role in the doping process. Although it is difficult to quantify the relationship between adsorption and rate of evaporation of solvent, it is clear that solvent vapor pressure correlates with the crystallinity and distribution of inorganic residue on the oxide surface. Crystallization must be avoided because of the exponential decrease in tunneling current with barrier thickness. The crystallinity of the doped material depends on both the solvent evaporation rate and dielectric constant. Thus, it is often possible to improve the quality of tunneling spectra by a factor of 10 with the proper choice of solvent properties. For example, no tunneling spectra of $KAu(CN)_2$ can be obtained by aqueous phase doping, but acetone solutions provide reasonable spectra.

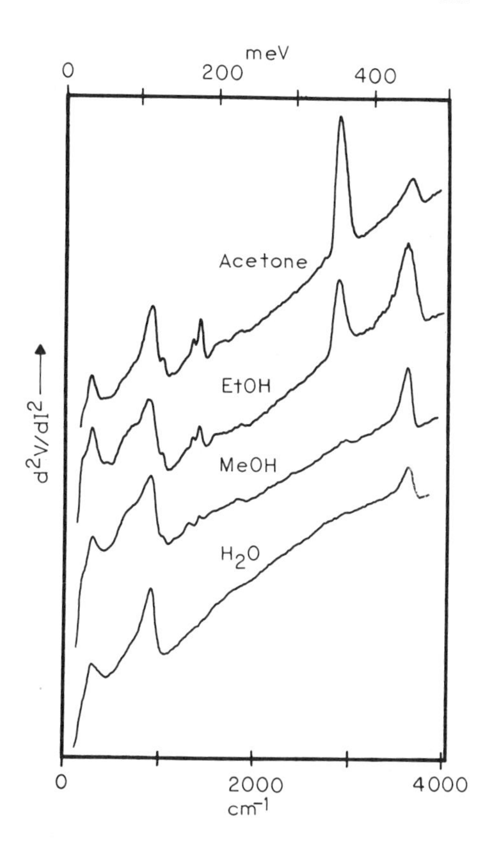

Figure 4. Tunneling spectra of some common solvents.

While a few inorganic ions are soluble only in water, most will dissolve in methanol or ethanol and some are soluble in acetone. The range of dielectric constants [78.5, 32.6, 24.3, and 20.7, respectively] and room temperature vapor pressures [24, 126, 59, and 230 Torr, respectively] provided by these solvents have proved useful in establishing various degrees of inorganic ion coverage on the oxide surface. Tunneling spectra of these solvents are collected in Figure 4. All were recorded under identical conditions of 2 mV modulation amplitude, 2 sec time constant, 4.2 K, and 1 μV lock-in amplifier scale.

None of these solvents provide significant interference in the metal–ligand motion region of the spectrum [$\leqslant 800\ cm^{-1}$] or in the CN^- or CO regions. We have successfully used these solvents with many types of inorganic ions and recommend their use. It should be noted, however, that acetone cannot be used in those cases requiring prolonged exposure to the solvent since it will react with alumina to produce prohibitively resistive barriers.

3.2. AlO_x and MgO supported O_ySiH_x Barriers

Eib *et al.* first observed that SiO deposited on a thin alumina barrier gave usable tunnel junctions showing a strong SiH stretch near 2215 cm^{-1}.[30] A spectrum of this type of barrier is shown in Figure 5b. Subsequent studies[31,32] have shown that the observed spectrum is due to supported O_ySiH units where $1.5 < y \leqslant 3$. Further, this species can be formed on virtually any insulator as is exemplified by the tunneling spectrum of O_ySiH on MgO shown in Figure 5a. Barriers of the type depicted in Figures 5a and 5b result when SiO is deposited under relatively anhydrous conditions. If water vapor is added to the vacuum system ($\geqslant 10^{-5}$ Torr) during SiO deposition, further processes occur as indicated by the tunneling spectrum, Figure 5c. The second SiH motion near 2000 cm^{-1} and other new bands are due to the presence of O_ySiH_x, where $y \leqslant 1$. Through deuterium

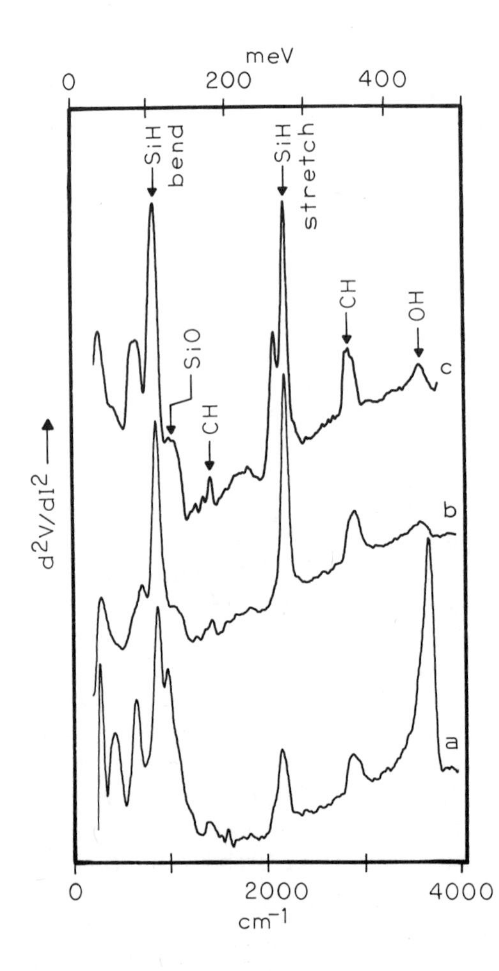

Figure 5. Tunneling spectra of composite barriers. (a) MgO–O_ySiH; (b) AlO_x–O_ySiH ("dry" preparation); (c) AlO_x–O_ySiH_z ("wet" preparation).

isotope substitution, it was possible to assign all the SiH stretching and bending modes.[31,32]

These barriers should probably not be visualized as molecularly dispersed or polymeric O_ySiH. Rather, they are more properly visualized as a mixture of SiO, SiO_2, Si, and O_ySiH groups bonded to either adjacent Si or Al.

Besides the intrinsic interest of these barriers and their chemistry of formation, there is also a practical interest. As indicated in Section 2, there is a strong need for chemically passive barriers which can be used as substrates in tunneling studies of inorganic ions. Although some questions about the passivity of the O_ySiH barrier remain, they are suitable for solution phase doping of inorganic ions. Figure 6 depicts representative spectra obtained from AlO_x/O_ySiH barriers doped with inorganic ions. Figure 6a is the tunneling spectrum of the thiocyanate ion adsorbed on the composite barrier. Figure 6b is that of the ferrocyanide ion. Clearly, the composite barrier is appropriate for the study of inorganic ions by tunneling spectroscopy. Whether marked differences exist in adsorptivity and/or reactivity of the composite with respect to alumina remains to be seen.

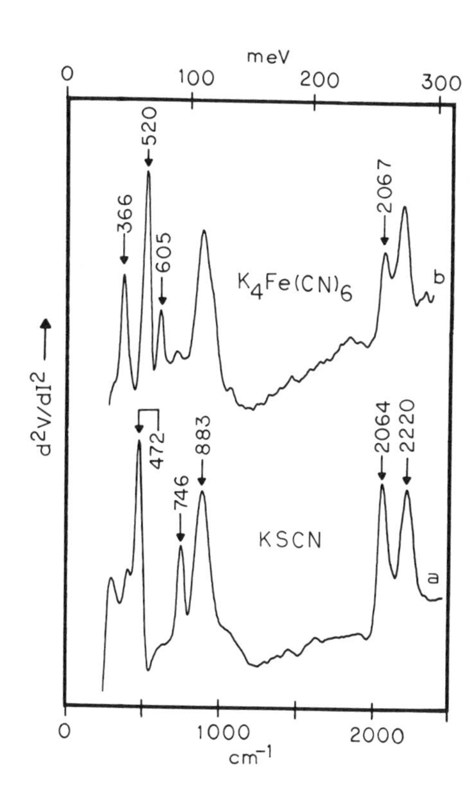

Figure 6. Tunneling spectra obtained from aqueous solution doped alumina–O_ySiH barriers. (a) KNCS doped, and (b) $K_4Fe(CN)_6$ doped.

4. Solution Phase versus Gas Phase Adsorption

While most inorganic ions are not stable in the vapor phase, there are a few inorganic acids which are stable and can be used to perform gas phase doping of tunnel diodes by inorganic ions. Notable among these is HNCO. The NCO^- ion is very stable in water and other polar solvents allowing for a comparison of gas and solution phase doping of this inorganic ion. This study is especially interesting in that NCO^- is formed by automotive exhaust catalysts and has been shown to adsorb strongly to alumina and alumina-silicate supports.[33–36] Further, it is worthwhile to investigate the bonding mode (Al–N vs Al–O) of this ligand.

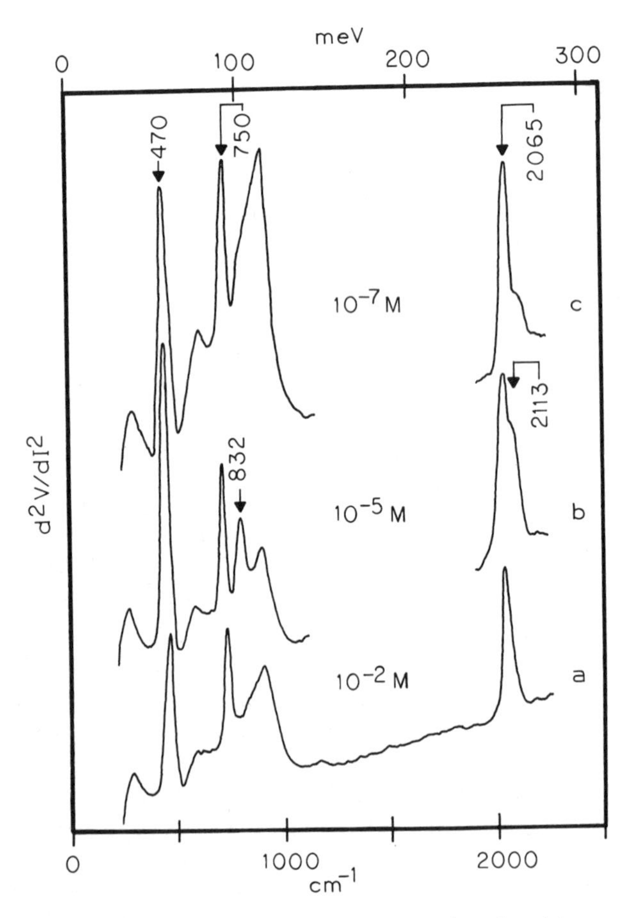

Figure 7. Tunneling spectra obtained from Al–AlO_x–Pb junctions doped from aqueous NCS^- solutions of the indicated concentrations.

Table 4. Peak Positions in cm^{-1} (mV) and Assignments

Assignment		Location[a]	
Species	Mode	cm^{-1}	(mV)
Al–NCS[b]	SCN bend	470	58.2
	SC stretch	750	93
	CN stretch	2065	256
Al–SCN[b]	SCN bend	470	58.2
	SC stretch	832	103
	CN stretch	2113	262
Al(OCN)[c]	OCN bend	630	78.1
	CN stretch	2146	266
Al(OCN)[d]	OCN bend	630	78.1
	CN stretch	2234	277

[a] Corrected for top metal superconductivity.
[b] From water solution.
[c] Ionic form—from water solution and gas phase.
[d] Covalent form—gas phase adsorption.

Tunnel diodes prepared by solution phase adsorption of NCO^- and the similar NCS^- ion have been reported.[37] The NCS^- ion displays the ability to bond through the nitrogen or the sulfur end of the molecule when adsorbed from water solution as indicated by the observed vibrational bands shown in Figure 7 and Table 4. The relative amounts of nitrogen and sulfur bonded species depend strongly on solution concentration, as shown in Figure 7. The tunneling spectra obtained from aqueous doping of NCO^-, on the other hand, always show only one species at all concentrations. The observed bands are very similar to those of KNCO, and using Solymosi's nomenclature,[33,34,36] we identify it as an ionic form of surface bound NCO^-. The tunneling spectrum of the ionic form is shown in Figure 8b.

The appearance of only one type of surface NCO^- in the case of aqueous doping is in sharp contrast with results obtained by gas phase adsorption of HNCO[38]: Figure 8 shows the NCO deformation and CN stretching regions of a tunneling spectrum obtained from a gas phase doped Al–AlO_x–Pb diode. Peak positions are reported in Table 4. The higher-frequency CN stretch has been observed in the infrared by Solymosi, and he identifies it as a coordinatively bound species. The relative amounts of ionic and coordinatively bound NCO^- in the tunnel diode depend strongly on the activation temperature of the barrier oxide and on the adsorption temperature.[38] Room temperature adsorption on a nonactivated oxide yields very little of the coordinatively bound species.

The observed dependence of the adsorption of OCN^- on surface preparation clearly correlates with the amount of surface water and/or hydroxyl present. Apparently some of the adsorption sites available on the

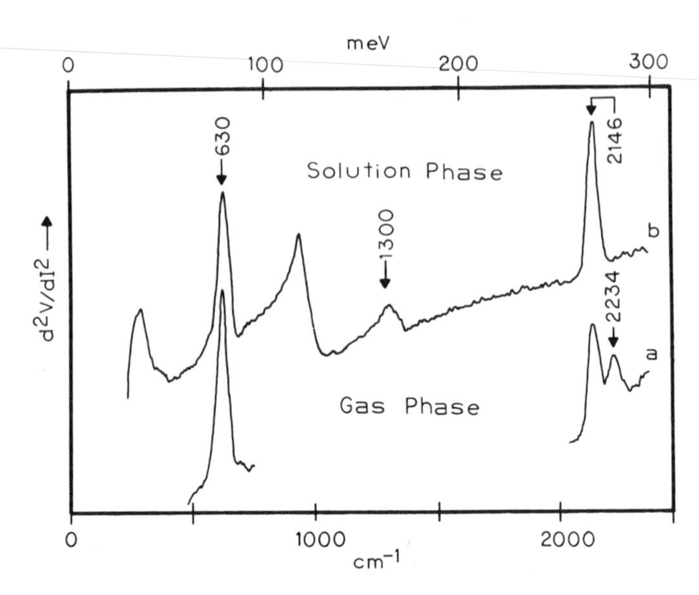

Figure 8. Tunneling spectra of the OCN^- ion. (a) gas phase (HNCO) adsorption on alumina at 200°C, and (b) aqueous solution (KNCO) doped.

alumina surface are blocked by the presence of water. Whether this is due to a physical blockage of surface sites by water or OH^-, or is due to a stabilization of the ionic form by two-dimensional hydration, requires further investigation. Clearly, however, solution phase adsorption of ionic species need *not* yield results similar to gas phase adsorption.

5. Representative Spectra

This section is a sampling of tunneling results obtained from inorganic ions. Most of the spectra shown here have been analyzed in published papers, and we will not present details of experimental procedure or analysis. Rather, attention will be directed towards the significance of results with respect to spectroscopic and surface chemical information obtained. Advantages resulting from the use of tunneling spectroscopy relative to ir or Raman will also be given.

5.1. Metal Cyanide Complexes

Of all the inorganic ions which have been studied by tunneling spectroscopy, the ferrocyanide ion is the most thoroughly studied.[6,7,39,40] The high symmetry, high stability, and relatively small size of this metal complex make it an especially interesting subject for tunneling studies. The spectrum obtained by aqueous phase doping of an Al–AlO$_x$–Pb junction with the $Fe(CN)_6^{-4}$ ion is presented in Figure 9. This figure also displays the ir spectrum obtained from a KBr pellet of $K_4Fe(CN)_6 \cdot 3H_2O$ and the Raman obtained from powdered $K_4Fe(CN)_6 \cdot \frac{1}{2}H_2O$.

As was indicated in Section 2, the Raman spectrum of the tri-hydrate salt differs but little from that of the ion in aqueous solution. The half-hydrate salt, on the other hand, has a significantly modified Raman spectrum. Note, for example, that both of the t_{1u} metal–carbon motions appear in the Raman spectrum. The appearance of these motions clearly indicates the absence of a center of symmetry for the ion. As is often the case, the marked increase in intensity of these modes is *not* accompanied by a significant change in band location. This is further emphasized by the small (1%) splitting of the t_{2g} band near 520 cm^{-1}. Previously reported bands and assignments, as well as peak positions determined from Figure 9, are given in Figure 10. The t_{1g} mode (ν_5) was assigned on the basis of combination bands; it cannot be observed as a fundamental in either ir or Raman, even in the highly distorted half-hydrate salts. Band a, observed at 479 cm^{-1} in the Raman of the half-hydrate and at 480 cm^{-1} in the

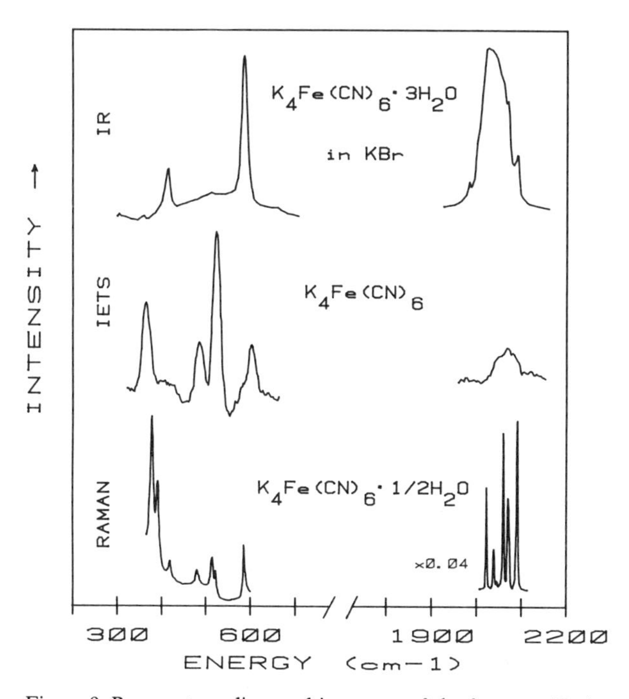

Figure 9. Raman, tunneling, and ir spectra of the ferrocyanide ion.

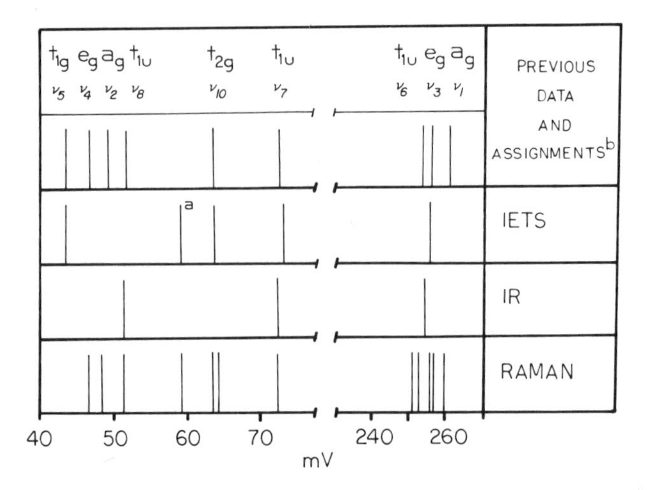

Figure 10. Peak positions from Figure 8 and previous assignments. (b) Taken from reference 5. Reprinted with permission from reference 7. Copyright (1980) The American Chemical Society.

tunneling spectrum, is not present in the Raman of the tri-hydrate or anhydrous salt.[(5)]

Figure 9 clearly demonstrates that the tunneling spectrum of the ferrocyanide ion provides more strong bands than the ir spectrum. Further, the tunneling and Raman spectra are different. The most exciting feature of the tunneling spectrum is the presence of bands at 358 and 480 cm^{-1}. The 480 cm^{-1} band is photon active only in the very distorted half-hydrate lattice, while the 358 cm^{-1} band is photon forbidden in all lattices studied. Presumably, the 358 cm^{-1} band is due to the doubly forbidden t_{1g} fundamental, while the 480 cm^{-1} band is due to one of the doubly forbidden t_{2u} modes. This assignment is contingent upon verification of the ferrocyanide ion as the surface species actually present in the tunnel diode.

That $Fe(CN)_6^{-4}$ is the species observed in solution doped Al–AlO_x–Pb barriers was verified in two ways.[(7)] Symmetrical isotopic substitution of the ferrocyanide ion was performed to produce $Fe(^{12}C^{14}N)_6^{-4}$, $Fe(^{13}C^{14}N)_6^{-4}$, and $Fe(^{12}C^{15}N)_6^{-4}$ potassium salts. The Raman, ir, and tunneling spectra of all three species were measured and the results are reported in Table 5 and Figure 11. It is clear from these results that peak positions and isotopic shifts are consistent with the ferrocyanide ion being the species observed by inelastic electron tunneling spectroscopy (IETS). The second corroborative method was based on the likely reaction of ferrocyanide with alumina in water, the formation of a mono-aquo species. Figure 12 contrasts the spectra obtained from Al–AlO_x–Pb junctions doped with aqueous solutions of $Na_4Fe(CN)_6$ and $Na_3Fe(CN)_5H_2O$. The enhanced CN stretching mode

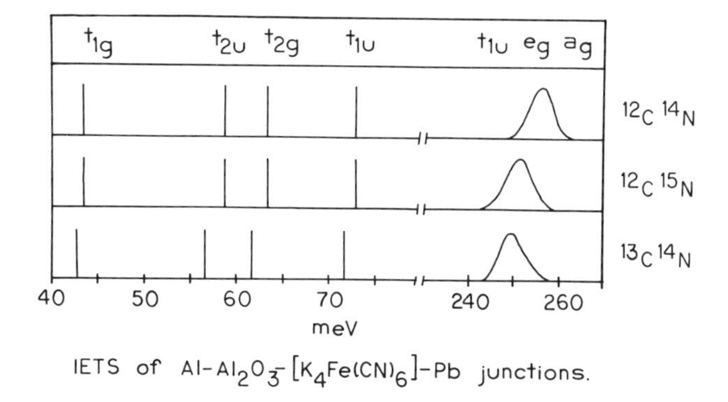

Figure 11. Assignments and isotopic shifts of bands observed in the tunneling spectrum of the ferrocyanide ion. Reprinted with permission from reference 7. Copyright (1980) The American Chemical Society.

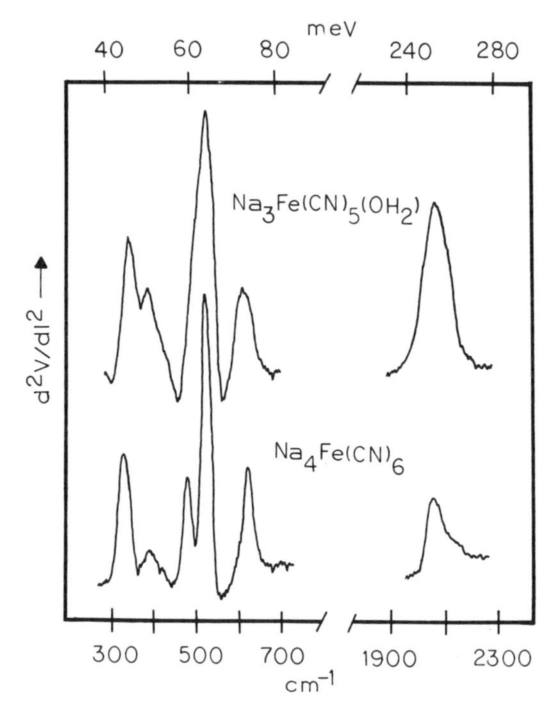

Figure 12. Tunneling spectra obtained from Al–AlO_x–Pb junctions doped with the indicated compounds from aqueous solution. Reprinted with permission from reference 7. Copyright (1980) The American Chemical Society.

Table 5. Prominent Bands Observed in the Tunneling[a] and Raman Spectra of Symmetric Isotopes of the Ferrocyanide Ion

$^{12}C^{14}N$		$^{13}C^{14}N$		$^{12}C^{15}N$	
IETS	Raman	IETS	Raman	IETS	Raman
358	—	347	—	358	—
—	377	—	370	—	370
—	390	—	381	—	383
480	479	462	462	480	478
—	514	—	—	—	512
516	—	500	500	516	—
—	522	—	—	—	522
594	584	587	576	594	583
2061	2091–2024	2010	2050–1985	2024	2059 –1992

[a] 8 cm^{-1} was subtracted from the observed positions to account for the superconductivity of the top metal.

intensity seen in the spectrum of the mono-aquo species is a characteristic of many mono-substituted complexes. Further, the metal–ligand region of the spectrum also allows the hexacyanide to be distinguished from the mono-aquo species. In summary, $Fe(CN)_6^{-4}$ is adsorbed by a *nondissociative* mechanism and exists as a unit on the oxide surface.

Having verified the existence of the ferrocyanide ion within the junction, we now review the significance of the vibrations observed. Of paramount importance is the observation of the t_{1g} fundamental. This band has not been observed as a fundamental in any photonic spectra. Even in the highly distorted half-hydrate powders, no band is seen. The intensity of this mode is, therefore, probably due to a dynamic interaction; i.e., it is due to the relaxed selection rules present in electron scattering processes. One of the forbidden t_{2u} modes is also observed. While this band is not present in the tri-hydrate and anhydrous salt spectra, it is present in the Raman of the half-hydrate powders. The relative intensities of the t_{2u} mode and the t_{2g} doublet in the Raman are also similar to the intensities seen in the tunneling spectrum. A static mechanism, therefore, can account for the appearance in the tunneling spectrum of this doubly forbidden mode. Apparently, static and dynamic mechanisms both may play a significant role in determining the tunneling intensities observed in the case of high-symmetry complexes.

An additional interesting feature of the assignments proposed in Figure 11 is that all the metal–ligand modes observed are primarily metal–ligand bending modes. The conspicuous absence of metal–ligand stretching motions is very suggestive in terms of the orientation of the ion on the surface, and fully deserves further theoretical study. The theoretical prediction of tunneling spectra obtained from octahedral and tetrahedral complexes is an exciting area which we are presently exploring.[39]

5.2. Metal Glycinates

In aqueous solution, glycine can exist in any of three forms.[41,42] Above pH 10, the anionic form is the principal species in solution. A dipolar anion is the major component in the pH 3–9 range, while a cationic form dominates below pH 3. The nature of glycine adsorbed on alumina may, therefore, exhibit pH-dependent structural changes. Above pH 7, glycine appears to adsorb primarily as a bidentate aluminum glycinate with a lesser amount of carboxylate bound zwitterion. As the solution pH is lowered to about four, the bidentate species is still present but there is increased indication of NH_3^+.[43,44] The tunneling spectra of glycine adsorbed from pH 8.6 and pH 4 aqueous solutions are presented in Figures 13 and 14.

If the adsorbate is a metal–glycine complex ion, a prototype speciation problem results. At low pH, hydrated metal ions and glycine cations are the major solution components. At very high pH, the metal ions in solution are primarily present as metal hydroxides. With the addition of an alumina surface in contact with the metal glycinate solution, the relative stability of

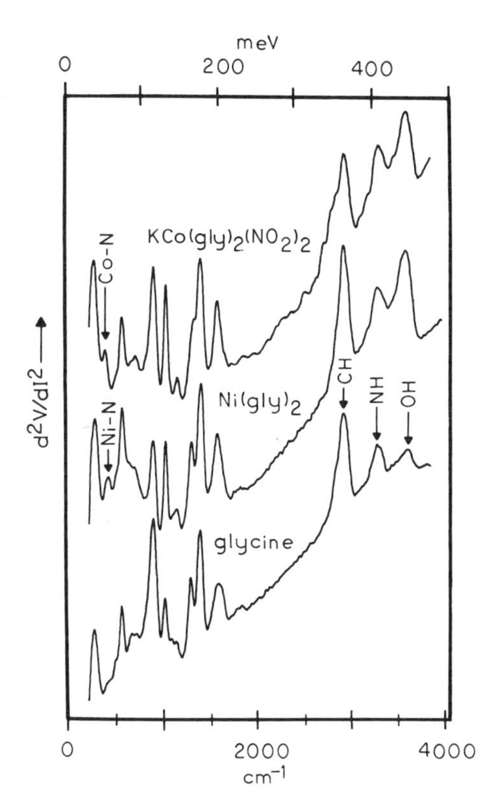

Figure 13. Tunneling spectra obtained by pH 8.6 solution phase doping of the indicated compounds.

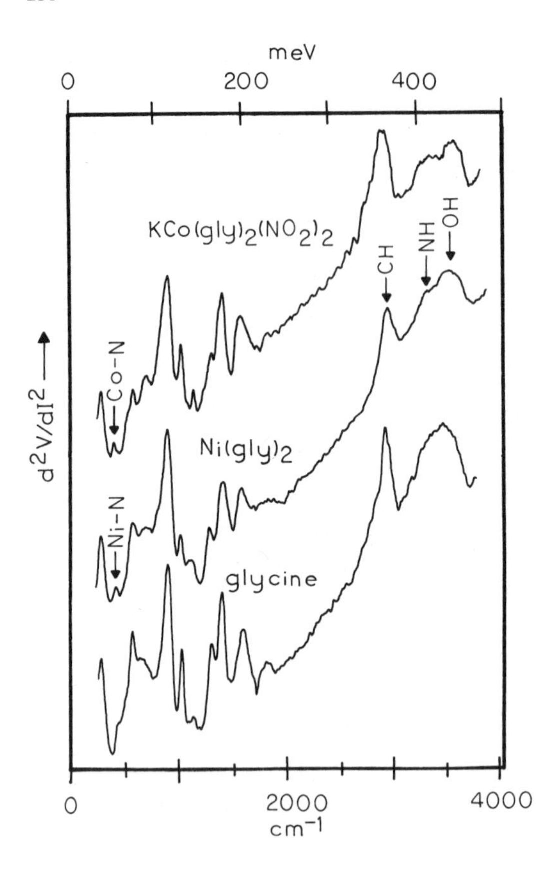

Figure 14. Tunneling spectra obtained by pH 4 solution phase doping of the indicated compounds.

surface complexes must also be considered. For some metal glycinates alumina stabilizes the metal glycine complex, while for others it plays a destabilizing role.[43] Certain nickel(II) and cobalt(III) glycine complexes are stabilized on the oxide surface.[43]

Figures 13 and 14 depict tunneling spectra obtained from Al–AlO_x–Pb junctions doped from pH 8.6 and pH 4 solutions prepared from $Ni(gly)_2$ and $KCo(gly)_2(NO_2)_2$. At pH 8.6 the $Ni(gly)_2$ and $KCo(gly)_2(NO_2)_2$ doped junctions give spectra that are clearly distinguishable from each other and from that of glycine. The most significant distinction is the presence of a well-defined metal–nitrogen stretch mode at 410 cm^{-1} (Co^{+3}) and 440 cm^{-1} (Ni^{+2}). The 1200–1600 cm^{-1} region also provides a means of distinguishing the cobalt, nickel, and aluminum species. While the cobalt complex probably binds to alumina by displacement of the NO_2^- groups in favor of surface oxide(s), the attachment mode of the nickel complex is not known. Spectra reported to date have not included the Ni–O stretching region, and it is this region which is most helpful in discriminating between the mono- and di-glycine complexes.

Adsorption from pH 4 solution (Figure 14) occurs less readily than at pH 8. It is clear, however, that the pH 8 nickel surface complex is still present at pH 4—well below the range of stability of the complex in solution. Some cobalt complex is also present on the alumina surface at pH 4, but there are indications in the spectrum of an additional species being adsorbed.[43]

The results presented here indicate that tunneling studies can be applied to speciation problems involving relatively complex equilibria. While the results reported to date are insufficient to assign the coordination sphere of the surface nickel–glycine complex, extension of the range of tunneling energies studied to the nickel–oxygen region may provide this assignment. Further, in the case of ligands having simple coordination and solution chemistry (SO_4^{-2}, CN^-, IO_3^-, etc.), tunneling should provide assignments of surface complex structures. Used in concert with surface ir and elution studies, tunneling spectroscopy can assist in analyzing very complex speciation problems.

5.3. Other Inorganic Systems

Although most of the inorganic systems studied to date are anions, this does not mean that neutral and cationic species cannot be studied. Figure 15, for example, gives the tunneling spectra obtained from the tetramethylammonium salt of the thiocyanate ion doped from acetone and ethanol. Both ions adsorb quite strongly, as indicated by the presence of thiocyanate (marked with an asterisk in the figure) and tetramethylammonium ion bands. It is also interesting to note that both types of surface bonded NCS^- (see Section 4) are present over a much wider concentration range than is the case of aqueous solution doping of KSCN. The tetraethylammonium cation also may be doped onto the oxide surface from nonaqueous solution.

Another important aspect of the study of the inorganic ions by tunneling spectroscopy is the occurrence of dissociative adsorption. Figure 16 reproduces the spectrum of $K_2Co(NCS)_4$ doped into an Al–AlO_x–Pb junction from acetone solution. This spectrum is identical to that of KNCS doped from acetone and indicates that the thiocyanate ion prefers to bond to alumina sites relative to Co^{+2}. There is no indication of any bands due to a cobalt species of any kind. It is not unusual to have the metal ion "disappear" from the junction in this way. A few of the systems for which this is the case are $Fe(oxalate)_3$, $Co(gly)_2$, $Fe(NO_2)_3$, and $Fe(acetylacetone)_3$. This behavior is not understood and could possibly be reversed by changes in solution parameters and adsorption conditions.

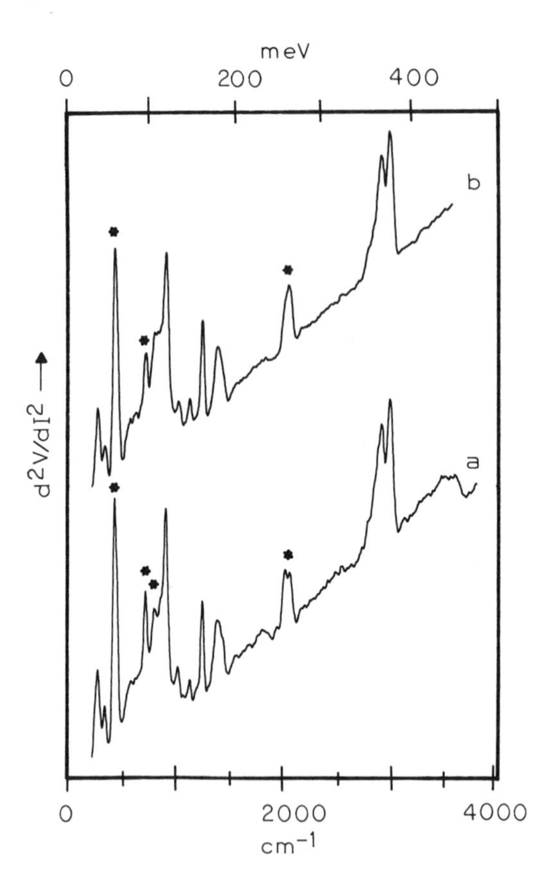

Figure 15. Low-resolution tunneling spectra of tetramethylammonium thiocyanate. Asterisk indicates NCS^- band. (a) ethanol doped; (b) acetone doped.

An additional type of dissociative adsorption in which the complex is only partly destroyed also occurs. Figure 17 contrasts the low-resolution tunneling spectra obtained from iron(II) complex, $Na_4Fe(CN)_6$, and the iron(III) nitroprusside complex $[Na_2Fe(CN)_5NO]$. The nitroprusside ion apparently undergoes reductive-dissociation during the diode fabrication process. Thus, there are no modes observed which correlate with NO motions and the CN stretch appears in the region associated with iron(II) cyanides. Reduction processes are discussed in Section 7.

Partial dissociation is expected to occur whenever one or more of the ligands of a transition metal complex is labile. Aquo, amine, and halide substitutions can lead to stronger absorption of a metal ion at the expense of the stability of the complex.(45) Surface displacement reactions of this type are of practical interest in catalyst formation and chromatography.

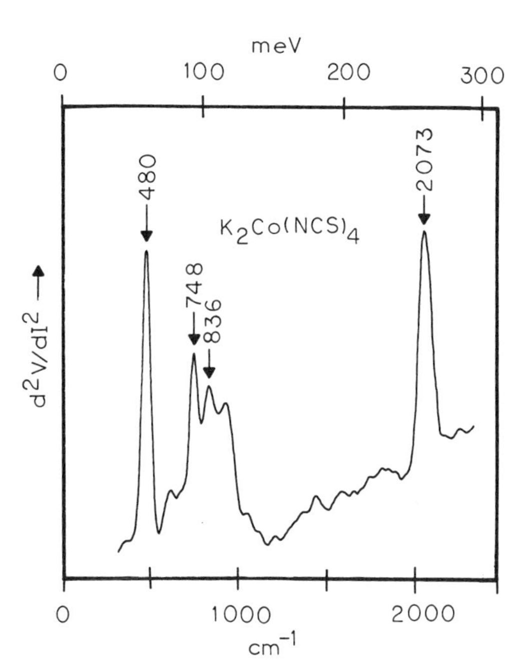

Figure 16. Low-resolution tunneling spectrum obtained from an Al–AlO_x–Pb junction doped with a $K_2Co(NCS)_4$ solution. No correction for Pb superconductivity has been made.

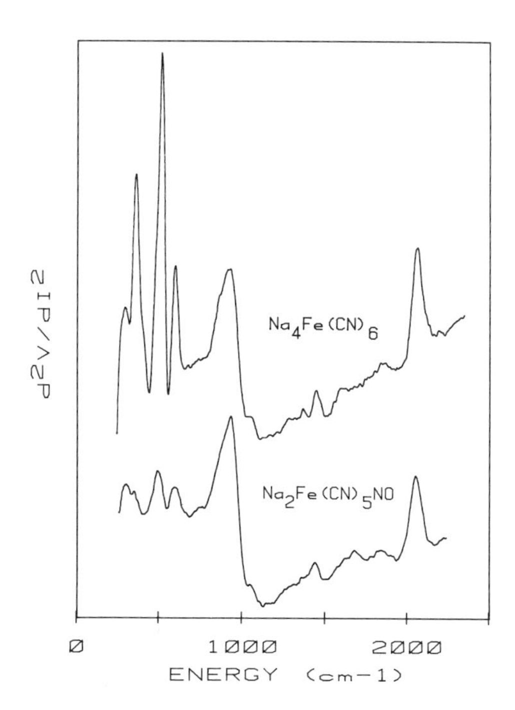

Figure 17. Medium-resolution tunneling spectra obtained from aqueous solution doped Al–AlO_x–Pb junctions. (a) $Na_2Fe(CN)_5(NO)$; (b) $Na_4Fe(CN)_6$.

6. The Role of Counterions

Since the surface of an oxide in contact with an aqueous solution is generally charged, there is no *a priori* reason to expect a balanced uptake of positive and negative ions from solution. In the case of ferrocyanide adsorption, for example, one can propose a simple ion exchange process in which surface OH^- groups are exchanged for ferrocyanide ions. In this picture, the counterions (K^+, H^+, Na^+, Li^+, or Cs^+) would play a minor role in the adsorption process. If adsorption of the cation did occur, its adsorption sites would correlate poorly with the ferrocyanide ion sites.

The above conclusions are in conflict with experimental evidence. The tunneling spectra obtained from five salts of the ferrocyanide ion show marked differences in metal–ligand band positions as a function of cation.[7] Table 6 presents the shifts relative to $K_4Fe(CN)_6$ for selected tunneling bands and for two ir active bands as measured by conventional means. The shifts of metal–ligand bands as a function of cation observed by tunneling spectroscopy are of the same order of magnitude, and generally larger, than the shifts observed by ir for the solid salts. The strength of the complex–cation interaction, therefore, indicates that cations are adsorbed in the immediate vicinity of the complex at a distance of the order of that present in crystalline solids. The distribution of cation locations must, furthermore, be relatively sharp. Were this not the case, the strong interactions observed would lead to very broad metal–ligand bandwidths.

These observations suggest the possibility of small crystallite formation as opposed to molecular adsorption. One can demonstrate, however, that the tunneling results are *not* due to crystallite formation. One method of determining the distribution of material on the surface is to measure the effective barrier width relative to undoped junctions. This was done by analyzing the current–voltage characteristics of a large number of ferro-

Table 6. Observed Shifts of Tunneling and ir Active Modes as a Function of Cation Relative to the Potassium Salt

		Salt shift (cm^{-1})			
Type	Mode	H^+	Li^+	Na^+	Cs^+
ir	t_{1u} (bend)	−3	+7	+2	−10
Tunneling	t_{1u} (bend)	+15	+14	+11	−1
Tunneling	t_{2g} (bend)	+5	+5	+1	−2
ir	CN stretch	+40	+15	+10	−5
Tunneling	CN stretch	~0	~0	~0	~0

cyanide doped junctions and no suggestion of crystallite formation was found.[7] Alternately, one may rely on purely vibrational data. If crystallization was the dominant role of ferrocyanide incorporation in tunnel diodes, all the modes of the molecule would be affected. Consideration of the CN stretching motions reported in Table 6 indicates that this is not the case.

Although all the ferrocyanide metal–ligand bands observed by tunneling spectroscopy show cation shifts, the CN stretching band does not. This can be partially explained by visualizing the adsorbed species as a $M_x\text{Fe(CN)}_6^{-4+x}$ surface salt.[7] The cations (M^+) being distributed symmetrically in the plane of the barrier, with the pseudo-fourfold axis of the complex normal to that plane.

Although cation dependence studies have so far been limited to the ferrocyanide ion, these results demonstrate the viability of tunneling spectroscopy for cation dependence studies. Of special practical interest is the role of counterions in the formation of impregnation catalysts. By combining adsorption isotherm, tunneling spectroscopic, and other conventional techniques, a really detailed picture of the role of counterions in the adsorption processes may result. It should be noted that tunneling spectroscopy is presently the only vibrational method which can be used to study metal–ligand motions of supported complexes of this type.

7. Oxidation and Reduction Processes

There are several stages in the preparation of doped tunnel junctions at which reduction and/or oxidation (redox) processes can occur. During adsorption of the species of interest, interaction of the adsorbate with the barrier can result in redox processes.[13,14] Following the adsorption step, exposure of the adsorbate to the ambient atmosphere of the laboratory can lead to redox processes. The evacuation step required for top metal deposition may induce a redox reaction. During deposition of the top metal, reactions between the adsorbate and top metal may occur. Finally, redox reactions may result during the measurement process as a consequence of the high internal fields produced by the bias voltage. Generally, it has been assumed that redox processes occur only during the adsorption step. While this may be true in the case of TCNQ,[46] and Fe(CN)_6^{-3},[7] there is evidence that indicates that later stages may also play a role in certain systems.[40]

Several attempts to measure the tunneling spectra of iron(III) cyanide complexes have resulted in the observation of iron(II) surface complexes.[7] Ferricyanide [Fe(CN)_6^{-3}] doped Al–AlO_x–M barriers (with $M = \text{Pb}$, Sn, Ag) all yield spectra identical with ferrocyanide doped junctions. Even with gold top metal barriers, there is no indication of iron(III) being present in the tunnel diode. Presumably, therefore, reduction of ferricyanide to ferrocyanide is due to reaction with the alumina barrier.

In order to test this hypothesis, the adsorption of ferricyanide and ferrocyanide on bulk alumina (12 m^2/g) was studied by transmission ir spectroscopy. Aluminas were generated in acid, neutral, and basic state by equilibration with NaOH or HCl solution and the potassium salts were adsorbed from similar pH solutions. Although substrate ir absorption precludes the observation of the metal–ligand motions, the CN stretching frequencies of iron(II) and iron(III) cyanides occur in the window region of alumina. Figure 18 shows a few of the ir spectra obtained. The ferrocyanide ion is stable on alumina throughout the pH 4–9 range. The ferricyanide ion, however, is only stable in acid solution. An increasing amount of ferrocyanide CN stretch is observed with increasing pH, with a significant amount being present at pH 7. Thus, at least part of the reduction observed on the alumina barrier of tunnel diodes is due to reaction with the alumina barrier.

Tunneling studies of the ferricyanide ion have also shown that redox reactions with the top metal can occur.[40] Gold undergoes a redox reaction with alumina supported iron hexacyanide to produce gold(I) dicyanide. Figure 19c is the tunneling spectrum obtained from an Al–AlO_x–Au junction that was doped with ferricyanide prior to gold deposition at room temperature. Figure 19b is the spectrum obtained from an Al–AlO_x–Pb junction doped with $KAu(CN)_4$. The latter spectrum also provides an example of a redox reaction, since the surface species CN stretch is in the gold(I) [and not the gold(III)] region of the spectrum. The similarity of Figures 19b and 19c is striking. It is especially interesting that the fer-

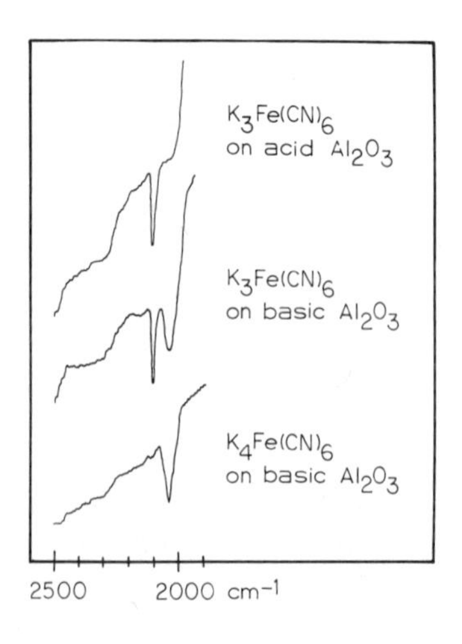

Figure 18. ir transmission spectra of complexes supported on 12 m^2/gram alumina.

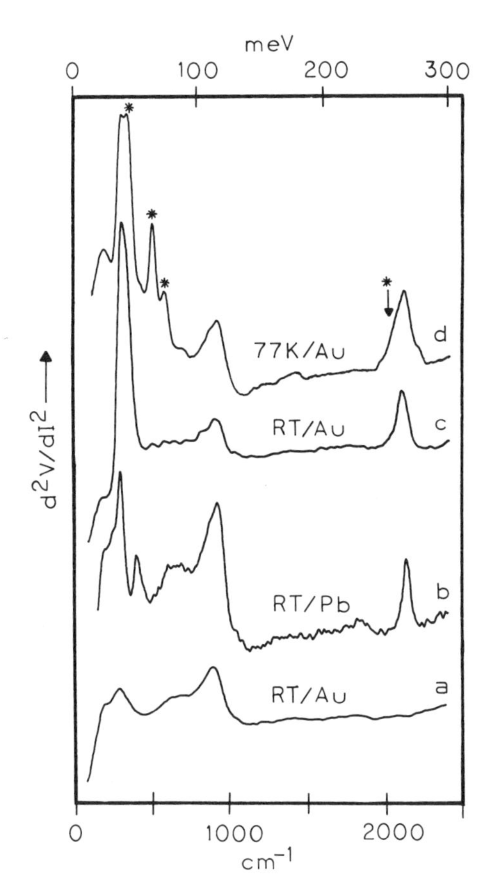

Figure 19. Low-resolution tunneling spectra. (a) Al–AlO_x–Au blank. (b) Al–AlO_x–Pb junction doped with an $Au(CN)_4^{-1}$ solution. (c) Al–AlO_x–Au junction doped with $K_3Fe(CN)_6$. (d) Al–AlO_x–Au junction doped with $K_3Fe(CN)_6$ and cooled to 77 K prior to and during Au deposition. Asterisk indicates bands due to $Fe(CN)_6^{-4}$.

ricyanide doped barrier shows no indication of the presence of iron, presumably owing to migration of reduced iron into the top metal.

Figure 20 presents the tunneling spectra obtained from junctions of the type Al–AlO_x–$KAu(CN)_2$–M, where $M =$ Au or Pb. Using the 118-mV Al–O motion as an intensity standard, it appears that an unexpected intensity enhancement occurs for the gold topped junctions. Whether this is due to chemical or physical interaction mechanisms is presently not known.

This top-metal-dependent redox process is due to thermal reaction of ferrocyanide with incident gold atoms during the early stages of the deposition process. This was demonstrated by the production of a ferricyanide doped gold top metal barrier at 77 K. Figure 19d shows that cooling the substrate during gold deposition reduces the extent of the $Au + Fe(CN)_6^{-4}$ reaction. To our knowledge, this is the only documented case of complete chemical modification of the adsorbate due to top metal reaction.

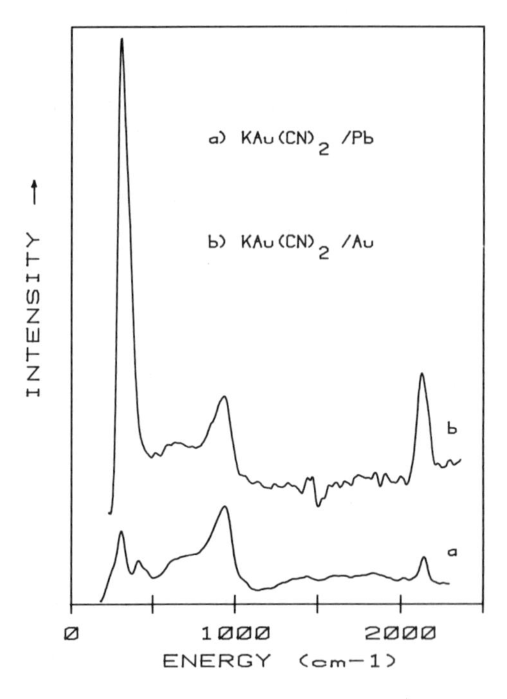

Figure 20. Al–AlO_x–$KAu(CN)_2$ [from acetone]—*M* junction tunneling spectra. They are drawn to make the Al–O motion intensities approximately equal. Note the tremendous intensity enhancement in the case of a gold top layer.

8. What's Next?

There is a clear and pressing need for less reactive barrier materials. The extent and complexity of transition metal complex reactions with alumina presently limits the range of systems that can be studied by tunneling. It would be especially convenient to have a barrier material which was soluble in water, ethanol, or acetone and could be "spin-dried" onto the substrate. This would allow the molecular system of interest to be intimately mixed in the barrier. Materials of this type are presently known, but their spectra dominate the entire vibrational region. For catalytic studies, a silica or alumina-silicate barrier is especially desirable. Further, the lower redox reactivity of silica also makes it an attractive barrier for molecular spectroscopic studies.

Theoretical calculations of inorganic ion spectra are needed. The tunneling spectra obtained from these ions clearly contain information about ion orientation. Extracting this information requires detailed theoretical calculations. In the future, the comparison of experimental data and theoretical calculation should become more common.

The handling procedures used during the doping process must be improved. Many spectroscopically and catalytically interesting systems cannot be vapor deposited or exposed to air. In order to study these systems an integrated inert atmosphere glove box and deposition system is required. Ideally, the substrate holder in the deposition chamber could be cooled to 4 K to allow for total control of construction and measurement conditions.

The study of inorganic ions has just begun. The few systems studied provide a tantalizing glimpse of the possible applications of tunneling, but they do not comprise a representative sampling. Many more systems must be studied before the value of tunneling spectroscopy to the study of inorganic ions can be assessed.

9. Conclusions

(1) Tunneling spectroscopy can be used to observe the photon forbidden vibrational transitions of high-symmetry metal complexes.

(2) Dissociation and reduction reactions of metal complex ions on alumina frequently occur. Surface species identification, therefore, is an important aspect of the study of inorganic complexes by tunneling spectroscopy.

(3) Tunneling is, potentially, a useful method for studying low-lying electronic states of transition metal complexes.

(4) Tunneling spectroscopy can be used to study counterion participation in the adsorption process.

(5) Tunneling spectroscopy can be used to study impregnation (adsorption) catalysts.

(6) Substrate cooling during top metal deposition reduces the probability of adsorbate–top-metal reaction.

References

1. L. H. Jones, *Inorganic Vibrational Spectroscopy*, Vol. 1, Marcel Dekker, New York (1971).
2. K. Nakamoto, *Infrared and Raman Spectra of Inorganic and Coordination Compounds*, 3rd Ed., John Wiley and Sons, New York (1978).
3. F. A. Cotton, *Chemical Applications of Group Theory*, Wiley-Interscience, New York (1971).
4. J. S. Griffith, *The Theory of Transition-Metal Ions*, Cambridge University Press, London (1971).
5. W. P. Griffith and G. T. Turner, Raman spectra and vibrational assignments of hexacyano-complexes, *J. Chem. Soc. A* **1970**, 858–862.
6. K. W. Hipps, Ursula Mazur, and M. S. Pearce, A tunneling spectroscopy study of the adsorption of ferrocyanide from water solution, *Chem. Phys. Lett.* **68**, 433–436 (1979).
7. K. W. Hipps and Ursula Mazur, An IETS study of some iron cyanide complexes, *J. Phys. Chem.* **84**, 3162–3172 (1980).

8. A. Leger, J. Klein, M. Belin, and D. Defourneau, Electronic transitions observed by inelastic electron tunneling spectroscopy, *Solid State Commun.* **11**, 1331–1335 (1972). See also Chapter 4 of this book.
9. J. Kirtley and P. K. Hansma, An experimental test of symmetry selection rules in IETS, *Surf. Sci.* **66**, 125–130 (1977).
10. S. Sugano, Y. Tanabe, and H. Kamimura, *Multiplets of Transition-Metal Ions in Crystals*, Academic Press, New York (1970).
11. R. E. DeSimone, Electron spin resonance studies of low spin d^5 complexes, *J. Am. Chem. Soc.* **95**, 6238–6244 (1973).
12. J. A. Stanko, H. J. Peresie, R. A. Bernheim, and R. Wang, Trigonal field splitting in tris(ethylenediamine) complexes, *Inorg. Chem.* **12**, 634–639 (1973).
13. H. Hosaka, N. Kawashima, and K. Meguro, Estimation of the surface properties of metal oxides by the use of TCNQ, *Bull. Chem. Soc. Jpn.* **45**, 3371–3375 (1972); H. Hosaka, T. Fugiwara, and K. Meguro, *Bull. Chem. Soc. Jpn.* **44**, 2616–2619 (1971).
14. M. Che, S. Coluccia, and A. Zecchina, Electron paramagnetic resonance study of electron transfer at the surface of alkaline earth oxides, *J. Chem. Soc. Faraday Trans. 1* **74**, 1324–1328 (1978).
15. T. A. Dorling, B. W. J. Lynch, and R. L. Moss, The structure and activity of supported metal catalysts, *J. Catal.* **20**, 190–201 (1971).
16. F. Bozon-Verduraz, A. Omar, J. Escard, and B. Pontvianne, Chemical state reactivity of supported palladium, *J. Catal.* **53**, 126–134 (1978).
17. J. P. Brunelle, Preparation of catalysts by metallic complex adsorption on mineral oxides, *Pure Appl. Chem.* **50**, 1211–1229 (1978).
18. B. Delmon, P. Grange, P. Jacobs, and G. Poncelet, *Preparation of Catalysts II*, Elsevier Press, New York (1979).
19. J. G. Grasselli, M. K. Snavely, and B. J. Bulkin, Applications of Raman spectroscopy, *Phys. Rep.* **65**, 231–344 (1980).
20. G. B. McVicker and M. A. Vannice, The preparation, characterization, and use of supported potassium–group VII metal complexes as catalysts for CO hydrogenation, *J. Catal.* **63**, 25–34 (1980).
21. J. O. Leckie and R. O. James, *Aqueous Environmental Chemistry of Metals* (J. A. Rubin, ed.), Ann Arbor Science, Ann Arbor (1974).
22. W. Stumm and J. J. Morgan, *Aquatic Chemistry*, J. Wiley and Sons, New York (1970).
23. A. E. Martell, The influence of natural and synthetic ligands on the transport of metal ions in the environment, *Pure Appl. Chem.* **44**, 81–113 (1975).
24. M. H. Cheng, *J. Water Pollut. Control Fed.* **47**, 302–310 (1975).
25. J. A. Rubin, *Chemistry of Waste Water Technology*, Ann Arbor Science, Ann Arbor (1978).
26. A. C. Bourg and P. W. Schindler, Effect of ethylenediaminetetraacetic acid on the adsorption of copper(II) on silica, *Inorg. Nucl. Chem. Lett.* **15**, 225–229 (1979).
27. J. A. Davis and J. O. Leckie, Speciation of adsorbed ions at the oxide/water interface, *ACS Symp. Ser.* **93**, 299–317 (1979).
28. L. G. Sillen and A. E. Martell, *Stability Constants of metal–Ion Complexes*, Metcalfe and Cooper Limited, London (1964).
29. I. N. Levine, *Physical Chemistry*, McGraw-Hill, New York (1978).
30. N. K. Eib, A. N. Gent, and P. N. Henriksen, Formation of SiH bonds when SiO is deposited on alumina, *J. Chem. Phys.* **70**, 4288–4290 (1979).
31. U. Mazur and K. W. Hipps, An IETS study of alumina and magnesia supported SiH_x and their uses for inorganic vibrational spectroscopy, *Chem. Phys. Lett.* **79**, 54–58 (1981).
32. U. Mazur and K. W. Hipps, Characterization of alumina supported O_ySiH by inelastic electron tunneling spectroscopy, *J. Phys. Chem.* **85**, 2244–2249 (1981).

33. F. Solymosi, L. Volgyesi, and J. Sarkany, The effect of support on the formation and stability of surface isocyanate on platinum, *J. Catal.* **54**, 336–344 (1978).
34. F. Solymosi, J. Kiss, and J. Sarkany, On the reaction of surface isocyanate over platinum catalyst, *Proceedings of the Seventh International Vacuum Congress and Third International Conference on Solid Surfaces*, (R. Dobrozemsky, ed.) Vienna (1977).
35. R. A. Della Betta and M. Shelef, Isocyanates from the reaction of CO+NO on supported noble-metal catalysts, *J. Mol. Catal.* **1**, 431–434 (1976).
36. F. Solymosi and T. Bansagi, Infrared spectroscopic study of the adsorption of isocyanic acid, *J. Phys. Chem.* **83**, 552–553 (1979).
37. U. Mazur and K. W. Hipps, An IETS study of the adsorption of NCS^- OCN^-, and CN^- from water by Al_2O_3, *J. Phys. Chem.* **83**, 2773–2777 (1979).
38. R. K. Knockenmuss and K. W. Hipps, work in progress.
39. S. D. Williams and K. W. Hipps, work in progress.
40. U. Mazur, S. D. Williams, and K. W. Hipps, A novel top metal–adsorbate reaction observed by tunneling spectroscopy, *J. Phys. Chem.*, **85**, 2305–2308 (1981).
41. K. P. Anderson, W. O. Greenhalgh, and E. A. Butler, Formation constants, enthalpy, and entropy values for the association of nickel(II) with glycinate, alanate, and phenylalanate ions, *Inorg. Chem.* **6**, 1056–1058 (1967).
42. K. Krishnan and R. A. Plane, Raman study of glycine complexes of zinc(II), cadmium(II), and the formation of complexes in solution, *Inorg. Chem.* **6**, 55–60 (1967).
43. K. W. Hipps and U. Mazur, pH Dependence of surface species formed by the adsorption of cobalt and nickel glycinates, *Inorg. Chem.* **20**, 1391–1395 (1981).
44. M. G. Simonsen and R. V. Coleman, High resolution IETS of macromolecules and adsorbed species with liquid phase doping, *J. Chem. Phys.* **61**, 3789–3799 (1974).
45. R. L. Burwell, R. G. Pearson, G. L. Haller, P. B. Tjok, and S. P. Chock, The adsorption and reaction of coordination complexes on silica gel, *Inorg. Chem.* **4**, 1123–1128 (1965).
46. C. S. Korman and R. V. Coleman, Inelastic electron tunneling spectroscopy of single ring compounds adsorbed on alumina, *Phys. Rev. B* **15**, 1877–1893 (1977).

9

Studies of Electron-Irradiation-Induced Changes to Monomolecular Structure

M. Parikh

1. Introduction

1.1. Why Study Irradiation-Induced Molecular Structure Changes?

The study of irradiation-induced changes to molecular structure has been pursued by radiation physicists and chemists for many decades. In recent years with the advent of techniques to probe the structure of molecules adsorbed on surfaces, there is a great need to quantitatively understand the changes induced in such structures due to the action of the probe. For example, in atomic resolution electron microscopy,[1] the deleterious action of energetic electrons needs to be understood and eventually minimized; in electron spectroscopy[2] of adsorbed molecular films, the degrading action of energetic electrons on the energy-loss spectrum needs to be quantified, in order to completely characterize the spectra; in electron diffraction[3] studies of molecular crystals, the deterioration of intensity due to structure changes needs to be convolved with the true diffraction pattern; in Auger electron spectroscopy,[4] the sensitivity of the spectrum to small

M. Parikh • Hewlett-Packard Laboratories, 3500 Deer Creek Road, Palo Alto, California 94304.

changes in submonolayer quantities of the adsorbed molecules needs to be understood. In brief, the action of the probe on a monolayer molecular specimen needs to be understood and quantified. It is hoped that such fundamental understanding will lead to "tailoring" of molecules that are more resistant (or more sensitive) to radiation action, thereby opening new avenues for applications of radiation-induced adsorbed molecular chemistry.

1.2. Why Use Tunneling Spectroscopy?

Radiation chemistry of molecular films and bulk specimens has been studied[5,6] extensively using conventional spectroscopies. The radiation chemistry of adsorbed monolayers of molecules is a more difficult study. Energetic radiation (generally electrons) has been used to characterize the action of the radiation itself. Changes in molecular electronic structure,[2] changes in molecular crystallinity,[7] or changes in mass[8] yield information about the electron exposure necessary for particular changes to occur but cannot inherently yield information about the changes in molecular bonding or structure. Vibrational spectroscopies, such as infrared or Raman, can provide molecular structure and bonding information but are difficult[9] to apply to a monolayer of adsorbed molecules.

Tunneling spectroscopy, on the other hand, is quite well suited to the radiation chemistry of monolayers of adsorbed molecules. This is due to the following reasons: (1) It is a vibrational spectroscopy of adsorbed molecules, sensitive to both ir and Raman modes. (See Chapters 1 and 2 for details.) (2) It is extremely sensitive[10]; in fact, it is sensitive to submonolayers of adsorbed molecules. (3) It is quantitative: theoretically the intensity of a peak from the tunneling spectrum is linearly proportional to the number of molecules contributing to that mode. This relative ease of quantification makes possible the application of tunneling spectroscopy to irradiation studies. (4) It seems to be relatively insensitive to the overlayer, in the sense that the overlayer does not radically deform molecular structure (the vibrational frequencies from tunneling spectroscopy correspond quite well to those from neat infrared and Raman spectra.)

1.3. Scope of this Chapter

This chapter is aimed towards providing a coherent overview on the results and interpretation of the state-of-the-art experiments involving electron irradiation. Discussion on experiments involving uv exposure are presented in Chapter 7; here we concentrate on electron irradiation experiments. Finally and possibly most importantly, suggestions are made for more advanced experiments that may yield new insights into fundamental radiation physics and chemistry of adsorbed molecules.

2. Present State-of-the-Art Experiments

2.1. Electron Irradiation Experiments

The earlier experiments that involved the use of tunneling spectroscopy as a tool for radiation-induced structural changes to molecular monolayers, were performed[12] by Paul Hansma and the author in 1975. These and subsequent experiments,[12–14] which will be designated to be of zeroth order in complexity, involved a routine preparation[12] of the tunnel junction (with the molecules under study), irradiation in a commercial scanning electron microscope, and subsequent routine[12] measurement of the tunneling spectra of the unirradiated as well as the irradiated junctions. The differences between the spectra were attributed to the action of the energetic electrons on the sandwiched molecules. In this section, a summary will be presented of the experimental conditions, the underlying assumptions, the results for a family of molecules, and the general interpretations and trends.

The zeroth-order experiments were performed (see Chapter 1) by (1) preparing tunnel junctions with molecules of interest. This was achieved by evaporating aluminum strips (2000 Å thickness, typically) onto a clean glass slide. The aluminum strips were allowed to oxidize in air, thereby forming a thin (~ 20 Å) layer of alumina. A liquid solution containing the molecule of interest was spun on the alumina strip, thereby "doping" the insulator. The junction is completed by evaporating several strips of lead (typical thicknesses of ~ 2000–2800 Å); this yields several junctions. By placing the junctions in close geometrical proximity to each other (typically less than 1 mm apart), approximately similar characteristics for the tunneling spectra are generally obtained. This can be checked with low-power ohmmeter measurements of the junction resistance. (2) Irradiation of the junctions (and thus the sandwiched molecules) is performed in a commercial scanning electron microscope. Generally, different junctions on a given glass substrate were irradiated with different incident electron exposures, while one junction was left unexposed. Typical irradiation parameters were as follows: electron beam voltage, 30 keV; beam current, 0.1–30 nA; exposure time, 2–10 min; pressure, 10^{-5}–10^{-7} Torr; temperature, ~ 300 K. (3) The spectra were measured in a routine manner by immersing the junctions in liquid helium and then taking second-harmonic measurements of the tunnel junction voltage. The peaks in the spectra, as is well known, can be correlated to the vibrational modes (and thus the structure) of the molecule.

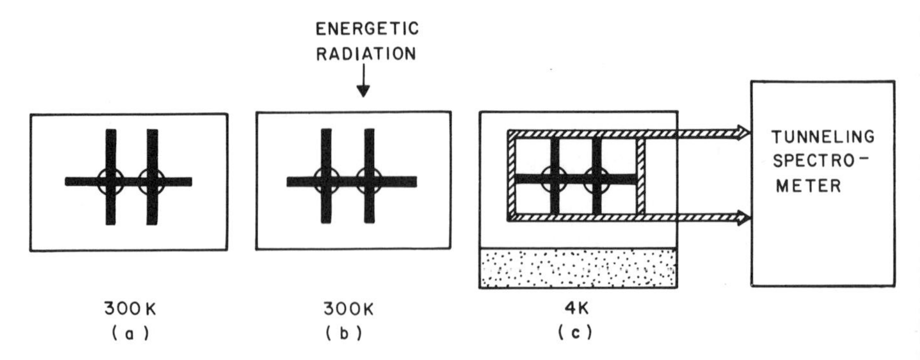

Figure 1. Schematic depicting the three steps involved in the zeroth-order experiments. (a) Completed junction at 300 K, consisting of a metal electrode, an organic monolayer (depicted by circles), and two metal electrodes. (b) Irradiation of the junction at 300 K. (c) Measurement of the tunneling spectrum of the junction at 4 K.

These three steps comprising the zeroth-order experiments are depicted schematically in Figure 1.

As an example, the results[12,13] of such experiments with β-D-fructose are shown in Figure 2. The effect of electron irradiation is to produce changes in intensity of various bands in the tunneling spectrum. These can be summarized as follows: The bands at 1100 cm^{-1} (denoted by △) and at 1260 cm^{-1} (▼) can be associated with the COH functional group in fructose; the former is assigned to the CO stretching vibration, the latter to the OH bending vibration. Both these seem to decrease in intensity

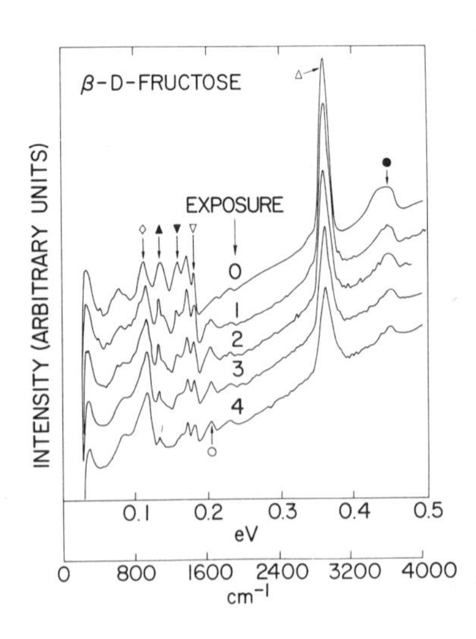

Figure 2. Tunneling spectra of junctions containing β-D-fructose irradiated by 30-keV electrons for different incident electron exposures in a scanning electron microscope. Curves are labeled by a unit of exposure $E_0 \sim 14$ mC/cm². Each curve was traced in approximately 100 min with a lock-in sensitivity of 2 μV (full scale) and a time constant of 3 sec. Wave numbers: ▲, 1100; ▼, 1260; ▽, 1460; △, 2900; and ●, 3580 cm^{-1}.

as a function of exposure, at approximately the same rate. The bands at 1460 cm^{-1} (denoted by ▽) and at 2920 cm^{-1} (△) can be associated with the various stretching modes of the CH, CH_2 functional groups in fructose. These too seem to decrease in intensity as a function of exposure; however, at a different rate than those for the COH functional group. In contradistinction a peak grows at 1638 cm^{-1} (denoted by 0) as a function of electron exposure; this position is characteristic of the C═C stretching vibrations. This suggests that as the COH and CH, CH_2 groups are destroyed by the action of energetic electrons, double bonds are formed between C atoms. Finally, the transformation of the symmetrical peak at 900 cm^{-1}, into an asymmetrical peak around 940 cm^{-1}, leads to the speculation of the persistence of the ring-stretching mode even at the high exposure values.

2.2. Underlying Assumptions

The association of the change in the spectra with electron irradiation requires the justification of the following assumptions: (1) spectral changes are *not due to specimen heating* caused by the electron beam. This was verified by heating the tunnel junctions to temperatures as high as 150°C for a period of ~60 sec. Results[13] of such experiments showed no discernible change in the tunneling spectrum. (2) Spectral changes are *not due to any exposure rate effects*. Experiments[13-15] were performed on two molecules wherein nearly identical spectra were obtained for cases when the total exposure was kept the same, but the exposure rate was varied by a factor of 5. These experiments led to a conclusion that the changes in spectra due to electron irradiation can be attributed to the action of electrons on the molecules in the junction.

The quantitative application of the tunneling spectra to irradiation studies of molecular monolayers necessitates additional assumptions. These are as follows: (1) Similar junctions have essentially identical spectra. Experimentally, this is generally the case for junctions prepared on the same glass substrate. Such an assumption is imperative (especially in the zeroth-order experiments) in making comparisons possible between different junctions that have been subjected to different exposures. (2) Peak areas (after appropriate background subtraction) can be correlated with the number of molecules in the junction. Theoretically (see Chapter 1 of this volume), the peak height is expected to be linearly proportional to the number of molecules participating in the particular vibrational mode. Experimentally,[11] the relationship between peak heights and the number of molecules in the junction is measured using radiographic techniques; the result is that an approximately linear relationship exists. The peak shapes approximate[13] Gaussian functions; this facilitates relatively simple fit

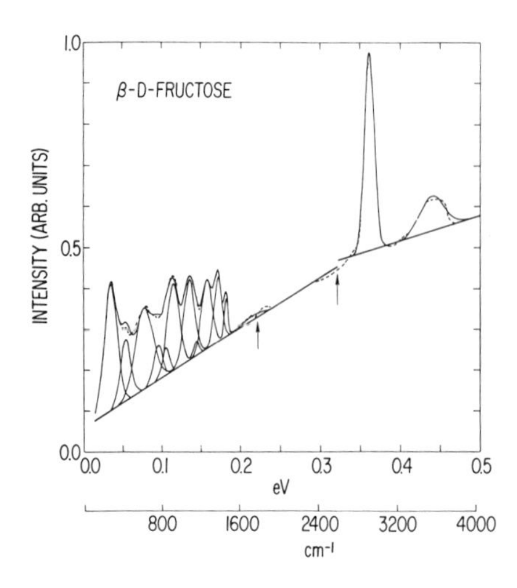

Figure 3. Nonlinear least-squares fit of spectrum (dashed curve) of undamaged β-D-fructose to a series of Gaussians (solid curves) with adjustable parameters. The background is determined by a linear least-squares fit of data in the region included by two arrows.

(Figure 3) of a spectrum with a collection of Gaussian functions. The linearly rising background is due to elastic tunneling; this is a good assumption (and theoretically justifiable) for values of $V \leqslant 0.5$ V. In Figure 3 the background due to elastic tunneling is determined by fitting a straight line to the data between 1800 and 2600 cm^{-1}.

2.3. Determination of "Damage" Cross Sections

The changes in the peak intensities of the tunneling spectra of irradiated molecules can be quantitatively related to the damage cross sections of these molecules. This relationship has been worked out in detail elsewhere.[13] Here it can be summarized through the equation

$$n(E) = n_0 \exp(-\sigma_D E) \tag{1}$$

where n_0 is the concentration of the undamaged molecules, $n(E)$ is the concentration of the damaged molecules after an exposure E (in units of electrons per unit area), and σ_D is the "damage" cross section of the molecule for the incident electrons.

The simple expression in Eq. (1) belies some fundamental assumptions: The incident electron exposure is related to the effective flux of electrons that appears in the vicinity of the molecules. Figure 4 shows a schematic of the geometry involved in the zeroth-order experiments; the contributions due to forward, secondary, and backward scattered electrons are also depicted. It is apparent that the flux of electrons appearing at the molecules is different from that at the surface of the junction. Detailed Monte Carlo

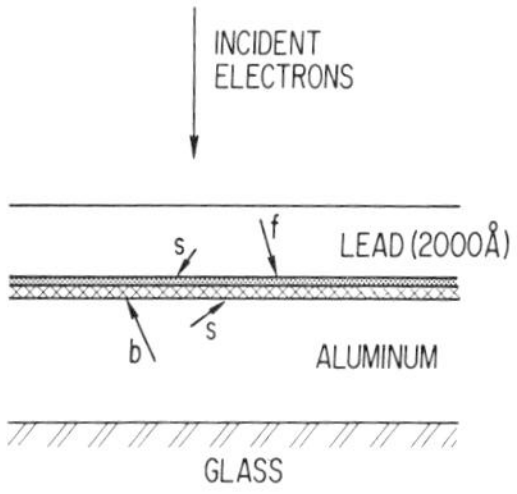

Figure 4. Schematic illustrating the geometry of the Al–Al_2O_3–organic-Pb junction on a glass substrate. The incident electrons generate forward-scattered (f), backscattered (b), and secondary electrons (s), as shown schematically.

simulation[13] has been performed on the geometry shown in Figure 4. Results of calculations for energy distributions of electrons traversing the lead–aluminum interface are shown in Figure 5, for the case of 30-keV electrons incident on different thicknesses of lead. For lead thickness 3000 Å, the distribution is highly skewed towards the incident electron energy. Typically over 50% of the electrons traversing the junction have energy within 10% of that of the incident electrons. Thicker lead overlayer (~6000 Å) leads to a more dispersed distribution. Finally note, through the factor Δ_0, that the number of electrons that traverse the junction may be greater than that incident. Taking into account the nonmonoenergetic

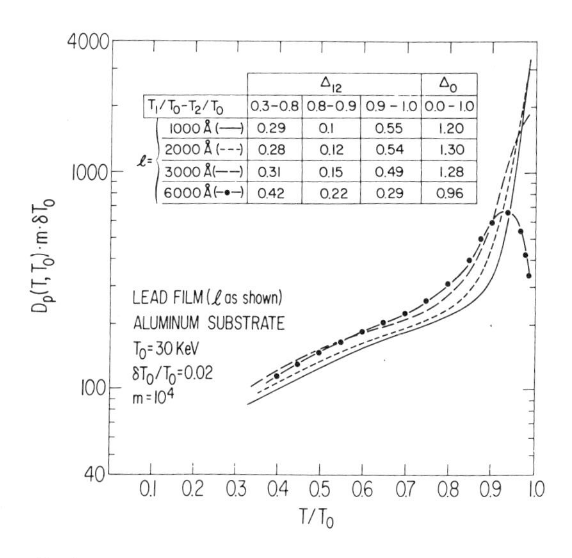

Figure 5. Energy distribution of scattered electrons at the interface of a lead film (thickness as shown) and aluminum substrate due to 30-keV incident electrons. Δ_0 is the average number of crossings of the interface region by a primary electron; while Δ_{12} is equal to the probability of an electron with energy between T_1 and T_2 crossing the interface.

nature of the electron distribution at the interface, the increased number of electrons traversing the interface, and the generation of slow secondary electrons, the relationship between the incident electron exposure and the electron influence at the sandwiched molecules is complicated. Reference 14 has derived some proportionality factors that can simplify the complications mentioned above.

The second assumption inherent in the exponential form of Eq. (1) is that the energetic radiation has an independent action on each molecule or a vibrational mode. In other words there is little intermolecular coupling between molecules or "bonds" during the action of the energetic electrons. In the absence of such a condition, nonexponential decay curves for vibrational mode intensity are observed.[13] Clearly, exponential decay is obtained if only the "initial" action of the radiation action is of interest. This has been the case in the experiments reported in references 13 and 14.

In summary, the straightforward association [given by Eq. (1)] of damage cross sections with measured changes in the intensity of tunneling spectra is possible, especially if experimental conditions are chosen so as to justify the above-mentioned assumptions. Finally, it is worth noting that damage cross section at energy T_1 can be estimated from measured cross section at energy T_2 via the general relationship[13,14]

$$\sigma_D(T_1) \leqslant T_1/T_2\sigma_D(T_2) \tag{2}$$

under the condition that $T_2 > T_1 > B$, where B is a constant, typically less than ~ 2 keV.

2.4. General Trends

Two sets of comprehensive experiments have been reported in the literature[13,14,16] that can help define some generalizations. The experiments[17] of Clark and Coleman on uv-induced damage to nucleotides has been discussed in Chapter 7. Here we will concentrate on the experiments[15] of Hall *et al.* on electron-induced damage to a family of saturated and conjugated hydrocarbons.

Results for a set of junctions doped with solutions of hexanoic acid, sorbic acid, and benzoic acid are shown in Figure 6; the acids react with alumina to form hexanoate, 2, 4-hexadienoate and benzoate ions.

Detailed analysis[15] of the spectra in Figure 6 yields the following generalizations: (1) Benzoate ions are the most resistant to damage; hexanoate ions, the least resistant. (2) "Initial" damage cross sections [as calculated through Eq. (1), using exposure values where peak intensities decrease exponentially], listed in Table 1, are generally smaller for the symmetric

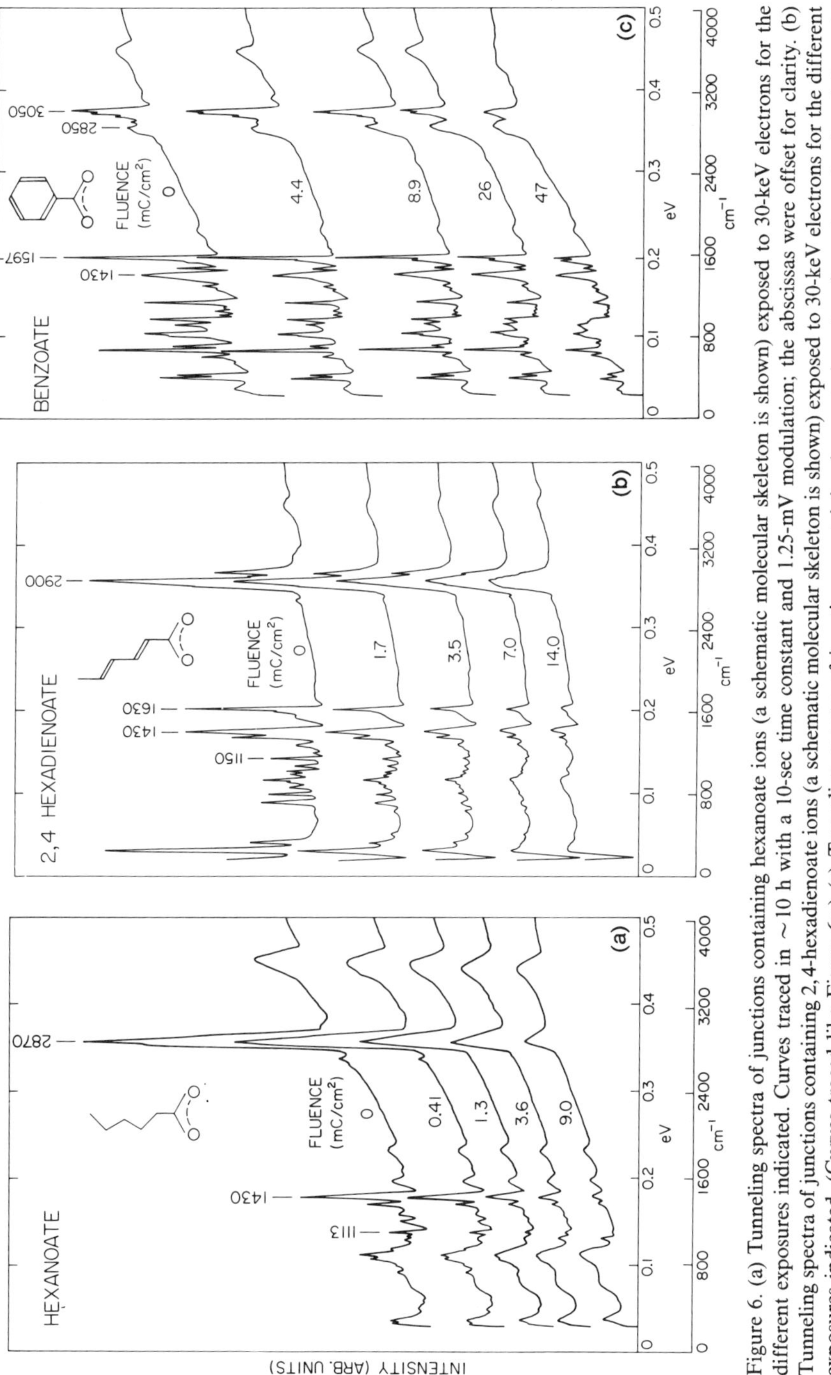

Figure 6. (a) Tunneling spectra of junctions containing hexanoate ions (a schematic molecular skeleton is shown) exposed to 30-keV electrons for the different exposures indicated. Curves traced in ~10 h with a 10-sec time constant and 1.25-mV modulation; the abscissas were offset for clarity. (b) Tunneling spectra of junctions containing 2,4-hexadienoate ions (a schematic molecular skeleton is shown) exposed to 30-keV electrons for the different exposures indicated. (Curves traced like Figure 6a.) (c) Tunneling spectra of junctions containing benzoate ions (a schematic molecular skeleton is shown) exposed to 30-keV electrons for the different exposures indicated. (Curves traced like Figure 6a).

Table 1. Initial Damage Cross Sections (Å^2/Incident 30-keV Electron)[a]

Mode	Hexanoate	2, 4-Hexadienoate	Benzoate
$\nu(C—O_2)_{sym}$	0.53±0.16	0.37±0.06	0.017±0.006
ν(C—H)	1.3 ±0.4	0.26±0.04	0.038±0.011
ν(C—C)	0.9 ±0.3	0.70±0.30	
η(C=C)		0.56±0.13	
ν(ring)			0.038±0.013

[a]See reference 15 for details.

carboxylate stretch $(C—O_2^-)_{sym}$ than for the C—C, C=C, or the ring modes in the three ions. This indicates that disintegration of the carbon chain or the ring precedes the disassociation of the ion/fragments from the alumina substrate. (3) The analysis of the behavior of the C—H bonds is complicated by the presence of any possible "impurity" hydrocarbons in the junctions, which also leads to a ν(C—H) peak at approximately the same locations. Results shown in Table 1 indicate the frailty of the C—H bonds in the saturated molecules; with increasing conjugation of the molecules, the C—H bonds become surprisingly resistant. This has been attributed to increased π electron delocalization and thus an increased probability for converting the excitation energy into nondissociative channels. The work of Clark and Coleman[17] illustrates such a behavior for nucleotides.

3. Suggestions for Future Experiments

3.1. Review of Zeroth-Order Experiments

A significant reason for the success of the zeroth-order experiments, described in the preceding sections, has been the ease of implementation. Within limits imposed by the mechanics of tunneling spectroscopy, the zeroth-order experiments are possible on a variety of molecules, with different radiation types and energies, and with different substrate and overlayer materials and geometries. However, several shortcomings exist. These are (1) The overlayer complicates the spectrum of the incident radiation. In addition, it plays a complex role in the direct or catalytic reactions with or amongst trapped fragments or radicals. (2) Irradiation at room temperature leads to significant reaction rates for most radicals. (3) Use and comparison of data between different junctions complicates analysis. This is especially the case when small differences in the spectra are sought; these are masked by uncertainties and the differences in the spectra between junctions.

In the succeeding sections, we will outline some suggestions for future experiments. These are grouped according to whether irradiation is performed through the overlayer (first-order experiments) or in the absence of the overlayer, before evaporation of the metal overlayer and the completion of the junction (second-order experiments). This grouping is *not* chosen to imply an ordering in terms of increasing difficulty. In fact, as has been recently noted,[18] second-order (phase B) experiments have been easier to perform than some of the first-order experiments.

3.2. First-Order Experiments

The family of the suggested first-order experiments (see Figure 7) have the following aspects in common. Irradiation is performed through the overlayer and at low temperature; the complications due to the presence of the overlayer remain. These experiments can be envisioned in three phases.

Phase A. After completion of the junction, with the sandwiched monolayer of molecules, the junction is transferred to an irradiation apparatus with a low-temperature stage. After irradiation, the junction is brought up to room temperature and subsequently transferred to the tunneling spectrometer for measurement. This procedure provides results that are essentially comparable to those obtained via the zeroth-order experiments under identical conditions. Small differences if any could be attributed to lowered radical reaction rates during the irradiation process. However, the room-temperature transfer may end up reactivating reaction paths, thereby yielding results that may be very similar to the case where the irradiation was performed at room temperature.

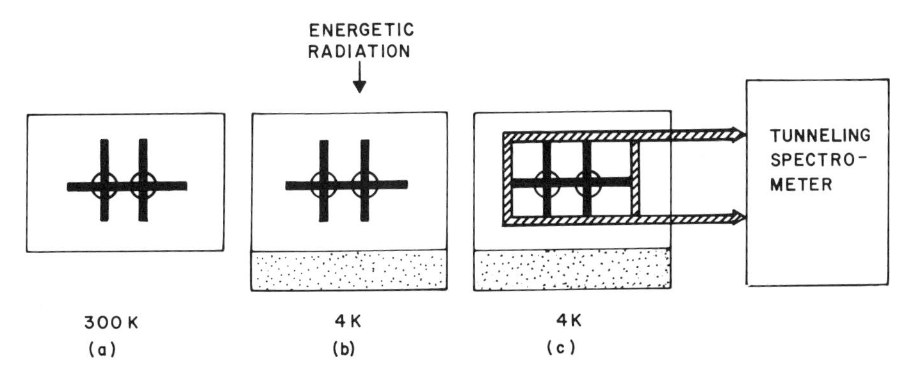

Figure 7. Schematic depicting the three steps involved in the suggested first-order experiments. Unlike the zeroth-order experiments, here the irradiation is performed at 4 K. Text discusses different phases of the suggested experiments.

Phase B. The transfer between the low-temperature irradiation and the spectrum measurement could be performed via a specially designed evacuated low-temperature "shuttle." Such an apparatus could, to a first approximation, yield "frozen" radicals. Comparison between results from the zeroth-order experiments and these experiments, for identical conditions, should demonstrate the effects of reduced radical reactions with other species in the junction as well as with the alumina and the overlayer. However, since comparison has to be made between different junctions, there may be some error due to differences between the characteristics of individual junctions.

Phase C. This phase involves the irradiation and the spectrum measurement in the same apparatus; consequently, it may be in terms of its construction requirements even simpler than phase B. The apparatus needs to have the following features: It should accommodate the completed junctions on a special holder than provides good thermal contact with a low-temperature stage (liquid He cooled or a closed cycle refrigerator). In addition, electrical contacts need to be made with the junction to facilitate tunneling spectrum measurements. An *in situ* electron gun or a window for an energetic photon source needs to be provided. Experiments performed in such an apparatus should allow for *repeated* irradiation and spectrum measurement on the *same* junction. This facility, in combination with the possibility of changing the substrate temperature between irradiation and/or spectrum measurements, provides one with a wide variety of *accurate* data on temperature dependencies of trapped fragments and radicals generated by varying levels of radiation action.

3.3. Second-Order Experiments

The suggestions for second-order experiments (see Figure 8) involve irradiation of the molecules before the overlayer has been evaporated; such experiments, by necessity, require comparison between different junctions. These experiments, therefore, can lead to insights regarding the influence of the overlayer on trapped radicals and fragments. In addition, the absence of the overlayer can make it possible to induce, via energetic radiation, the direct polymerization or bonding to the substrate; this is especially interesting in the case of molecules which do not ordinarily adhere physically or chemically to alumina. It is possible thus to study the initial phase of radiation-induced nucleation.

Phase A. These experiments would involve irradiation as an intermediate step during the completion of a junction during the zeroth-order experiment. Clearly, such experiments are no more difficult than the simple experiments described in Section 2. The difficulty, however, lies in the

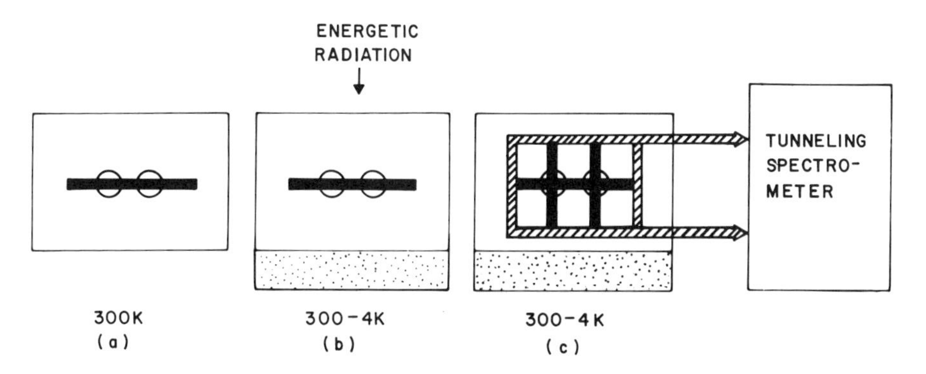

Figure 8. Schematic depicting the three steps involved in the suggested second-order experiments. (a) Oxidized metal electrode with molecules (depicted by circles). (b) Irradiation at temperatures ranging from 300 to 4 K. (c) Completion of junction via evaporation of overlayer and measurement of the tunneling spectrum at 4 K.

validity of results obtained from such experiments. This is because the alumina surface between irradiation and evaporation of the overlayer is subject to atmospheric action; this can lead to altering the true effects of radiation action. The extent to which this is a serious concern, needs to be investigated with model compounds. In the case of molecules that either polymerize or strongly bond to alumina, after radiation action, the effect of atmosphere may be minimal. In brief, phase A experiments may be a "quick and dirty" way to investigate molecules capable of polymerizing under the action of energetic radiation.

Phase B. These experiments involve maintaining the irradiation and the evaporation to complete the junction in the same vacuum system. This can be achieved either by "shuttling" the junctions via an evacuated "shuttle" between the systems or by building an apparatus that can irradiate and complete evaporation in the same chamber. The latter apparatus is now a rudimentary part of most research laboratories; it is an electron gun evaporator with minor modifications. If a low-temperature substrate holder can also be installed, interesting temperature variation effects can also be studied. In general, the utility of such experiments is evident after comparison with phase A experiments; atmospheric contamination concern is alleviated.

Phase X. This "ultimate" phase of the second-order experiments is based on a low-temperature "shuttle" that can transport to the measurement apparatus without significant change in the substrate temperatures. Clearly such an experimental scheme will provide information on trapped or frozen radicals that were created in the absence of the overlayer. If comparison with equivalent experiment of a first-order phase C experiment is possible, some insight into the influence of the overlayer on the radicals may be obtained.

4. Conclusions

This chapter has attempted to provide an overview of an application of tunneling spectroscopy: the study of radiation-induced changes to monolayers of organics. The strength of this technique, besides its relative ease of implementation, is in its applicability to monolayers and in its quantitative nature. Experiments on a variety of molecules have yielded interesting new results; "damage" cross sections for various bonds have been obtained.

This chapter has also outlined some suggestions for new generations of experiments. These could yield new insights into the influence of overlayer on trapped fragments or radicals generated by radiation actions; the effect of substrate temperature on radicals and other short-lived species; the creation and detection of polymerized or strongly bonded (to the substrate) molecules, due to the action of energetic radiation.

Acknowledgments

Without an extremely fruitful association and collaboration with Professor Paul K. Hansma, this work would not have been possible. Most of this work was performed at IBM T. J. Watson Research Center between 1974 and 1977; I am grateful for the support I received then from my associates and management. Thanks are due to Nancy Parikh for moral support and Ms. Judie Fekete and Ms. Mady Barrett for typographical assistance.

References

1. J. R. Breedlove, Jr. and G. T. Trammell, Molecular microscopy: Fundamental limitations, *Science* **170**, 1310–1312 (1970).
2. A. V. Crewe, J. Wall, and J. Langmore, Visibility of single atoms, *Science* **168**, 1338–1340 (1970).
3. M. Isaacson, D. Johnson, and A. V. Crewe, Electron beam excitation and damage of biological molecules. Its implications for specimen damage in electron miscroscopy, *Radiat. Res.* **55**, 205–224 (1973).

4. P. J. Estrup and E. G. McRae, Surface studies by electron diffraction, *Surf. Sci.* **25**, 1–52 (1971).
5. C. C. Chang, Auger electron spectroscopy, *Surf. Sci.* **25**, 53–79 (1971).
6. L. G. Christophorou, *Atomic and Molecular Radiation Physics*, Wiley, New York (1970).
7. J. B. Birks, *Photophysics of Aromatic Molecules*, Wiley, New York (1970).
8. R. M. Glaeser, Limitations to significant information in biological electron microscopy as a result of radiation damage, *Ultrastructure Res.* **36**, 466–482 (1971).
9. J. Wall, Ph.D. thesis, University of Chicago, 1971, unpublished.
10. J. Pacansky and H. Coufal, Electron beam induced Wolff rearrangement, *J. Am. Chem. Soc.* **102**, 410–412 (1980).
11. J. Langan and P. K. Hansma, Can the concentration of surface species be measured with inelastic electron tunneling?, *Surf. Sci.* **52**, 211–216 (1975).
12. P. K. Hansma and M. Parikh, A tunneling spectroscopy study of molecular degradation due to electron irradiation, *Science* **188**, 1304–1305 (1975).
13. M. Parikh, P. K. Hansma, and J. Hall, Quantitative tunneling spectroscopy study of molecular structural changes due to electron irradiation, *Phys. Rev. A* **14**, 1437–1446 (1976).
14. M. Parikh, P. K. Hansma, and J. Hall, Quantitative tunneling spectroscopy study of molecular structural changes due to electron irradiation, IBM Research Report RC 5935 (1976).
15. J. Hall, P. K. Hansma, and M. Parikh, Electron beam damage of chemisorbed surface species: A tunneling spectroscopy study, *Surf. Sci.* **65**, 552–562 (1977).
16. J. Klein, A. Leger, M. Belin, D. Defourneau, and M. J. L. Sangster, Inelastic-electron-tunneling spectroscopy of metal—insulator–metal junctions, *Phys. Rev. B* **7**, 2236–2348 (1973).
17. J. M. Clark and R. V. Coleman, Inelastic electron tunneling study of uv radiation damage in surface-adsorbed nucleotides, *J. Chem. Phys.* **73**, 2156–2178 (1980).
18. P. K. Hansma and A. Baymen, private communications (1981).

10

Study of Corrosion and Corrosion Inhibitor Species on Aluminum Surfaces

Henry W. White

1. Introduction

1.1. General Remarks

This chapter discusses applications of tunneling spectroscopy to the study of corrosion and corrosion inhibitor species on aluminum.

Destructive corrosion of metals can usually be associated with a failure of a passivating oxide layer. This protective oxide may be ruptured by mechanical strain, dissolved, or otherwise detrimentally altered by chemical or electrochemical processes. Active corrosion of the bare metal can result when the protective oxide is damaged. Often the synergistic effects of mechanical and chemical factors lead to catastrophic failure of a metal as, for example, in stress corrosion cracking.

Corrosion of aluminum alloys by organic liquids is a significant commercial problem. Of particular importance in the aerospace, electronic, and chemical industries is the problem of corrosion by chlorinated solvents. These solvents are employed extensively for vapor degreasing and metal cleaning. Large-scale cleaning operations using these solvents encounter major problems arising from the corrosion of aluminum by the solvents and contamination of the solvents by the corrosive reaction.

Henry W. White • Physics Department, University of Missouri, Columbia, Missouri 65211.

Inhibitor additives are used to help prevent corrosive attack but current inhibitors are not entirely satisfactory at reflux temperatures. The basic mechanisms by which typical industrial solvents such as 1,1,1-trichloroethane or trichloroethylene attack aluminum and its alloys are not well understood and neither are the mechanisms of inhibition. Knowledge of the surface reactions, reaction intermediates, and products is needed for improved prevention of corrosion and for the design of more effective inhibitors.

The technical objectives of our tunneling spectroscopy program are to provide fundamental information about the corrosion of aluminum and its alloys by chlorinated solvents, and the mechanisms by which chemical additives inhibit surface reactions, and to relate such information to the problems of improved inhibitors and more resistant aluminum alloys.

A program to investigate corrosion and corrosion inhibition of aluminum in acid media has also been started.

1.2. Corrosion of Aluminum in Organic Media

The subject of corrosion of metals in organic media has been reviewed recently by Heitz.[1] Corrosion phenomena are more complex in organic solvents than in the more extensively studied aqueous solutions. While there are many similarities, numerous corrosion processes occur in organic solvents which have no analog in aqueous corrosion.

Organic media can be classified as nonpolar aprotic (e.g., hydrocarbons), dipolar aprotic (e.g., acetone), or protic (e.g., alcohols) depending upon the nature of the bonding of the solvation shell. The protic or aprotic character of a solvent is dependent upon its ability to provide protons. Nonpolar solvation forces include van der Waals or dispersion forces, and π-complexing. Under "dipolar" are included ion–dipole and dipole–dipole forces.

Organic corrosion media may also be classified as one-component or multicomponent systems. One-component systems provide their own oxidizing groups. For example, the aggressive elements of pure alcohol are the solvated protons and the $-\overset{|}{\underset{|}{C}}-OH$ groups. Multicomponent systems may contain corrosive agents as solutes.

In many instances water has a profound effect upon corrosion rates in organics. Water in an organic medium may act as an inhibitor or stimulator of corrosion depending upon the medium and other variables. The reaction of aluminum with HBr or HCl in a hydrocarbon solvent occurs readily only in the presence of water[2] (or other protic component). In acetic acid the corrosion rate of aluminum is a minimum for the pure acid but increases with the addition of H_2O or acetic acid anhydride.[3] Similar stimulation of corrosion is found for organic amines.[4] With boiling solvents, however, water may act as a corrosion inhibitor.[5] For example, 1% water inhibits corrosion of aluminum in boiling methanol, ethanol, glycol, and monomethyl ester.

It is evident that the role of H_2O in corrosive processes in organic solvents is an important one, but not a simple one. In any attempt to construct explanations it must be considered that the bulk concentration of water in solution may not represent the effective concentration at the substrate surface. In fact, water may be produced at the interface as a corrosion product when corrosion is by oxygen in the presence of a proton donor.

Regarding the mechanisms of corrosion, the question of whether a process is primarily "electrochemical" or "chemical" is often debated. Electrochemical processes generally require two charge transfer reactions (anodic and cathodic) while for the chemical mechanism only one is needed. Many corrosion processes, especially reactions of metals with protic media, are assumed to proceed by electrochemical mechanisms. The anodic partial reaction is the dissolution of a metal according to the equation

$$Me \rightarrow Me^{n+} + ne^- \tag{1}$$

The cathodic partial reaction, i.e., the reduction of acidic hydrogen of a proton donor, is represented by

$$\mathrm{H}A + e^- \rightarrow \tfrac{1}{2}\mathrm{H}_2 + A^- \tag{2}$$

Most of the processes involve the direct reaction of the nondissociated proton donor with the metal.[6]

Many organics are known to form surface complexes[7] with an aluminum oxide, and such reactions play an important role in corrosion. Even for systems where corrosion occurs by an electrochemical mechanism it is often the case that surface chemical reactions are also involved. Organic species, not evident in the net electrochemical reaction, may exist as intermediate species.

Chemical destruction of the passivating oxide is often involved in corrosion of aluminum in organic media. Boies and Northan[8] reported studies of aluminum corrosion in ethylene glycol and distilled water. They concluded that there is an initial period in which the oxide is destroyed, hydrogen is evolved, and an alkaline environment is formed. The oxide layer is apparently self-repairing until the oxygen in solution is exhausted. They found that the anodic reaction of aluminum dissolution to form ions occurred at the interface between the metal and the amorphous oxide layer. Aluminum ions diffuse into solution driven by a concentration gradient and electrons flow outward due to the potential difference between anodic and cathodic sites. Reaction of the aluminum with the water to form a nonprotective oxide and hydrogen proceeds at a rapid rate once the compact passivating oxide layer has been damaged. The mechanisms involved in the interface attack of the passivating oxide layer are not known in detail.

1.3. Corrosion by Chlorinated Hydrocarbons

The reaction between aluminum and chlorinated solvents appears to yield $AlCl_3$ and a dimer of the chlorinated solvent. Thus corrosion by carbon tetrachloride is expected to yield C_2Cl_6 as a principal product. However, unsaturated compounds such as perchloroethylene and hexachlorobutadiene have also been identified as the dimer reacts further with the surface. The presence of water in the solvent gives complex products. It has been thought that the solvent hydrolizes to yield HCl. On the other hand, water may merely facilitate the removal of corrosion products from the metal surface.

A most interesting reaction occurs when 2024 aluminum is scratched while immersed in uninhibited 1,1,1-trichloroethane. A bleeding red color which is thought to be due to a complex urthan aluminum chloride corrosion product appears instantly. It is proposed[23] that Cl^- is abstracted by the Al_2O_3 surface to leave a dichloroethane carborium ion which loses a proton to give 1,1-dichloroethylene according to the following reaction:

$$CH_3{-}CCl_3 \xrightarrow[Al_2O_3]{} CH_3CCl_2^+ + Al_2O_3Cl^- \xrightarrow[H_2O]{} CH_2{=}CCl_2 + Al(OH)_2Cl \tag{3}$$

The aluminum hydroxychloride salt is sufficiently soluble to be removed by the solvent. In addition there is evidence that the following cleavage mechanism proposed for CCl_4 with Al exists

$$2Al + 6CH_3CCl_3 \rightarrow 3CH_3CCl_2CCl_2CH_3 + 2AlCl_3 \tag{4}$$

The $AlCl_3$ produced in the reaction appears to complex with the solvent to give the red color and to degrade it by HCl elimination via the sequence

$$AlCl_3 + CH_3CCl_3 \rightarrow \text{complex} \rightarrow Cl_2C{=}CH_2 + HCl \tag{5}$$

Further dehydrahologenation occurs to eventually turn the entire system into a tarry mass.[9]

Additional information on the corrosion reactions of aluminum by carbon tetrachloride and trichloroethylene is presented later in Sections 2 and 4, respectively.

1.4. Corrosion Inhibitors for Aluminum in Chlorinated Solvents

Inhibitors for the above corrosion reactions are typically electronegative molecules capable of interacting with the electron deficient aluminum sites. These molecules can effectively complex with $AlCl_3$ in the solvent, or form barrier films on active sites on the aluminum oxide surface. While the exact mechanism of a given inhibitor's action is not known at present, it does appear that a surface reaction is often very important. Multifunctional inhibitor molecules appear to work best. In terms of donor atoms nitrogen is more effective than sulfur, and sulfur is better than oxygen. Cyclic nitrogen structures work well, especially those with additional electronegative substitutions.

The problem remains, though, that many of the inhibitors do not provide long protection at reflux temperatures, nor do many of the inhibitors work well on the most common alloys. The copper–aluminum alloy 2024 is especially troublesome in the presence of CH_3CCl_3.

Archer[10] has established some of the relationships between structure and inhibition activity for a number of molecules which inhibit the reaction of aluminum by 1,1,1-trichloroethane. However, the better 1,1,1-trichloroethane inhibitors are not always the best structures for inhibiting other reactions such as the methylene chloride–aluminum reaction. Other factors, such as the solubility of the resultant metal chloride–inhibitor complex at the metal surface, probably help determine inhibitor performance. Analytical methods to determine the solubility of the complex are needed along with a means of detecting the stepwise sequence of inhibition.

1.5. Corrosion and Inhibitor Surface Species

There are a number of aspects of practical corrosion systems which makes their study and analysis particularly difficult. First, the systems of interest are necessarily complex, consisting of a metal substrate, on oxide film, an adsorbed heterogeneous layer of reactive molecular and ionic species, and in some cases a conducting liquid which serves as an electrolyte. Obviously, complex chemical reactions are possible for such systems and in fact are probable. A complete analysis requires detailed information about the nature of the metal oxide surface, the adsorbed monolayer, and the various chemical species which are generated at the interface and how these species change with the relevant variables.

The mechanisms of corrosion and of corrosion inhibition depend upon the nature of the chemical species existing at the microscopic interface (one or two atomic layers in thickness) between a metal or metal oxide surface and the corrosive medium. Knowledge of the interface chemistry and how it is altered by variables such as interfacial water and impurity content, substrate alloy composition, and temperature are essential for the understanding of corrosion and corrosion inhibition.

Extensive work on the theory of the mechanisms of corrosion and inhibition is available in the scientific literature.[10–12] It is reasonable to suppose that corrosion processes could be reasonably well understood within the existing framework of chemical and electrochemical models if detailed knowledge of the actual chemical species existing at the interface could be determined with certainty. Very few experimental methods are available for monitoring the chemical constituents in an interface system which approximates the actual conditions present in a practical corrosion system.

Traditional electrochemical studies have contributed enormously to the understanding of corrosion but these types of studies cannot directly monitor the surface molecular species, and special problems arise for organic solvents which have low conductivity such as chlorinated hydrocarbon solvents.

Several methods of studying molecular adsorption have been developed which are applicable to corrosion studies. These include tunneling spectroscopy, SRS (surface Raman scattering), XPS (x-ray photoelectron spectroscopy), and ESD (electron-stimulated desorption). They can be used to determine the identity and nature of molecular species adsorbed on an oxide surface when it is exposed to organic solvents and inhibitors.

A brief review of the results obtained using tunneling spectroscopy is contained in the following sections. From the results it has been possible to identify several of the reaction products and to postulate likely mechanisms for corrosion reactions and the action of the inhibitors. Section 2 discusses the corrosion of aluminum by carbon tetrachloride. A discussion of the results obtained for the inhibitor formamide, including how it can be effective for the aluminum–carbon tetrachloride system, is given in Section 3. Section 4 presents results for the corrosion reactions and surface species for aluminum–trichloroethylene. Results obtained for two inhibitors, acridine and thiourea, which are effective in reducing corrosion of aluminum in hydrochloric acid, are given in Section 5. Section 6 summarizes the application of tunneling spectroscopy to these systems, and suggests its possible application to other corrosion problems.

2. Corrosion of Aluminum by Carbon Tetrachloride

2.1. Proposed Reactions

The corrosion of aluminum metal by carbon tetrachloride has been studied by several investigators. From weight loss measurements Stern and Uhlig[13] found the corrosion rate of 99.99% aluminum with boiling anhydrous CCl_4 to be very high (20 in./yr). The reaction was shown to be

$$2Al + 6CCl_4 \rightarrow 3C_2Cl_6 + 2AlCl_3 \tag{6}$$

with the production of aluminum chloride and hexachloroethane. The corrosion rates for the liquid and vapor phases were similar.

For this reaction Stern and Uhlig proposed a free radical mechanism which is initiated with homolytic cleavage of the C—Cl bond to produce the free radicals $^{\cdot}Cl$ and $^{\cdot}CCl_3$. Production of aluminum chloride, $AlCl_3$, allows the subsequent formation of the complex $CCl_3^+[AlCl_4]^-$. This complex can later dissociate into free radicals taking part in the free radical chain reaction. The $^{\cdot}CCl_3$ free radicals can dimerize to form the hexachlorothane. Stern and Uhlig interpreted results of their work to exclude an electrochemical mechanism and concluded that the reaction was a typical direct chemical reaction. Minford, Brown, and Brown[14] and Brown, Cook, Brown, and Minford[15] have also investigated the reaction of aluminum with CCl_4. Brown *et al.*[15] concluded that the reaction was electrochemical rather than chemical, and that the rate-determining step in the reaction does not involve free radicals.

More recently, Archer and Harter[16] have reported studies involving the reaction of aluminum and magnesium in CCl_4 and in CCl_4 with hydroxides present. They concluded that their results substantiate the existence of the free radical $^{\cdot}CCl_3$.

2.2. Tunneling Spectroscopy Studies

2.2.1. *Experimental Procedure*

Tunneling spectroscopy has been used to study the chemical species formed on aluminum oxide exposed to carbon tetrachloride.[17] Aluminum–aluminum-oxide–carbon-tetrachloride–lead tunnel junctions were fabricated. The aluminum electrode was evaporated to a thickness of about 100 nm in an ion-pumped vacuum system. This step was followed immediately by the growth of an oxide layer about 2.0 nm thick using a glow discharge in a partial pressure of oxygen. Approximately one monolayer of spectroscopically pure carbon tetrachloride was deposited on the oxide by placing several drops on the electrode, and spinning off the excess. The doped substrate was quickly returned to the vacuum chamber and a lead counter electrode evaporated over the molecules at approximately 10^{-6} Torr.

To obtain spectra the junctions were placed in a spectrometer and cooled to 4.2 K. A dc bias across the junction was swept slowly from 30 to 500 mV using the ramp output of a 2048 channel electronic signal averager. An ac modulation signal of 1000 Hz and about 1 mV (rms) was applied continuously. The amplitude of the second-harmonic signal (proportional to d^2V/dI^2) generated by the tunnel junction was recorded as a function of the dc bias. Peaks in this signal located the energies of the vibrational modes as read from the measured bias voltage (energy) scale.

2.2.2. *Surface Species*

The tunneling spectrum of a junction doped with spectroscopically pure CCl_4 is shown in Figure 1.[17] The energy range is from about 30 to 500 meV (240 to 4000 cm^{-1}). The ordinate is in arbitrary units of d^2V/dI^2. For reasons which are not understood it was not possible to obtain good signal-to-noise ratios on junctions doped with CCl_4, especially at the higher bias voltage values. The significant peaks in the spectrum have been labeled by their wave number values in units of cm^{-1}. These are listed in Table 1 and, for convenience, have been given labels A_1 through A_{11} and B_1 through B_4. These peak positions have not been corrected for peak shifts due to the superconducting lead electrode. Those labeled B_1-B_4 refer to peaks in the background spectrum shown in Figure 3 of reference 17. These are B_1 (aluminum oxide phonon), B_2 (C—H bend), B_3 (C—H stretch), and B_4 (O—H stretch). It is not clear whether or not the aluminum phonon mode near 294 cm^{-1} in the background spectrum is present in the CCl_4 spectrum. The frequencies for the peaks labeled A_1-A_{11} were compared with accepted values for infrared and Raman frequencies associated with the molecular species CCl_4, $^{\cdot}CCl_3$, C_2Cl_6, AlCl, $AlCl_2$, $AlCl_3$, the complex $CCl_3^+[AlCl_4]^-$, and those which might reflect oxygen bonding to adsorbed CCl_4. The species listed occur in the reaction sequence proposed by Stern and Uhlig.[13]

The assignments listed in Table 1 were made on the basis of the agreement between optical and tunneling values for a particular mode. The validity of the assignments is necessarily limited by the small number of lines available for association with a given species, and by the fact that in an effort of this type the list of species considered in the matching process cannot be exhaustive. Recognizing these limits, the assignments in Table 1

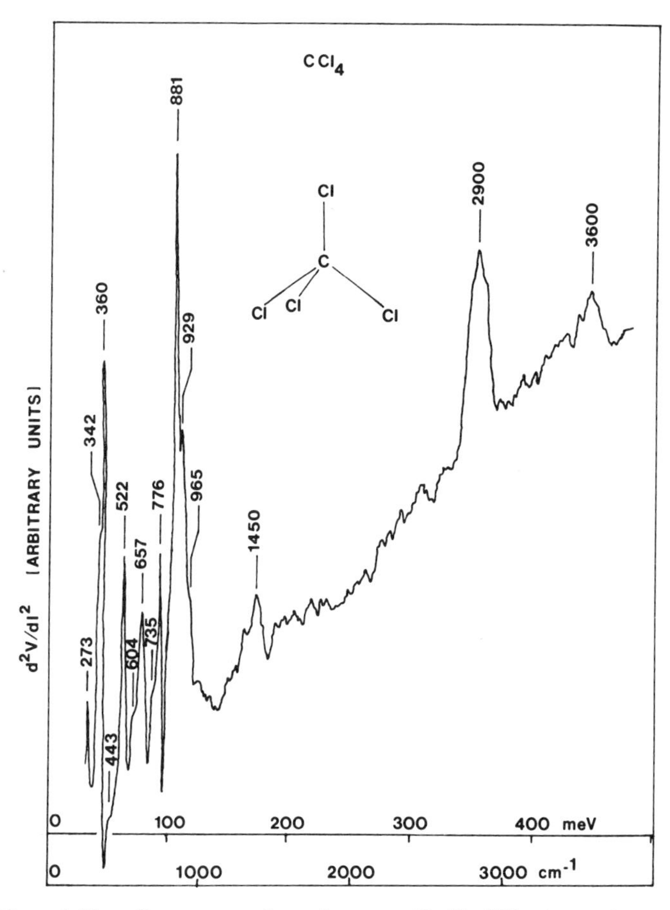

Figure 1. Tunneling spectrum for carbon tetrachloride CCl_4 (from reference 17).

are valid to the extent that close agreement exists between the observed tunneling and optical frequencies, and that the species selected for comparison are reasonable. Modes for the species AlCl, $AlCl_2$, $AlCl_3$, $^{\cdot}CCl_3$, C_2Cl_6, CCl_4, and $CCl_3^+[AlCl_4]^-$ can be matched with reasonable agreement. None of the observed lines could be associated with a C—O stretch which would occur at about 1050 cm^{-1}. The closest observed mode in this region was A_{11} at 965 cm^{-1}, which was assigned as C—Cl stretch in C_2Cl_6, a likely species. The absence of modes reflecting oxygen bonding to the CCl_4, and the predominence of modes associated with aluminum–chlorine species, indicates that the reaction of CCl_4 with the barrier layer of aluminum oxide occurs at sites of exposed aluminum. These exposed sites could occur at oxygen vacancies and other imperfections such as grain boundaries.

Table 1. Frequencies and Mode Assignments for the Peaks in the Tunneling Spectrum of Carbon Tetrachloride, the Infrared and Raman Frequencies Associated with the Designated Species are Listed

Peak label	IETS (meV)	IETS (cm^{-1})	ir (cm^{-1})	Raman (cm^{-1})	Mode	Species
A_1	33.8	273 (m)[a]		284	Al—Cl stretch	Al_2Cl_6?
A_2	42.4	342 (m)		340	Al—Cl stretch	Al_2Cl_6
				345	Al—Cl stretch	$AlCl_2$
A_3	44.6	360 (s)	380		Al—Cl stretch	$AlCl_2$
A_4	55.0	443 (w)	443		Al—Cl stretch	AlCl
A_5	64.7	522 (m)			C—Cl stretch	$CCl_3^+ [AlCl_4]^-$
A_6	74.9	604 (w)	610		Al—Cl stretch	$AlCl_3$
			625		Al—Cl stretch	Al_2Cl_6
A_7	81.4	657 (m)			Al—C stretch	$CCl_3^+ [AlCl_4]^-$
A_8	91.1	735 (w)	761		C—Cl stretch	CCl_4
A_9	96.2	776 (m)	787		C—Cl stretch	CCl_4
A_{10}	109.2	881 (s)	898		C—Cl stretch	CCl_3
A_{11}	119.6	965 (w)		975	C—Cl stretch	C_2Cl_6
B_1	115.2	929			lattice phonon	
B_2	180	1450			C—H bend	
B_3	360	2900			C—H stretch	
B_4	450	3600			O—H stretch	

[a] s = strong, m = medium, and w = weak peak intensities.

The species identified are those in the free radical sequence proposed by Stern and Uhlig. The results do not make it clear, however, whether the reaction mechanism is chemical, electrochemical, or a combination of these. It is also not possible to conclude that the model proposed by Stern and Uhlig is correct in all respects. The tunneling results should be interpreted from the point of view that new information on a molecular level has been obtained regarding the species which occur during the corrosion process.

Other spectroscopic data on this and related solvent and metal systems would be useful in order to obtain more complete information on the molecular species present, and the types of bonding and bond breaking which occur.

3. Inhibition of Corrosion by Formamide

3.1. Surface Species

Tunneling spectroscopy has been used to determine the molecular species formed on aluminum oxide exposed to the inhibitor formamide.[18]

Small concentrations of formamide are known to inhibit the corrosion of aluminum by carbon tetrachloride.[19] Junctions were prepared which were doped with formamide, $HCONH_2$, the deuterated forms $DCONH_2$ and $HCOND_2$, and mixtures of CCl_4 and formamide. A tunneling spectrum of formamide adsorbed on the oxide surface of a tunnel junction is shown in Figure 2. The peak positions have not been corrected for peak shifts due to the superconducting lead electrode. A strong C—O (single bond) mode is observed at 1048 cm^{-1}. A very weak peak located at 506 cm^{-1} was interpreted as an Al—N stretching mode. Other important peaks showed up at 629 cm^{-1} and at 1370 cm^{-1} due to O—C—N skeletal deformation and C—N stretching modes, respectively.

No peaks due to N—D stretching or bending vibrations were detected in spectra for $HCOND_2$. A spectrum is shown in Figure 3. The fact that

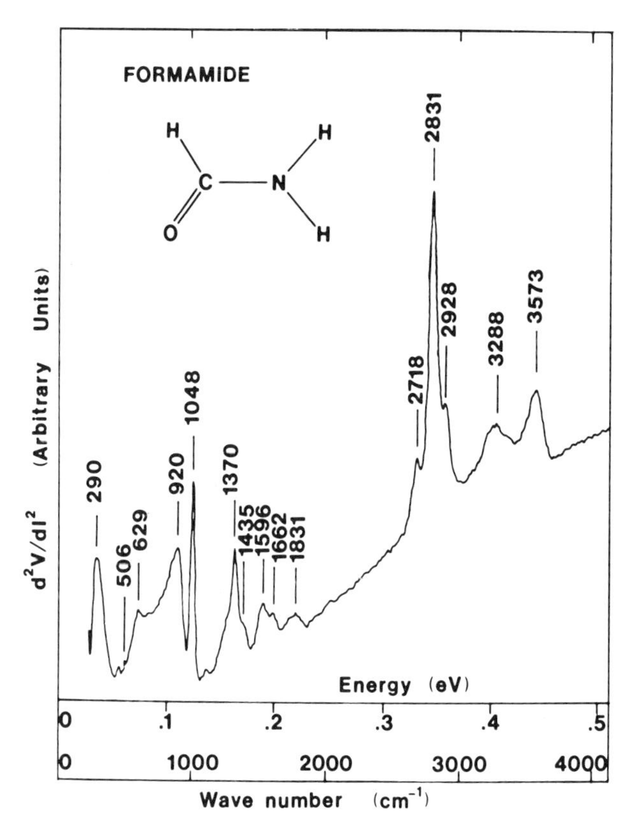

Figure 2. Tunneling spectrum for formamide $HCONH_2$. The wave numbers of significant peaks are labeled in units of cm^{-1}. (From reference 18.)

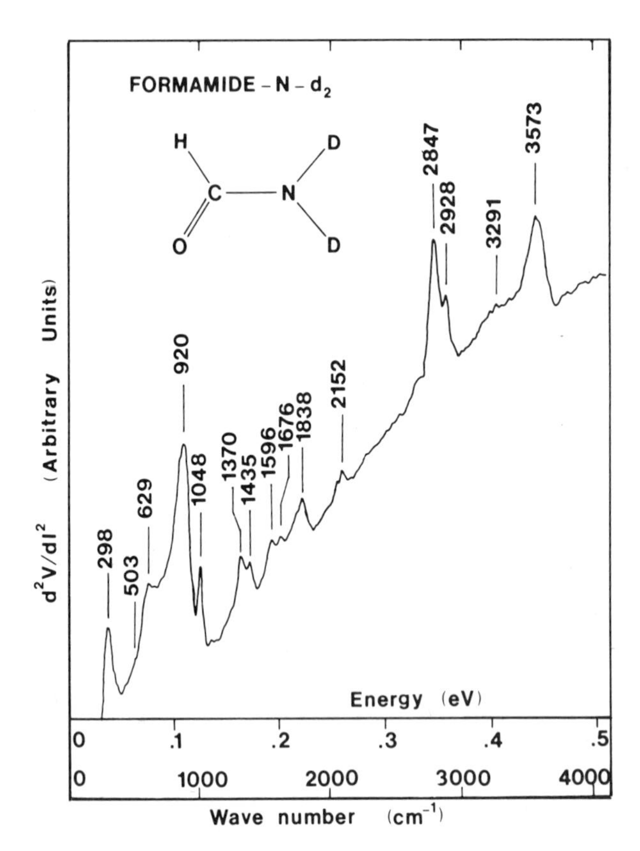

Figure 3. Tunneling spectrum for formamide $HCOND_2$. The wave numbers of significant peaks are labeled in units of cm^{-1}. (From reference 18.)

N—H stretching modes were instead observed (near 3291 cm^{-1}) implied that the second deuteron attached to the nitrogen underwent an exchange reaction with a surface hydrogen. The absence of a peak which could be associated with an O—D stretching mode suggested that the source of the hydrogen involved in the exchange was probably not a surface hydroxyl group. From discussions on the types of surface sites which can be associated with various surface reactions it appears that the most reasonable configuration for the surface site which takes part in the H—D exchange is that of an oxygen vacancy occupied by a hydrogen atom, i.e., a Bronsted site.[20]

The tunneling results indicated that formamide molecules preferentially adsorbed to the aluminum oxide surface, and that the surface species had features which were different from those of nonadsorbed molecules. A very tentative model was presented which is consistent with the observed facts.[18] The most notable of the observed features were the change in the carbon–oxygen bond from having a partial double bond to a single bond, the picking up of a hydrogen (or deuteron) by the carbon atom, the H–D exchange on the amide group, and the Al–N bonding. The discussion also includes the types of surface sites with which the formamide molecule could react, a sequence by which the necessary changes could occur, and a final orientation.

The tunneling spectrum for a dilute mixture of $DCONH_2$ in carbon tetrachloride showed the same features observed in the tunneling spectra of $DCONH_2$ in the absence of CCl_4. This result indicated that the formamide molecules were preferentially chemisorbed on the oxide surface in such a way as to passify the active sites involved in the corrosion reaction. This determination of preferential adsorption does not rule out the possibility that the adsorbed formamide species could also be neutralizing or binding aggressive species in solution. An adsorbed species which plays such a dual role may indeed be more effective as an inhibitor.

3.2. Inhibition Mechanism

The question arises as to whether or not the bonding of this surface species could reasonably be expected to inhibit the corrosion of aluminum by CCl_4. The fact that the tunneling work on the corrosion of aluminum by CCl_4 showed a preponderance of Al—Cl species, and the absence of any evidence for oxygen bonding to the CCl_4 molecule, indicates that the attack occurs at the exposed aluminum sites. Active sites involved in catalytic reactions of this type must have at least a double oxygen vacancy (three or more, or a chain, are even better). These are most likely to occur along grain boundaries.[20] Hence, according to this model the chemisorption of formamide could block the reaction of CCl_4 with aluminum at such sites, and thereby inhibit the corrosion of aluminum by CCl_4.

This model contains reasonable features but obviously has not been proved in detail. Clearly, a great deal more information is required in order to unambiguously determine the surface chemistry of chlorinated solvents on aluminum oxide. Data from other spectroscopies will be of great value in validating the results that are obtained by tunneling measurements. Of particular interest will be the study of aluminum alloys, which are more difficult to study by tunneling because of the problems of fabricating good junctions.

4. Corrosion of Aluminum by Trichloroethylene

4.1. Reaction with Aluminum

Tunneling spectroscopy has been used to study the reaction of trichloroethylene with aluminum oxide to determine the intermediate species and reaction products formed, and to see if these results could be correlated with a reasonable reaction sequence.[21] There were several reasons for choosing trichloroethylene for study. First, trichloroethylene has been used extensively as a commercial metal cleaning agent for many years, although its use has decreased during the past few years as a result of various health regulations. Second, there is very little published work on the reaction mechanism for the corrosion of aluminum metal by trichloroethylene. Archer[22] has discussed the reactions of several commercially important chlorinated solvents which are currently used in large quantities. Among these are 1,1,1-trichloroethane and methylene chloride. The reaction of trichloroethylene with aluminum can also be quite rapid, especially if a large amount of small aluminum particles is present. It is, however, considered to be less rapid than that of carbon tetrachloride, although no quantitative comparison is available.

Archer[23] proposed the following reactions for trichloroethylene in the presence of aluminum chloride, $AlCl_3$

$$2CHCl{=}CCl_2 \xrightarrow{AlCl_3} CCl_2{=}CH{-}CCl_2{-}CHCl_2 \tag{7}$$

$$CCl_2{=}CH{-}CCl_2{-}CHCl_2 \rightarrow CCl_2{=}CH{-}CCl{=}CCl_2 + HCl \tag{8}$$

Reactions of this type where aluminum chloride products are present can lead to autocatalytic degradation of the solvent.

Aluminum–aluminum oxide–trichloroethylene–lead tunnel junctions were fabricated as discussed in Section 2.2.1. Reagent grade trichloroethylene was supplied by PPG Industries, Inc. The phenolic stabilizer in the solvent was removed by three extractions with 5% NaOH, followed by washing with water and distillation under nitrogen. Approximately one monolayer of freshly distilled, uninhibited trichloroethylene was deposited on the oxide by placing several drops on the electrode and spinning off the excess. The doped substrate was quickly returned to the vacuum chamber, and a lead counter electrode evaporated.

4.2. Surface Species and Reactions

A tunneling spectrum of a junction doped with trichloroethylene is shown in Figure 4. Spectra were obtained for a large number of junctions. Several junctions could be made during one preparation cycle, and junctions

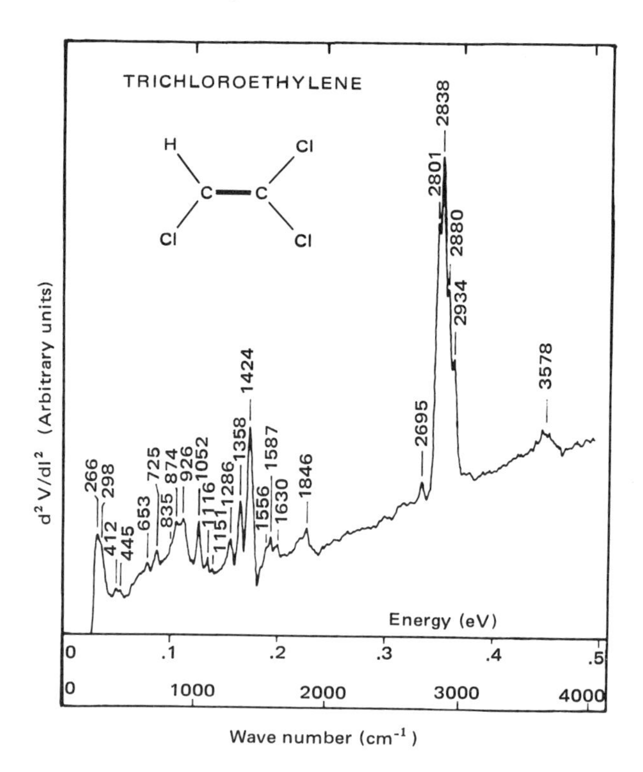

Figure 4. Tunneling spectrum for trichloroethylene $CHClCCl_2$. The wave numbers of significant peaks are labeled in units of cm^{-1}. (From reference 21.)

were selected from more than ten sets. This procedure was necessary to assure reproducibility and minimize the possibility of contamination since other hydrocarbons present in the atmosphere can compete with trichloroethylene for adsorption sites. Based on the signal-to-noise ratios observed, it is likely that coverage was sometimes less than saturation. The significant peaks in the spectrum have been labeled by their wave number values in units of cm^{-1} and listed in Table 2. These peak positions have not been corrected for peak shifts due to the superconducting lead electrode. Several of these refer to peaks in the background spectrum which is similar to that shown in Figure 3 of reference 17.

Several of the peaks in Figure 4 indicated the existence of an Al—O—C—C backbone with associated bonded hydrogen. A number of the strong peaks can be associated with symmetric and antisymmetric CH_2 and CH_3 rock, bend, wag, and stretch modes, which could be part of the Al—O—C—C species. A close comparison of the modes assigned to the Al—O—C—C structure, or portions thereof, and those associated with the C—H,

Table 2. Frequencies and Mode Assignments for the Peaks in the Tunneling Spectrum of Trichloroethylene, the Values are in Units of cm^{-1}

Peak position	Approximate character
266 (m)[a]	Al—Cl stretch or in-plane skeletal deformation
298 (m)	Al phonon
412 (w)	Al—Cl stretch?
445 (w)	C—C, C—O, or C—C—O torsion
653 (w)	C—Cl or Al—Cl stretch
725 (m)	CH_2 rock or bend
835 (w) 874 (m)	CH_3 rock, C—C stretch
926 (m)	Oxide phonon
1052 (s)	Al—O—C stretch or C—H bend
1116 (m) 1151 (w)	C—C—O skeletal deformation
1286 (s) 1358 (s) 1424 (s)	CH_2 wag, CH_3 deformations
1556 (w) 1587 (m)	C=C stretch
1630 (w)	(?)
1846 (m)	Overtone of 926
2695 (w)	Combination? (1116+1587)
2801 (s) 2838 (s) 2880 (s) 2934 (s)	C—H stretch (symmetric and asymmetric for CH_2 and CH_3)
3578 broad	O—H stretch

[a] s = strong, m = medium, and w = weak peak intensities.

CH_2, and CH_3 groups indicates the presence of a surface species which is, or has close resemblance to, an aluminum ethoxide species. Such a surface structure is compatible with a reaction sequence in which trichloroethylene reacts with the aluminum oxide surface as follows

$$:AlOH + CHCl{=}CCl_2 \rightarrow :AlOCH{=}CCl_2 + HCl \tag{9}$$

The $AlOCH{=}CCl_2$ species has an enolate type bond. At Type II sites[20] the $AlOCH{=}CCl_2$ species can undergo a reaction with trichloroethylene to form a dimer, as follows

$$:AlOCH{=}CCl_2 + CHCl{=}CCl_2 \rightarrow CCl_2{=}CH{-}CCl{=}CCl_2 + :AlO^{\cdot} + {}^{\cdot}H \tag{10}$$

A reaction of this type has been postulated by Archer.(23) Such a reaction occurring at a type-II site is illustrated in Figure 5a. The notation regarding sites is such that type-I sites are represented by oxygen bonding to one aluminum atom, and type II for oxygen bonding to two aluminum atoms. The dimer formed by such a reaction could leave the site and is therefore not shown in the third step. Figure 5b illustrates a reaction more likely to occur at type-Ia sites with the production of an aluminum ethoxide species. For the type-I site reaction the product in reaction (9) plus hydrogens from the surface (or from the dimer reaction) yields

$$:AlOCH{=}CCl_2 + 4{}^{\cdot}H \rightarrow :AlOCH_2{-}CH_3 + 2{}^{\cdot}Cl \tag{11}$$

From these reactions ${}^{\cdot}Cl$, and possibly HCl, are formed. Different reactions can be expected to occur at the two sites since the C—O bond for type-Ia

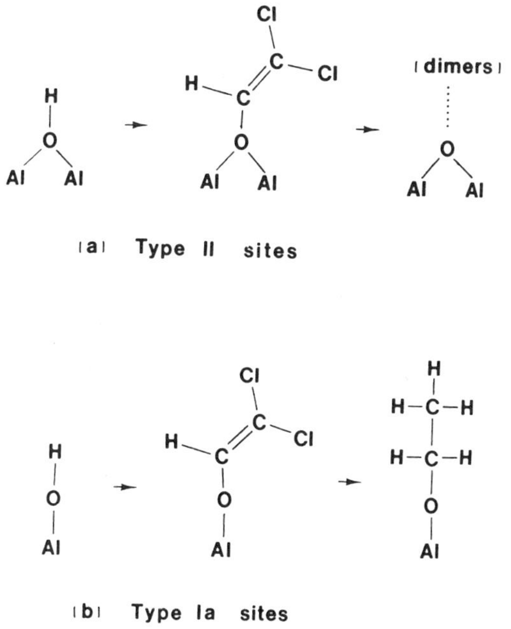

Figure 5. Schematics of possible reaction steps following the chemisorption of trichloroethylene on an aluminum oxide surface. (a) The sites can leave the surface by forming dimers. (b) Termination occurs at type-Ia sites by formation of an aluminum ethoxide species. The degree of hydrogenation for the species is not known. (From reference 21.)

sites is stronger than that for the type-II sites.[20] Accordingly, the enolic bond structure at type-Ia sites can transform to an ethoxide structure, while dimers are formed at type-II sites. These dimers would be free to leave the surface. The presence of these species can precipitate aluminum corrosion by mechanisms in which the chlorine free radical reacts with exposed aluminum atoms at oxygen vacancies.

The sequence for reaction (9) would involve successive removal of $^{\cdot}Cl$ radicals from the $AlOCH{=}CCl_2$ product. The extent to which this reaction proceeds is not known; therefore, the ratio of the aluminum ethoxide species to species of the form with Cl instead of H atoms is not known.

The possibility of trichloroethylene reacting directly with an aluminum atom has not been discussed in this model. Such a process could conceivably occur by homolytic cleavage of a C—Cl bond, as was presented in Section 2 for carbon tetrachloride. The relatively few spectral lines in the trichloroethylene data which could be assigned to Al—Cl modes compared with many in carbon tetrachloride spectra indicated that the reaction mechanisms for the two solvents with aluminum oxide were different.

4.3. Corrosion Mechanism

The results and proposed mechanism do offer an explanation as to why the corrosive attack of carbon tetrachloride on aluminum could be initiated more rapidly than that of trichloroethylene. In both cases the chlorine free radicals can initiate attack by reaction with an exposed aluminum atoms at oxygen vacancies. For trichloroethylene the chlorine free radical is a product of the reaction of a trichloroethylene molecule with an AlO (or AlOH) surface site. The chlorine free radical produced can then react with an exposed aluminum at an oxygen vacancy site. It is reasonable to assume that this two-step process could give a slower initiation of corrosive attack than the one-step mechanism reported for carbon tetrachloride.

5. Corrosion Inhibitors for Aluminum in Hydrochloric Acid

5.1. Acridine Surface Species

Acridine is the effective inhibitor for the corrosion of aluminum by hydrochloric acid. Jenckel and Woltmann[24] have determined that acridine has an inhibitor efficiency of 99% for 0.5 g/l acridine in 3 *N* hydrochloric acid in contact with aluminum. Sundararajan and Rama Char[25] determined acridine to be a good inhibitor even at concentrations as low as 0.02 g/l in 1 *N* acid. When inhibitors are effective in such low concentrations the mechanism of inhibition is probably by adsorption.

Tunneling spectroscopy has been used to study the surface species of acridine adsorbed on aluminum oxide.[26] The junctions were prepared by the liquid doping technique. A tunneling spectrum for acridine is shown in Figure 6. The peak locations have not been corrected for shifts due to the lead superconducting electrode. Spectra were also obtained for the deuterated compound $C_{13}ND_9$. The observed frequencies for both compounds were compared with optical data for acridine, anthracene, quinoline, and pyridine. The general agreement between the locations of infrared and Raman modes for the nonadsorbed three-ring molecules with the locations of those observed by tunneling spectroscopy indicated that the three-ring structure of the acridine molecule is preserved in the adsorbed species on the oxide surface. Although quinoline and pyridine spectral mode assignments were used for the acridine assignments, the ring-stretching frequencies between 1300 and 1700 cm^{-1} have medium to strong relative intensities and are thereby very similar to the ring stretching modes seen for anthracene by Califano[27] in the infrared and to the lines seen by Brigodiot *et al.*[28] and Fialkovskaya *et al.*[29] for both anthracene and acridine in both infrared and Raman spectra. The presence of modes at 450 and 1326 cm^{-1}, whose assignments are based in part on similarities with features in quinoline spectra, could be used to argue against the continued presence of the three-ring structure. Although a breakup of the rings into a quinoline form is not deemed impossible, such a breakup is considered to be unlikely on the basis of the tunneling data.

The results were also examined to determine whether information could be obtained regarding the orientation of the adsorbed species relative to the surface. Such information is of interest since steric hindrance is considered

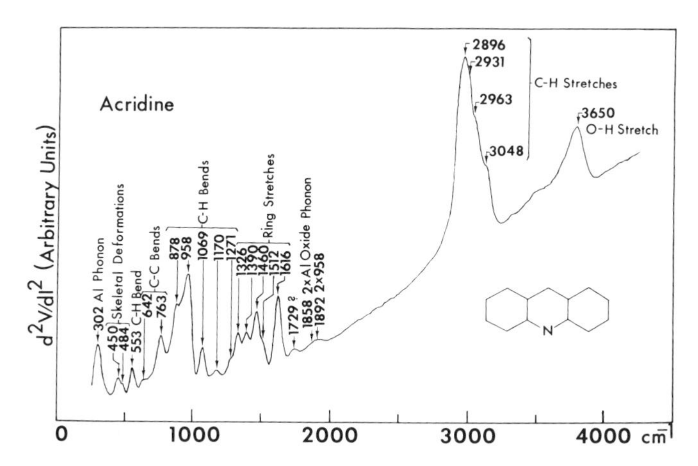

Figure 6. Tunneling spectrum for acridine $C_{13}NH_9$. The wave numbers of the significant peaks are labeled in units of cm^{-1}

to be an important mechanism by which some adsorbed surface species inhibit the breakdown of the passive layer. The presence of the relatively strong ring breathing modes suggests that the rings are not lying flat on the surface but rather are standing erect. If the rings were lying flat on the surface, the intensity of the ring breathing modes would probably be reduced considerably from those observed. This orientation suggests that the bonding is through the nitrogen atom and is probably significant.

5.2. Orientation of Thiourea on Aluminum Oxide

Comparison of calculated and experimental tunneling intensities for thiourea have provided information on the orientation of the surface species formed.[30] Thiourea in low concentrations is also an effective inhibitor for the corrosion of aluminum by hydrochloric acid.[25]

Tunneling spectra were obtained on tunnel junctions doped with thiourea using the liquid doping method described. Modes were assigned using optical data and force constant calculations. The relative experimental peak intensities were measured for seven modes which are listed in Table 3. Peak intensity was defined as the area under a peak, rather than just peak height. The peak intensities were normalized to the intensity associated with the SCNN deformation mode.

A transfer-Hamiltonian formalism with a partial charge approximation was used to calculate relative peak intensities for each of the seven modes for a number of orientations of the thiourea molecule relative to the oxide surface. This model was developed by Kirtley, Scalapino, and Hansma,[31] who used it to show the existence of an orientational dependence. Modes oriented perpendicular to the surface have a larger intensity than those oriented parallel to the oxide. The procedures used for the calculations were similar to those used by Godwin, White, and Ellialtioglu[32] to determine the orientation of the surface species associated with formic acid adsorbed on

Table 3. Comparison of Relative Experimental and Calculated Perpendicular and Parallel Tunneling Intensities for Thiourea. The Intensities are Normalized to That for the SCNN Deformation.[a]

Mode	Experimental	Perpendicular	Flat
NH_2 deformation*	0.42	2.35	1.23
CN asymmetrical stretch	0.57	0.54	0.48
CS stretch	0.94	0.82	0.19
NH_2 deformation*	0.53	24.8	0.18
CN symmetrical stretch	1.66	1.89	0.73
SCNN deformation*	1.89	8.54	9.52
CN_2 deformation	1.00	1.00	1.00

[a] Asterisk denotes that mode was not used.

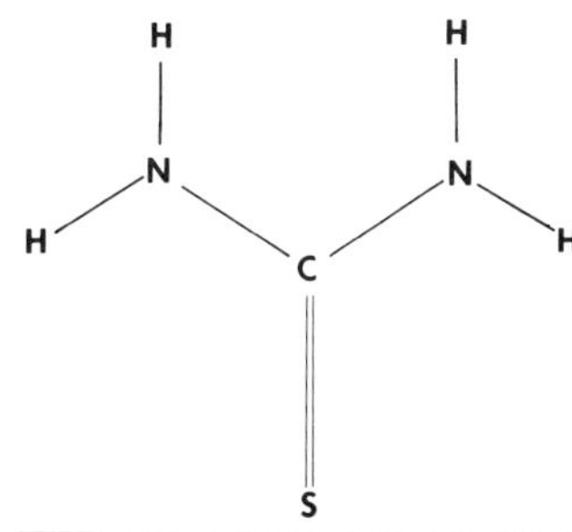

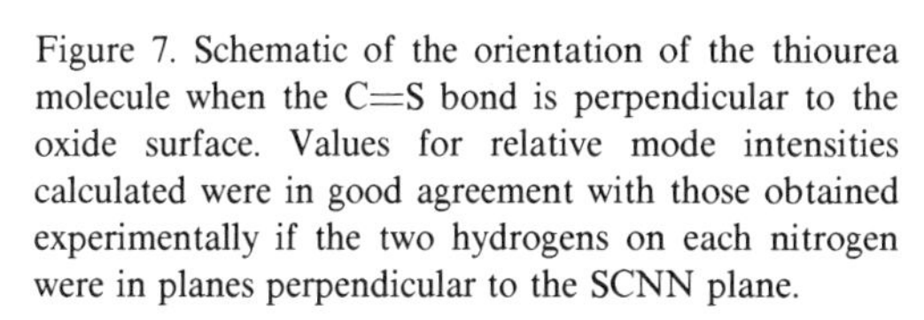
Figure 7. Schematic of the orientation of the thiourea molecule when the C=S bond is perpendicular to the oxide surface. Values for relative mode intensities calculated were in good agreement with those obtained experimentally if the two hydrogens on each nitrogen were in planes perpendicular to the SCNN plane.

aluminum oxide. Relative experimental and calculated intensities were obtained for each of seven modes of the thiourea species. A comparison of numerous sets of calculated intensities with those obtained experimentally revealed several very interesting results. First, the best agreements occurred for the orientation in which the C=S bondline was perpendicular to the oxide surface, as illustrated in Figure 7. Other orientations, such as flat, edgewise, or tilted, gave poorer agreement. Second, by using only completely planar structures for the thiourea species it was not possible to get a reasonable separation between the thiourea molecule and the second (lead) electrode. This electrode forms an image plane so that the relative intensities of the modes depend on the distance between the image plane and the molecule. It was necessary to alter the structure from planar to one in which the SCNN atoms were planar, but the hydrogens on each nitrogen were in planes perpendicular to the SCNN plane. This structure, with C=S perpendicular to the surface, gave the separation between the lead electrode and the nearest hydrogen to be 1.5 Å. Otherwise, if the structure was constrained to be planar, the best agreement between calculated and experimental intensities occurred when this separation distance was near 6 Å. A distance near 1 Å is reasonable; whereas, a distance of 6 Å is much too large.

The information thus obtained indicated that thiourea adsorbs to aluminum oxide relatively intact and that the bonding to the oxide occurs through the sulfur. It is reasonable to expect sulfur bonding to exposed aluminum, which could account, at least in part, for the inhibiting properties of thiourea for aluminum in hydrochloric acid.

6. Conclusions

Tunneling spectroscopy has been used to study the corrosion and corrosion inhibition of aluminum in organic and acid media. Based on these and related studies the following specific conclusions can be made regarding the application of tunneling spectroscopy to the study of corrosion:

(1) The aluminum oxide layer in tunnel junctions is structurally and chemically similar to that of the thin passivating "barrier" oxide which quickly forms on aluminum and helps protect it from corrosion.

(2) Tunneling spectroscopy can be a useful tool for obtaining information on the molecular species produced on the aluminum oxide surface during reaction with corrosive organics. The sensitivity of tunneling spectroscopy allows microscopic investigation of this interfacial layer.

(3) Tunneling spectroscopy can be used to obtain information on the surface species associated with molecules which inhibit corrosion of aluminum by formation of an adsorbed layer. This information can be correlated with corrosion species results to gain more insight regarding the mechanisms of corrosion and corrosion inhibition.

(4) Comparison of relative calculated peak intensities with those obtained experimentally can provide information on the orientation of adsorbed inhibitor species with respect to the oxide layer. These results can also be used to assist in determining the nature of the bonding to the oxide surface.

References

1. E. Heitz, in *Advances in Corrosion Science and Technology* (M. G. Fontana and R. W. Staehle, eds.), Vol. 4, pp. 149–243, Plenum Press, New York (1974).
2. J. J. Demo, Effect of low concentrations of acid and water on the corrosion of metals in organic solvents, *Chem. Eng. World* **7**, 115–124 (1972).
3. K. Fischbeck, On the application of the warburg apparatus for corrosion testing, *Werkst. Korros.* **15**, 59–63 (1964).
4. A. Bukowiecki, Investigations on the corrosion behavior of aluminum and its alloys in organic and inorganic bases, *Werkst. Korros.* **10**, 91–105 (1959).
5. K. Brookman, Investigations on the corrosion behavior of aluminum in organic solvents, *Aluminum* **34**, 30–35 (1958).
6. E. Heitz and C. Kyriazis, Reactions during corrosion of metals in organic solvents, *Ind. Eng. Chem. Prod. Res. Dev.* **17**, 37–41 (1978).
7. P. K. Hansma, Inelastic electron tunneling, *Phys. Rep.* **30C**, 145–206 (1977).
8. D. B. Boies and B. J. Northan, Aluminum corrosion: Possible mechanisms of inhibition, *Mater. Prot.* **7**, 27–30 (1968).
9. W. L. Archer, Paper presented at T-3 Symposium, Conf. of National Association of Corrosion Engineers, Houston, Texas, 1978 (unpublished).
10. W. L. Archer, Comparison of chlorinated solvent–aluminum reaction inhibitors, *Ind. Eng. Chem. Prod. Res. Dev.* **18**, 131–135 (1979).
11. J. O'M. Bockris and A. K. N. Reddy, *Modern Electrochemistry* Vols. 1 and 2, Plenum Press, New York (1973).
12. *Corrosion Inhibitors* (C. C. Nathan, ed.) National Association of Corrosion Engineers, Houston, Texas (1973).
13. M. Stern and H. H. Uhlig, Mechanism of reaction of aluminum and aluminum alloys with carbon tetrachloride, *J. Electrochem. Soc.* **100**, 543–551 (1953).

14. J. D. Minford, M. H. Brown, and R. H. Brown Reaction of aluminum and carbon tetrachloride. I, *J. Electrochem. Soc.* **106**, 185–191 (1959).
15. R. H. Brown, E. H. Cook, M. H. Brown, and J. D. Minford, Reaction of aluminum and carbon tetrachloride. II, *J. Electrochem. Soc.* **106**, 192–199 (1959).
16. W. L. Archer and M. K. Harter, Reactivity of carbon tetrachloride with a series of metals, *Corrosion* (Houston) **34**, 159–162 (1978).
17. R. W. Ellialtioglu, H. W. White, L. M. Godwin, and T. Wolfram, Study of the corrosion of aluminum by CCl_4 using inelastic electron tunneling spectroscopy, *J. Chem. Phys.* **72**, 5291–5296 (1980).
18. R. M. Ellialtioglu, H. W. White, L. M. Godwin, and T. Wolfram, Study of the corrosion inhibitor formamide in the aluminum–carbon tetrachloride system using IETS, *J. Chem. Phys.* **75**, 2432 (1981).
19. E. Plueddemann and R. Rathman, U.S. Pat. 2,423,343 (1947).
20. H. Knozinger and P. Ratnasamy, Catalytic aluminas: Surface models and characterization of surface sites, *Catal. Rev.* **17**, 31–70 (1978).
21. H. W. White, R. Ellialtioglu, and J. E. Bauman, Jr., Study of the corrosion of aluminum by trichloroethylene using inelastic electron tunneling spectroscopy, *J. Chem. Phys.* **75**, 3121 (1981).
22. W. L. Archer and V. L. Stevens, Comparison of chlorinated, aliphatic, aromatic, and oxygenated hydrocarbons as solvents, *Ind. Eng. Chem. Prod. Res. Dev.* **16**, 319–325 (1977).
23. W. L. Archer, private communication.
24. V. E. Jenckel and F. Woltmann, Study of the reaction rates of aluminum in acid solutions of pyridine derivatives (in German), *Z. Anorg. Allg. Chem.* **233**, 236–256 (1937).
25. J. Sundararajan and T. L. Rama Char, Inhibition of corrosion of aluminum in acid solutions, *J. Sci. Ind. Res.* **17B**, 387–388 (1958).
26. R. J. Graves and H. W. White, Study of the corrosion inhibitor acridine adsorbed on aluminum oxide using tunneling spectroscopy (to be published).
27. S. Califano, Infrared spectra in polarized light and vibrational assignment of the infrared active modes of anthracene and anthracene–d_{10}, *J. Chem. Phys.* **36**, 903–909 (1962).
28. M. Brigodiot and J. M. Lebas, Vibrational study of acridine and the acridinium ion by infrared and Raman spectroscopy: Comparison with anthracene, *J. Chim. Phys.* **69**, 964–971 (1972).
29. O. V. Fialkovskaya and A. V. Nefedov, A study of vibrational processes in molecules with different symmetries but similar masses and structures, *Opt. Spectrosc.* **25**, 428–429 (1968).
30. Wen-Jung Yang and H. W. White, Comparison of experimental and theoretical intensities of inelastic tunneling spectra for thiourea, *Surf. Sci.* (in press).
31. J. Kirtley, D. J. Scalapino, and P. K. Hansma, Theory of vibrational mode intensities in inelastic electron tunneling spectroscopy, *Phys. Rev. B* **14**, 3177–3184 (1976).
32. L. M. Godwin, H. W. White, and R. Ellialtioglu, Comparison of experimental and theoretical inelastic electron tunneling spectra for formic acid, *Phys. Rev. B* **23**, 5688–5699 (1981).

11

Adsorption and Reaction on Aluminum and Magnesium Oxides

D. G. Walmsley and W. J. Nelson

1. Introduction

The usefulness of inelastic electron tunneling as a vibrational spectroscopy is now widely recognized. The customary arguments in its favor as a probe of surface adsorbates include its high sensitivity, wide spectral range, and good resolution. Even a cursory inspection of the chapter titles in this volume will show the breadth of its potential applications. As an introduction to such studies it is instructive to consider the outcome of some basic experiments in which relatively simple organic molecules are examined. From the results much may be learned that has more general application. Along the way we shall find many helpful indicators to how the method might profitably be further extended.

2. Clean Aluminum Oxide

With few exceptions the tunneling spectra arise from the study of species incorporated within aluminum–aluminum-oxide–lead (Al–I–Pb) sandwiches. Aluminum oxide has two desirable properties in this context, one physical and one chemical: it offers a high potential barrier to the tunneling electrons and it contains hydroxyl species. The first property

D. G. Walmsley and W. J. Nelson • School of Physical Sciences, New University of Ulster, Coleraine BT52 1SA, Northern Ireland.

allows the thin (20-Å) oxide tunnel barrier to withstand the application of ~0.5 V dc bias without disastrous increase of the tunnel current and the second gives rise to interesting surface chemistry. Let us begin by looking at the spectrum from an undoped Al–I–Pb sandwich as displayed in Figure 1. This will be present as background in all other spectra from such sandwiches. It is important to emphasize that this same spectrum has been consistently reported by all workers using tunneling spectroscopy. Thus it would seem that the oxide layer, whether grown thermally in laboratory air or by plasma oxidation in an oxygen glow discharge, shows no detectable variation in composition. It does not necessarily follow that the chemical reactivity of the oxide surface is identical in all cases. In Figure 1 three features are clear: a peak at 300 cm^{-1} (1 meV = 8.065 cm^{-1}) due to aluminum metal phonons in the electrode, a broad asymmetric band at 940 cm^{-1} due mainly to vibrational modes of the aluminum oxide, and another

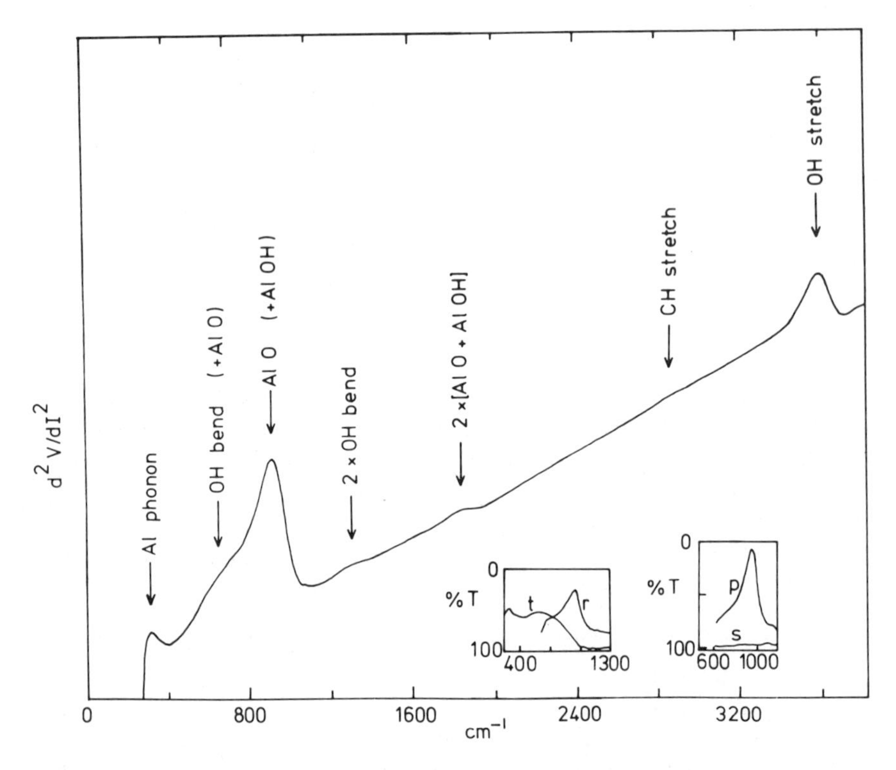

Figure 1. Tunneling spectrum of undoped Al–I–Pb sandwich. First inset shows infrared reflectance and transmission spectra of electrolytically anodized aluminum. Second inset shows polarization dependence of reflectance spectrum.

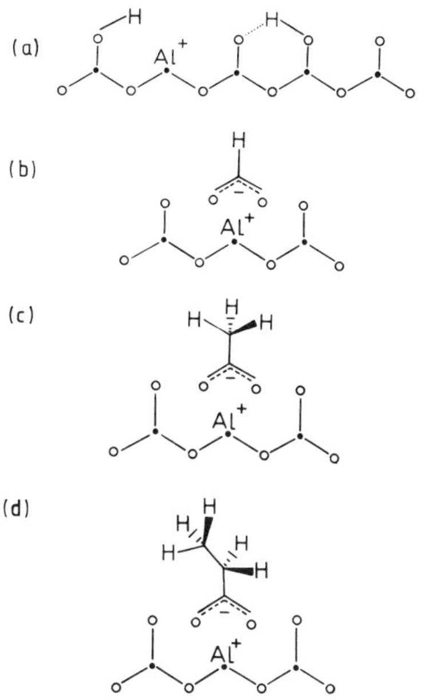

Figure 2. (a) Idealized representation of plasma-grown aluminum oxide. (b) Formate anion on aluminum oxide after doping with formic acid. (c) Acetate anion on aluminum oxide after doping with acetic acid. (d) Propionate anion on aluminum oxide after doping with propionic acid.

fairly prominent band at 3600 cm^{-1} which arises from stretching vibrations of surface hydroxyl species. The latter is quite reproducible in position, shape, and strength; it reflects the need for water vapor to be present during oxidation of aluminum. An idealized representation of the surface might be something like that shown in Figure 2a. The aluminum oxide band at 940 cm^{-1} agrees well with the infrared reflectance spectrum but not the transmission spectrum (Figure 1, first inset) of air-oxidized or electrolytically anodized aluminum. In fact only the *p*-polarized component of the reflected radiation shows[1] the absorption (Figure 1, second inset). This suggests that the relevant dipolar oscillations in the oxide are perpendicular to the surface which is the best configuration for excitation by tunneling electrons.[2] There is very little signal due to CH stretch modes. We conclude that organic contamination is negligible. The remaining assignments have been discussed in a number of papers interpreting the tunnel spectrum of clean aluminum oxide barriers.[3-5]

3. Dirty Aluminum Oxide

The study of adsorbates on aluminum oxide as on any surface has one major pitfall, contamination. During sample fabrication there is a critical stage immediately after the oxide has been grown. The freshly prepared

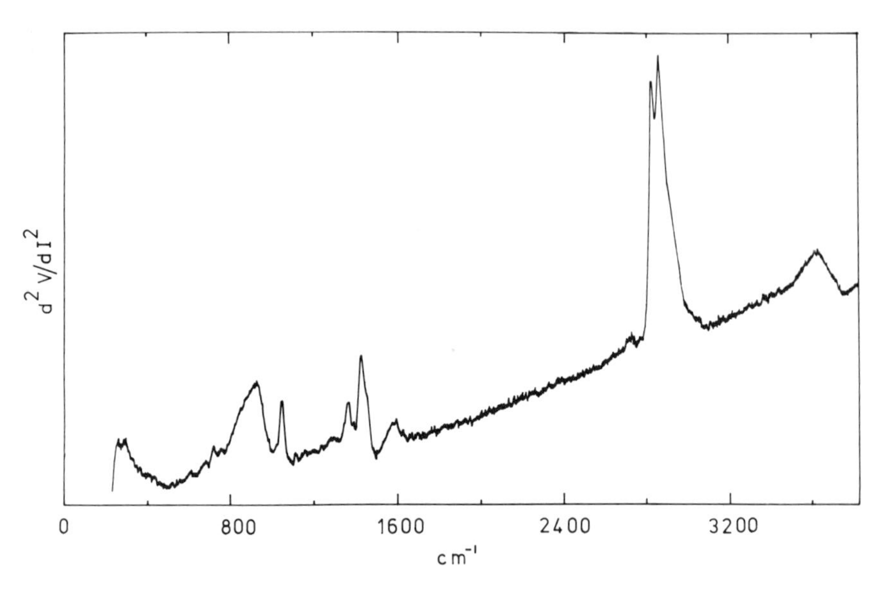

Figure 3. Tunneling spectrum of Al–*I*–Pb sandwich accidentally exposed to atmospheric contamination during fabrication.

oxide is quite reactive and will rapidly pick up any stray vapor to which it is exposed. Therefore it is important to use an oil with very low vapor pressure in the diffusion pump of the sample preparation chamber. If vapor doping is used this measure should suffice. However, with liquid doping or infusion doping, which require the partially completed sandwich to be removed from the preparation chamber, great care must also be taken with the ambient atmosphere in which the doping procedure is carried out. A clean air laminar-flow bench is then desirable. A straightforward test of whether the particular arrangement is adequate is to go through the complete exercise of making a sample but omitting just the doping step. If liquid doping is being used then the sample should be removed from the preparation chamber and mounted in a spinner just as if the doping had taken place. Only by going through such a procedure and examining the spectrum of the undoped sample can it be seen whether or not the sample is still clean. This test is very important. Figure 3 shows an example of a dirty sample. It was intended as a sample liquid doped with carbon tetrachloride, CCl_4. At the time of preparation there was a very low concentration of acetaldehyde vapor in the laboratory atmosphere as a result of some other work in progress. The spectrum obtained shows clearly the characteristic features of adsorbed acetate ion and some additional aliphatic hydrocarbon. Any structure due to adsorbed carbon tetrachloride is lost in the contamination. Having sounded the warning let us go on to consider spectra obtained under more scrupulous conditions. Vapor doping has been used in most cases.

4. Doped Aluminum Oxide

Where do we start? Though not limited to organic adsorbates these must be foremost in our thinking since vast areas of scientific and technological enterprise including work on catalysis, corrosion, adhesion, and lubrication rest very heavily on surface adsorbed organics and it is in these fields of activity that tunneling spectroscopy should be able to make its mark.

A quick look through an introductory textbook on organic chemistry gives the classifications of compounds into families such as acids, alcohols, amines, and so on. These are characterized by a functional group, for example the COOH group of an acid, with the remainder of the molecule inert. When a molecule is chemically adsorbed on a surface it is the functional group of the molecule that reacts in its own characteristic fashion. The same reaction behavior is expected for a series of compounds in a given family, the only difference being the size of the inert tail of the molecule. The vibrational spectrum of the molecule should reflect this state of affairs. Certain features derived from the reacted functional group should recur for all compounds of a given family but the more atoms there are in the inert tail, the more structure will be expected in the spectrum.

We have interested ourselves for some time in looking at the tunneling spectra of different families of compounds adsorbed on aluminum oxide in the hope that we could decipher the nature of the reactions in each case. In general we have preferred to work with small molecules because they have less complex spectra. If the reaction can be understood for a small molecule it is expected that the same reaction will take place for the same functional group in a large molecule—always provided of course that there are no geometrical difficulties in accommodating the molecule at a potential reaction site on the adsorbent surface.

The following sections relate the outcome of these studies. Molecules studied include acids and the closely related aldehydes, acid chlorides, and acid anhydrides. Also hydrocarbons, alcohols, amines, and a few bifunctional species. The treatment is illustrative rather than rigorous. Original papers are referenced and should be consulted if more information is required.

4.1. Formic Acid

Small reactive molecules are most likely to give a high-density adsorbate and strong spectral signals. Larger molecules may not pack so well on the surface or they may partially block some potential surface reaction sites while unreactive molecules, especially small ones, tend to come off the surface again under the vacuum conditions required for depositing a top electrode. With these considerations in mind let us examine the tunneling spectrum in Figure 4 obtained when a small reactive molecule, formic acid with formula HCOOH, is adsorbed on aluminum oxide. Klein *et al.*[6] interpreted this spectrum as showing a reaction of formic acid on the aluminum oxide surface to produce adsorbed formate ion, $HCOO^-$. See Figure 2(b). The clue to interpretation lies in the absence of the C=O and C—OH stretch modes of the unreacted acid which are expected at 1780 and

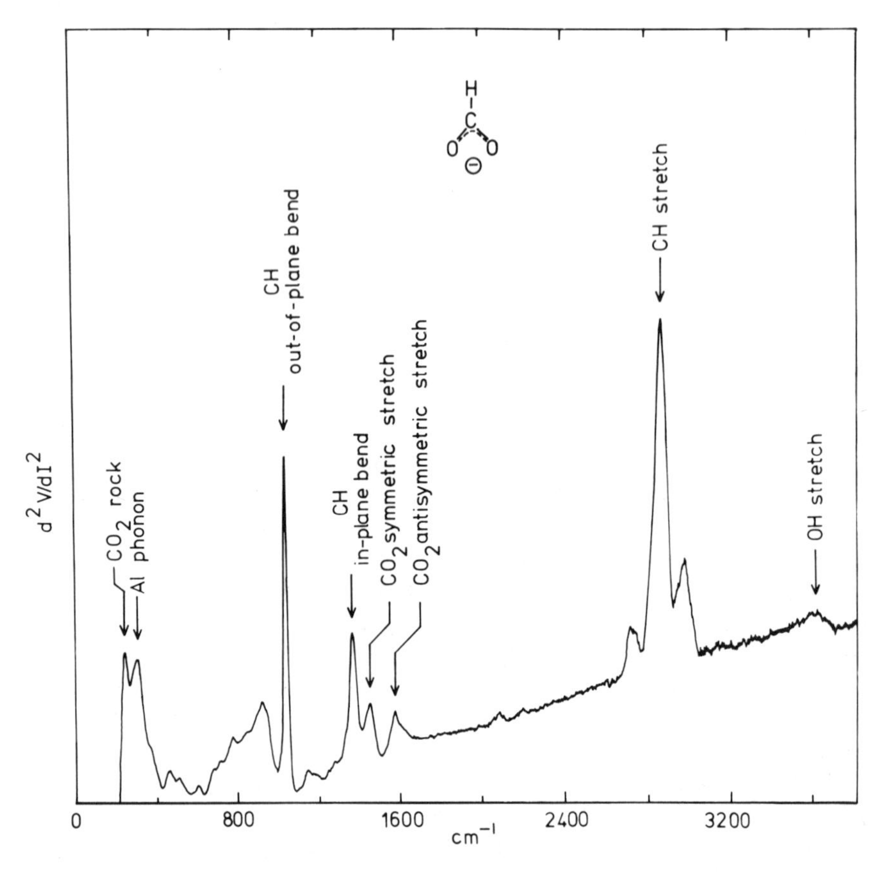

Figure 4. Tunneling spectrum of Al–*I*–Pb sandwich doped with formic acid.

1100 cm^{-1}, respectively. Instead there are two bands characteristic of the anionic carboxylate, namely, the CO_2^- antisymmetric stretch mode at 1580 cm^{-1} and the CO_2^- symmetric stretch mode at 1456 cm^{-1}.

General considerations dictate that a molecule or ion with N atoms should have $3N-6$ normal modes of vibration. Hence we expect six modes in all for the formate ion. The CH stretch mode is clearly seen at 2875 cm^{-1} and there are two CH bend modes. That corresponding to bending of the CH bond in the CO_2^- plane is higher is energy at 1370 cm^{-1} than the out-of-plane bend mode, with its lower restoring forces, at 1038 cm^{-1}. The sixth mode is a CO_2^- rock down at 236 cm^{-1}.

These assignments come about in large measure through consulting the existing literature of vibrational spectroscopy. That literature has developed in step with the use of infrared spectroscopy, and more recently Raman spectroscopy, of bulk materials. There are two main lines of attack in general use for making assignments. The experimental one seeks to determine which features of a molecular spectrum are sensitive to substitution of one particular atom by another. A favorite substitution is deuterium for hydrogen. More subtle effects arise if a polyatomic group occurs at a different location in a molecule. The other important approach is calculation of the frequencies of the normal modes of vibration of the molecule using interatomic potentials derived *ab initio* or by iterative approximation to obtain a satisfactory fit with observed data. Both the calculational and experimental approaches are complementary and must eventually aim for compatibility of conclusions. Both rely heavily on the symmetry of the molecule to provide simplification of the problem. Much careful work along these lines has built up a substantial body of knowledge which is invaluable for the interpretation of vibrational spectra.

In addition, compilations of actual infrared spectra of a very large range of compounds are available[7,8] for molecule identification but these do not normally bear mode assignments. They are extremely useful for checking out the positions of lines observed in the tunneling spectrum if the species requires identification or confirmation. Unfortunately the relative intensities of the lines in tunneling spectra do not correspond closely to those in infrared or Raman spectra. A good illustration of this, and convenient bonus, is the relatively high strength of lines associated with CH bend and stretch modes in the tunnel spectra of most molecules. The difference hinges on the nature of the interaction between the tunneling electron and the molecule. Dr. Kirtley has discussed this question in Chapter 2. Although the intensities differ between infrared, Raman, and tunneling spectroscopies there is no intrinsic difficulty in using tunneling spectroscopy to identify a species once its tunneling spectrum has been determined. The tunneling intensities are reproducible.

An advantage of tunneling spectroscopy is that the resolution obtained for an adsorbate at liquid helium temperatures, in the range 1–4 K, often far outstrips that seen in an infrared spectrum of the same species in bulk at room temperature. Indeed it is true of the formate ion which we have just considered. The spectrum of Figure 4 and a comparable spectrum from the formate ion adsorbed on magnesium oxide have enabled us to make firm assignments. These spectra show conclusively that the CO_2^- symmetric stretch mode is higher in energy than the CH in-plane bend mode.[9,10] Previously this has been totally unclear and the balance of argument, based on poorly resolved vibrational spectra, pointed in the opposite direction. So tunneling spectroscopy has sorted out an uncertainty of assignment in a very simple ion.

4.2. Acetic Acid and Closely Related Molecules

After formic acid the next simplest member of the aliphatic acid family is acetic acid, CH_3COOH, which differs by the addition of a CH_2 group. Like its lower homologue it is a liquid of moderate vapor pressure under laboratory conditions and can readily be introduced to a tunnel sandwich by the vapor doping technique. The spectrum is seen in Figure 5a. A good starting point for its interpretation is to notice that the OH stretch mode at ~ 3600 cm^{-1} is less strong than in the undoped sandwich spectrum in Figure 1. This observation suggests a Lewis-acid–Lewis-base reaction in which a proton is transferred from the acetic acid to the hydroxyl species on the aluminum oxide surface

$$CH_3COOH + Al(OH)_x \rightarrow CH_3COO^- Al(OH)_{x-1}^+ + H_2O$$

The water released is pumped away by the vacuum system of the sample preparation chamber. Although we have not yet directly observed the water after release the reverse process can be induced by water infusion through the top electrode of a completed sandwich.[11] Upon infusion the 3600-cm^{-1} mode registers an increase in strength. Features characteristic of molecular water itself are not normally observed in tunneling spectra: free water has a stretch mode at 3300 cm^{-1} and a bend mode at 1630 cm^{-1}. See Figure 2c.

Other features in the spectrum confirm the reaction. The CO_2^- antisymmetric stretch mode is found at 1583 cm^{-1} while the identification of the 1463 cm^{-1} line as the CO_2^- symmetric mode is confirmed by its shift to lower energy, 1445 cm^{-1}, when adsorbed on magnesium oxide. For purposes of interpretation, the acetate ion may be viewed as two triatomic groups joined by a C—C bond which has its stretch mode at 945 cm^{-1}. The CO_2^- group displays a scissor mode at 678 cm^{-1} and two rock modes, one

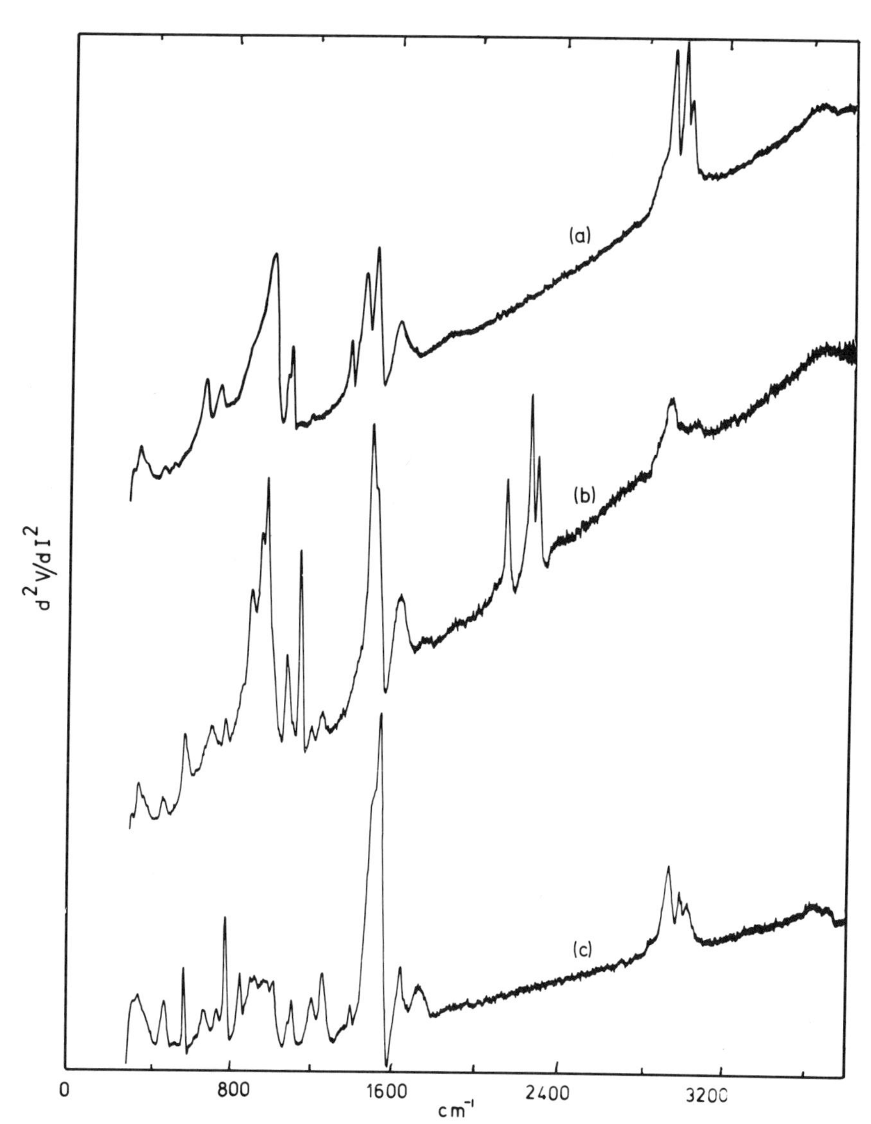

Figure 5. Tunneling spectra of Al–*I*–Pb sandwiches doped with (a) acetic acid, (b) perdeutero-acetic acid, (c) trifluoroacetic acid.

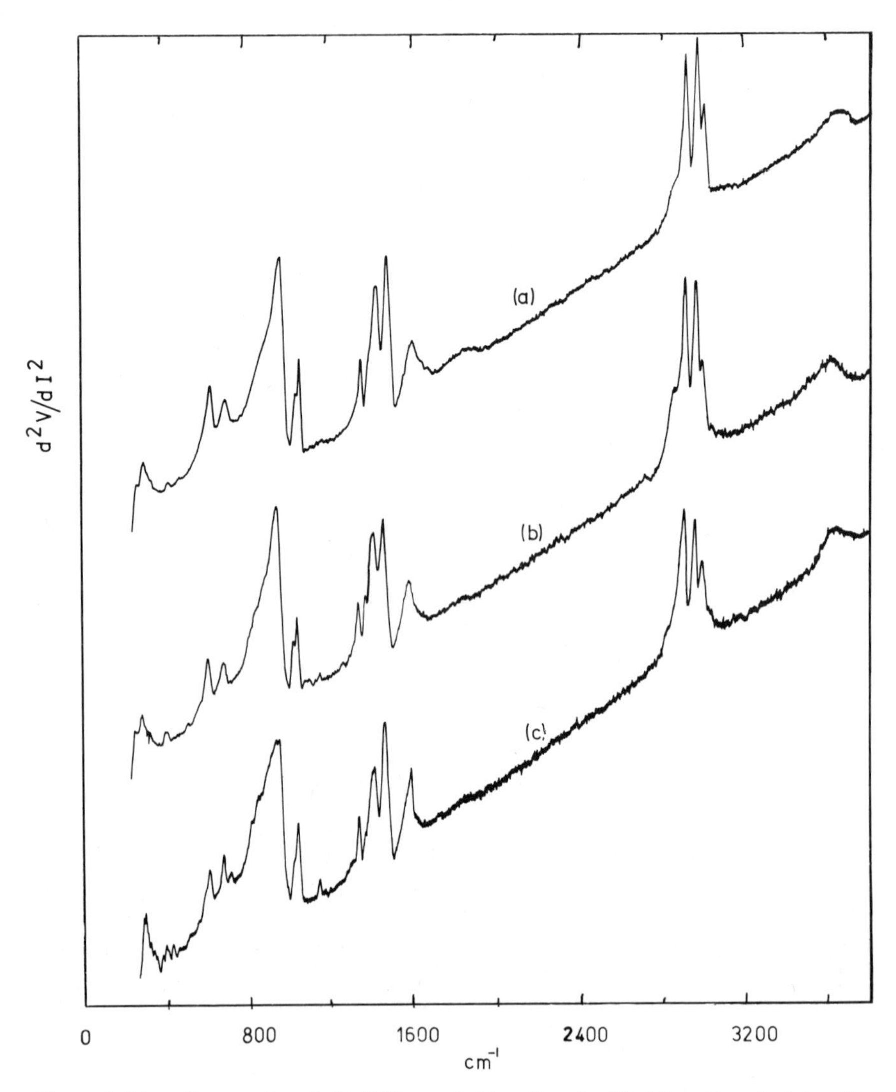

Figure 6. Tunneling spectra of Al–I–Pb sandwiches doped with (a) acetic anhydride, (b) acetaldehyde, and (c) acetyl chloride.

in plane at 468 cm^{-1} and one out of plane at 605 cm^{-1}. The role reversal here of the in-plane and out-of-plane modes as compared with the CH bend modes of the formate ion in the previous section comes about as follows. Here it is the surface environment which governs the CO_2^- behavior in contrast with the intramolecular potential which is the determining influence on the CH bend modes of the formate ion. As for the remaining modes in the acetate ion spectrum, the CH_3 group accounts for those remaining in the 1000–1500-cm^{-1} region and its stretch modes are seen at 2900–3000 cm^{-1}.[12,13]

Figure 5b is the spectrum of perdeuteroacetic acid, CD_3COOD. It shows clearly the expected $1/\sqrt{2}$ isotopic downshift in the CD stretch modes and careful study of it helps in confirming the assignments of other modes in the undeuterated species. Some CH stretch modes attributable to contamination persist. Figure 5(c) is obtained from trifluoroacetic acid, CF_3COOH. As well as showing the lower energy of the CF stretch modes it exhibits not one but two sets of CO_2^- modes, which suggests that two different surface sites are occupied. These may be a reflection of the greater strength of trifluoroacetic acid.

Next is shown, in Figure 6, a comparison of the spectra obtained by adsorbing (a) acetic anhydride $(CH_3CO)_2O$ (b) acetaldehyde CH_3CHO, and (c) acetyl chloride CH_3COCl on plasma grown aluminum oxide. In all cases the outcome is the same; it is the spectrum that we have identified as arising from adsorbed acetate ion. Possible reaction mechanisms have been discussed elsewhere[12] and will not be repeated here except to note that the results are substantially different from those found in bulk alumina adsorption studies. If we wish to study adsorbed carboxylate species it would seem unimportant whether we use the acid, aldehyde, anhydride, or acid chloride as starting material. In practice the aldehyde has some advantages. For example, it tends to be more volatile than the hydrogen-bonded acid, an advantage when vapor doping, and it gives less spurious structure in the spectrum than does the highly reactive acid chloride. Comparison of the spectra in Figures 6a, 6b, 6c and Figure 5a shows rather well the reproducibility that can be achieved from sample to sample in tunneling spectroscopy.

4.3. Higher Acids

The spectra of the higher monobasic aliphatic acids are more complicated as a result of the additional CH_2 groups in the corresponding adsorbed C_n-carboxylates, $CH_3(CH_2)_{n-2}CO_2^-$. Figure 7 shows the spectra from sandwiches doped with propionic acid (C_3), butyric acid (C_4), and caprylic acid (C_8). As may be seen from Table 1 all dopants are liquid at

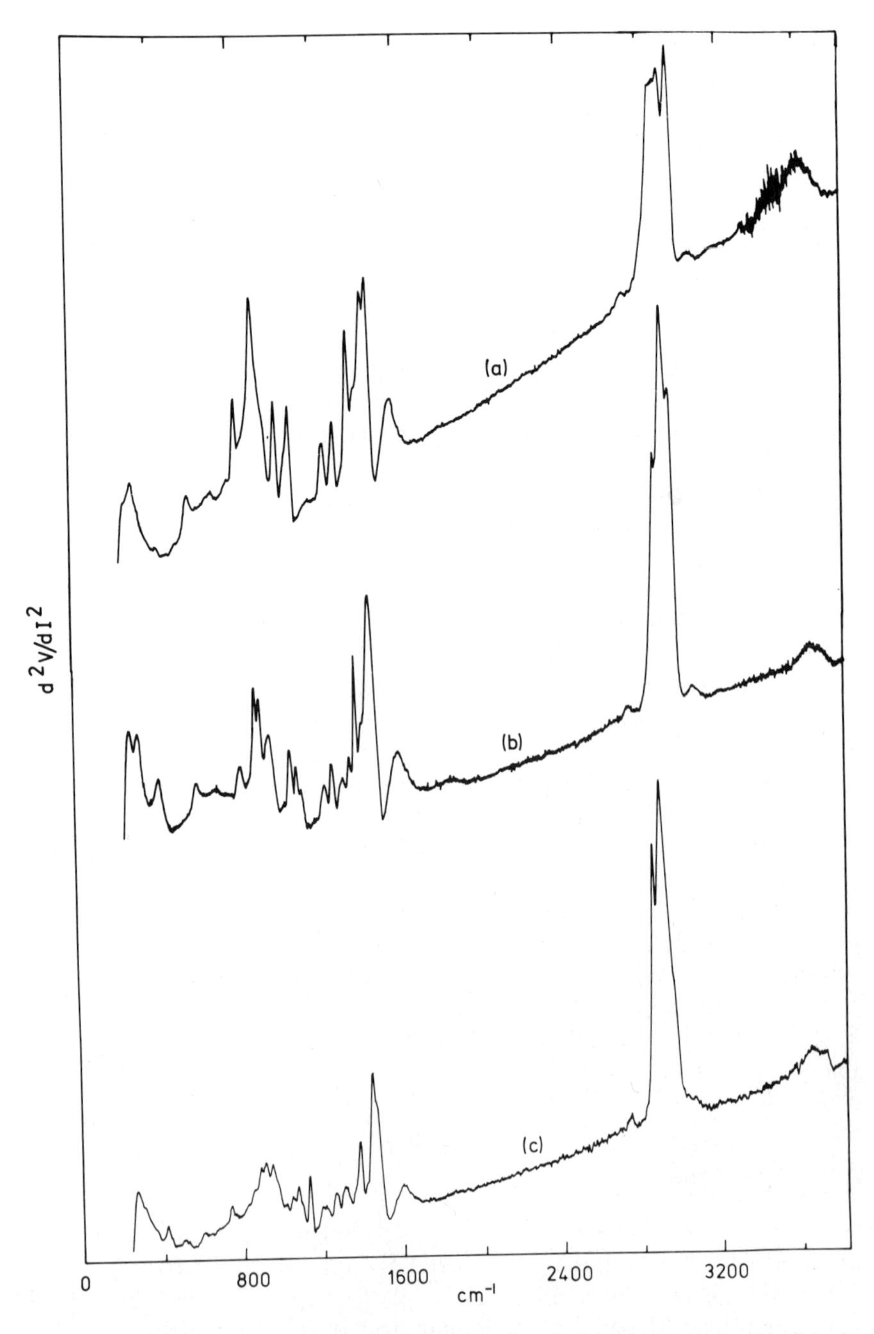

Figure 7. Tunneling spectra of Al–I–Pb sandwiches doped with (a) propionic acid (C_3), (b) butyric acid (C_4), and (c) caprylic acid (C_8).

Table 1. Melting Points and Boiling Points of Some Dopants

Dopant	Melting point (°C)	Boiling point (°C)
Formic acid	8.4	101
Acetic acid	16.6	118
Perdeuteroacetic acid		116
Trifluoroacetic acid	−15.3	72.4
Acetic anhydride	−73	140
Acetaldehyde	−121	20.8
Acetyl chloride	−112	50.9
Propionic acid C_3	−20.8	141
Butyric acid C_4	−4.3	164
Caprylic acid C_8	16.5	239
Lauraldehyde C_{12}	44.5	185
Palmitaldehyde C_{16}	34	200
Stearaldehyde C_{18}	55	261

room temperature. Caproic acid (C_6) has been studied by Cass et al.[14] using a liquid solution doping technique. Figure 8 shows the spectra from sandwiches doped with lauraldehyde (C_{12}), palmitaldehyde (C_{16}), and stearaldehyde (C_{18}). Though introduced by vapor doping these required heating and in some instances had to be entrained in a flow of warm nitrogen gas. As the results show the vapor doping procedure can, with a little modification, cope with fairly involatile species.

In each spectrum the CH stretch modes remain close in frequency at around 2900 cm^{-1} but give increasing intensity as the series is ascended. The increase is partly concealed in the figures by the reduced amplification used for detecting the higher members of the series. A rough but not quantitatively rigorous guide to the sensitivity of spectrometer operation is the background slope on which the spectral lines are superposed. When there are strong signals the instrument gain is kept low and the background remains relatively flat. Conversely, when little dope adsorbs, the signals will be weak, the spectrometer gain will be turned up, and the background slope will be seen as fairly steep. Figure 9 demonstrates the increase in strength of the CH stretch mode relative to the observed strength of the CO_2^- antisymmetric stretch mode at ~1585 cm^{-1} with increase in the number of carbon atoms in the molecule. In each case the surface oxide is thought to be site saturated and the carboxylate mode is expected to be a reliable reference.

Analysis of the low-energy part of the spectra shows clearly the CO_2^- antisymmetric mode at ~1585 cm^{-1} in every case. Thus we conclude all are adsorbed as carboxylate anions as in Figure 2d. The CO_2^- symmetric mode may be identified at 1435 ± 10 cm^{-1}. Of the remainder, the most interesting

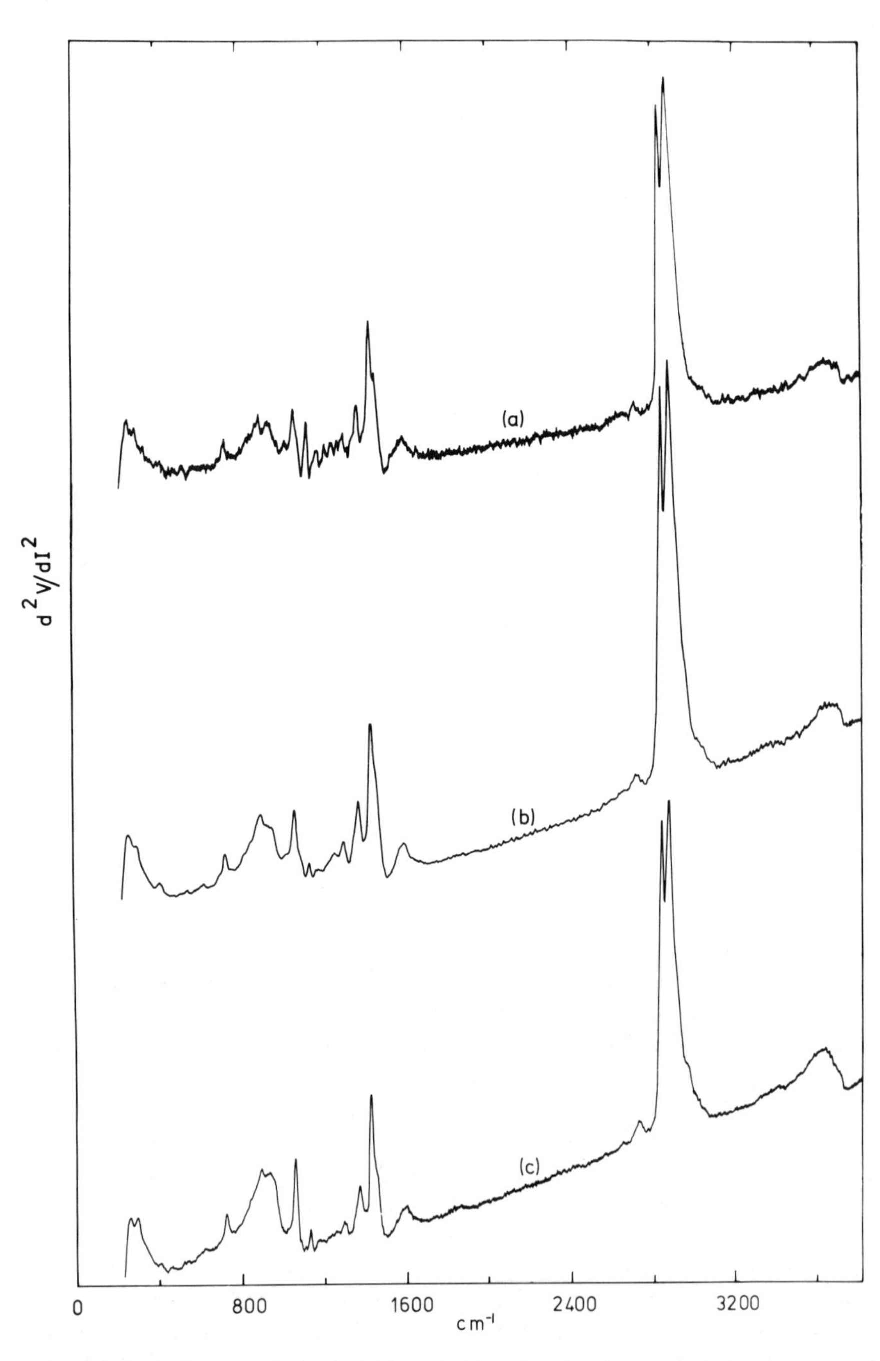

Figure 8. Tunneling spectra of Al–I–Pb sandwiches doped with (a) lauraldehyde (C_{12}), (b) palmitaldehyde (C_{16}), and (c) stearaldehyde (C_{18}).

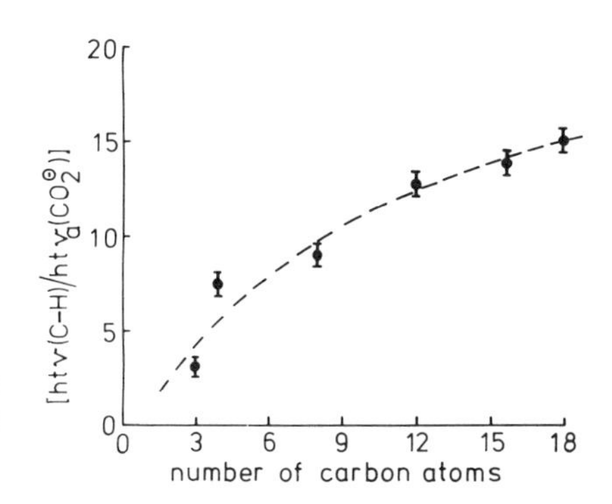

Figure 9. Variation of intensity of CH stretch modes in tunneling spectra of monobasic C_n-aliphatic acids as a function of n.

bands are those associated with the collective motions of the CH_2 groups. They lie in the 1160–1350-cm^{-1} range. Their number, positions, and intensities depend on how many adjacent CH_2 groups there are in the molecule. Their intensity is enhanced by the presence of a polar terminal substituent. They are taken to indicate that the hydrocarbon chains are in regular *trans* zig-zag configuration. The tunneling spectrum of the laurate anion (C_{12}) shows the bands most clearly and they are in good agreement with infrared data. Fairly good agreement is found for the other spectra too. Assignment of the remaining bands is discussed elsewhere.(15)

The spectra of the higher aliphatic carboxylates illustrate the possibility of using tunneling spectroscopy for quite large molecules. Indeed one might expect to be able to obtain very detailed information by the tunneling method. The limit on resolution is set only by thermal smearing and thus one might expect to resolve all $3N-6$ modes of the largest molecules simply by going to low enough temperatures. Alas, not so. We have examined the spectra of a number of adsorbates at temperatures down to less than 100 mK in a helium dilution refrigerator and found no lines with natural width less than 1 meV (8 cm^{-1}). It would therefore be impossible to see resolved all the lines of a molecule having more than 50 or 100 atoms. Such is the nature of the internal coupling of the dipoles within a molecule and the redundancy in bond type that broad bands unevenly spaced are likely to defeat the objective long before this limit is reached.

There is actually little to be learned about large molecules by tunneling spectroscopy that can not more easily be determined from studies of the same species in bulk by standard infrared or Raman methods. The real power of tunneling is in seeing surface reactions and, for these, quite small molecules serve just as well as large ones. Furthermore, the spectra are then much easier to interpret. Exceptions to this general rule will arise when there is an interest in how a specific large molecule, such as one of biological importance, behaves on a surface(16) or permeates a membrane.(11)

4.4. Unsaturated Acids

Before considering small molecules with different functional groups we first look at the spectra of unsaturated acids. For Figure 10a propiolic acid was adsorbed and once again the CO_2^- modes of the $HC{\equiv}C{-}CO_2^-$ anion are seen. Also clearly evident are the terminal $\equiv C{-}H$ stretch mode at 3272 cm^{-1} and the $C{\equiv}C$ stretch mode at 2108 cm^{-1}. Though pleasantly resolved these come as no surprise. There is, however, much unexpected structure which bears a close resemblance to the spectrum obtained when acrylic acid, $H_2C{=}C{-}COOH$, is similarly adsorbed as shown in Figure 10b. To take one particular example, the $=C{-}H$ mode at 3075 cm^{-1} is closely mirrored at the same energy in Figure 10a. We have concluded[17] from a detailed comparison of the spectra that some of the propiolic acid, as well as adsorbing as an anion, has reacted to acrylate on the surface. A similar argument applies to the acrylic acid, which shows spectral evidence of further reaction at the site of its double bond but the final product is not clearly identifiable as propionate. An amusing comparison is the spectrum obtained by Jaklevic and Gaerttner[11] when propiolic acid is infused into a clean sandwich. Here the spectrum shows no sign of these further reactions and the result is much cleaner; the CH stretch region of the spectrum is particularly simple. The third spectrum from an unsaturated acid, Figure 10c, is the result of introducing dimethyl acrylic acid. Here we expected the methyl groups to shield the double bond from direct exposure to the reactive aluminum oxide surface. The spectrum indicates by its relative simplicity that in large measure we have succeeded.

In the propiolic and acrylic acid examples there is reaction at two sites in the adsorbate. The formation of an anionic species has been discussed in connection with saturated acids where it was suggested to arise by proton transfer from the acid to the surface hydroxyl species. The protonation of the unsaturated carbon chains is postulated to originate in surface hydroxyl species of high Brønsted acidity. These will readily donate protons to the adsorbate. A dual role is thus being proposed for the hydroxyl species at the aluminum oxide surface. Possible configurations are suggested in Figure 11.

4.5. Unsaturated Hydrocarbons

After seeing these reactions of unsaturated acids a next step is to investigate unsaturated hydrocarbons which have no functionalized group. Figure 12 shows that spectra may be obtained[18] after exposing aluminum oxide to three alkenes: 1-hexene, 1-heptene, and 1-octene. In each of these there is a double bond between the first two carbon atoms of the chain. The spectra are weak, particularly for the heavier molecules, and the structure

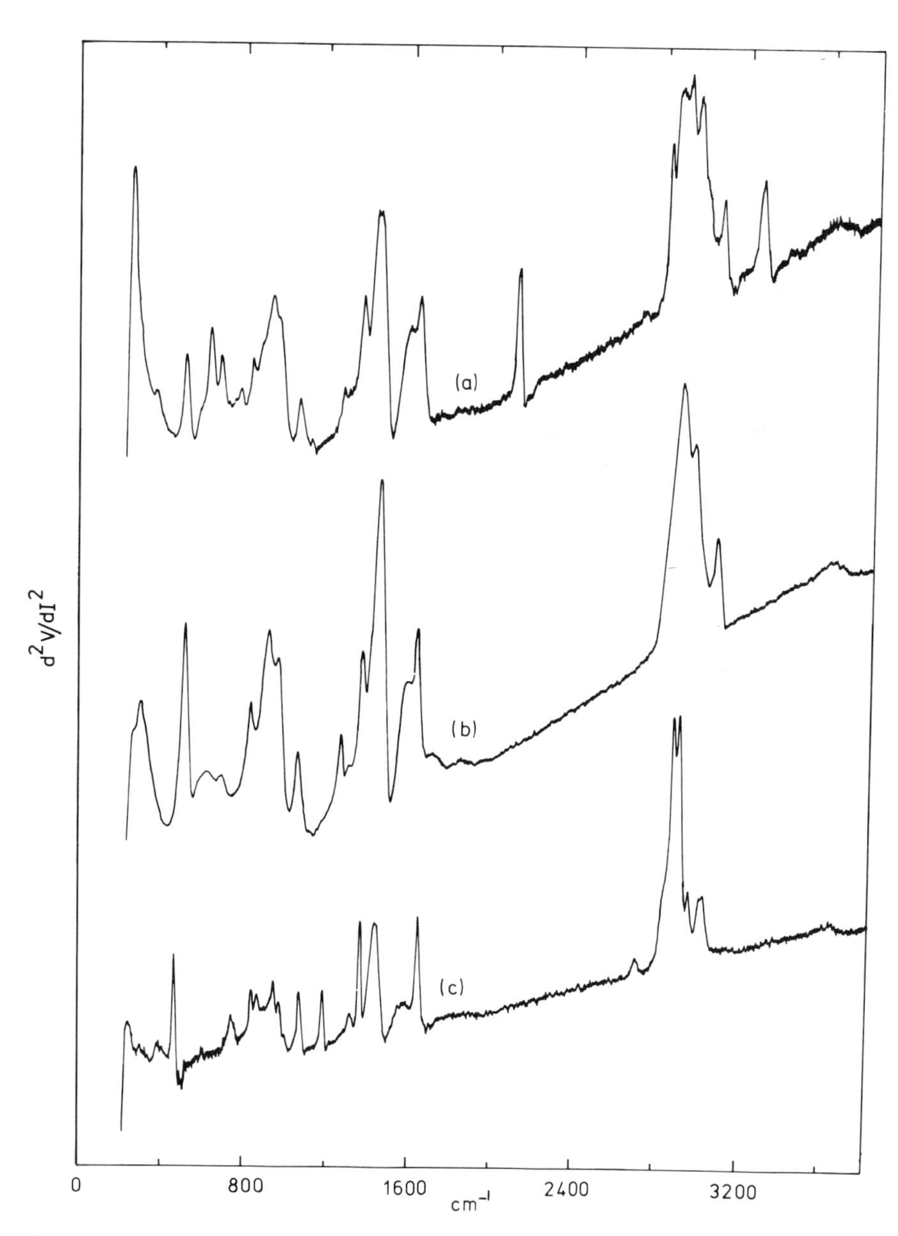

Figure 10. Tunneling spectra of Al–*I*–Pb sandwiches doped with (a) propiolic acid, (b) acrylic acid, and (c) dimethylacrylic acid.

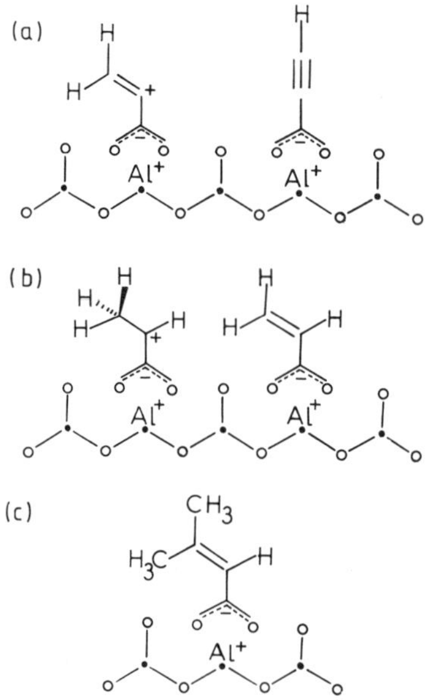

Figure 11. Possible configurations of (a) propiolate anion, (b) acrylate anion, and (c) dimethyl acrylate anion on aluminum oxide.

seen is simple. At the high-energy end of the spectrum the CH stretch modes are not sufficiently resolved to be highly informative. The C=C stretch mode is present at $\sim 1610\ cm^{-1}$, which is about $30\ cm^{-1}$ lower than the mode in free alkenes, and is rather broad in all three cases. Since the surface OH stretch band is not greatly different from that seen in undoped samples protonation of the double bonds by acidic surface hydroxyls is in question. However, the total amount of adsorbate is low and may be limited by the number of hydroxyl species of sufficient acidity on the surface. The alternative explanation is a Lewis base interaction of the double bond with surface Lewis acid sites such as exposed aluminum cations. In Lewis-base–Lewis-acid reactions electron rather than proton transfer is involved. The weak band seen at $\sim 2225\ cm^{-1}$ in all cases is close to the C≡C stretch mode of the corresponding 3-alkynes. No contamination by alkynes could be detected in the starting materials. Possibly a reaction of the type $2\,(R{=}CH_2CH{=}CH_2) \rightarrow R{-}CH_2CH_2CH_3 + R{-}C{\equiv}C{-}CH_3$ takes place with the unsaturated product retained on the surface and the unreactive saturated product pumped away at the time of fabrication of the sample. The spectral bands near $1200\ cm^{-1}$ are CH_2 wag modes and the band near $1050\ cm^{-1}$ arises from C—C stretch and CH_2 rock modes.

Three adsorbed alkynes, 1-hexyne, 3-hexyne, and 1-heptyne gave the spectra shown in Figure 13. The ≡C—H stretch of the 1-alkynes shows up

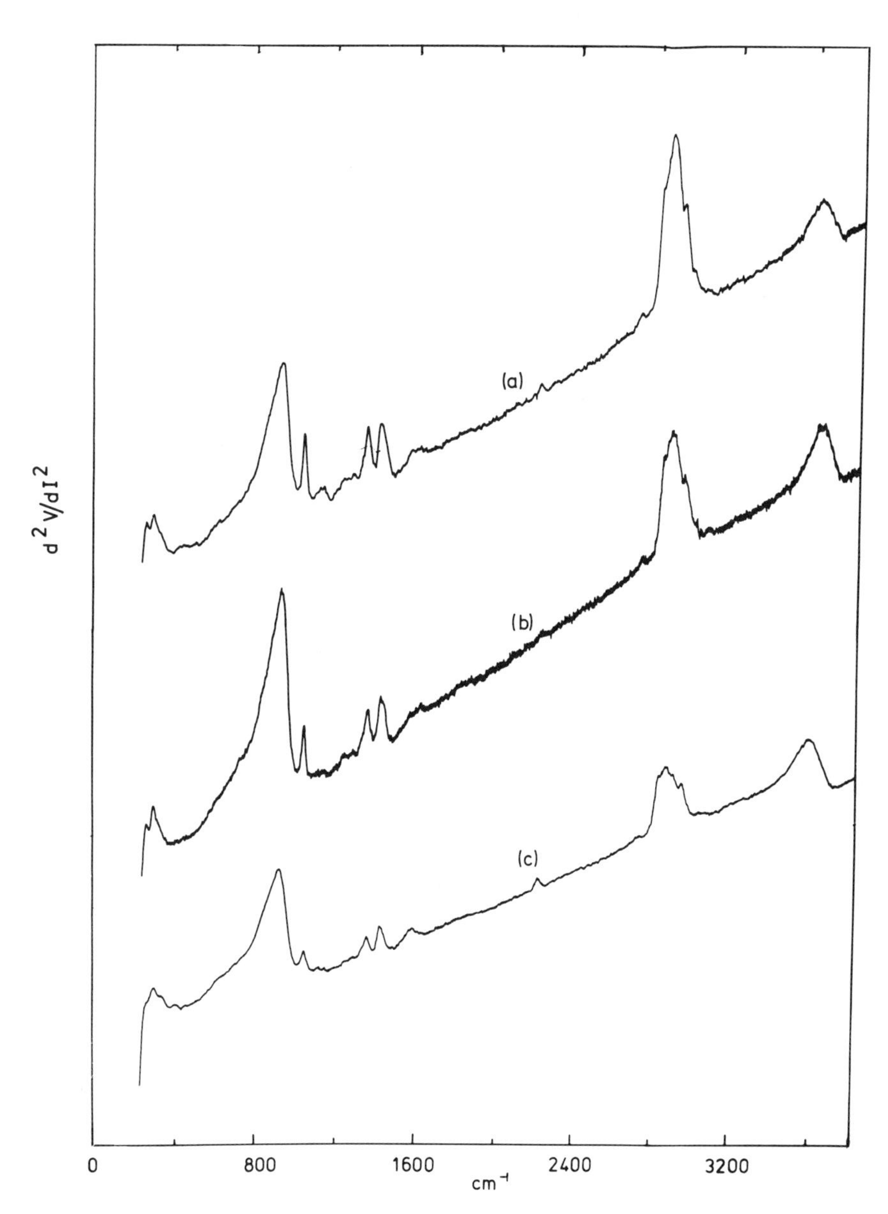

Figure 12. Tunneling spectra of Al–*I*–Pb sandwiches doped with (a) 1-hexene, (b) 1-heptene, and (c) 1-octene.

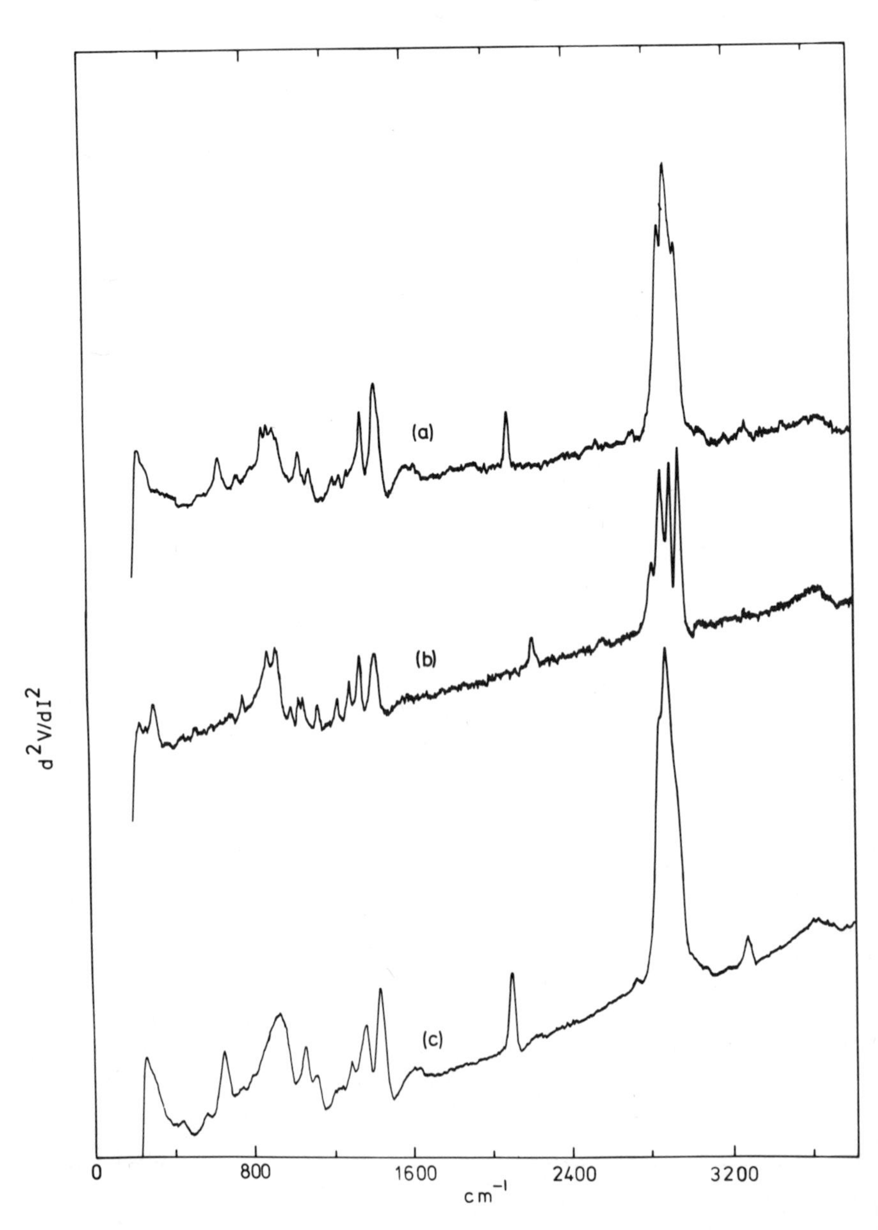

Figure 13. Tunneling spectra of Al–*I*–Pb sandwiches doped with (a) 1-hexyne, (b) 3-hexyne, and (c) 1-heptyne.

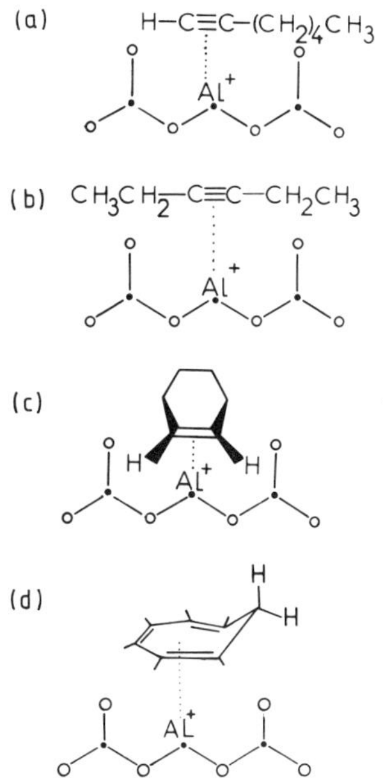

Figure 14. Possible configurations of (a) 1-heptyne, (b) 3-hexyne, (c) cyclohexene, and (d) cycloheptatriene on aluminum oxide.

well resolved at $\sim 3290\ \text{cm}^{-1}$ while of course the 3-alkynes have no acetylenic proton and the mode is missing. Similarly, the positions of the C≡C stretch modes of the 1-alkynes are close near $2100\ \text{cm}^{-1}$ while the 3-heptyne mode is at $2232\ \text{cm}^{-1}$. These modes are all near the values found in the free species. It would seem that the adsorption causes little or no change to the structure of the alkynes. However, there is also some indication of a broad band close to $1600\ \text{cm}^{-1}$ which we may associate with a C=C bond and interpret as due to the partial reduction, possibly by protonation, of the triple bond. The relative weakness of the OH stretch mode supports the idea of its involvement in the protonation. Possible configurations are shown in Figures 14a and 14b.

Two unsaturated cyclic species, cyclohexene and 1,3,5-cycloheptatriene, gave the spectra shown in Figure 15 after adsorption. The olefinic CH stretch and C=C stretch modes are seen in both. By contrast with the 1-alkenes adsorption seems not to cause major perturbation. Possible configurations are shown in Figures 14c and 14d. Vapor doping with other closely similar species failed in many cases to produce proper adsorption or satisfactory spectra. This is not understood.

4.6. Phenols

Of the many tunneling spectra that have been reported in the literature, those from phenol, C_6H_5OH, and benzoic acid, C_6H_5COOH, are the most

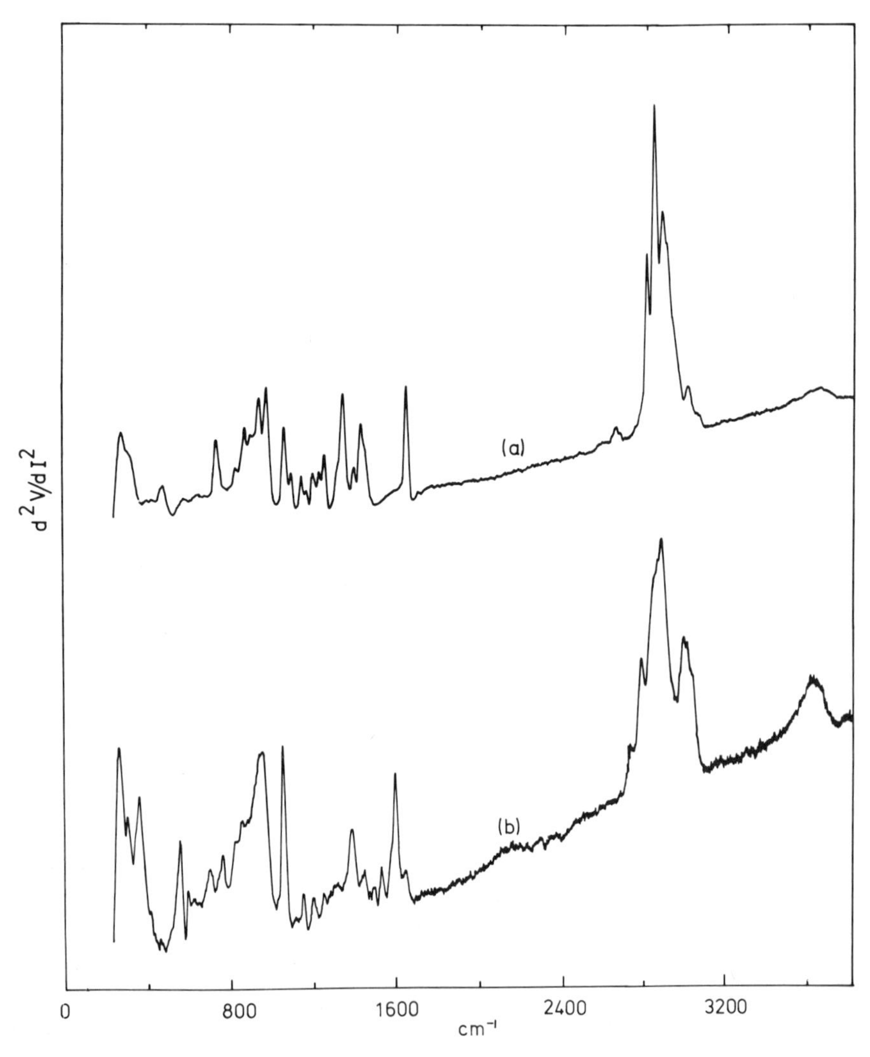

Figure 15. Tunneling spectra of Al–I–Pb sandwiches doped with (a) cyclohexene and (b) cycloheptatriene.

elegant. Several advantageous conditions combine in these compounds. They readily react onto oxidized aluminum as seen in Figure 16; for reasons that are in part obvious, and in part obscure, surface reaction by the adsorbate is almost a necessity for obtaining good spectra. They have a moderate number ($\sim$12) of atoms in the molecule so that upon adsorption there is a fairly rich but not overbusy spectrum. The single aromatic ring

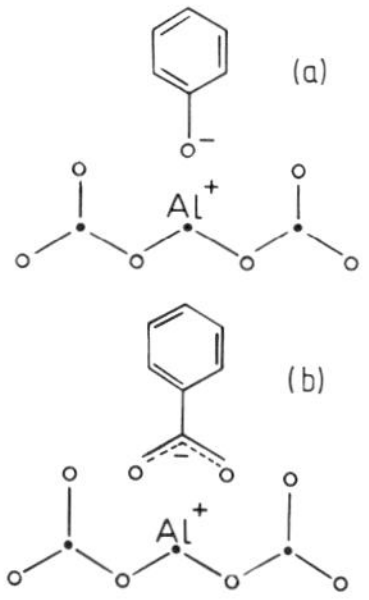

Figure 16. Possible configurations of (a) phenolate anion and (b) benzoate anion on aluminum oxide.

structure has a set of normal vibrational modes that are well spaced in energy. Together these factors make phenol and benzoic acid ideal test compounds when matters basic to the tunneling technique are being investigated. For example linewidth[19] and sensitivity[20] have been explored using them. Anyone wishing to test out sample-making techniques or spectrometer performance could well be recommended to try reproducing the spectrum of adsorbed benzoate or adsorbed phenolate anion.

We show in Figure 17 the spectra of three phenols with Cl, Br, and NO_2 groups substituted at the opposite side of the benzene ring from the reactive hydroxyl group which characterizes phenols. The interpretation of the first two along with that of *p*-fluorophenol has been presented in some detail before.[21] The most striking feature is the similarity of the spectra, which is unsurprising in view of the similarity of the adsorbed phenolate ions. Only a careful scrutiny of the positions of the few substituent-sensitive bands allows the spectra to be distinguished. The characteristic frequency (~ 3000 cm^{-1}) of the aromatic CH stretch modes lies between that of the saturated compound —CH stretch modes and that of the =CH modes in unsaturated compounds. From each spectrum the phenolic OH modes are missing and it is deduced that a proton liberated from the phenol combines with a surface hydroxyl group to form water. The water vapor is pumped away at the time of sample fabrication. The spectrum of *p*-nitrophenol in Figure 17c is rather different from the others. It does not have a strong narrow C—C ring mode near 1600 cm^{-1}. Also, it shows additional bands around 2900 cm^{-1} which are attributable to stretch modes of the NO_2 group.

4.7. Aromatic Alcohols and Amines

Two important families of organic compounds are the alcohols and amines. Their adsorption onto aluminum oxide may be monitored by tunneling spectroscopy. The interpretation of the spectra from adsorbed

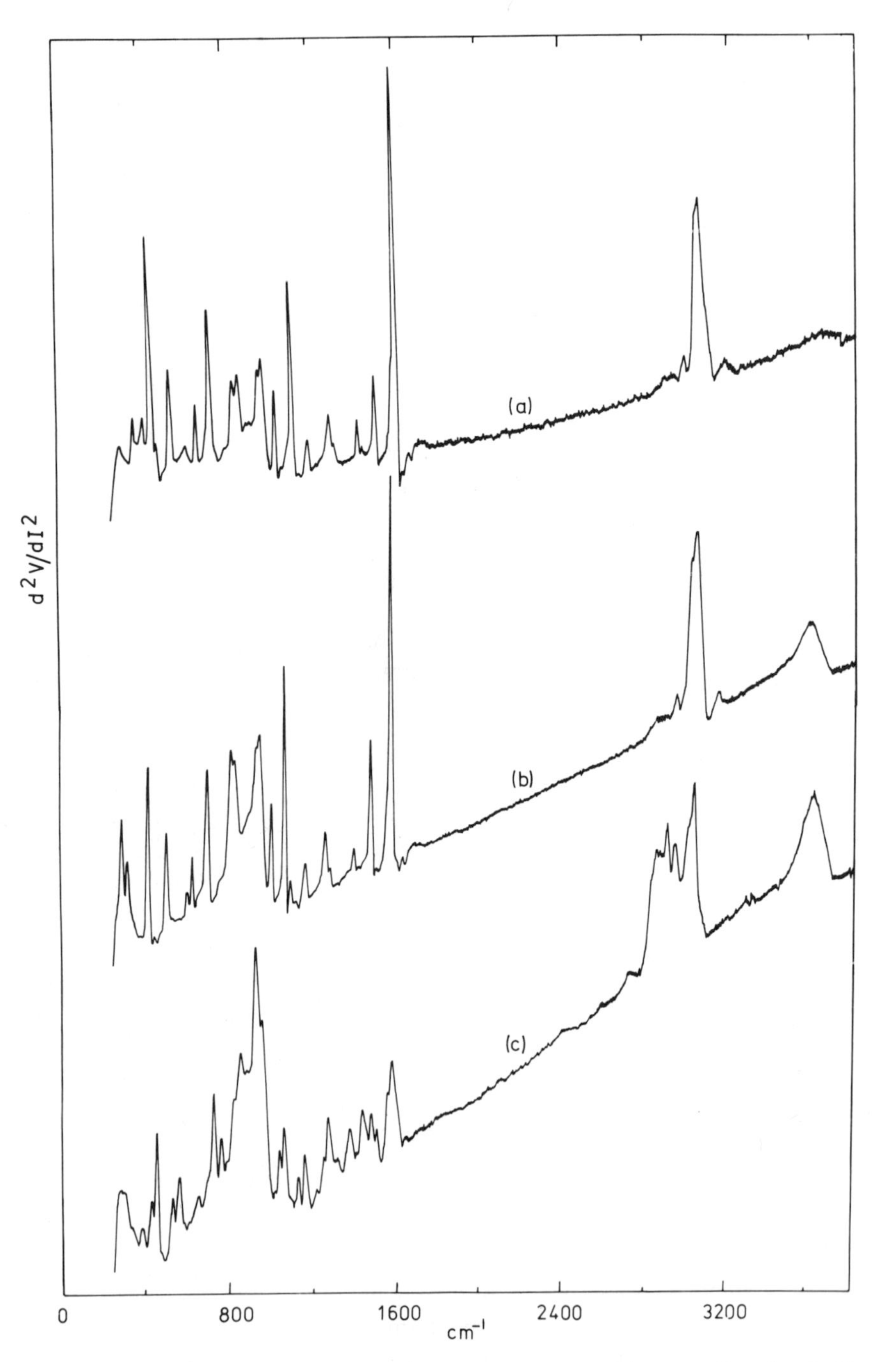

Figure 17. Tunneling spectra of Al–I–Pb sandwiches doped with (a) *p*-chlorophenol, (b) *p*-bromophenol, and (c) *p*-nitrophenol.

benzyl alcohol $C_6H_5CH_2OH$ and adsorbed benzylamine $C_6H_5CH_2NH_2$, seen in Figure 18, caused us some problems. The two spectra are quite similar and the modes have been assigned elsewhere. The unexpected feature in both is a broad band centered near 2900 cm^{-1}. The band is wider in the amine spectrum and more intense in that of the alcohol. It is well known that both OH and NH_2 groups are subject to hydrogen-bond interaction with their environment. The effect is to reduce the frequencies of OH and NH internal stretch modes. Local variation of the environment means the frequency shift may be different for different molecules so that,

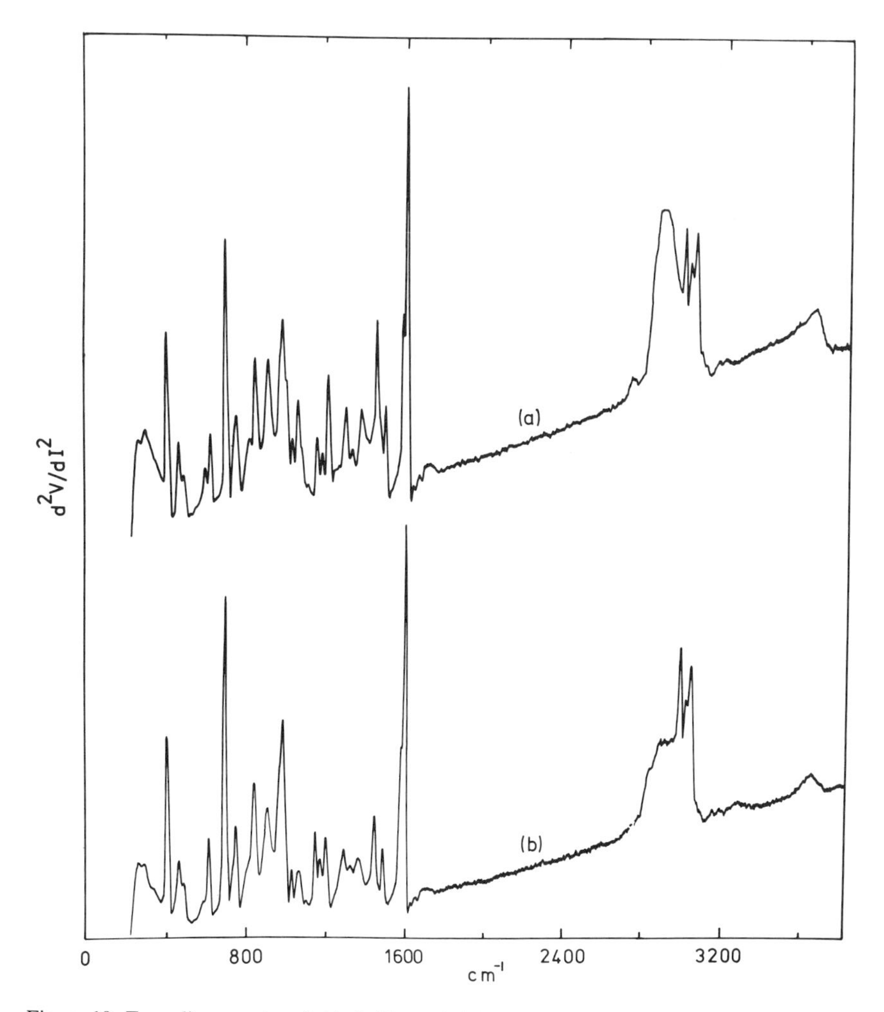

Figure 18. Tunneling spectra of Al–*I*–Pb sandwiches doped with (a) benzyl alcohol and (b) benzylamine.

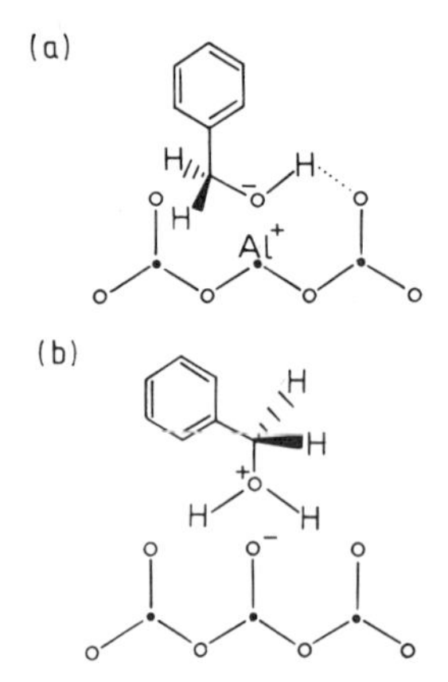

Figure 19. Possible configurations of benzyl alcohol on aluminum oxide showing (a) Lewis-acid–Lewis-base coordination bond of adsorbate to oxide and (b) Brønsted acid protonation of adsorbate by surface.

as well as shifting, the mode is often broadened. Normally, hydrogen-bonded OH and NH stretch modes appear in the range 3100–3500 cm^{-1}. We can therefore rule out such a mechanism here. Instead we offer two choices to account for what we at first identified as grossly downshifted OH and NH modes at 2900 cm^{-1}. Either there is (a) Lewis-acid–Lewis-base coordination of the adsorbed species at an exposed surface aluminum cation with substantial charge redistribution to leave the oxygen and nitrogen centers more positively charged as in Figures 19a and 20a; simultaneously strong hydrogen bonds are formed with adjacent oxide centers; or (b) Brønsted acid protonation of the adsorbate by the surface hydroxyl to give a surface bound benzyl oxonium $C_6H_5CH_2\overset{+}{O}H_2$ or ammonium species as in Figures 19b and 20b. Further details of the argument and supporting literature from related studies are to be found in our original paper.[22] More recently, however, we have examined deuterated alcohols and amines and conclude that the 2900-cm^{-1} peak may be largely due to methylene group stretch modes and not downshifted OH and NH vibrations.

Two amines closely related to benzylamine gave the spectra of Figure 21. These are 2-phenylethylamine, $C_6H_5CH_2CH_2NH_2$, and diphenylamine, $(C_6H_5)_2NH$. The first differs only by the presence of an additional methyl-

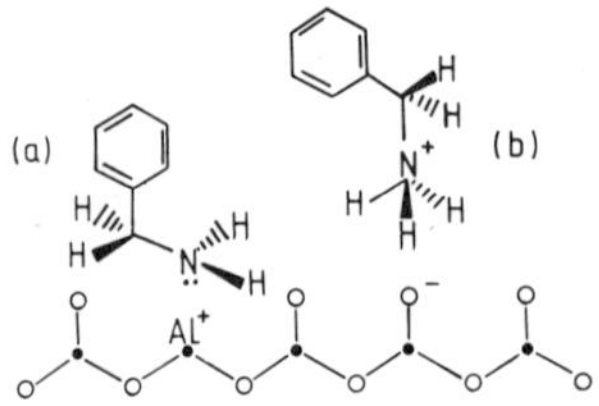

Figure 20. Possible configurations of benzylamine on aluminum oxide showing (a) Lewis-acid–Lewis-base coordination bond of adsorbate to oxide and (b) Brønsted acid protonation of adsorbate by surface.

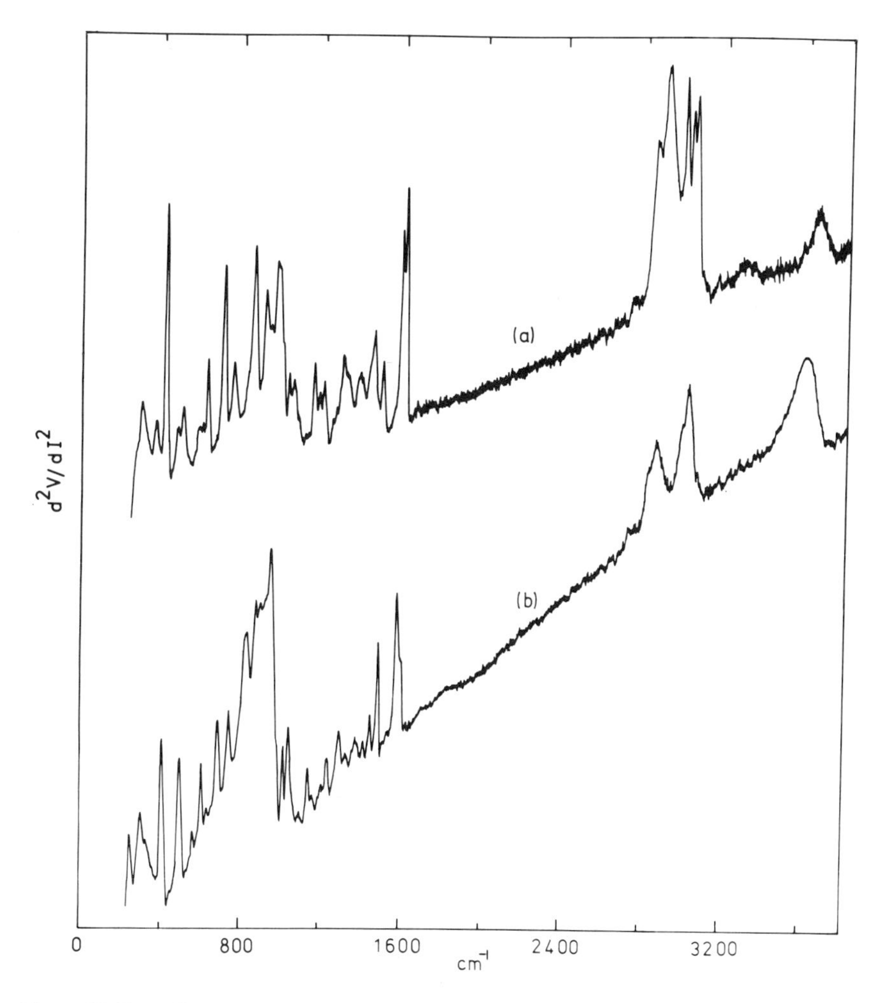

Figure 21. Tunneling spectra of Al–*I*–Pb sandwiches doped with (a) 2-phenylethylamine and (b) diphenylamine.

ene group and its spectrum is closely similar to Figure 18b. Possible configurations are shown in Figure 22. A weak mode at $\sim 3300\ cm^{-1}$ is attributed to physisorbed amine. The second has two benzene rings and its spectrum is weaker overall probably as a result of some potential reaction sites being blocked by the unreactive parts of those molecules which adsorb first. Since this molecule contains no methylene groups the identification of the band at $\sim 2900\ cm^{-1}$ with downshifted NH stretch modes is supported.

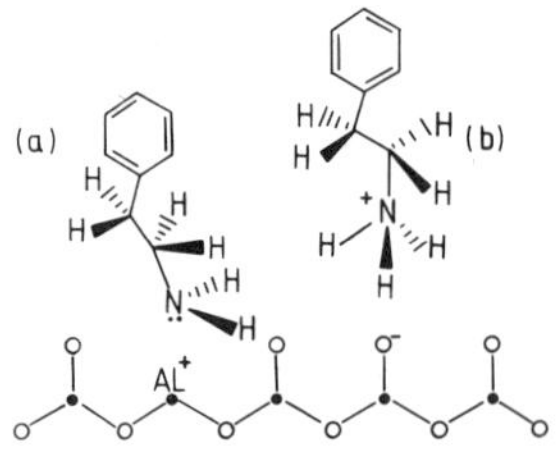

Figure 22. Possible configurations of 2-phenylethylamine on aluminum oxide showing (a) Lewis-acid–Lewis-base coordination bond of adsorbate to oxide and (b) Brønsted acid protonation of adsorbate by surface.

A more detailed study of deuterated alcohols and amines would be a helpful check on these interpretations and is now in hand.

4.8. Bifunctional Molecular Species

Having examined the reactions of molecules with different functional groups on plasma-grown aluminum oxide there arises the question of what happens to molecules with two functional groups. Much variety is clearly possible within such terms of inquiry. We present four illuminating spectra. Figure 23a is the result of adsorbing ethylene glycol, $C_2H_4(OH)_2$, which has two potentially reactive OH groups. There is no evidence of OH stretch modes in the 3100–3800-cm^{-1} range. The band at 2900 cm^{-1} is due mainly to methylene stretch modes. The two OH bonds of the adsorbate have been broken and the protons released have probably combined with surface hydroxyl groups to form water which has been pumped away during sample preparation. When ethanolamine, $HOCH_2CH_2NH_2$, is adsorbed Figure 23b is the outcome. Here we see something quite elusive in tunnel spectra: a strong band at 3300 cm^{-1}. It is the NH stretch mode, and its presence suggests that the amine group is not strongly hydrogen-bonded to the surface. The hydroxyl group has, presumably, reacted once again to produce the strong chemisorption.

In Figure 24 are shown the spectra from adsorbed 1,3-propanediol $HO(CH_2)_3OH$, and 1,3-propane diamine $NH_2(CH_2)_3NH_2$. In (a) the methylene CH stretch modes at 2900 cm^{-1} provide a beautiful, narrow, strong band. Again both OH groups have probably reacted with loss of a proton but there is a possibility that the low-energy tail may be due to a surface-bonded OH group in each molecule. In the latter case, if the bonding is to the same surface aluminium cation a chelation-stabilized five-membered ring will be formed. The diamine in (b) appears to have one unreacted NH_2 group as witnessed by the $\sim$3300-cm^{-1} band and possibly one surface-bonded NH_2 group with a stretch mode at 2850 cm^{-1}. These interpretations are tentative until more studies have been completed. The opportunities for studying reactivity at adjacent surface sites are almost limitless. Deuteration of the adsorbates should be particularly helpful in unraveling the processes.

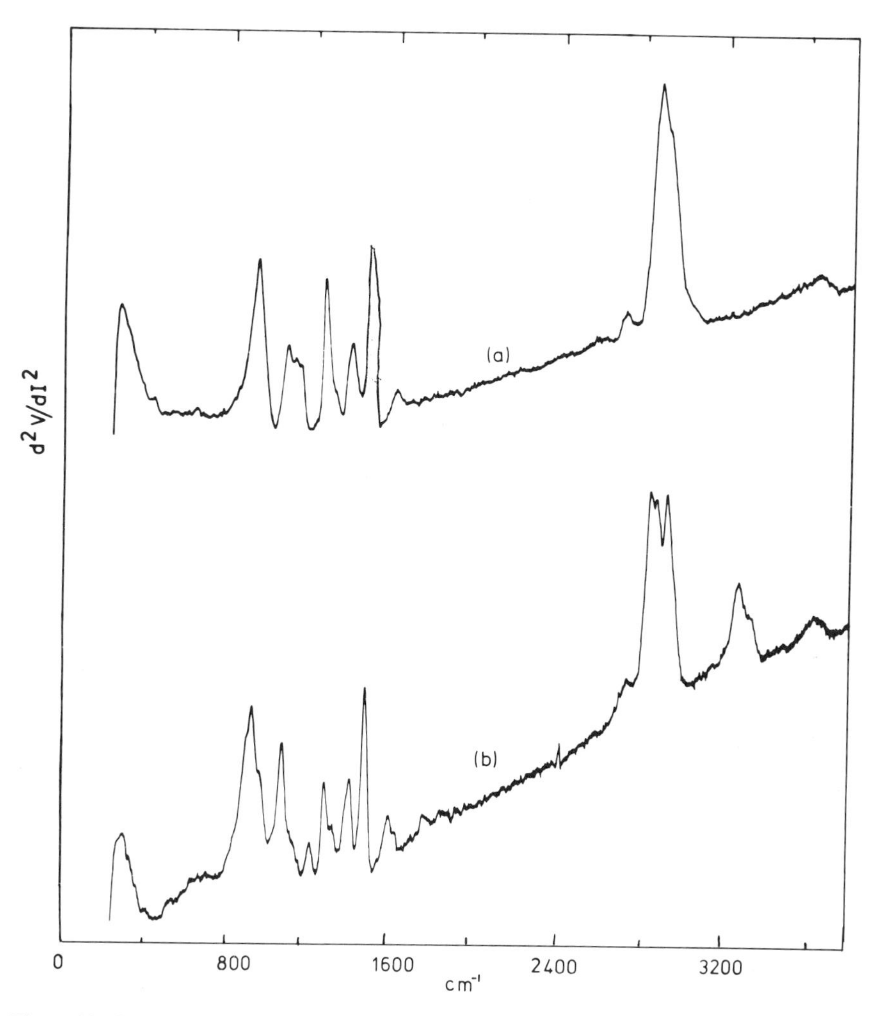

Figure 23. Tunneling spectra of Al–I–Pb sandwiches doped with (a) ethylene glycol and (b) ethanolamine.

4.9. Chemical Mixtures

There has been very little investigation of chemical mixtures by tunneling spectroscopy. The spectrum of a commercially available mixture which enjoys widespread popular usage is shown in Figure 25b. It is Old Bushmills whiskey, a product of the oldest distillery in the world, licensed in 1608 and situated in the town of Bushmills some 12 km from our laboratory. For comparison we show in Figure 25a a spectrum from adsorbed ethanol which

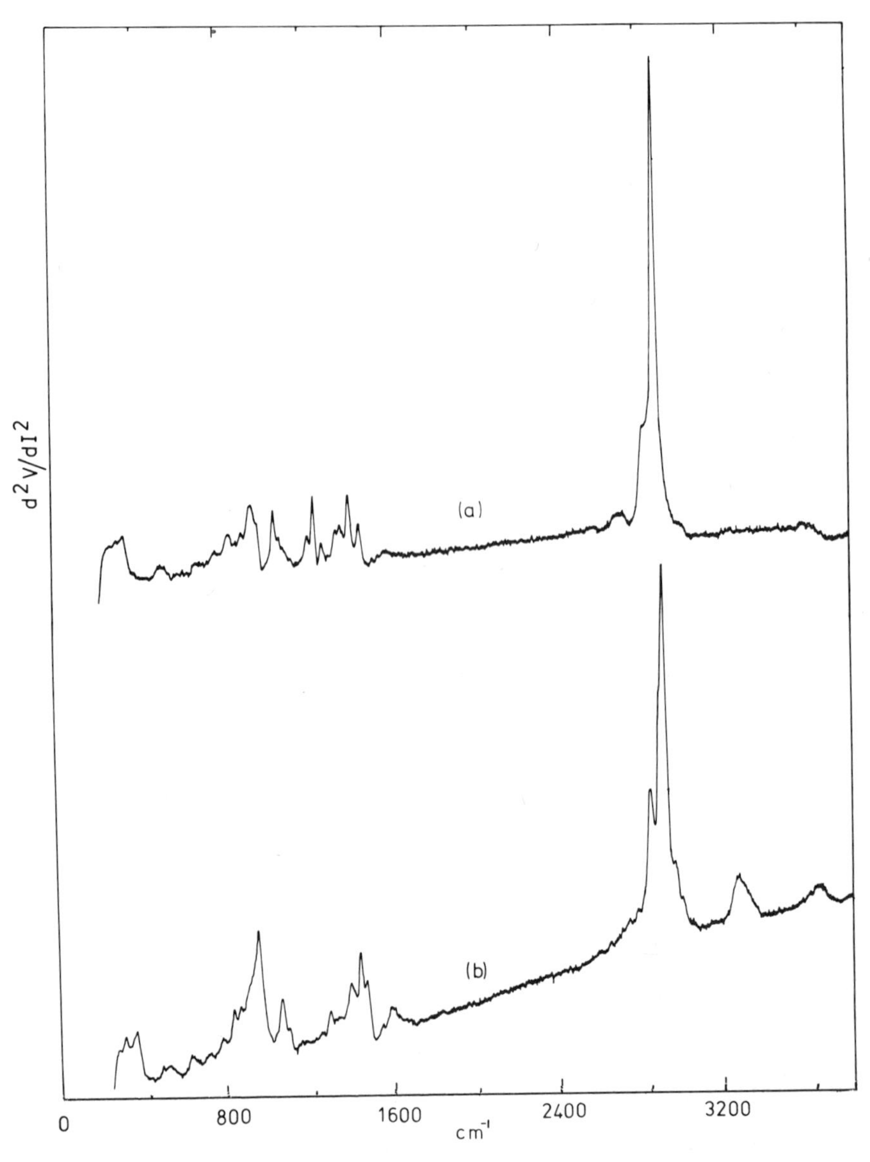

Figure 24. Tunneling spectra of Al–*I*–Pb sandwiches doped with (a) 1,3-propanediol and (b) 1,3-propane diamine.

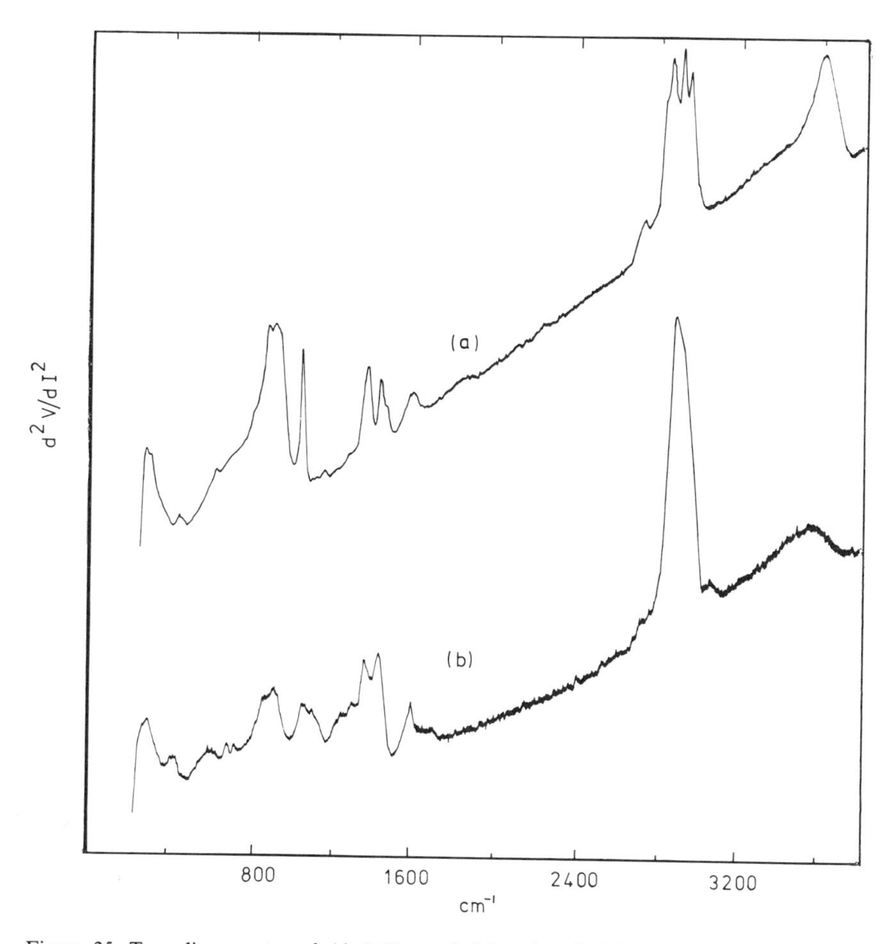

Figure 25. Tunneling spectra of Al–I–Pb sandwiches doped with (a) ethanol and (b) Old Bushmills whiskey (Black Bush).

has been the subject of earlier work.[22–24] The similarity of the spectra is not unexpected. Interpretation of the differences is left as an exercise for the reader. Other mixtures for investigation may suggest themselves to the enthusiast.

5. Clean Magnesium Oxide

Tunneling spectroscopy of adsorbates cannot prosper as a useful technique if it is limited to aluminum oxide substrates. Fortunately there are other possibilities. Metal particles supported by aluminum oxide are now

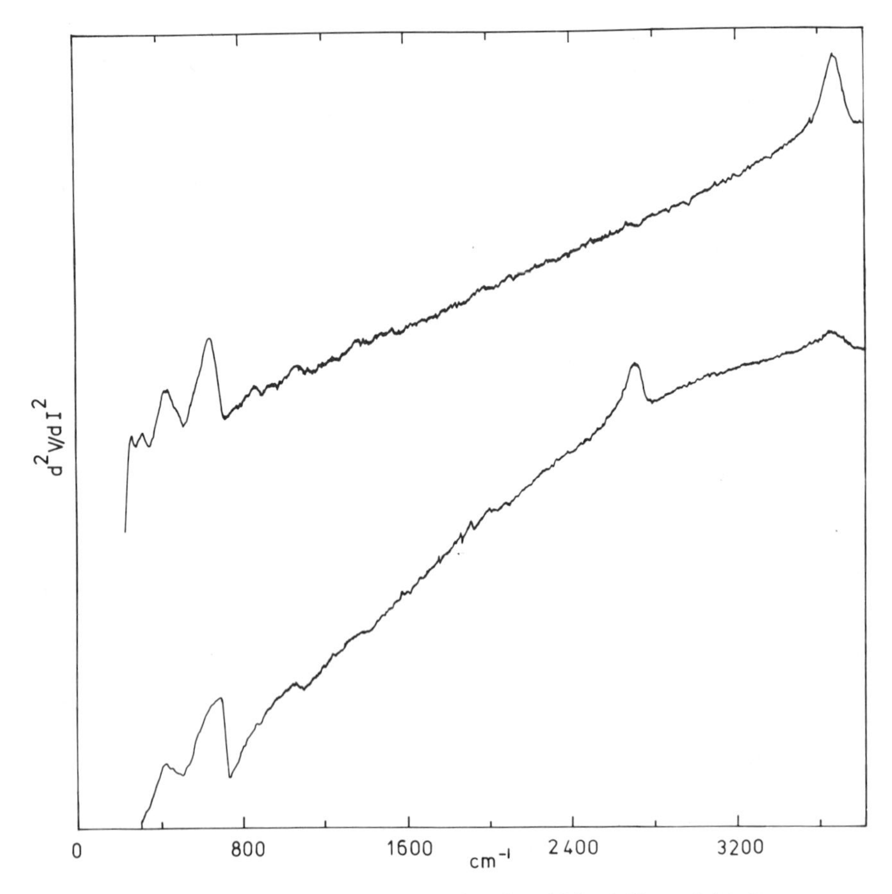

Figure 26. Upper curve: Tunneling spectrum of undoped Mg–I–Pb sandwich. Lower curve: Tunneling spectrum of Mg–I–Pb sandwich with oxide insulator grown in presence of D_2O.

receiving attention. Prompted by a study of the vibrational properties of clean magnesium oxide by Klein *et al.*,[6] we have attempted to adsorb various species on magnesium oxide, with a view to looking at their spectra.[9] The results are promising.

First we show in Figure 26 the spectrum of clean magnesium oxide in a Mg–I–Pb sandwich. There is wide variation in the appearance of such spectra from sample to sample in contrast with those from aluminum oxide, which are nearly perfectly reproducible. For example, the OH stretch mode at 3675 cm^{-1} is highly sensitive to the partial pressure of water vapor during oxide growth. The two peaks at 430 cm^{-1} and 645 cm^{-1} vary somewhat in overall shape from one sample to another. Klein *et al.*[6] identified structure in this region as due to MgO modes. The second spectrum shown in Figure

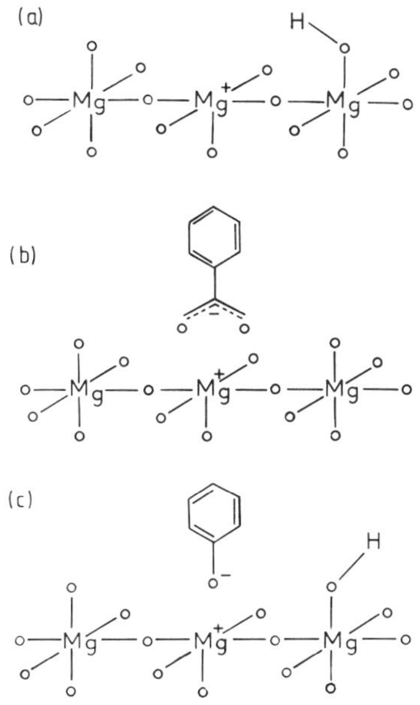

Figure 27. (a) Idealized representation of plasma-grown magnesium oxide. (b) Benzoate anion on magnesium oxide after doping with benzaldehyde. (c) Phenolate anion on magnesium oxide after doping with phenol.

26 comes from an oxide grown in the presence of D_2O. As expected, the peaks at 430 and 645 cm^{-1} remain unshifted but the OD stretch mode is down in energy by a factor of approximately $1/\sqrt{2}$ (actually 1.36) at 2700 cm^{-1}. The assignments are thus confirmed. The OH mode can be almost completely replaced by OD substitution. A schematic representation of the surface is shown as Figure 27a.

6. Doped Magnesium Oxide

Already we can be optimistic about magnesium oxide as a suitable host for adsorbate spectroscopy. The tunnel barrier is sufficiently high to tolerate applied dc biases out to 0.5 V without catastrophic growth of the elastic background current and there is evidence of hydroxyl species which hold promise of chemical interest.

6.1. Benzaldehyde

The quartz microbalance in the preparation chamber showed that benzaldehyde vapor readily adsorbed to magnesium oxide in the same familiar way as with aluminum oxide. The resulting tunnel spectrum is

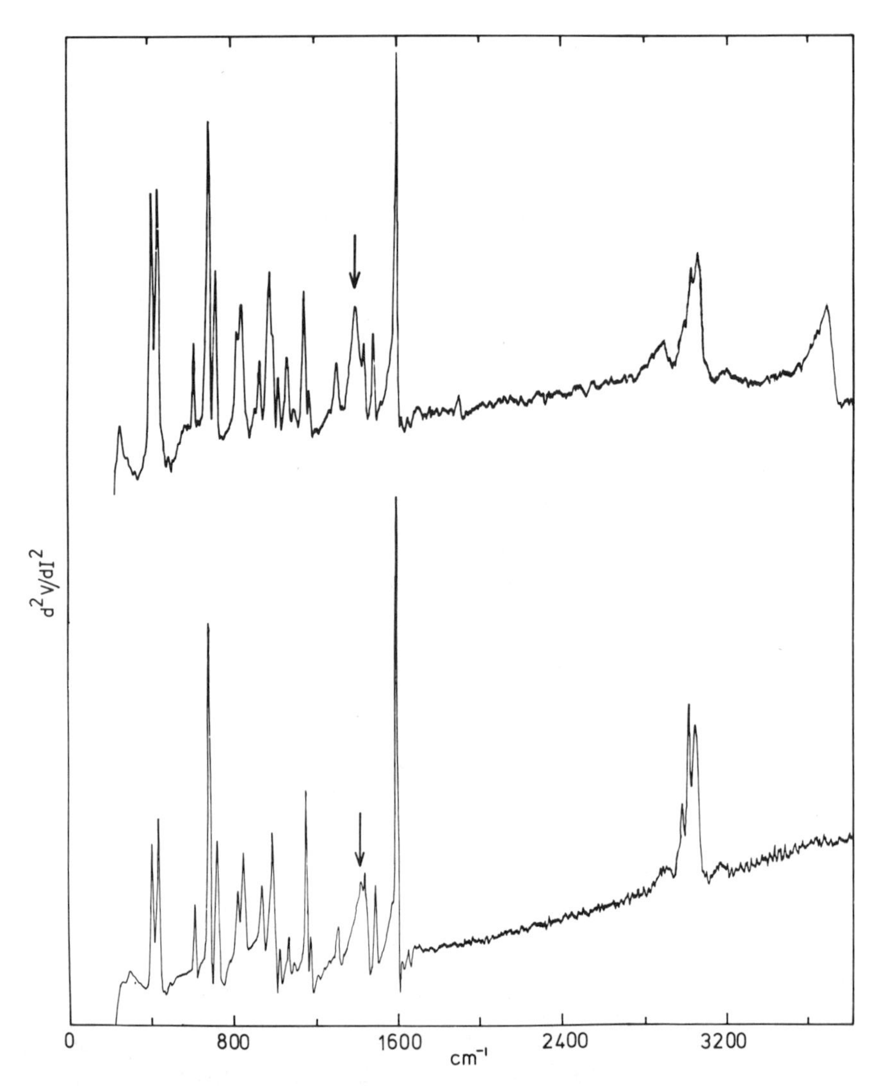

Figure 28. Upper curve: Tunneling spectrum of Mg–I–Pb sandwich doped with benzaldehyde. Lower curve: Tunneling spectrum of Al–I–Pb sandwich doped with benzaldehyde.

shown in Figure 28, and below it for comparison is the spectrum from the same species adsorbed on aluminum oxide. The outcome is remarkably similar in both instances. Overall the line positions, linewidths, and line intensities are comparable but there is one rather nice exception. It is marked with an arrow on the two spectra and may be identified with the CO_2^- symmetric stretch mode of the adsorbed benzoate anion.[9] This mode has shifted down slightly on magnesium oxide so that it is better resolved than on aluminum oxide. We should emphasize that the reproducibility of features in tunnel spectra is very high, perhaps positions vary by $\sim 5\ cm^{-1}$ at most, and this shift of $\sim 32\ cm^{-1}$ from 1428 to 1396 cm^{-1} is clear-cut in each sample we have studied. It convincingly demonstrates a real difference between the two adsorbing oxides. A suggested configuration for adsorbed benzoate anion is shown in Figure 27b.

6.2. Formic, Acetic, and Propionic Acids

When propionic acid is adsorbed the result is rather similar, as seen in Figure 29. Again the CO_2^- symmetric mode of the propionate anion is lower

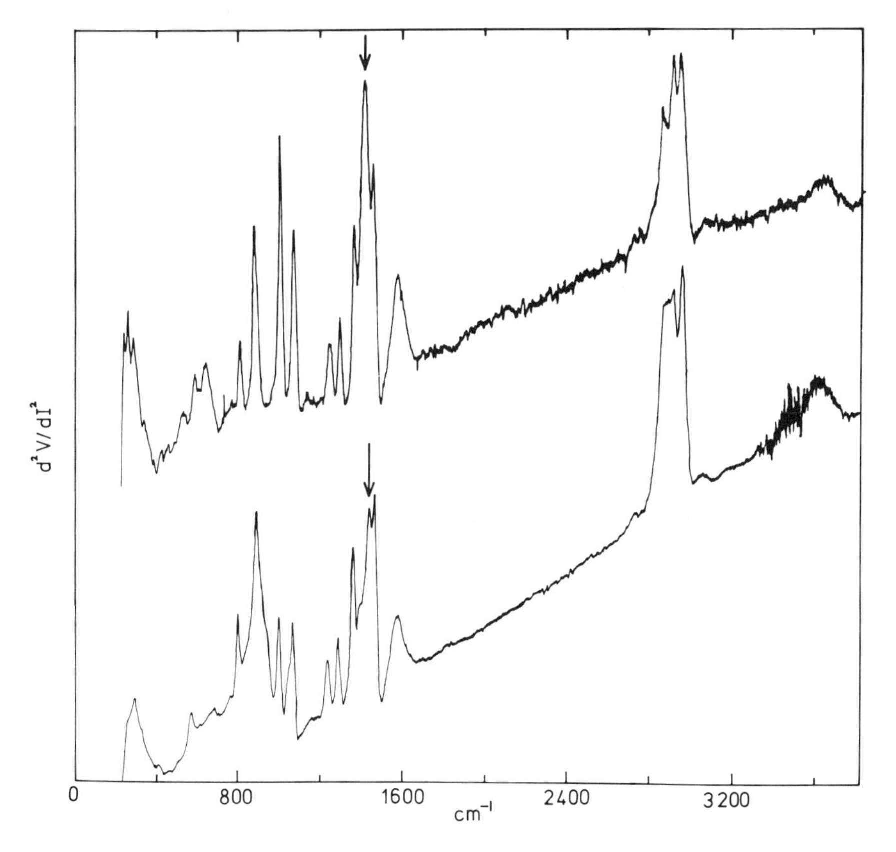

Figure 29. Upper curve: Tunneling spectrum of Mg–I–Pb sandwich doped with propionic acid. Lower curve: Tunneling spectrum of Al–I–Pb sandwich doped with propionic acid.

in energy on magnesium oxide. The spectrum also illustrates rather well an important advantage of having two oxides available. The mode at 879 cm^{-1} is readily distinguishable from the background AlO mode by the line shape of the overall structure but it stands out ever so much more clearly on MgO where the background is featureless in this region.

The formate ion adsorbed on magnesium oxide shows a greater downshift (>100 cm^{-1}) in the position of its CO_2^- symmetric mode, as compared with the same mode on aluminum oxide, than any other species we have examined. Two modes are now coincident: the CO_2^- symmetric stretch and CH in-plane bend modes. Figure 30 displays the spectra.

We have also examined the acetate ion in both oxides, and again found a lower energy for the CO_2^- symmetric mode on magnesium oxide. The

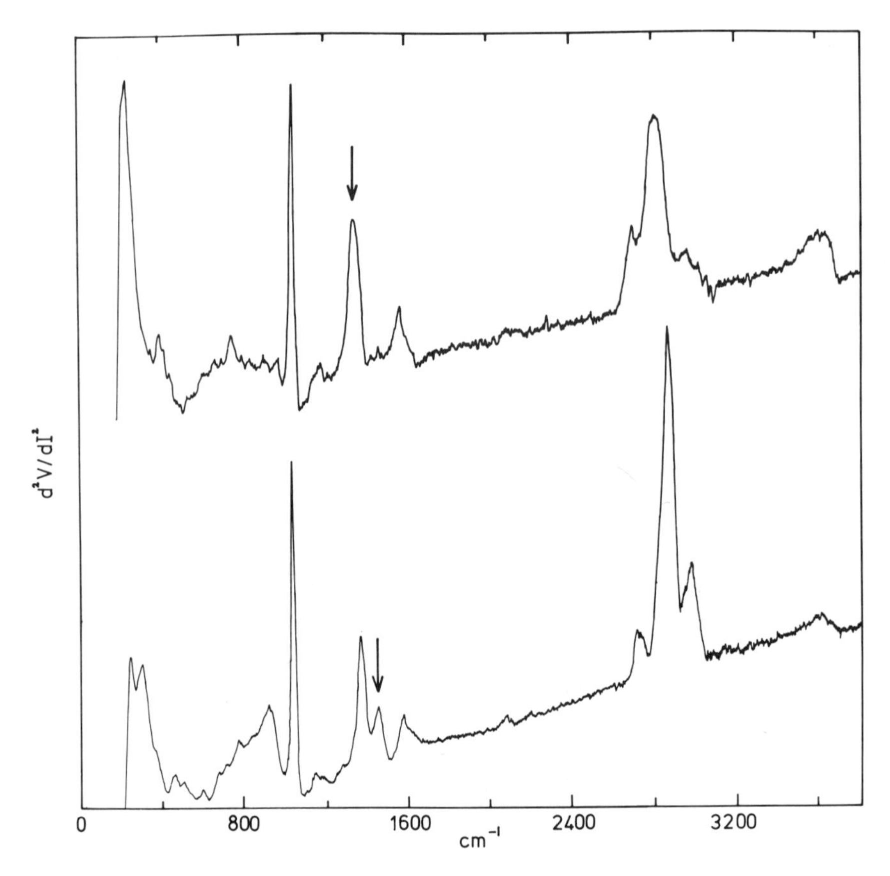

Figure 30. Upper curve: Tunneling spectrum of Mg–I–Pb sandwich doped with formic acid. Lower curve: Tunneling spectrum of Al–I–Pb sandwich doped with formic acid.

Table 2. CO_2^- Symmetric Stretch Mode Positions

Adsorbed species	Host	
	Magnesium oxide	Aluminum oxide
Benzoate	1396 cm^{-1}	1428 cm^{-1}
Propionate	1417 cm^{-1}	1444 cm^{-1}
Acetate	1445 cm^{-1}	1463 cm^{-1}
Formate	1348 cm^{-1}	1456 cm^{-1}

results are collected in Table 2 and suggested configurations for the adsorbates on magnesium oxide are seen in Figure 31.

At the higher end of the energy range there is a distinct tendency for the intensities of spectral lines to be weaker for adsorbates on magnesium oxide than on aluminum oxide. This is due to the more rapid increase of the conductance of magnesium oxide with increasing bias which in turn influences spectrometer sensitivity.[25,26]

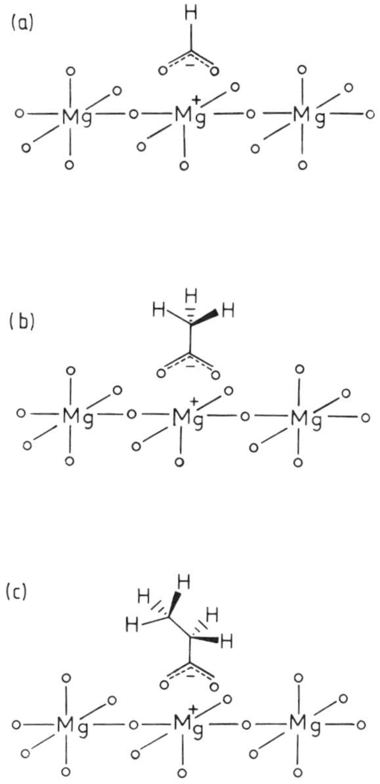

Figure 31. Possible configurations of (a) formate anion, (b) acetate anion, and (c) propionate anion on magnesium oxide.

6.3. Phenol

When phenol is adsorbed on magnesium oxide the resultant spectrum differs little from that on aluminum oxide. There are no obvious shifts of line position. As usual there is a falloff in intensity at the high-energy end of the spectrum and the three lines indicated with arrows in Figure 32 are significantly stronger relative to their neighbors. One has developed from simply being a shoulder on the side of another line to being a well-resolved line itself. The lowest is a C—O stretch mode of the bond which is nearest to the oxide and the others are C—C stretch modes of the benzene ring. The shape of the OH stretch mode suggests that the proton liberated when the phenolate ion adsorbs on the magnesium oxide surface is held on the oxide. It may adsorb at a surface oxygen anion site to enhance the surface hydroxyl concentration and give the high-energy sharp peak seen in the OH

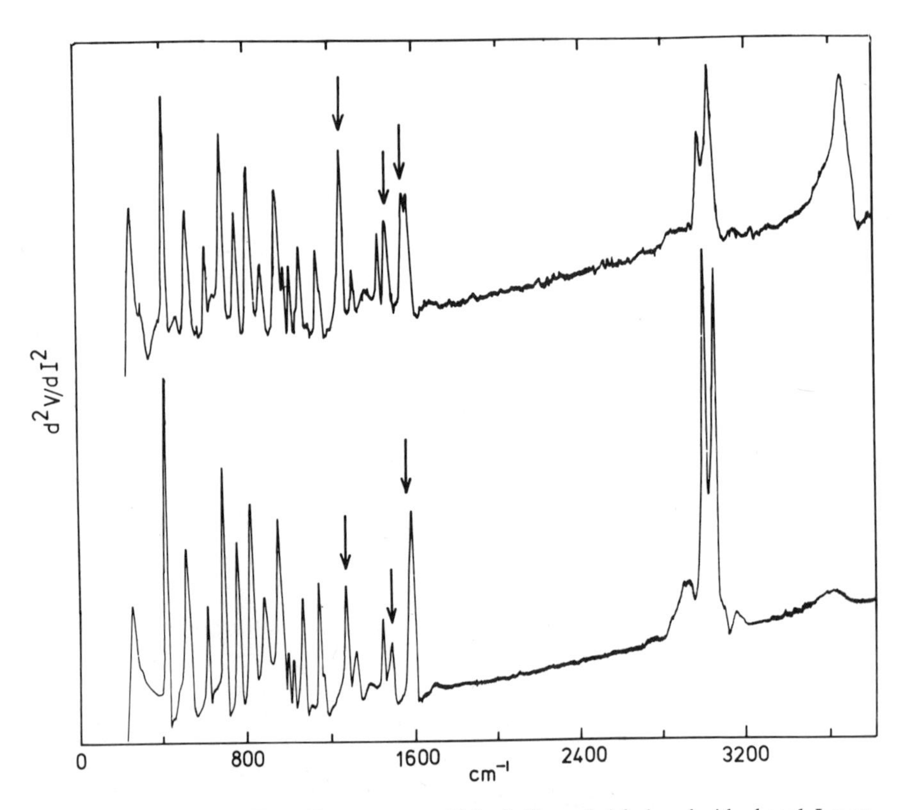

Figure 32. Upper curve: Tunneling spectrum of Mg–I–Pb sandwich doped with phenol. Lower curve: Tunneling spectrum of Al–I–Pb sandwich doped with phenol.

stretch mode. With deuterated phenol this could be fully confirmed. Perhaps it would be useful to mention in passing that phenol appears sometimes to adsorb on aluminum oxide with the OH bond intact in a fraction of the adsorbate. This shows up as a broad mode at $\sim 2925\ cm^{-1}$ on the lower side of the CH stretch modes. It is thus substantially downshifted by its bonding to the oxide surface. The corresponding OH bend mode can be seen as a weak broad band at $\sim 1390\ cm^{-1}$ with a strength which closely correlates from sample to sample with the intensity of the 2925-cm^{-1} stretch band. Whether this also happens on magnesium oxide is less clear. Phenolate anion adsorbed on magnesium oxide is shown in Figure 27c. Again deuterated adsorbates would be helpful in clarifying these issues.

6.4. Carboxylate Mode Shift

The shift of the CO_2^- symmetric stretch mode in the carboxylates to lower energy on magnesium oxide is interesting because that mode has a net dipole moment perpendicular to the surface. It is therefore sensitive to the strength of the adsorption bond. A stronger $Mg^+ —CO_2^-$ bond will tend to cause a greater charge polarization in the carboxylate group and hence reduce the strength of the internal CO bond along the surface normal direction. This is our interpretation of the shift. By contrast, the charge distribution in the CO bond of the phenolate ion is much less susceptible to perturbation by the surface electric fields because of its compact geometry.

6.5. Benzyl Alcohol

When benzyl alcohol is adsorbed on magnesium oxide the spectrum of Figure 33 results. For comparison the spectrum from the same adsorbate on an aluminum oxide host is also shown. The broad mode in both at $\sim 2850\ cm^{-1}$ may be the OH stretch mode of the benzyl alcohol perturbed downwards in energy but not broken by adsorption on the oxide. The band is lower in energy and broader on magnesium oxide. This implies a stronger interaction. Figures 34a and 34b show possible configurations. There is also evidence from the surface OH stretch mode, which is stronger than on an undoped sample prepared under the same conditions, that some of the adsorbate yields up a proton to the surface where it stays and contributes to the narrow sharp peak at the high-energy side of the surface OH stretch band at $\sim 3650\ cm^{-1}$, just as we found for adsorbed phenol. Figure 34c attempts to represent the configuration. Once more deuterated adsorbates would help considerably in establishing these interpretations and determining the contribution of methylene modes to the $2850\ cm^{-1}$ band.

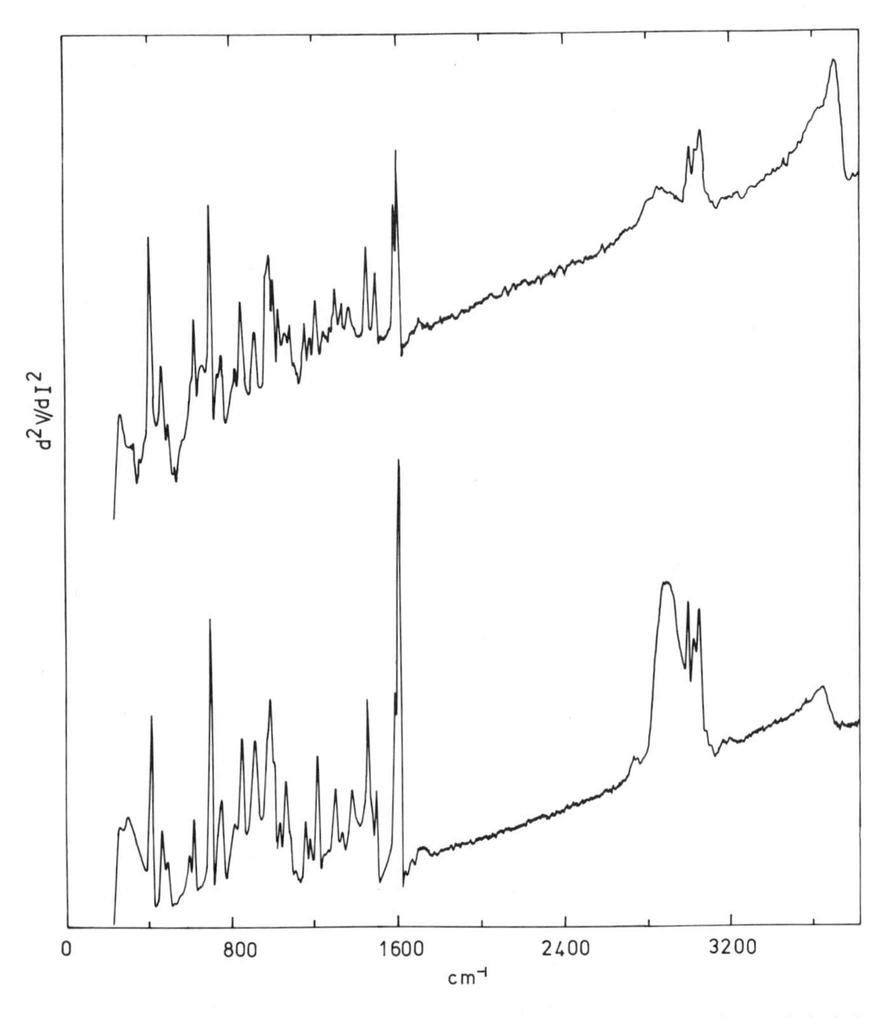

Figure 33. Upper curve: Tunneling spectrum of Mg–I–Pb sandwich doped with benzyl alcohol. Lower curve: Tunneling spectrum of Al–I–Pb sandwich doped with benzyl alcohol.

6.6. Unsaturated Hydrocarbons

Adsorption of an unsaturated hydrocarbon, l-heptyne, on magnesium oxide gave a good spectrum which showed marked differences from that on aluminum oxide. See Figure 35. The triple bond has almost totally disappeared with only a trace of the C≡C stretch mode at ~ 2100 cm^{-1} still visible and no clear sign of the ≡C—H stretch mode at ~ 3300 cm^{-1}. Probably a surface-bound C=C band is denoted by the broad feature at

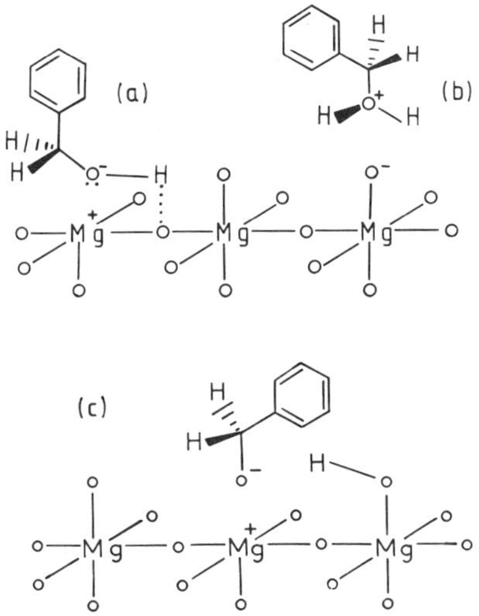

Figure 34. Possible configurations of benzyl alcohol on magnesium oxide showing (a) Lewis-acid–Lewis-base coordination bond of adsorbate to oxide, (b) Brønsted acid protonation of adsorbate by surface, and (c) benzyloxy species and surface bound proton.

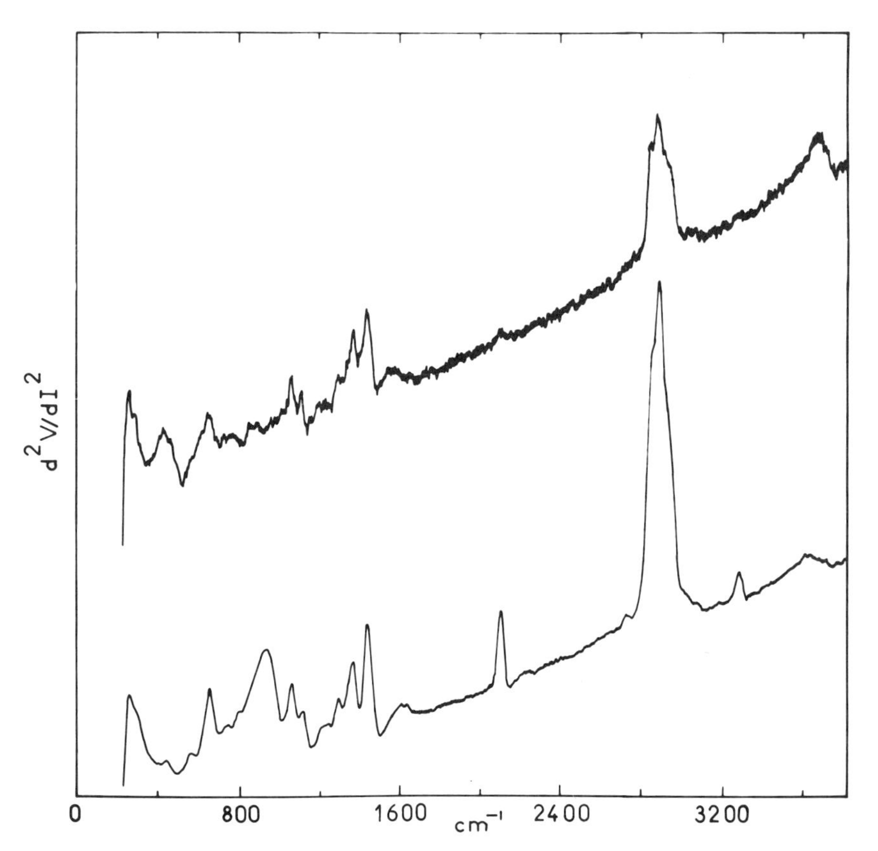

Figure 35. Upper curve: Tunneling spectrum of Mg–I–Pb sandwich doped with 1-heptyne. Lower curve: Tunneling spectrum of Al–I–Pb sandwich doped with 1-heptyne.

$\sim 1600\ cm^{-1}$. We conclude that there are more sites on the plasma-grown magnesium oxide which can produce an apparent partial reduction, probably by protonation of the triple bond, than on aluminum oxide.[27]

The adsorption of 3-hexyne on the two oxides follows a roughly similar course: the triple bond is much weaker on magnesium oxide. That it survives at all suggests that some adsorption sites need not seriously perturb the molecule. See Figure 36. The high intensity of the C—C stretch modes of adsorbed 3-hexyne on magnesium oxide may imply a change in orientation on this oxide. The stronger bonding to the magnesium oxide surface may cause the C—C bonds away from the adsorption site to assume an

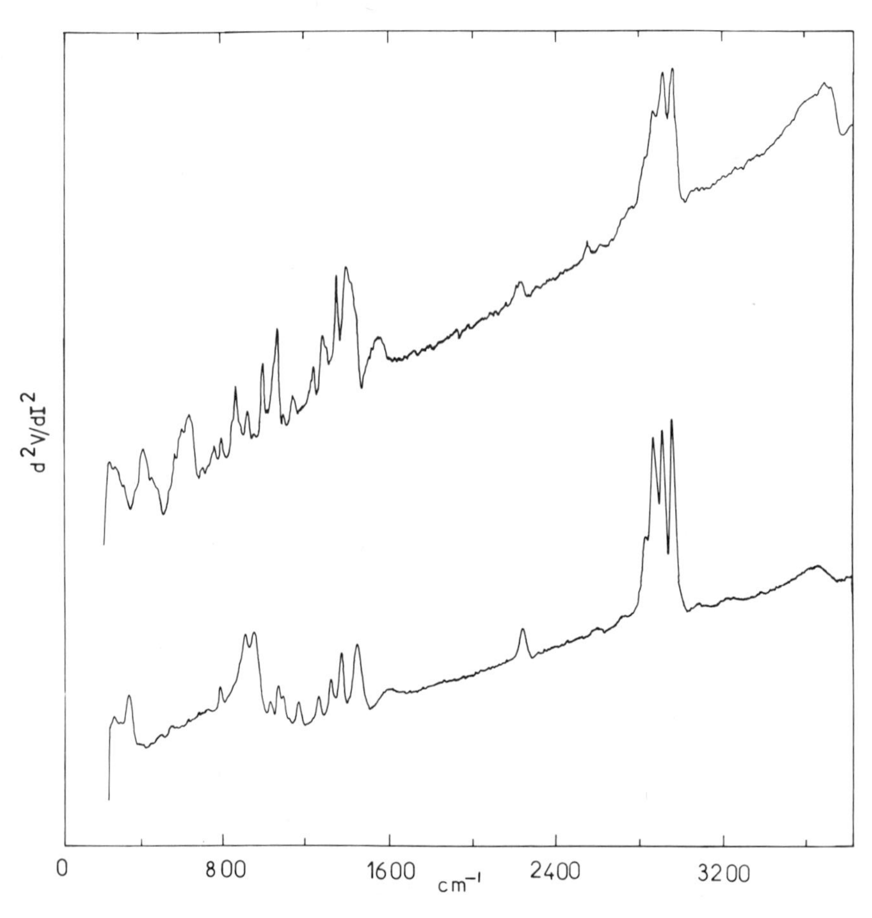

Figure 36. Upper curve: Tunneling spectrum of Mg–I–Pb sandwich doped with 3-hexyne. Lower curve: Tunneling spectrum of Al–I–Pb sandwich doped with 3-hexyne.

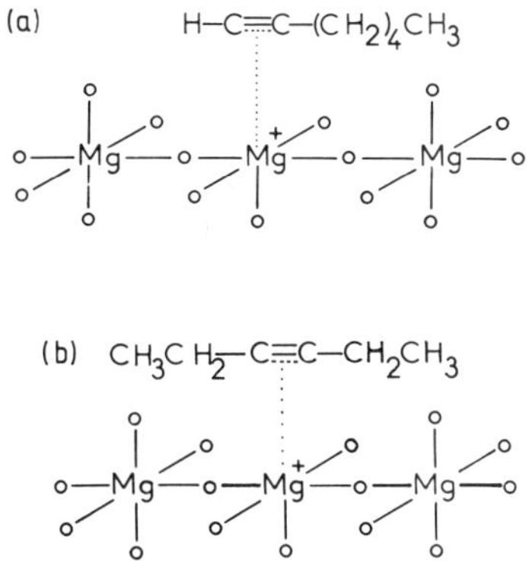

(a) H—C≡C—$(CH_2)_4CH_3$

(b) CH_3CH_2—C≡C—CH_2CH_3

Figure 37. Possible configurations of (a) 1-heptyne and (b) 3-hexyne on magnesium oxide.

orientation closer to the surface normal. That would increase the detected signal if the dipole interaction between the tunneling electrons and the adsorbate(2,3) is operative. Another feature confirming the stronger adsorption on magnesium oxide is the somewhat lower frequency (by $\sim 50\ cm^{-1}$) of the C=C modes seen in adsorbates on magnesium oxide. Possible configurations of the unsaturated adsorbates are shown in Figure 37.

6.7. Diketone

Acetylacetone $CH_3(C{=}O)CH_2(C{=}O)CH_3$ is an example of a bifunctional ketone. The spectra from it on magnesium oxide and aluminum oxide in Figure 38 show that it is not simply physisorbed in unreacted ketone form. Rather the lines correspond to a chemisorbed chelating anion at the surface-exposed metal cation sites, as represented in Figure 39. Studies of a family of such compounds will appear in published form soon.(28)

Hopefully too we shall shortly see spectra from other host oxides. Then tunneling spectroscopy will truly have arrived.

7. Technical Postscript

Nearly all the spectra shown in this chapter were obtained by following a routine procedure for sample preparation in which the oxide was exposed at room temperature to vapor of the compound being studied. Spectrometer operation was likewise routine with a 2-mV peak-to-peak (0.7 mV rms) ac modulation, ~40-min dc sweep, and sample temperature in the range 1.2–2 K. The full spectra are plotted on graph paper 38 cm wide at a sensitivity of 12.5 mV/cm; this means that each cm of the paper corresponds rather

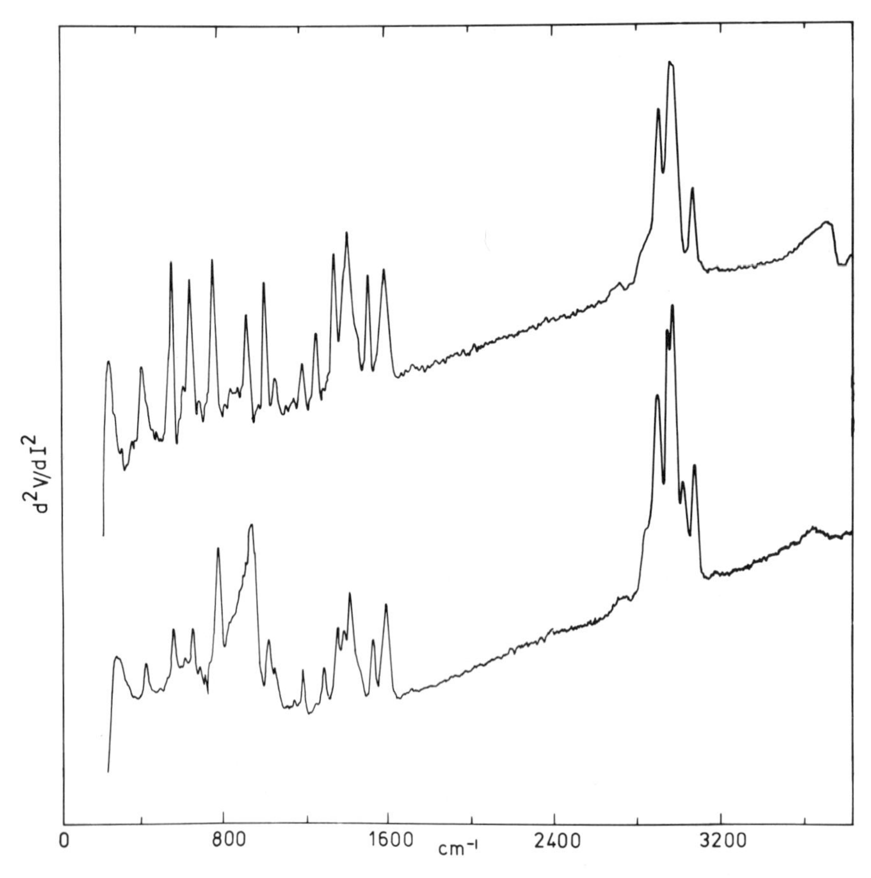

Figure 38. Tunneling spectrum of Mg–*I*–Pb sandwich doped with acetyl acetone. Lower curve: Tunneling spectrum of Al–*I*–Pb sandwich doped with acetyl acetone.

(a) H_3C, H, CH_3, Mg, Mg^+, Mg

(b) H_3C, H, CH_3, Al^+

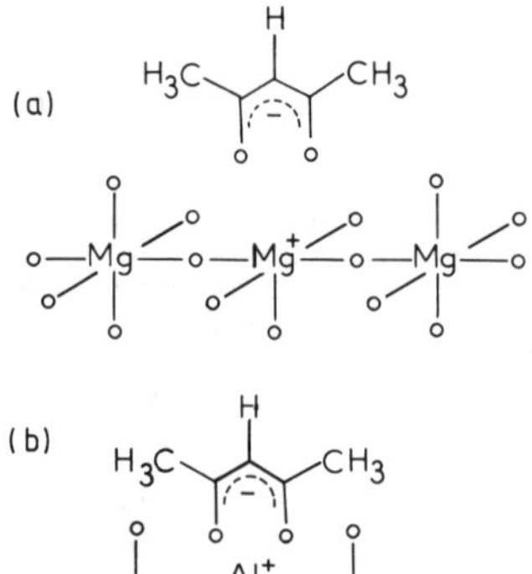

Figure 39. Possible configuration of acetyl acetone on (a) magnesium oxide and (b) aluminum oxide.

closely to 100 cm^{-1} (1 meV $\equiv$ 8.065 cm^{-1}) and makes quick inspection of the data easy. All peak positions have been corrected for peak shifts due to the superconducting lead electrode by subtracting 8 cm^{-1} from the measured positions.

8. Conclusions

(1) Tunneling spectroscopy of organic adsorbates on aluminum oxide is now open to widespread development.

(2) Species which chemisorb are easiest to study. This is not a significant limitation since it is the reaction itself which tunneling explores to great advantage.

(3) Acids, unsaturated hydrocarbons, alcohols, and amines are examples of compounds that have been successfully investigated. Bifunctional molecules are at a very early stage of exploration.

(4) Magnesium oxide is a suitable host for tunneling spectroscopy studies of adsorbates. Its availability opens up opportunities for clarifying details of surface reactions by comparison with results from the same adsorbates on aluminium oxide.

(5) Magnesium oxide is more reactive than aluminum oxide. Line shifts in adsorbed carboxylates and chemical reduction of unsaturated hydrocarbons support this statement.

(6) Although the availability of magnesium oxide represents a breakthrough in that it doubles the number of oxides which may be used, there is a pressing need for the investigation of other possible adsorbate hosts.[29,30]

(7) If you are a beginner try to produce a spectrum from adsorbed phenol or benzaldehyde. Should it fall short of the required standard, contact one of the established research groups and get advice. Happy tunneling!

Acknowledgments

This work is the outcome of our good fortune in having a dedicated, but good-humored group of colleagues who have been unstinting in their efforts. Ian McMorris, Bill Timms, Robert Floyd, and Albert Turner made everything work. Norman Brown has determined the direction of much of this sequence of studies, supplied us with chemicals of reliable purity, and provided the interpretative skills so necessary in analyzing most of the spectra.

References

1. A. J. Maeland, R. Rittenhouse, W. Lahar, and P. V. Romano, Infrared reflection–absorption spectra of anodic oxide films on aluminum, *Thin Solid Films* **21**, 67–72 (1974).
2. D. J. Scalapino and S. M. Marcus, Theory of inelastic electron–molecule interactions in tunnel junctions, *Phys. Rev. Lett.* **18**, 459–461 (1967).
3. J. Lambe and R. C. Jaklevic, Molecular vibration spectra by inelastic electron tunneling, *Phys. Rev.* **165**, 821–832 (1968).
4. A. L. Geiger, B. S. Chandrasekhar, and J. G. Adler, Inelastic electron tunneling in Al–Al-Oxide–Metal systems, *Phys. Rev.* **188**, 1130–1138 (1969).
5. W. M. Bowser and W. H. Weinberg, The nature of the oxide barrier in inelastic electron tunneling spectroscopy, *Surf. Sci.* **64**, 377–392 (1977).
6. J. Klein, A. Léger, M. Belin, D. Défourneau, and M. J. L. Sangster, Inelastic-electron-tunneling spectroscopy of metal–insulator–metal junctions, *Phys. Rev. B* **7**, 2336–2348 (1973).
7. C. J. Pouchert, *The Aldrich Library of Infrared Spectra*, Aldrich Chemical Company Inc., Milwaukee (1970).
8. *Sadtler Standard Spectra*, Heyden, London (1970).
9. D. G. Walmsley, W. J. Nelson, N. M. D. Brown, and R. B. Floyd, Development of inelastic electron tunneling spectroscopy: Comparison of adsorbates on aluminum and magnesium oxides, *Appl. Surf. Sci.* **5**, 107–120 (1980).
10. D. G. Walmsley, Inelastic electron tunneling spectroscopy, Chapter 5 in *Vibrational Spectroscopy of Adsorbates*, Springer Series in Chemical Physics 15, Springer-Verlag, Berlin (1980).
11. R. C. Jaklevic and M. R. Gaerttner, Inelastic electron tunneling spectroscopy. Experiments on external doping of tunnel junctions by an infusion technique, *Appl. Surf. Sci.* **1**, 479–502 (1978).
12. N. M. D. Brown, R. B. Floyd, and D. G. Walmsley, Inelastic electron tunneling spectroscopy (IETS) of carboxylic acids and related systems chemisorbed on plasma-grown aluminum oxide. Part 1.—Formic acid (HCOOH and DCOOD), Acetic acid (CH_3COOH, CH_3COOD and CD_3COOD), Trifluoroacetic acid, Acetic anhydride, Acetaldehyde and Acetylchloride, *J. Chem. Soc. Faraday Trans. 2* **75**, 17–31 (1979).
13. D. G. Walmsley, W. J. Nelson, N. M. D. Brown, S. de Cheveigné, S. Gauthier, J. Klein, and A. Léger, Evidence from inelastic electron tunneling spectroscopy for vibrational mode reassignments in simple aliphatic carboxylate ions, *Spectrochimica Acta* **37A**, 1015–1019 (1981).
14. D. A. Cass, H. L. Strauss, and P. K. Hansma, Vibrational spectroscopy of chemisorbed fatty acids with inelastic electron tunneling, *Science* **192**, 1128–1130 (1976).
15. N. M. D. Brown, R. B. Floyd, and D. G. Walmsley, Inelastic electron tunneling spectroscopy (IETS) of carboxylic acids and related systems chemisorbed on plasma-grown aluminum oxide. Part 3.—Propanoic acid, Butanoic acid and Butanal, Octanoic acid, Dodecanal, Hexadecanal and Octadecanal, *J. Chem. Soc. Faraday Trans. 2* **75**, 261–270 (1979).
16. P. K. Hansma and R. V. Coleman, Spectroscopy of biological compounds with inelastic electron tunneling, *Science* **184**, 1369–1371 (1974).
17. N. M. D. Brown, W. J. Nelson, and D. G. Walmsley, Inelastic electron tunneling spectroscopy (IETS) of carboxylic acids and related systems chemisorbed on plasma-grown aluminum oxide. Part 2.—Propynoic acid, Propenoic acid and 3-methyl-but-2-enoic acid, *J. Chem. Soc. Faraday Trans. 2* **75**, 32–37 (1979).
18. N. M. D. Brown, W. E. Timms, R. J. Turner, and D. G. Walmsley, Inelastic electron tunneling spectroscopy (IETS) of simple unsaturated hydrocarbons adsorbed on plasma-grown aluminum oxide, *J. Catal.* **64**, 101–109 (1980).

19. D. G. Walmsley, R. B. Floyd, and S. F. J. Read, Inelastic electron tunneling spectral lineshapes below 100 mK, *J. Phys. C* **11**, L107–L110 (1978).
20. J. D. Langan and P. K. Hansma, Can the concentration of surface species be measured with inelastic electron tunneling?, *Surf. Sci.* **52**, 211–216 (1975).
21. I. W. N. McMorris, N. M. D. Brown, and D. G. Walmsley, A study of the chemisorption of phenol and its derivatives on plasma-grown aluminum oxide by inelastic electron tunneling spectroscopy, *J. Chem. Phys.* **66**, 3952–3961 (1977).
22. N. M. D. Brown, R. B. Floyd, W. J. Nelson, and D. G. Walmsley, Inelastic electron tunneling spectroscopy of selected alcohols and amines on plasma-grown aluminum oxide, *J. Chem. Soc. Faraday Trans. 1* **76**, 2335–2346 (1980).
23. H. E. Evans and W. H. Weinberg, The reaction of ethanol with an aluminum oxide surface studied by inelastic electron tunneling spectroscopy, *J. Chem. Phys.* **71**, 1537–1542 (1979).
24. H. E. Evans and W. H. Weinberg, A comparison of the vibrational structures of ethanol, acetic acid and acetaldehyde adsorbed on alumina, *J. Chem. Phys.* **71**, 4789–4798 (1979).
25. D. G. Walmsley, W. E. Timms, and N. M. D. Brown, Explanation of asymmetry in tunnel spectra, *Solid State Commun.* **20**, 627–629 (1976).
26. J. G. Adler and J. E. Jackson, System for observing small nonlinearities in tunnel junctions, *Rev. Sci. Instrum.* **37**, 1049–1054 (1966).
27. W. J. Nelson and D. G. Walmsley, The adsorption of unsaturated hydrocarbons on aluminium oxide and magnesium oxide: An inelastic electron tunneling spectroscopy study, in *Proceedings of the Conference on "Vibrations at Surfaces"* Namur, Belgium, 10–12 Sept. 1980, pp. 471–481, Plenum Press, New York (1982).
28. N. M. D. Brown, W. J. Nelson, R. J. Turner, and D. G. Walmsley, Inelastic electron tunneling spectroscopy (IETS) of 1,3-diketones and related molecules on plasma-grown aluminum oxide, *J. Chem. Soc. Faraday Trans. 2* **77**, 337–353 (1981).
29. D. G. Walmsley, E. L. Wolf, and J. W. Osmun, Conductance of niobium oxide tunnel barriers, *Thin Solid Films* **62**, 61–66 (1979).
30. V. Mazur and K. W. Hipps, An IETS study of alumina and magnesia supported SiH_x and their uses as substrates for inorganic vibrational spectroscopy, *Chem. Phys. Lett.* **79**, 54–58 (1981).

12

The Structure and Catalytic Reactivity of Supported Homogeneous Cluster Compounds

W. Henry Weinberg

1. Introduction

Although tunneling spectroscopy has been applied heretofore to only a limited extent in the investigation of homogeneous cluster compounds supported on oxide surfaces, it is quite possible that it will prove to be one of the more valuable techniques that can be used to elucidate the structure and catalytic reactivity of this important class of catalysts. The principles of tunneling spectroscopy have been described in Chapter 1 of this book as well as in the previous review articles.[1–5] Examples of the use of tunneling spectroscopy to investigate the catalytic properties of metal crystallites supported on oxide surfaces have been described in previous review articles,[1–5] and, in particular, in Chapter 13 of this text. This application is closely related to the study of supported homogeneous cluster compounds. In the former case, one is concerned with aggregates of reduced metallic atoms attached to an oxide "support," while in the latter case, the topic of this chapter, one is concerned with the attachment and catalytic reactivity of cluster compounds (which may or may not have lost one or more of their ligands) "supported" on an oxide surface.

W. Henry Weinberg • Division of Chemistry and Chemical Engineering, California Institute of Technology, Pasadena, California 91125. Work supported by the National Science Foundation under grant No. CPE-8024597.

The generalized reaction of a homogeneous cluster compound with an oxide surface may be written as

$$M_xL_y + \text{surface} \rightarrow M_xL_{y-n}(\text{surface}) + L_n(\text{surface})$$

where n may vary between zero and y. In the case of $n=0$, there is molecular adsorption of the complex onto the surface. This type of adsorption would normally be rather weak, and consequently this type of complex would not be expected to have broad catalytic activity since at even moderate temperatures, where catalytic conversions proceed rapidly, the complex would be expected either to desorb from the surface or lose one or more of its ligands to become anchored more strongly to the oxide support surface. If all ligands are lost from the complex ($n=y$), then the bare metallic cluster M_x remains. Depending on the catalytic reaction conditions, this cluster can either maintain its molecular structure or sinter by combining with adjacent bare metallic clusters to become in essence a supported metallic catalyst (although possibly a quite highly dispersed one).

The fate of the homogeneous cluster compound, once it is adsorbed on the oxide support as the M_xLy-n entity, depends upon a number of factors such as the temperature of the surface, the catalytic reaction in question (i.e., the total pressure and the partial pressures of reactants and products of reaction), and the nature of the surface of the oxide support (e.g., the acid–base properties of the surface, the degree of hydroxylation of the surface, the density of defects, etc.). Insofar as tunneling spectroscopy is concerned, all homogeneous cluster compounds that have been investigated to date have been supported on plasma grown alumina surfaces in aluminum–alumina–lead tunneling junctions. This fabrication procedure results in alumina surfaces with hydroxyl groups present on them.[6,7] The catalytic activity of these surfaces most nearly mimics that of γ-alumina.[8]

Only two classes of homogeneous cluster compounds supported on alumina have been investigated with tunneling spectroscopy. These are the following: (1) A $Zr(BH_4)_4$ complex which serves as a model for a hydrocarbon polymerization catalyst[9–14] and (2) multimetallic carbonyl complexes of ruthenium,[14,15] rhodium,[14,16] and iron.[17] The importance of tunneling spectroscopy to determine the structure and to examine the reactivity of these supported homogeneous cluster compounds cannot be overemphasized. Such catalysts combine the activity and selectivity exhibited by homogeneous systems with the stability and ease of separation characteristic of heterogeneous catalysts, and they frequently exhibit activities an order of magnitude or more greater than the unsupported systems.[18] Characterization of supported homogeneous cluster catalysts has been quite poor traditionally. Prior to the application of tunneling spectroscopy, the structure of no supported cluster catalyst had been determined unequivocally, at least employing a vibrational spectroscopy. The major reason for this is the inability of optical spectroscopies to probe the low-frequency vibrational spectra of supported systems. The ability of tunneling spectroscopy to examine the vibrational spectral region from below 250 cm^{-1} to over 4000 cm^{-1} without significant interference from the oxide support and without perturbing the cluster is critical both in determining the structure of the supported cluster catalysts and in examining their catalytic properties.

To summarize, tunneling spectroscopy is ideally suited to determine both the structure and the catalytic properties of supported homogeneous cluster catalysts as a consequence of its high sensitivity (on the order of 1% of a monolayer on a surface the total area of which is typically 1 mm^2 or less), its rather high resolution, and especially its wide dynamic range (from below 250 cm^{-1} to over 4000 cm^{-1}). In Section 2, the experimental procedures employed in tunneling spectroscopy, especially as they pertain to an examination of supported cluster compounds, are reviewed. In Section 3, the previous results of the application of tunneling spectroscopy to supported cluster compounds are presented and discussed. Finally, in Section 4, a preview of probable future directions of tunneling spectroscopy in this area and a summary of the major conclusions based on the work to date are presented.

2. Experimental Procedures

In most of the tunneling spectroscopic studies of supported cluster catalysts,[9-16] the preparation of the tunneling junctions has consisted of the following steps. First, a strip of aluminum the thickness of which is approximately 1000 Å is evaporated onto a clean glass substrate in an oil diffusion pumped bell jar the base pressure of which is below 10^{-7} Torr. Then the top few atomic layers of the aluminum are oxidized in a plasma discharge either of pure O_2 or O_2 with traces ($<1\%$) of water vapor present at a total pressure of 0.12–0.16 Torr. The resulting thin (approximately 20 Å) layer of Al_2O_3 forms the insulating barrier necessary for the tunneling spectroscopic measurements, and, in addition, serves as the alumina substrate used for supporting the cluster compounds.

Adsorption of the gaseous $Zr(BH_4)_4$ complex onto the support was accomplished by exposing the Al_2O_3 films to 5×10^{-2} Torr of $Zr(BH_4)_4$ for 15 min (a surface area of approximately 1 mm^2 being probed during the measurement). This exposure is sufficient to produce a saturation coverage of the complex on the alumina, and evacuation of the gaseous $Zr(BH_4)_4$ from the bell jar results in the desorption of any weakly, reversibly held complex (including multilayers) from the alumina support, both observations being verified by subsequent tunneling spectroscopic measurements. The supported zirconium complex was then heated ($\leqslant 475$ K for up to 30 min) either in vacuum or in the presence of hydrogen (5 Torr) or water vapor (1 Torr) to ascertain its stability in vacuum and in the presence of reducing and oxidizing atmospheres. Finally, the supported zirconium complex was allowed to interact with ethylene, propylene, acetylene, cyclohexene, 1,3-cyclohexadiene, and benzene with the effects of exposure and temperature examined for each of the hydrocarbon reactants. In these experiments involving the adsorption and the reaction of hydrocarbons, the surface temperature was varied between 300 and 575 K, and the exposure of the alumina to the hydrocarbon was varied between 15 Torr sec (5×10^{-2} Torr for 300 sec) and 6000 Torr sec (5 Torr for 1200 sec). Heating of the alumina substrate was accomplished by means of a technique due to Bowser and Weinberg[(19)] which employs resistive heating with simultaneous temperature measurement.

Adsorption of both the $Ru_3(CO)_{12}$ and the $[RhCl(CO)_2]_2$ complexes onto the alumina support was accomplished by *in situ* sublimation of the solid complexes,[(14–16)] at 408 K in the case of $Ru_3(CO)_{12}$ and at 353–363 K in the case of $[RhCl(CO)_2]_2$. In order to enhance the adsorption of these two complexes, the alumina was cooled via conduction to a liquid-nitrogen reservoir to temperatures as low as 180 K. The adsorption of the $Fe_3(CO)_{12}$ complex was not carried out *in situ* from the vapor phase, but rather from a benzene solution after removal of the alumina substrate from the vacuum system.[(17)]

The final step in the synthesis of all tunneling junctions was the evaporation of Pb cross-strips on top of the samples, the Pb and the underlying Al layers serving as the two metal electrodes required for the tunneling measurements. Measurements were carried out over the spectral range from 240 to 4000 cm^{-1} with several samples examined at each set of conditions in order to ensure reproducibility. The spectral resolution was typically 30 cm^{-1}.

The tunneling spectra were measured with the samples immersed in liquid He at 4.2 K. With the exception of the $Fe_3(CO)_{12}$ complex, the desired second derivative was obtained by a modulation and harmonic detection scheme, using a modulation frequency of 50 kHz. The electronics were controlled, and the spectra recorded in digital form, by a PDP 11/10 laboratory minicomputer. The modulation amplitudes used in the measurement were below 2.0 meV rms (as measured at a bias voltage of 250 meV), depending on the resolution desired and the signal-to-noise characteristics of the junction. In the case of $Fe_3(CO)_{12}$, the measurement scheme was similar, in general, to that described above, but different in detail.[(20)]

3. Results and Discussion

3.1. $Zr(BH_4)_4$ on Al_2O_3 at 300 K[9, 10, 14]

The tunneling spectrum of $Zr(BH_4)_4$ adsorbed irreversibly on Al_2O_3 at 300 K together with an identification of the peak positions are shown in Figure 1.[9–11] The mode assignment of each peak in Figure 1 is listed in Table 1. In $Zr(BH_4)_4$, the boron atoms are arranged tetrahedrally around

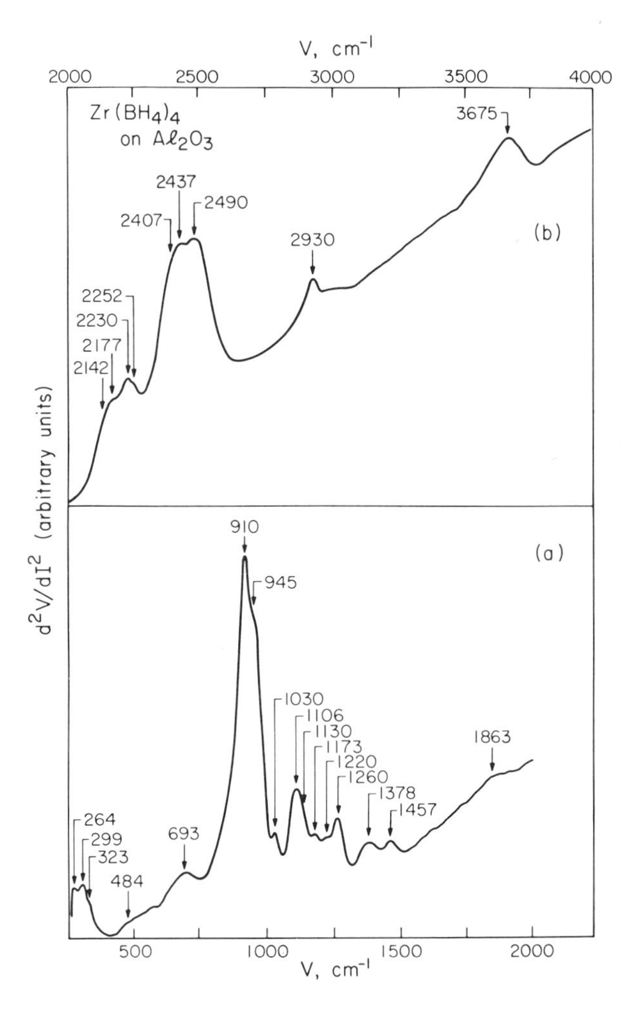

Figure 1. Tunneling spectrum for $Zr(BH_4)_4$ adsorbed on Al_2O_3 at 300 K over the spectral ranges (a) 240–2000 cm^{-1}, and (b) 2000–4000 cm^{-1}.

Table 1. Peak Positions (in cm^{-1}) for $Zr(BH_4)_4$ Supported on Al_2O_3[a]

In vacuum		Exposed to D_2		Exposed to D_2O or H_2O		
300 K	475 K	300 K	475 K	300 K	475 K	Assignments
264	264	264	264	—	—	BH_4–Zr–BH_4 bend or Zr–BH_4 torsion
299	299	299	299	299	299	Al phonon
323	323	323	323	323	323	Metal oxide or Zr–BH_4 torsion
480–	480–	—	—	—	—	Zr–BH_4 stretch or Zr–O modes
580	580					
693	685	705	693	693	695	Zr–O stretch
910	910	910	910	possible sh near 910		Zr–O stretch
945	945	945	945	942	942	Bulk Al–O stretch
1030	—	1056	1056	1048	1048	CH bend (contamination)
1106	1106	1114	1114	—	—	BH deformation
1130	1121	—	—	—	—	
1173	1165	1165	1170	—	—	
1220	1214	—	—	—	—	
1260	1252	1257	1260	1258	1258	BH deformation and B–O stretch
		(possible weak feature near 1300)				
1378	1378	1378	1385	1374	1380	B–O modes
1457	1457	1457	1457	1450	1450	
1870	1873	1868	1870	1870	1870	Harmonic of 945 cm^{-1}
2142	—	—	—	—	—	Bridging B–H stretch
2177	2137–2258 (weak–broad)	—	—	—	—	
2230	—	—	—	—	—	
2252	—	—	—	—	—	
2407	—	—	—	—	—	Terminal B–H stretch
2437	2455 (broad)	2455 (broad)	2455 (broad)	—	—	
2490	—	—	—	—	—	
—	—	—	—	2670	2670 (D_2O only)	O–D stretch
2930	2930	2930	2930	2930	2930	CH stretch (contamination)
—	—	—	—	3615	3615(D_2O)	O–H stretch for D_2O exposure only
3675	3675	3669	3669	3640	3640(H_2O)	O–H stretch for all other samples

[a]After subtracting 8.15 cm^{-1} to correct for the effect of the Pb superconducting gap (reference 44).

the central zirconium atom, each boron being bonded to the zirconium through three bridging hydrogens (H_b). There is also a single terminal hydrogen (H_t) on each boron. Among zirconium complexes, this type of tridentate bonding is uniquely restricted to this complex with four BH_4 ligands. In related complexes such as $(C_5H_5)_2Zr(BH_4)_2$ and $(C_5H_5)_2Zr(H)(BH_4)$, the bonding between the boron and zirconium atoms is bidentate, occurring through two bridging hydrogen atoms.[21-23]

During the adsorption of $Zr(BH_4)_4$, one or more of the BH_4 ligands are displaced in order to accommodate zirconium-surface bonding. For supported complexes, the surface becomes an effective ligand and can be expected to contribute both electronically and sterically to exert an influence on the remainder of the complex. Consequently, it would not be expected *a priori* that the bonding between the zirconium atoms and the remaining BH_4 ligands would retain an unperturbed tridentate form. On the contrary, bidentate bonding would appear to be more probable. The observed frequency range for $Zr(BH_4)_4$ adsorbed on Al_2O_3 is compared in Figure 2 with the observed transitions for the tridentate $Zr(BH_4)_4$, as well as for a number of complexes with bidentate structures. The frequencies observed for the adsorbed $Zr(BH_4)_4$ indicate a bidentate structure. Moreover, the observation of more than two B–H_t stretching modes as well as more than two B–H_b stretching modes (at least four at 2142, 2177, 2230, and 2252 cm^{-1}) suggests the presence of more than a single type of adsorbed complex on the surface.

The presence of surface complexes with zirconium atoms both singly and multiply coordinated to oxygen atoms at the surface, as well as the presence of both Zr(III) and Zr(IV) and [and possibly Zr(II)] complexes, would almost certainly result in multiple types of BH_4 ligands, corresponding to the multiple bands observed in the tunneling spectrum of Figure 1 for both terminal and bridging boron–hydrogen stretching vibrations in the complexes formed when $Zr(BH_4)_4$ adsorbs irreversibly on alumina. The differences in zirconium–oxygen coordination are reflected also in the tunneling spectrum of Figure 1 (cf. Table 1). Although the additional peaks in the B–H stretching region can be accounted for by surface complexes with different types of Zr–O coordination (and perhaps different oxidation states of the zirconium), there are other possible surface species which deserve consideration. During the adsorption of $Zr(BH_4)_4$, displaced ligands might well remain on the surface, becoming associated with either Al or O atoms. For example, $Al(BH_4)_3$ is a known complex which is quite stable when not exposed to air and which possesses bidentate hydrogen bridge bonding.[22] It has also been shown that diborane (B_2H_6) adsorbs dissociatively on alumina, resulting in the formation of the following

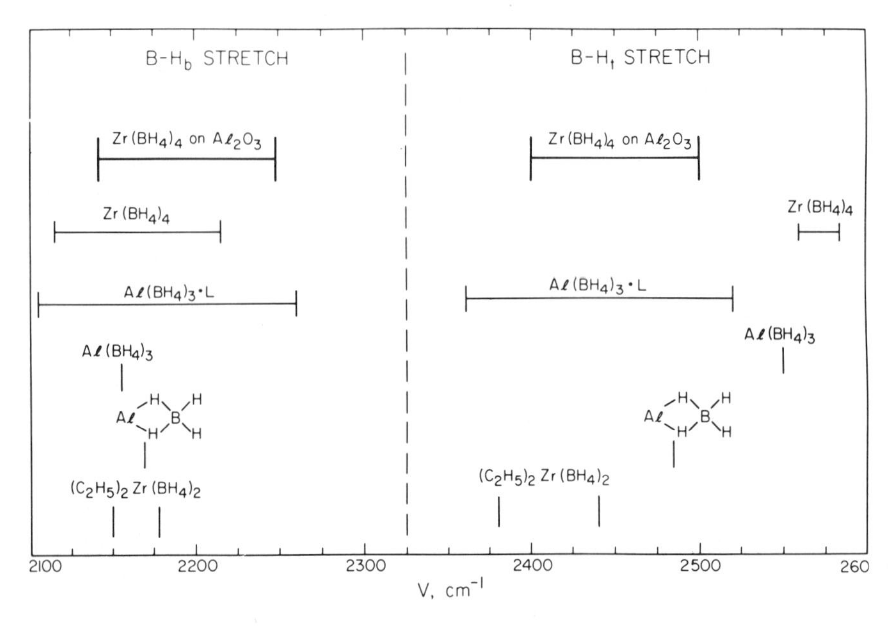

Figure 2. Comparison of both terminal (H_t) and bridging (H_b) B–H stretching frequencies for $Zr(BH_4)_4$ supported on Al_2O_3 with the corresponding vibrations for $Zr(BH_4)_4$ (reference 70); $Al(BH_4)_3 \cdot L$, where L is a coordination ligand (reference 22); $Al(BH_4)_3$ (reference 22); AlH_2BH_2, resulting from diborane adsorption on Al_2O_3 (reference 24); and $(C_5H_5)_2Zr(BH_4)_2$ (reference 25).

complexes on the surface[24]:

H H
\ /
B
/ \
H H
\ /
Al

and

H H
\ /
B
|
O

Similar surface complexes might well be formed from the BH_4 ligands which are displaced during the irreversible adsorption of $Zr(BH_4)_4$ on the alumina surface. The slow decomposition of $Zr(BH_4)_4$ producing B_2H_6 at room temperature presents another possible source for these surface species.[25]

3.2. $Zr(BH_4)_4$ on Al_2O_3 at 475 K[11]

As may be seen in Figure 2, heating the surface on which $Zr(BH_4)_4$ has adsorbed at 300–475 K does not destroy the catalyst completely. The decomposition of various zirconium borohydrides, which occurs only slowly at room temperature, proceeds rapidly above approximately 450 K, evolving mainly H_2.[25] The adsorption of $Zr(BH_4)_4$ on Al_2O_3 obviously stabilizes the complex somewhat. Examination of the tunneling spectrum of $Zr(BH_4)_4$ on Al_2O_3 heated to 475 K in Figure 3 indicates that although the B–H deformation and stretching modes are all reduced in intensity, some of the B–H bonds remain intact. The peak positions, as well as their assignments, for $Zr(BH_4)_4$ adsorbed irreversibly on Al_2O_3 at 475 K are summarized in Table 1.

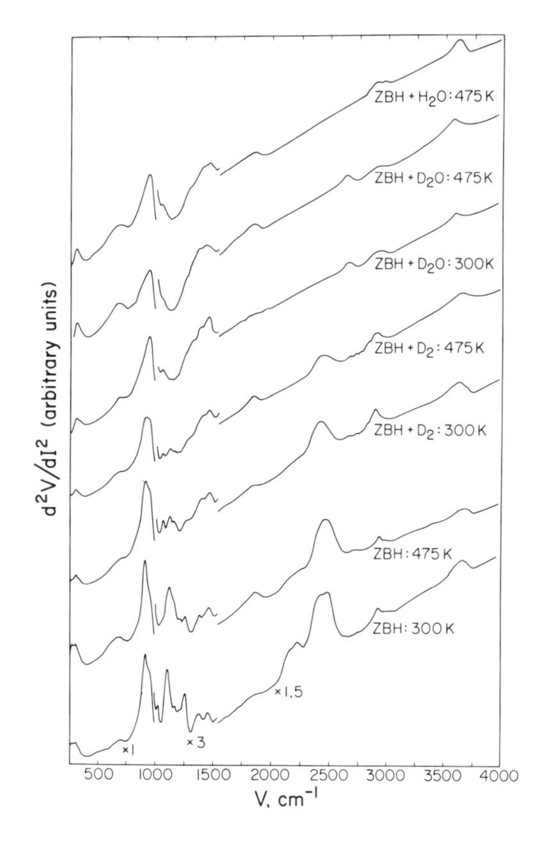

Figure 3. Tunneling spectra over the range 240–4000 cm^{-1} for $Zr(BH_4)_4$, designated "ZBH," adsorbed on Al_2O_3 at 300 and 475 K, and for the supported complex after exposure to D_2, D_2O, and H_2O at the same temperatures.

3.3. The Interaction of $Zr(BH_4)_4$ on Al_2O_3 with D_2, D_2O, and H_2O[11]

Exposure of the $Zr(BH_4)_4$ complex on Al_2O_3 to D_2 at 300 K results in similar changes compared to heating the complex in vacuum to 475 K, although both the increases and the decreases in intensity are more pronounced after exposure to D_2, as may be seen in Figure 3 and Table 1. Exposure of the $Zr(BH_4)_4$ complex adsorbed on Al_2O_3 to D_2 fails to produce any exchange with B–H groups. Although BH_4 ligands are being displaced, as evidenced by decreases in intensity in all regions of the tunneling spectrum in which B–H or Zr–BH_4 modes appear, there is no conclusive evidence for a new Zr–D or Zr–H stretching mode which would be expected near 1150 cm^{-1} [26] and 1625 cm^{-1},[18, 21, 27] respectively. It is important to note that, especially after exposure to D_2 at 475 K, some B–H_t bonds remain even though there are no B–H_b bonds left at the surface. This tends to confirm the presence of O–BH_2 species on the surface. That these groups are more stable with respect to both heating and chemical reaction than are other BH species on the surface is in agreement with previous work.[24]

Zirconium and aluminum borohydrides, as well as diborane and other B–H compounds, are hydrolyzed easily even at room temperature.[22] Tunneling spectra of the $Zr(BH_4)_4$ complex supported on Al_2O_3 which has been exposed to D_2O at both 300 K and 475 K are shown in Figure 3. These spectra are devoid of any B–H stretching features, nor are there any new features which might be attributed to B–D modes. The absence of isotopic exchange with the supported complex is confirmed by comparison to the tunneling spectrum in Figure 3 for exposure to H_2O at 475 K. This confirms quantitatively the expected sensitivity of the supported catalyst to water vapor.

3.4. The Interaction of $Zr(BH_4)_4$ on Al_2O_3 with C_2H_4, C_3H_6, and C_2H_2[12]

Tunneling spectroscopic measurements were performed for the $Zr(BH_4)_4$ catalyst supported on alumina after interaction with ethylene at exposures from 15 Torr sec (5×10^{-2} Torr of C_2H_4 for 300 sec) to greater than 6000 Torr sec (5 Torr of C_2H_4 for 1200 sec), and at various temperatures between 300 and 575 K. Very little adsorption, with no resolvable structure above the background, was observed at the lower exposures. Furthermore, no C_2H_4 (nor C_3H_6 nor C_2H_2) adsorbs irreversibly on the bare Al_2O_3 surface under these experimental conditions, as verified by the appropriate blank experiments. Saturation coverage of C_2H_4 on the $Zr(BH_4)_4$ catalyst supported on alumina was obtained with exposures of 3000 Torr sec at all temperatures, as judged by the lack of spectral changes

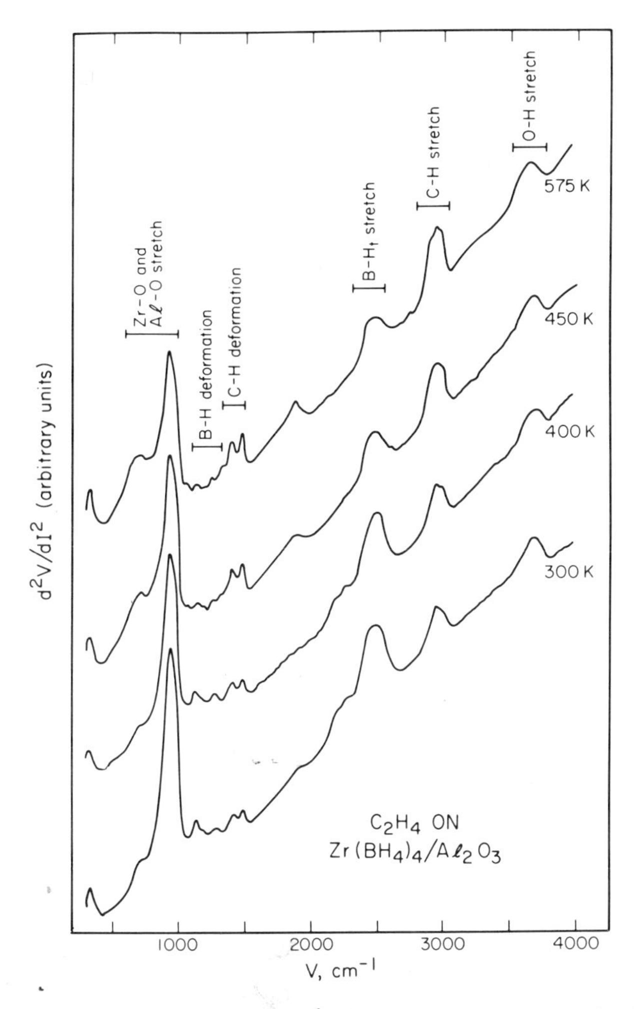

Figure 4. Tunneling spectra (240–4000 cm^{-1}) for saturation coverages of ethylene on $Zr(BH_4)_4/Al_2O_3$ at 300, 400, 450, and 575 K.

at higher exposures. Tunneling spectra for saturation coverage of C_2H_4 at 300, 400, 450, and 575 K are shown in Figure 4. No shifts in the positions of the peaks were observed as a function of coverage, and the peak positions as well as their assignments are listed in Table 2. Tunneling spectra corresponding to saturation exposures of propylene to the $Zr(BH_4)_4$ catalyst supported on Al_2O_3 at 300, 400, and 575 K are shown also in Figure 5 with the peak positions and the mode assignments listed in Table 2.

Table 2. Peak Positions (in cm^{-1}) for C_2H_2, C_2H_4, and C_3H_6 on Alumina-Supported $Zr(BH_4)_4$[a]

C_2H_2		C_2H_4		C_3H_6		
300 K	400 K	300 K	525 K	300 K	525 K	Assignments
264	264	264	264	264	264	BH_4–Zr–BH_4 bend or Zr–BH_4 torsion
299	299	299	299	299	299	Al phonon
323	323	323	323	323	323	Metal oxide mode or Zr–BH_4 torsion
485	485	~480	~480	~490	~490	Zr–BH_4 stretch, Zr–O modes; and (acetylene only) polymer skeletal deformations
		to	to	to	to	
573	573	580	580	580	580	
690	690	693	693	693	693	Zr–O stretch
910	910	910	910	910	910	Zr–O stretch
945	945	945	945	945	945	Al–O (bulk) stretch
1030	1030	1044?	1040?	1030	1030	CH modes
1071	1071					
1106	1106	1106	1106	1102	1102	BH deformations
1130	1130	1130	1130	—	—	
1173	1173	1169	—	1165	—	
1220	1220	1218	—	1213	—	
1260	1260	1256	1244	1258	1258	BH or B–O mode
1351	1351	1302	1302	1342	1342	C–C stretch (C_2H_2 only), CH deformations and B–O modes
1380	1380	1383	1380	1378	1378	
1452	1452	1455	1455	1449	1449	
1597	1597	—	—	—	—	C=C stretch?
1870	1870	1870	1874	1870	1876	Harmonic of 945 cm^{-1}
2150	2150	2142	—	2134	—	Bridging BH stretch
2181	2177	2177	—	2177	—	
2236	2233	2229	—	2229	—	
2258	2250	2256	—	2252	—	
2410	2410	2406	—	2410	2410	Terminal BH stretch
2434	2434	2435	2440 (broad)	2427	2427	
2484	2492	2486	—	2473	2473	
2505	—	—	—	2501	2490	
2871	2871	2879	2879	2863	2867	CH stretch
—	—	2913	2920	2903	2899	
2935	2922	2930	2938	2930	2930	
2981	2960	2960	2958	2952	2950	
3677	3677	3677	3641	3669	3625	OH stretch

[a]After subtracting 8.15 cm^{-1} to correct for the effect of the Pb superconducting gap (reference 44).

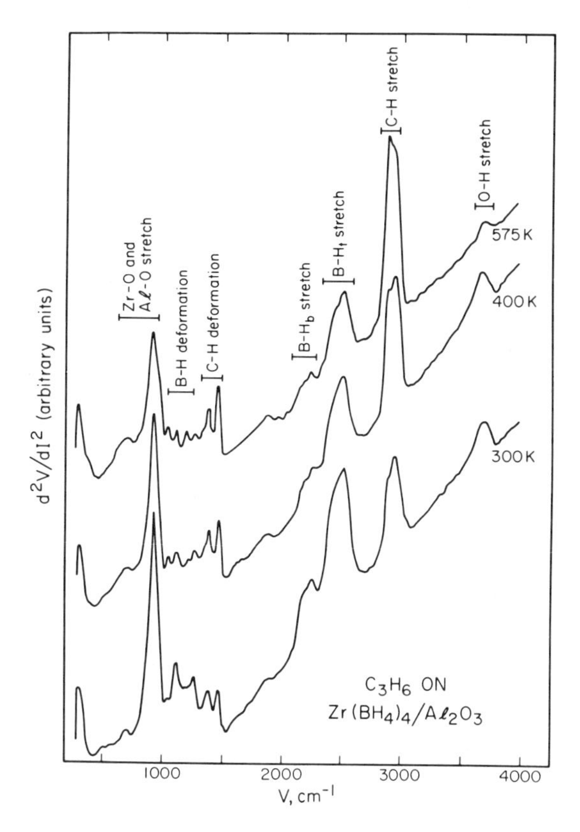

Figure 5. Tunneling spectra for saturation coverages of propylene on $Zr(BH_4)_4/Al_2O_3$ at 300, 400, and 575 K.

The adsorption of acetylene on the supported $Zr(BH_4)_4$ catalyst exhibited qualitatively different behavior from either ethylene or propylene. Tunneling spectra for an exposure of 300 Torr sec of acetylene at 300 and 400 K are shown in Figure 6 with the peak positions and the mode assignments listed in Table 2. *At higher temperatures, the concentration of hydrocarbon species on the surface increased rapidly with no apparent saturation limit.* Even at the lowest exposures of acetylene which could be measured reliably, 5 Torr sec, the resistances of the resulting tunneling junctions, synthesized at temperatures of adsorption above 400 K, were immeasurably high.

Unsaturated hydrocarbon molecules can be expected to interact with many, if not all, of the various types of surface species formed after the adsorption of $Zr(BH_4)_4$ on alumina as discussed in Sections 3.1 and 3.2. This has been confirmed for the AlH_2BH_2 and OBH_2 species in a study of

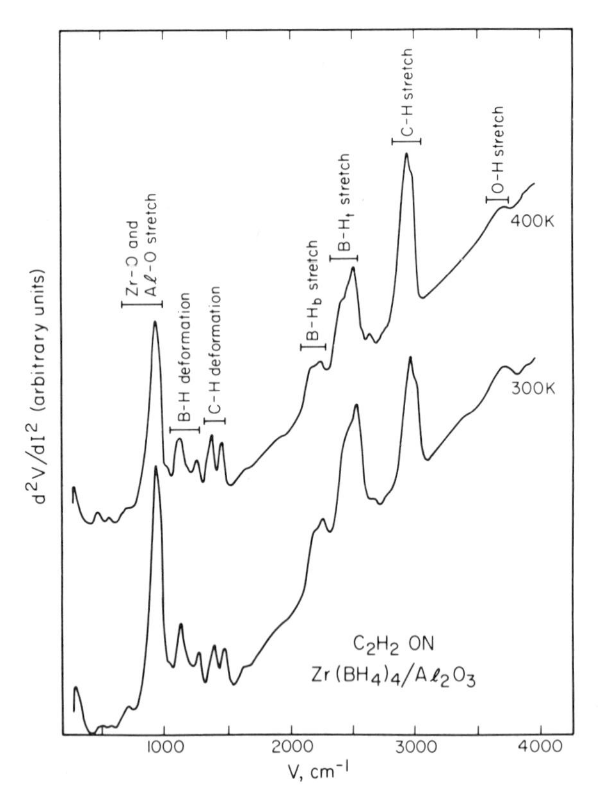

Figure 6. Tunneling spectra of acetylene (300 Torr sec exposure) on $Zr(BH_4)_4/Al_2O_3$ at 300 and 400 K.

olefins interacting with adsorbed diborane.[24] Supported zirconium complexes, particularly those supported on Al_2O_3, are known to be active for the polymerization of olefins. A weak Lewis-acid–Lewis-base interaction permits the formation of nonlinear π complexes as olefins become coordinated to zirconium atoms. This coordination, without the displacement of additional ligands from the supported complex, is generally considered to be the first step in the polymerization reaction. The next step is the insertion of the hydrocarbon between the zirconium atom and one of its remaining ligands in the complex. Repeated coordination and insertion of other olefin molecules create a polymer chain. It has been postulated that only H ligands form "active centers" for polymerization,[28] but the BH_4 groups could also have several functions in such a reaction scheme, e.g., providing a source of hydrogen to form the Zr–H sites, providing a variable coordination sphere for the zirconium, or exerting some activating influence on other ligands.

The observed behavior of acetylene exposed to the supported $Zr(BH_4)_4$ catalyst can be explained consistently by the formation of polyacetylene on the surface. There are no correspondingly conclusive indications of polymer formation after exposure of the catalyst to either ethylene or propylene. These conclusions are based both on the observed characteristics of adsorption and on a comparison of the measured tunneling spectra to published results for saturated and unsaturated hydrocarbons,[29, 30] polyethylene,[31–33] polypropylene,[33–35] and polyacetylene.[36] Perhaps the most obvious indication of polymer formation in the case of acetylene is its continuous adsorption without achieving saturation coverage at temperatures above 400 K, even at rather low pressures of acetylene (below 1 Torr). On the other hand, the observed rapid attainment of saturation coverage for ethylene and propylene indicates no extensive formation of polymers at the surface. Finally, it is of considerable interest to note in the case of acetylene, a high surface concentration of hydrocarbon can accumulate with very little, if any, accompanying loss of BH groups (cf. Figures 1 and 6). This is consistent with the view that the formation of the polymer at a few active centers, probably Zr–H sites, is the mechanism for incorporating additional acetylene molecules into the surface phase.

3.5. The Interaction of $Zr(BH_4)_4$ on Al_2O_3 with Cyclohexene, 1,3-Cyclohexadiene, and Benzene[13]

A preliminary investigation of the interaction of cyclohexene (C_6H_{10}), 1,3-cyclohexadiene (C_6H_8), and benzene (C_6H_6) with $Zr(BH_4)_4$ supported on Al_2O_3 has appeared recently.[13] Saturation exposures of these three hydrocarbons on the supported $Zr(BH_4)_4$ at 300 K give rise to the tunneling spectra shown in Figure 7. Blank experiments were carried out which demonstrated that none of the hydrocarbons adsorbed irreversibly on the Al_2O_3 surface in the absence of the $Zr(BH_4)_4$ cluster catalyst. A comparison of the tunneling spectra in Figure 7 suggests that cyclohexene perturbs the supported catalyst less than either benzene or 1,3-cyclohexadiene based on the relative intensities of the $B–H_t$ and $B–H_b$ stretching and deformation modes. Consequently, 1,3-cyclohexadiene, and to a much less extent benzene, interact with the supported $Zr(BH_4)_4$ primarily via a displacement of the BH_4 ligands. The tunneling spectra demonstrate that this displacement occurs readily at room temperature in the case of the interaction of 1,3-cyclohexadiene with the supported catalyst. Recent results employing tunneling spectroscopy suggest that at elevated temperatures benzene does not adsorb appreciably on the catalyst, whereas 1,3-cyclohexadiene polymerizes.[37]

The tunneling spectra of the $Zr(BH_4)_4$ catalyst supported on Al_2O_3 after saturation exposure to cyclohexene at 300, 400, and 480 K are shown in Figure 8 for the spectral range between 2000 and 4000 cm^{-1}. The cyclohexene does not polymerize on the surface at these temperatures and at

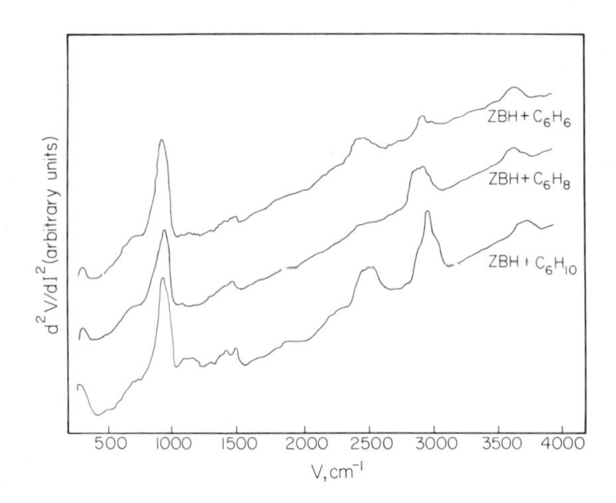

Figure 7. Tunneling spectra (240–4000 cm^{-1}) of cyclohexene (C_6H_{10}), 1,3-cyclohexadiene (C_6H_8), and benzene (C_6H_6) on $Zr(BH_4)_4/Al_2O_3$ (ZBH) at 300 K.

pressures below approximately 5 Torr. At 400 K, the B–H stretching modes are reduced in intensity relative to the C–H stretching modes, indicating a greater extent of adsorption compared to 300 K. The tunneling spectra of Figure 8 indicate that the surface complex formed at 480 K may be a Zr–alkyl complex resulting from the hydrogenation of the cyclohexene. Previous work has demonstrated that zirconium hydrides are catalysts for the hydrogenation of olefins, including cyclohexene.[38]

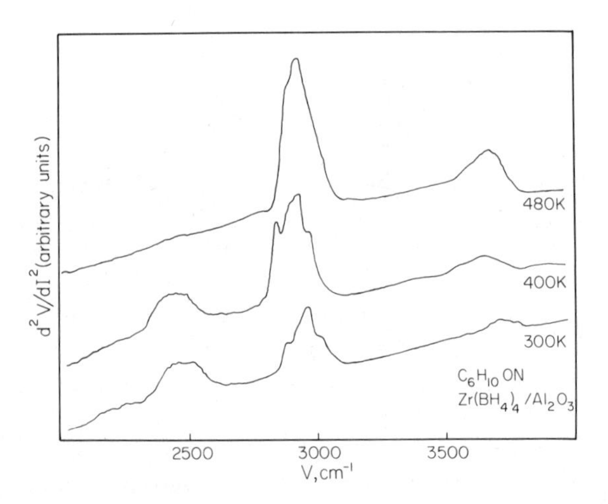

Figure 8. Tunneling spectra (2000–4000 cm^{-1}) of cyclohexene on $Zr(BH_4)_4/Al_2O_3$ at 300, 400, and 480 K.

3.6. $Ru_3(CO)_{12}$ on Al_2O_3[14, 15]

The $Ru_3(CO)_{12}$ complex is an air stable solid which does not decompose below 423 K. The structure of the complex, determined by x-ray crystallography, is shown in Figure 9.[39] The complex consists of an equilateral triangle of Ru atoms, each metal atom being bound directly to the other two metal atoms. Of the four CO ligands, two are *axial*, aligned perpendicularly to the Ru triangle, and the other two are *radial*, lying in the plane of the Ru triangle.

The $Ru_3(CO)_{12}$ complex is adsorbed weakly and does not dissociate on the Al_2O_3 surface at the low (vapor) pressures of adsorption examined using tunneling spectroscopy. A temperature of adsorption below 200 K was necessary to prevent desorption of the complex upon evacuation prior to the evaporation of the Pb overlayer. The vibrational modes present in the tunneling spectra were quite weak even for saturation coverage at 180 K, as may be seen in Figure 10. In order to enhance the visibility of those features due to adsorbed $Ru_3(CO)_{12}$, a digital background subtraction technique was employed as illustrated in Figure 10.[40] The tunneling spectrum in the frequency range between 240 and 2250 cm^{-1}, obtained by averaging the tunneling spectra from four separate junctions, each exposed to the $Ru_3(CO)_{12}$, is shown in Figure 10a. The tunneling spectrum of a clean junction in the same range of frequencies is shown in Figure 10b. The difference spectrum due only to the adsorbed $Ru_3(CO)_{12}$ is obtained by

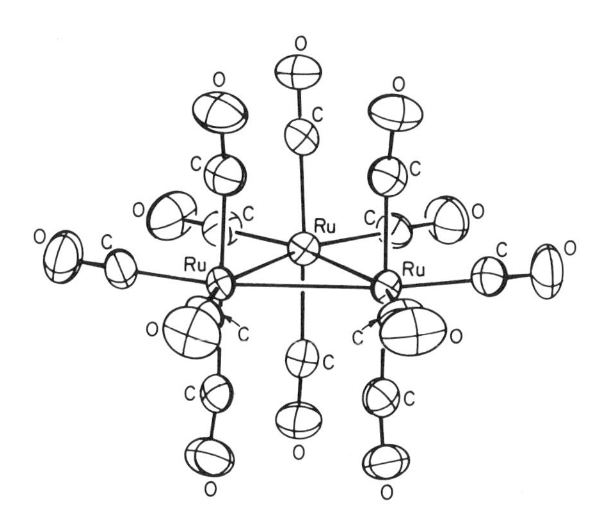

Figure 9. $Ru_3(CO)_{12}$, viewed from a direction 75° from the normal to the Ru_3 plane (reference 39).

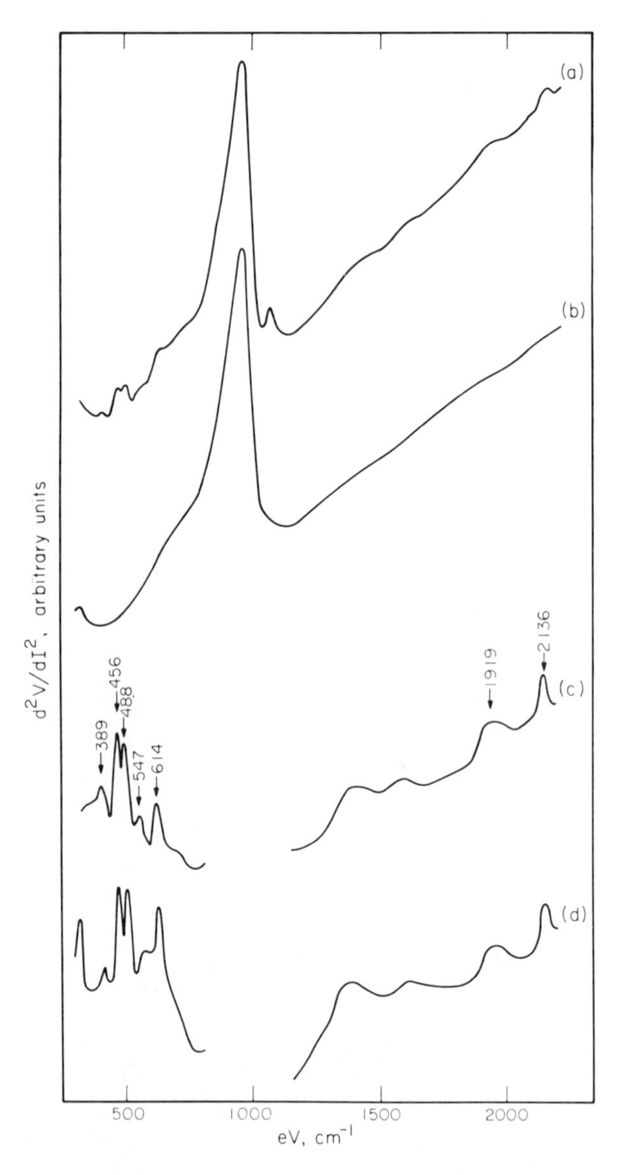

Figure 10. (a) The tunneling spectrum, from 240 to 2220 cm^{-1}, of $Ru_3(CO)_{12}$ adsorbed on aluminum oxide at approximately 180 K. (b) The tunneling spectrum of a clean sample, over the same energy range as spectrum (a). (c) The spectrum resulting from the digital subtraction of spectrum (b) from spectrum (a), expanded vertically by a factor of 3. (d) The difference spectrum, analogous to spectrum (c), for $Ru_3(CO)_{12}$ adsorbed on a deuteroxylated alumina surface.

subtracting spectrum (b) from spectrum (a) in Figure 10, and the result, expanded vertically by a factor of 3, is shown in Figure 10c. The tunneling spectrum shown in Figure 10d, is the same as that of Figure 10c except that the original surface contained OD groups rather than OH groups. The equivalence of spectra (c) and (d) hence shows that there is no chemical reaction between the $Ru_3(CO)_{12}$ complex and surface hydroxyl groups although there well could be polar bonding between the two. The low-intensity of the vibrational modes of $Ru_3(CO)_{12}$ on Al_2O_3 in the tunneling spectra of Figure 10 is probably due to a weak coupling of the tunneling electrons with the vibrational modes of the complex due to the large physical size of the latter.[15] The vibrational modes due to the adsorption of $Ru_3(CO)_{12}$ on the alumina surface as well as the mode assignments are given in Table 3.

It has been found that when $Ru_3(CO)_{12}$ is adsorbed at room temperature on high-surface-area powders both of silica[41,42] and γ-alumina,[41] it retains its molecular structure. Consequently, it would be expected that $Ru_3(CO)_{12}$ would adsorb molecularly also on the cooled alumina surface of a tunnel junction. This expectation is confirmed by the close agreement between the vibrational modes observed in the tunneling spectra and both the ir and Raman modes observed for solid $Ru_3(CO)_{12}$[43] and the ir modes of $Ru_3(CO)_{12}$ adsorbed molecularly on silica,[42] all of which are listed in Table 3.

The fact that the tunneling electrons couple best to those vibrational modes which are perpendicular to the oxide surface (parallel to the tunneling electrons) implies that the tunneling spectra probe the *axial* CO ligands in the complex preferentially (making the reasonable assumption that the weakly adsorbed $Ru_3(CO)_{12}$ is attached to the oxide with the Ru_3 ring parallel to the oxide surface). In view of this, it is clear that there is an *upshift* of 9 cm^{-1} for the high-frequency CO stretching mode in the tunneling spectra compared to the solid complex, and a *downshift* of between 109 cm^{-1} and 115 cm^{-1} for the low-frequency CO stretching mode in the tunneling spectra compared to the solid complex. To understand this, it should be recalled that the top Pb electrode required in tunneling spectroscopy has the effect of downshifting the frequencies and broadening the peaks of the vibrational modes,[44] an effect which is generally quite significant for chemisorbed CO.[45-47] In view of this, the reason for the *upshift* in the high-frequency CO stretching mode requires an explanation. This small upshift in frequency of 9 cm^{-1} is due to a change in the electronic environment of the complex upon adsorption on the alumina surface. Upshifts of a similar magnitude have been observed also for two of the three ir bands of $Ru_3(CO)_{12}$ adsorbed on silica,[42] as may be seen in Table 3. Furthermore, it is important to realize that the shifts due to the Pb

Table 3. Vibrational Frequencies for $Ru_3(CO)_{12}$, and Corresponding Mode Assignments, as Observed by Tunneling, Raman, and ir Spectroscopies

Tunneling[a]	Raman[b]	ir[b]	ir[c]		Mode assignments for $Ru_3(CO)_{12}$ with D_{3h} symmetry
2136	2127	—	—	A_1'	Symmetric axial CO stretching vibration
—	—	2062	2065	A_2''	Antisymmetric axial CO stretching vibration
—	—	2053	—		
—	—	2042	—		
—	2034	—	—	E''	Axial CO stretching vibrations
1919	2028	2026	2035	E'	
—	2011	—	—	A_1'	Symmetric radial CO stretching vibration
—	2004	2002	2010	E'	Radial CO stretching vibrations
—	1994	—	—	E'	
—	1989	1989	—	A_2'	Antisymmetric radial CO stretching vibration
614	607	606	—		Predominantly Ru–C–O deformation in character (no detailed assignment attempted)
—	596	594	—		
—	—	574	—		
547	546	546	—		
—	513	512	—		
488	489	—	—		Predominantly Ru–C stretching in character (no detailed assignment attempted)
456	458	466	—		
—	446	448	—		
—	—	400	—		
389	392	389	—		

[a] Inelastic electron tunneling spectrum of molecular $Ru_3(CO)_{12}$ bound to Al_2O_3 (reference 15). [After subtracting 8.15 cm^{-1} to correct for the effect of the Pb superconducting gap (reference 44).]
[b] Raman and ir spectra of solid $Ru_3(CO)_{12}$ (reference 43).
[c] ir spectrum of molecular $Ru_3(CO)_{12}$ bound to silica (reference 42).

electrode give important information as to which axial CO ligands are viewed with tunneling spectroscopy. According to a theory of Kirtley and Hansma,[44] an (unresolvable) downshift of 0.3 cm^{-1} is expected in the CO stretching frequency for those axial CO ligands adjacent to the alumina surface, whereas a downshift on the order of 100 cm^{-1} is expected in the CO stretching frequency for those axial CO ligands adjacent to the Pb surface.[15] Hence, it may be concluded that the 2136 cm^{-1} mode in the tunneling spectra corresponds to axial CO ligands in close proximity to the alumina surface, and the 1919 cm^{-1} mode corresponds to axial CO ligands in close proximity to the Pb surface at the surface alumina–Pb interface. An exact assignment of the low-frequency modes cannot be made due to the lack of data for isotopically substituted clusters. However, in general, those bands between 350 and 500 cm^{-1} are attributed to modes with predominantly metal–carbon stretching character, whereas the bands between 500 and 650 cm^{-1} are attributed to modes with metal–carbon–oxygen bending character.[43]

3.7. $[RhCl(CO)_2]_2$ on Al_2O_3[14, 16]

The rhodium chlorodicarbonyl dimer is an air-sensitive, crystalline solid at room temperature, sublimes readily at 353–363 K, and decomposes above 398 K.[48] The structure of the dimer, determined by x-ray diffraction, is shown in Figure 11.[49] The interaction of the rhodium chlorodicarbonyl dimer with an alumina surface is of interest for several reasons. First, the adsorption of CO on Rh crystallites supported on Al_2O_3 has been investigated extensively by both ir[50–54] and tunneling spectroscopy,[45, 46, 55] and, hence, direct comparisons can be made between this work and the results for $[RhCl(CO)_2]_2$ adsorbed on Al_2O_3. Also, this complex has been observed to adsorb from a liquid solution onto γ-alumina,[54] and, finally, it would be expected to form a more highly dispersed metallic catalyst, after decarbonylation, than larger rhodium carbonyl complexes which have three or more

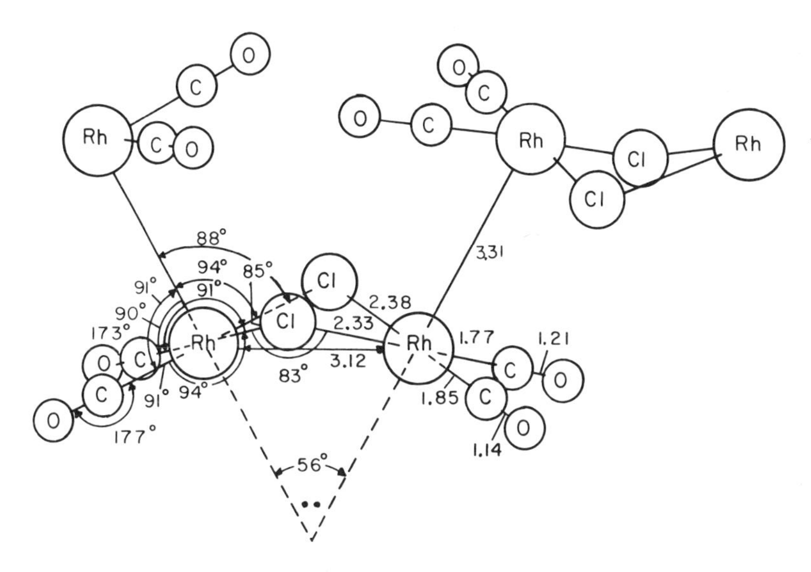

Figure 11. The structure of the $[RhCl(CO)_2]_2$ dimer as determined from x-ray crystallography (reference 49).

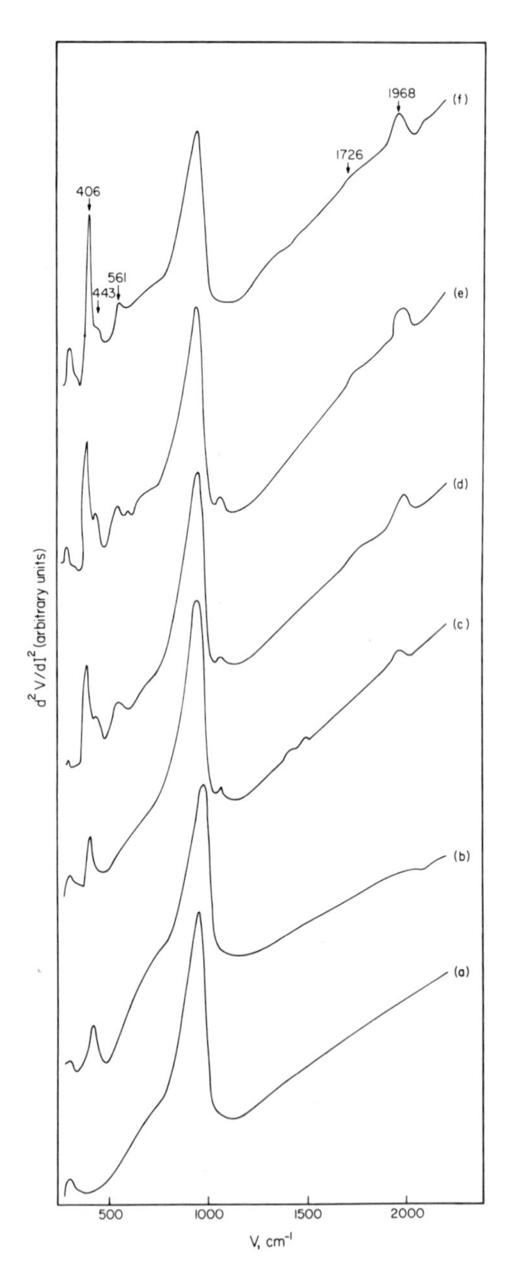

Figure 12. Tunneling spectra of alumina cooled to various temperatures and exposed to $[RhCl(CO)_2]_2$ from the gas phase. A clean surface spectrum is shown in (a). The other spectra represent the surface exposed to the complex at (b) room temperature (295 K), (c) 240 K, (d) 210 K, (e) 190 K, and (f) 180 K. The lead was evaporated on the cold surface.

metallic atoms bound together within the molecular complex. This is of interest since significant increases in the activity of some reactions have been related to high dispersion of the reduced metallic catalyst.[56, 57]

Tunneling spectra of the "clean" alumina surface (with trace amounts of hydrocarbon contamination) and alumina surfaces exposed to $[RhCl(CO)_2]_2$ at surface temperatures between 295 and 180 K are shown in Figures 12 and 13. Since no modes are observed which could be due to Rh–Cl, Al–Cl, or O–Cl vibrations, the $[RhCl(CO)_2]_2$ evidently decomposes to form $Rh_n(CO)_m$ species bound to the alumina surface. The frequencies of these five vibrational modes, listed in Table 4, agree quite well with the frequencies of the five peaks found in tunneling spectroscopic studies of CO chemisorbed on Rh crystallites which were deposited by evaporation onto an Al_2O_3 surface.[45, 46] This favorable comparison suggests that upon

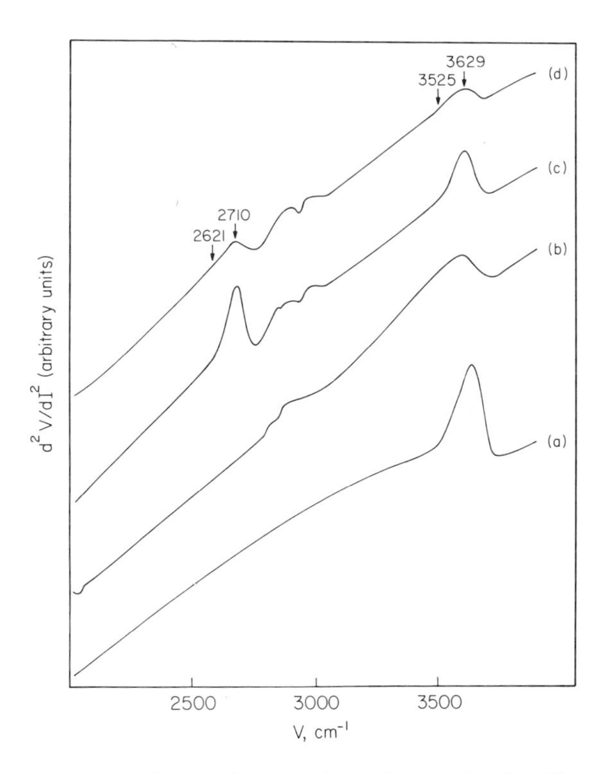

Figure 13. Tunneling spectra showing the OH and OD vibrational region. The spectra are for (a) a clean sample, (b) alumina exposed to $[RhCl(CO)_2]_2$ at 210 K corresponding to spectrum (d) of Fig. 12, (c) clean surface partially deuterated, and (d) deuterated surface exposed to $[RhCl(CO)_2]_2$ at 200 K.

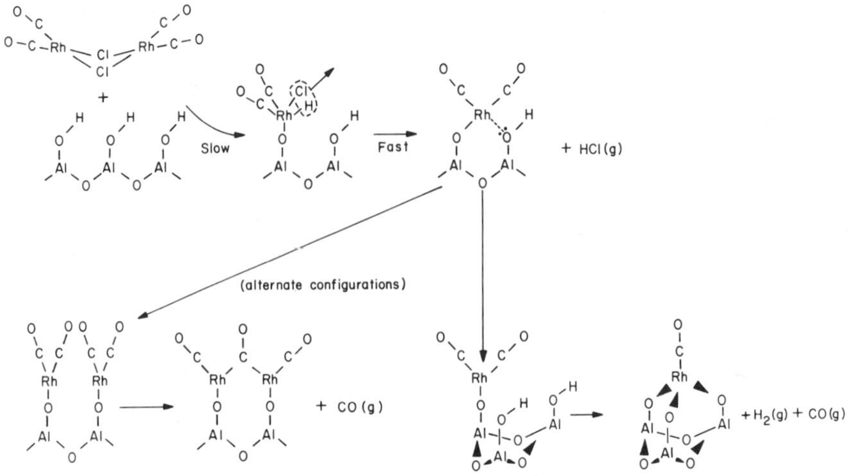

Figure 14. Postulated adsorption mechanism for $[RhCl(CO)_2]_2$ on a hydroxylated alumina surface.

adsorption on the alumina, the $[RhCl(CO)_2]_2$ decomposes losing the chlorine via HCl evolution leaving rhodium–CO species bound to the surface. This reaction is shown schematically in Figure 14. This scheme is consistent also with the results of Smith *et al.*[54] for the adsorption of $[RhCl(CO)_2]_2$ on alumina from the liquid phase. The increase in intensity of the vibrational modes in the tunneling spectra, which correspond to the dissociatively adsorbed $[RhCl(CO)_2]_2$ as the temperature of the alumina is decreased (cf. Figure 12) is due to a higher concentration of adsorbed dimer at the lower surface temperature. Then, the heat of condensation of the Pb overlayer provides the necessary thermal activation for the reaction scheme shown in Figure 14 to occur.[16]

In addition to the modes which are observed in the tunneling spectra due to the dissociatively adsorbed $[RhCl(CO)_2]_2$ complex, it is clear from Figure 13 that the hydroxyl groups on the alumina surface are perturbed strongly by the adsorption of the complex. The peak at 3646 cm^{-1} has been broadened significantly toward low frequency, and it appears that a second OH stretching mode at approximately 3525 cm^{-1} is present in addition to the original peak which has been downshifted to approximately 3629 cm^{-1}. If the hydroxyl groups on the surface are exchanged partially with D_2O to form OD groups on the surface (by exposure to D_2O at room temperature prior to cooling and exposure to the $[RhCl(CO)_2]_2$), the same type of perturbation of the OD stretching mode is observed (cf. Figure 13). In no

Table 4. Vibrational Modes Resulting from $[RhCl(CO)_2]_2$ Adsorption on Alumina

Frequency (cm^{-1})[a]	Surface specie	(Type of mode)
406	$Rh(CO)_2$	(M–C–O bending)
443	RhCO	(M–C–O bending)
561	Rh_2CO	(M–C–O bending or M–CO stretching)
1726	Rh_2CO	(CO stretching)
1968	$Rh(CO)_2 + RhCO$	(CO stretching)
2621	Coordinated OD	(OD stretching)
2710	Free OD	(OD stretching)
3525	Coordinated OH	(OH stretching)
3629	Free OH	(OH stretching)

[a]After subtracting 8.15 cm^{-1} to correct for the effect of the Pb superconducting gap (reference 44).

case, however, were modes corresponding to Rh–H or Rh–D bonds observed in the tunneling spectra.

3.8. $Fe_3(CO)_{12}$ on Al_2O_3[17]

Triiron dodecacarbonyl $[Fe_3(CO)_{12}]$ would be difficult to investigate with tunneling spectroscopy employing the same techniques as those described in Sections 3.6 and 3.7, namely, *in situ* sublimation of the solid carbonyl complex onto the alumina support. The reason for this is the possible decomposition of the $Fe_3(CO)_{12}$ to $Fe_2(CO)_9$ and especially $Fe(CO)_5$ during sublimation. The metal–metal bond strength is considerably less in $Fe_3(CO)_{12}$ than, for example, in $Ru_3(CO)_{12}$ and $Os_3(CO)_{12}$, and this is reflected also in the chemical reactivity of these clusters.[58–60] Only under quite mild conditions is the triiron structure retained by comparison to conditions in which the triruthenium and triosmium structures are retained.[61] Accordingly, Hipps and co-workers adsorbed $Fe_3(CO)_{12}$ onto the plasma grown Al_2O_3 surface of the barrier of their tunneling junctions from a benzene solution external to their vacuum system.[17] After adsorption from the liquid phase, the fabrication of the tunneling junction was completed via evaporation of the top Pb cross strips in vacuum. It would be difficult to postulate *a priori* the types of iron carbonyl species which would be expected on the alumina under these conditions. For example, the structure of the $Fe_3(CO)_{12}$ is different in the solid phase from that in solution. As a solid, the structure of $Fe_3(CO)_{12}$ is an isosceles triangle of

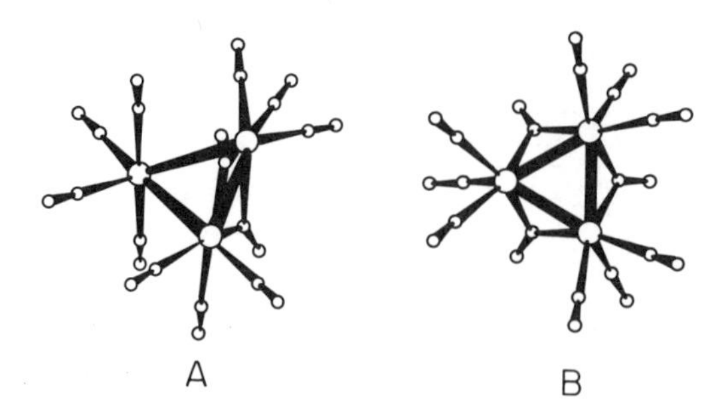

Figure 15. (A) The structure of solid $Fe_3(CO)_{12}$. (B) The structure of $Fe_3(CO)_{12}$ in solution.

iron atoms in which an $Fe(CO)_4$ group is symmetrically coordinated by only iron–iron bonds to an $Fe_2(CO)_8$ group containing two identical $Fe(CO)_3$ groups connected to one another by *two* bridging carbonyl groups and by an iron–iron bond [cf. Fig. 15(A)].[62] On the other hand, in solution, three identical $Fe(CO)_3$ groups are positioned at the vertices of an equilateral triangle and are linked in pairs to one another by a bridging carbonyl group and by an iron–iron bond, i.e., there are *three* symmetrical, bridging carbonyls [cf. Figure 15B].[62]

The tunneling spectrum obtained after the adsorption of $Fe_3(CO)_{12}$ from a benzene solution onto an alumina surface is shown in Figure 16.[17] The modes due to the benzene solvent are marked "B," while the modes due to the carbonyl complex(es) are marked by numerical frequencies (in wave numbers). The most important conclusion to be drawn from Figure 16 is that it is possible to examine supported homogeneous cluster compounds by tunneling spectroscopy after adsorption from a liquid solution. The conditions employed in this case, including transfer of the supported complex through the laboratory atmosphere, make a definitive structural determination of the surface species difficult due to probable decomposition and/or oxidation reactions of the supported catalyst.[63–65] However, one striking conclusion that can be drawn from the tunneling spectrum of Figure 16 is that there are an immeasurably small number of bridging carbonyl ligands on the surface species (especially if one recalls that the top Pb electrode has the effect of downshifting the CO stretching frequencies). The shift of bridging carbonyl ligands to terminal positions has been observed previously in the case of $Fe_2(CO)_9$.[66] In the presence of an $AlBr_2$ (Lewis acid) adduct, two of the three bridging CO ligands apparently shift to the terminal position, but the stretching frequency of the one remaining bridging CO ligand is downshifted strongly to 1557 cm^{-1} due to the probable formation of a $>$CO—$AlBr_3$ linkage.[66] In the case of $Fe_3(CO)_{12}$ adsorbed on alumina, however, this type of Lewis acid interaction does not occur due to the absence of any type of bridging carbonyl.

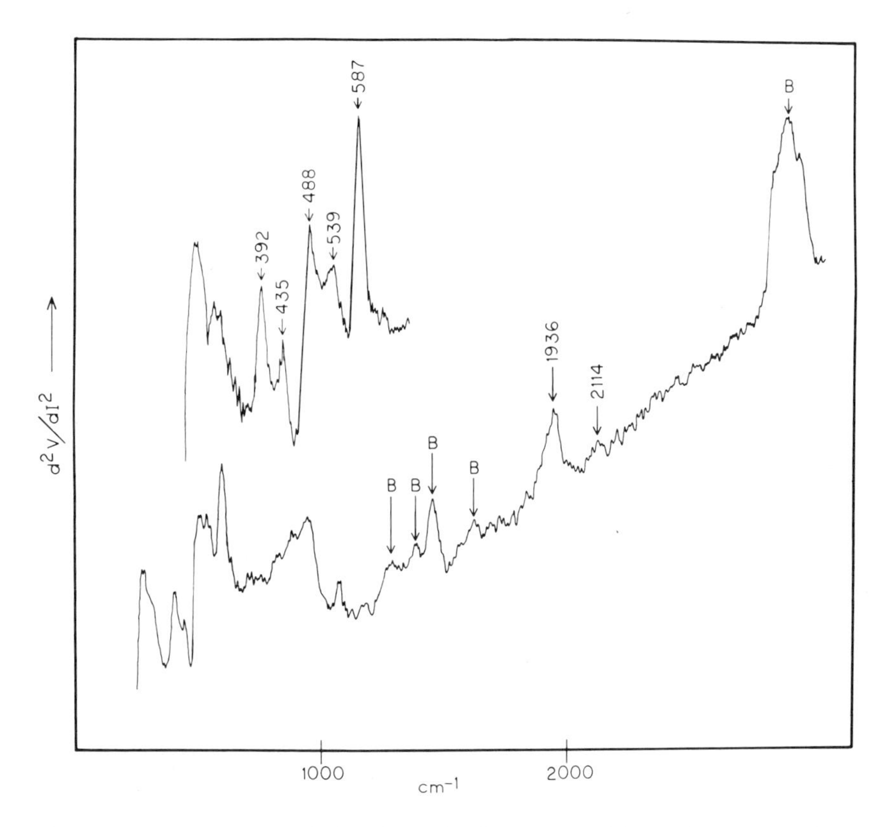

Figure 16. Tunneling spectra obtained from $Fe_3(CO)_{12}$ adsorbed from a benzene solution on alumina. Inset is a medium-resolution scan of the Fe–CO region (reference 17).

As may be seen in Table 5, there is a striking similarity between the tunneling spectra of $Fe_3(CO)_{12}$ and of $Ru_3(CO)_{12}$ adsorbed on alumina. However, it is almost certain that this apparent agreement is fortuitous, and that the complexes do not possess the same molecular structure on the alumina surface. Rather, it would appear much more likely that the $Fe_3(CO)_{12}$ decomposes on the surface to form predominantly $Fe(CO)_x$, where x may range from 1 to 4. The absence of any observed bridging carbonyls indicates also that the iron is very highly dispersed on the alumina surface, in agreement with recent, more traditional catalytic studies.[63] It would be of clear interest to repeat this tunneling spectroscopic investigation of supported iron carbonyl clusters under better characterized conditions, especially in the absence of an oxidizing atmosphere in all stages of the preparation of the tunneling junctions.

Table 5. Tunneling, ir, and Raman Bands (in Parentheses) for a Series of Metal Carbonyls in Units of cm^{-1}

Tunneling[a]		ir (Raman)			
$Fe_3(CO)_{12}$[b] (from benzene solution)	$Ru_3(CO)_{12}$[c] (from vapor phase sublimation)	$Fe_3(CO)_{12}$[d] Zeolite	$Fe(CO)_5$[e] liquid	$Fe_2(CO)_9$[f] solid	$Fe_3(CO)_{12}$[g] $AlBr_3$ adduct
2114(?)	2136(*m*)	2112(*w*)		(2109)	2124(*w*)
		2050			2081
1936	1919(*b*)	2030(*b*)	2022	2021	2070–2008
		1950	2000		1922
Solvent		1795		1828	
inter-		1760			1548
ference					
			754		
				677	
			637		
587	614	—	615	599	590
				562	568
539(*m*)	547(*m*)	—	554	524	540
488	488	—	475	(480)	500
435(*m*)	456	—	432	453	435
				(415)	410
392	389(*m*)	—	(380)	(390)	

[a] After subtracting 8.15 cm^{-1} to correct for the effect of the Pb superconducting gap (reference 44).
[b] Reference 17.
[c] Reference 15.
[d] Reference 67.
[e] Reference 68.
[f] Reference 69.
[g] References 17 and 66.

4. Conclusions

The following major conclusions may be drawn from the tunneling spectroscopic investigations that have been performed to date that are concerned with supported homogeneous cluster catalysts:

(1) A variety of surface species are formed when the polymerization catalyst $Zr(BH_4)_4$ is adsorbed on Al_2O_3. The BH_4 ligands which remain attached to zirconium atoms as well as those which migrate to the alumina support change their internal bonding geometry from a tridentate structure (three bridging hydrogen atoms) to a bidentate structure (two bridging hydrogen atoms). The existence of $O–Zr(BH_4)_x$, AlH_2BH_2, and $O–BH_2$ groups on the alumina surface was demonstrated for surface temperatures up to 475 K.

(2) Water vapor readily hydrolyzes the supported $Zr(BH_4)_4$ catalyst even at room temperature rendering it inactive. No isotopic exchange was observed to occur between the supported complex and either D_2 or D_2O in the range of temperatures investigated, 300–475 K.

(3) Both ethylene and propylene saturate upon adsorption on the supported $Zr(BH_4)_4$ forming surface species with similar types of $—CH_3$ and $—CH_2—$ groups.

(4) At surface temperatures on the order of only 400 K, acetylene is polymerized by the supported $Zr(BH_4)_4$ catalyst. The polyacetylene consists mainly of —CH— and $—CH_2$ groups while retaining some characteristics of unsaturated carbon–carbon bonds.

(5) Cyclohexene, 1,3-cyclohexadiene and benzene adsorb on the supported $Zr(BH_4)$ catalyst. It appears that benzene interacts rather weakly with the catalyst, cyclohexene forms a saturated hydrocarbon species on the surface, and 1,3-cyclohexadiene polymerizes at rather modest pressures and temperatures.

(6) The $Ru_3(CO)_{12}$ multinuclear carbonyl complex adsorbs molecularly on a hydroxylated alumina surface. The adsorption is reversible, and the concentration of $Ru_3(CO)_{12}$ on the alumina surface increases as the surface temperature is decreased from 300 to 180 K.

(7) The $[RhCl(CO)_2]_2$ adsorbs weakly on a hydroxylated alumina surface but can dissociate with only mild thermal activation. At low concentrations of the complex on the surface, the predominant surface species are $Rh(CO)_2$ and RhCO, whereas at higher surface concentrations bridging carbonyl ligands in Rh_2CO species are observed. In no case is chlorine retained by the alumina surface, but rather the desorption of HCl occurs where the hydrogen atom is derived from a hydroxyl group present on the alumina surface.

(8) The adsorption of $Fe_3(CO)_{12}$ on an alumina surface from a benzene solution was observed successfully with tunneling spectroscopy. The complex appears to adsorb dissociatively (with possible oxidation due to the experimental conditions employed). Since no bridging carbonyl ligands were observed, the predominant surface species appear to be $Fe(CO)_x$, where x may vary between 1 and 4.

To summarize, tunneling spectroscopy has already been shown to be a powerful spectroscopic probe of the molecular structure and catalytic reactivity of homogeneous cluster compounds supported on alumina surfaces. Now that its potential has been demonstrated explicitly, much additional work in this area can be expected to occur. The following research activities would seem to be particularly promising: (1) varying the degree of hydroxylation of the oxide surface, (2) varying the acid–base properties of the oxide surface via ligand attachment, (3) varying the oxide substrate (since only alumina has been examined to date), (4) adsorbing air-sensitive complexes from solution using a combination of established glove box and vacuum techniques, and (5) examining additional chemical reactions catalyzed by the supported complexes (since only polymerization reactions have been considered to date).

Acknowledgments

Our research in the development and utilization of tunneling spectroscopy has been generously supported by the National Science Foundation. The contributions of past and present co-workers, in particular, Bill Bowser, Howard Evans, and Lynn Forester, as well as useful conversations with Kerry Hipps are very much appreciated. An expanded review of this subject will appear in *Vibrational Spectra and Structure*, Volume II, Elsevier, New York (1982).

References

1. P. K. Hansma, Inelastic electron tunneling, *Phys. Rep.* **30C**, 145–206 (1977).
2. W. H. Weinberg, Inelastic electron tunneling spectroscopy: A probe of the vibrational structure of surface species, *Ann. Rev. Phys. Chem.* **29**, 115–139 (1978).
3. P. K. Hansma and J. Kirtley, Recent advances in inelastic electron tunneling spectroscopy, *Acc. Chem. Res.* **11**, 440–445 (1978).
4. *Inelastic Electron Tunneling Spectroscopy* (T. Wolfram, ed.), Springer-Verlag, New York (1978).
5. R. M. Kroeker and P. K. Hansma,Tunneling spectroscopy for the study of adsorption and reaction on model catalysts, *Catal. Rev.* **23**, 553–603 (1981).
6. W. M. Bowser and W. H. Weinberg, The nature of the oxide barrier in inelastic electron tunneling spectroscopy, *Surf. Sci.* **64**, 377–392 (1977).
7. H. E. Evans, W. M. Bowser, and W. H. Weinberg, An XPS investigation of alumina thin films utilized in inelastic electron tunneling spectroscopy, *Appl. Surf. Sci.* **5**, 258–274 (1980).
8. P. K. Hansma, D. A. Hickson, and J. A. Schwarz, Chemisorption and catalysis on oxidized aluminum metal, *J. Catal.* **48**, 237–242 (1977).
9. H. E. Evans and W. H. Weinberg, Inelastic electron tunneling spectroscopy of zirconium tetraborohydride supported on aluminum oxide, *J. Am. Chem. Soc.* **102**, 872–873 (1980).
10. H. E. Evans and W. H. Weinberg, Synthesis and catalytic reactions of $Zr(BH_4)_4$ on Al_2O_3 characterized by inelastic electron tunneling spectroscopy, *J. Vac. Sci. Technol.* **17**, 47–48 (1980).
11. H. E. Evans and W. H. Weinberg, A vibrational study of zirconium tetraborohydride supported on aluminum oxide. 1. Interactions with deuterium, deuterium oxide and water vapor, *J. Am. Chem. Soc.* **102**, 2548–2553 (1980).
12. H. E. Evans and W. H. Weinberg, A vibrational study of zirconium tetraborohydride supported on aluminum oxide. 2. Interactions with ethylene, propylene and acetylene, *J. Am. Chem. Soc.* **102**, 2554–2558 (1980).
13. L. Forester and W. H. Weinberg, A vibrational study of $Zr(BH_4)_4$ supported on alumina: Interactions with cyclohexene, 1,3-cyclohexadiene and benzene, *J. Vac. Sci. Technol.* **18**, 600–601 (1981).
14. W. H. Weinberg, W. M. Bowser, and H. E. Evans, Reduced metallic clusters and homogeneous cluster compounds "supported" on aluminum oxide as studied by inelastic electron tunneling spectroscopy, *Surf. Sci.* **106**, 489–497 (1981).
15. W. M. Bowser and W. H. Weinberg, An inelastic electron tunneling spectroscopic study of $Ru_3(CO)_{12}$ adsorbed on an aluminum oxide surface, *J. Am. Chem. Soc.* **102**, 4720–4724 (1980).

16. W. M. Bowser and W. H. Weinberg, An inelastic electron tunneling spectroscopic study of the interaction of $[RhCl(CO)_2]_2$ with an aluminum oxide surface, *J. Am. Chem. Soc.* **103**, 1453–1458 (1981).
17. S. D. Williams, M. K. Dickson, D. M. Roundhill, and K. W. Hipps, The vibrational spectrum of the surface species obtained by solution doping of alumina with $Fe_3(CO)_{12}$ as studied by inelastic electron tunneling spectroscopy, Washington State University (unpublished report).
18. V. A. Zakharov and Yu. I. Yermakov, Supported organometallic catalysts of olefin polymerization, *Catal. Rev.* **19**, 67–103 (1979).
19. W. M. Bowser and W. H. Weinberg, Sample heating and simultaneous temperature measurement in inelastic electron tunneling spectroscopy, *Rev. Sci. Instrum.* **47**, 583–586 (1976).
20. U. Mazur and K. W. Hipps, An inelastic electron tunneling spectroscopy study of the adsorption of NCS^-, OCN^- and CN^- from water solution on Al_2O_3, *J. Phys. Chem.* **83**, 2773–2777 (1979).
21. T. J. Marks, W. J. Kennelly, J. R. Kolb, and L. A. Shimp, Structure and dynamics in metal tetrahydroborates. II. Vibrational spectra and structures of some transition metal actinide tetrahydroborates, *Inorg. Chem.* **11**, 2540–2546 (1972).
22. B. D. James and M. G. H. Wallbridge, Metal tetrahydroborates, *Prog. Inorg. Chem.* **11**, 99–231 (1970).
23. T. J. Marks and J. R. Kolb, Covalent transition metal, lanthanide, and actinide tetrahydroborate complexes, *Chem. Rev.* **77**, 263–293 (1977).
24. T. Matsuda and H. Kawashima, An infrared study of hydroboration of lower olefins with diborane on γ-Al_2O_3, *J. Catal.* **49**, 141–149 (1977).
25. B. E. Smith, B. D. James, and J. A. Dilts, Bis(tetrahydroborate)bis (cyclopentadienyl)zirconium(IV): A synthetic, thermoanalytical and vibrational spectroscopic study, *J. Inorg. Nucl. Chem.* **38**, 1973–1978 (1976).
26. H. E. Flotow and D. W. Osborne, Heat capacities and thermodynamic functions of ZrH_2 and ZrD_2 from 5 to 350°K and the hydrogen vibrational frequency in ZrH_2, *J. Chem. Phys.* **34**, 1418–1425 (1961).
27. B. E. Smith, H. F. Shurwell, and B. D. James, Molecular vibrations of zirconium(IV) tetrahydroborate, a compound containing triple hydrogen bridges, *J. Chem. Soc. Dalton Trans.* **1978**, 710–722.
28. D. G. H. Ballard, Transition metal alkyl compounds as polymerization catalysts, *J. Polym. Sci.* **13**, 2191–2212 (1975).
29. N. L. Albert, W. E. Keiser, and H. A. Szymanski, *Theory and Practice of Infrared Spectroscopy*, Plenum Press, New York (1970).
30. R. T. Conley, *Infrared Spectroscopy*, Allyn and Bacon, Boston (1972).
31. S. Krimm, C. Y. Liang, and G. B. B. M. Sutherland, Infrared spectra of high polymers. II. Polyethylene, *J. Chem. Phys.* **25**, 549–562 (1956).
32. J. R. Nielsen and A. H. Woollett, Vibrational spectra of polyethylenes and related substances, *J. Chem. Phys.* **26**, 1391–1400 (1957).
33. D. O. Hummel, Infrared Spectra of Polymers, *Polymer Rev.* **14**, 8–97 (1966).
34. K. Abe and K. Yanagisawa, Infrared spectra of melted polypropylene films, *J. Polym. Sci.* **36**, 536–539 (1959).
35. J. P. Luongo, Infrared study of polypropylene, *J. Appl. Polym. Sci.* **3**, 302–309 (1960).
36. C. R. Ficker, Jr., M. Ozaki, A. J. Heeger, and A. G. MacDiarmid, Donor and acceptor states in lightly doped polyacetylene, $(CH)_x$, *Phys. Rev. B* **19**, 4140–4148 (1979).
37. L. Forester, unpublished results.
38. P. C. Wailes, H. Weingold, and A. P. Bell, Hydrido complexes of zirconium(IV): Reactions with olefins, *J. Organomet. Chem.* **43**, C32–C34 (1972).

39. M. R. Churchill, F. J. Hollander, and J. P. Hutchinson, An accurate redetermination of the structure of triruthenium dodecacarbonyl, $Ru_3(CO)_{12}$, *Inorg. Chem.* **16**, 2655–2659 (1977).
40. W. M. Bowser, PhD thesis, California Institute of Technology (1980).
41. J. R. Anderson, P. S. Elmes, R. F. Howe, and D. E. Mainwaring, Preparation of some supported metallic catalysts from metallic cluster carbonyls, *J. Catal.* **50**, 508–518 (1977).
42. J. Robertson and G. Webb, Catalysis by supported group VIII metal compounds. I. The interaction of *n*-butane with hydrogen over silica-supported ruthenium carbonyl catalysts, *Proc. R. Soc. London Ser. A* **341**, 383–398 (1974).
43. C. O. Quicksall and T. G. Spiro, Raman frequencies of metal cluster compounds: $Os_3(CO)_{12}$ and $Ru_3(CO)_{12}$, *Inorg. Chem.* **7**, 2365–2369 (1968).
44. J. R. Kirtley and P. K. Hansma, Vibrational mode shifts in inelastic electron tunneling spectroscopy: Effects due to superconductivity and surface interactions, *Phys. Rev. B* **13**, 2910–2917 (1976).
45. R. M. Kroeker, W. C. Kaska, and P. K. Hansma, How carbon monoxide bonds to alumina-supported rhodium particles: tunneling measurements with isotopes, *J. Catal.* **57**, 72–79 (1979).
46. J. Klein, A. Léger, S. DeCheveigné, C. Guinet, M. Belin, and D. Defourneau, An inelastic electron tunneling spectroscopy study of the adsorption of CO on Rh, *Surf. Sci.* **82** L288–L292 (1979).
47. H. E. Evans, W. M. Bowser, and W. H. Weinberg, The adsorption of ethanol on silver clusters supported on alumina, *Surf. Sci.* **85**, L497–L502 (1979).
48. C. W. Garland and J. R. Wilt, Infrared spectra and dipole moments of $Rh_2(CO)_4Cl_2$ and $Rh_2(CO)_4Br_2$, *J. Chem. Phys.* **36**, 1094–1095 (1962).
49. L. F. Dahl, C. Martell, and D. L. Wampler, Structure of and metal–metal bonding in $Rh(CO)_2Cl$, *J. Am. Chem. Soc.* **83**, 1761–1762 (1961).
50. A. C. Yang and C. W. Garland, Infrared studies of carbon monoxide chemisorbed on rhodium, *J. Phys. Chem.* **61**, 1504–1512 (1957).
51. H. Arai and H. Tominga, An infrared study of nitric oxide adsorbed on a rhodium-alumina catalyst, *J. Catal.* **43**, 131–142 (1976).
52. J. T. Yates, Jr., T. M. Duncan, S. D. Worley, and R. W. Vaughan, Infrared spectra of chemisorbed CO on Rh, *J. Chem. Phys.* **70**, 1219–1224 (1979).
53. D. J. C. Yates, L. L. Murrell, and E. B. Prestridge, Ultradispersed rhodium rafts: Their existence and topology, *J. Catal.* **57**, 41–63 (1979).
54. G. C. Smith, T. P. Chojnacki, S. R. Dasgupta, K. Iwatate, and K. L. Watters, Surface-supported metal cluster carbonyls. I. Decarbonylation and aggregation reactions of rhodium clusters on alumina, *Inorgan. Chem.* **14**, 1419–1421 (1975).
55. P. K. Hansma, W. C. Kaska, and R. M. Laine, Inelastic electron tunneling spectroscopy of carbon monoxide chemisorbed on alumina-supported transition metals, *J. Am. Chem. Soc.* **98**, 6064–6065 (1976).
56. G. Carturan and G. Strukul, Atomically dispersed palladium as a borderline case between heterogeneous and homogeneous hydrogenation of olefins, *J. Catal.* **57**, 516–521 (1979).
57. J. F. Hamilton and R. C. Baetzold, Catalysis by small metal clusters, *Science*, **205**, 1213–1220 (1979).
58. J. Lewis, A. R. Manning, J. R. Miller, and J. M. Wilson, Chemistry of polynuclear compounds. VII. The mass spectra of some polynuclear metal carbonyl complexes, *J. Chem. Soc. A* **1966**, 1663–1670.
59. R. B. King, Some novel features in the mass spectra of polynuclear metal carbonyl derivatives, *J. Am. Chem. Soc.* **88**, 2075–2077 (1966).
60. J. Lewis and B. F. G. Johnson, Mass spectra of some organometallic molecules, *Acc. Chem. Res.* **1**, 245–256 (1968).

61. M. I. Bruce and F. G. A. Stone, Dodecacarbonyltriruthenium, *Angew. Chem. Int. Ed. Engl.* **7**, 427–432 (1968).
62. C. H. Wei and L. F. Dahl, Triiron dodecacarbonyl: An analysis of its stereochemistry, *J. Am. Chem. Soc.* **91**, 1351–1361 (1969).
63. A. Brenner and D. A. Hucul, Catalysts of supported iron derived from molecular complexes containing one, two, and three iron atoms, *Inorg. Chem.* **18**, 2836–2840 (1979).
64. F. Hugues, A. K. Smith, Y. Ben Taarit, J. M. Basset, D. Commereuc, and Y. Chauvin, Surface-supported metal carbonyl clusters: Formation of $[HFe_3(CO)_{11}]^-$ by interaction of $Fe_3(CO)_{12}$ and $Fe(CO)_5$ with alumina and magnesia, *J. Chem. Soc. Chem. Commun.* **1980**, 68–70.
65. D. Commereuc, Y. Chauvin, F. Hugues, J. M. Basset, and D. Olivier, Catalytic synthesis of low molecular weight olefins from CO and H_2 with $Fe(CO)_5$, $Fe_3(CO)_{12}$ and $[HFe_3(CO)_{11}]^-$ supported on inorganic oxides, *J. Chem. Soc. Chem. Commun.* **1980**, 154–155.
66. J. S. Kristoff and D. F. Shriver, Adduct formation and carbonyl rearrangement of polynuclear carbonyls in the presence of group VIII halides, *Inorg. Chem.* **13**, 499–506 (1974).
67. D. Ballivet-Tkatchenko, G. Condurier, H. Mozzanega, and I. Tkatchenko, in *Fundamental Research in Homogeneous Catalysis* (M. Tsutsui, ed.), p. 257, Plenum Press, New York (1979).
68. M. Bigorgne, Étude spectroscopique Raman et infrarouge de $Fe(CO)_5$, $Fe(CO)_4L$, et trans-$Fe(CO)_3L_2$ ($L = PMe_3$, $AsMe_3$, $SbMe_3$). I. Attribution des bandes de $Fe(CO)_5$, *J. Organomet. Chem.* **24**, 211–229 (1970).
69. I. S. Butler, S. Kishner, and K. R. Plowman, Vibrational spectra of solid tri-μ-carbonyl-(hexacarbonyl)di-iron(O), $Fe_2(CO)_9$, *J. Mol. Struct.* **43**, 9–15 (1978).
70. N. Davies, M. G. H. Wallbridge, B. E. Smith, and B. D. James, Vibrational spectra of zirconium tetrahydroborate and related molecules, *J. Chem. Soc. Dalton Trans.* **1973**, 162–165.

13

Model Supported Metal Catalysts

R. M. Kroeker

1. Introduction

The modeling of supported metal catalysts to allow the identification of the chemical species present on the active surface is an exciting application of tunneling spectroscopy. Such catalysts are essential to a growing number of industrial processes, and while much is known about their general use little is known about the reaction intermediates formed on their catalytic surfaces. Since the chemistry of interest is in a surface monolayer in close proximity to an oxide (often aluminum oxide) modeling the essential features of these catalysts with tunneling junctions is a promising area of research.

The adsorption and reaction of carbon monoxide on supported metal particles is a model system that has been extensively studied for many years with many techniques.[1] Carbon monoxide is a small molecule, yet it bonds in many different ways to transition metals. Its reaction with hydrogen can produce such diverse products as carbon dioxide, methane, paraffin, and graphite. Thus much of the current work of modeling supported metal catalysts with tunneling junctions deals with the chemistry of carbon monoxide on the catalytic surface.

The study of supported metals with tunneling spectroscopy began with the observation of carbon monoxide chemisorbed on alumina supported rhodium.[2] Model catalysts have since been formed with at least eight metals (Fe, Co, Ni, Ru, Rh, Pd, Ag, and Pt) and two different supports (alumina and magnesia). It is expected that this list will continue to expand,

---

R. M. Kroeker • IBM San Jose Research Laboratory, 5600 Cottle Road, San Jose, California 95193.

and that tunneling will make a growing contribution to the understanding of the chemistry of supported catalysts.

2. Special Techniques

Supported metal catalysts cannot be directly examined via tunneling; modeling the catalysts is necessary to incorporate the two metal electrodes needed to form a tunneling junction. Figure 1 shows a schematic view of a model supported rhodium catalyst.[3] The aluminum film is deposited onto the substrate and oxidized in a standard manner. The oxide formed has been found to have the properties of α-alumina.[4] A small amount of metal is then evaporatively deposited onto the clean oxide surface. This evaporation is monitored with a quartz crystal microbalance; commercially available devices can measure an average thickness of less than 1 Å (0.1 nm). The metal forms small particles on the oxide surface. The size of the particles can be controlled to some extent by varying the substrate temperature and the evaporation rate. The key requirement is that the metal particles are formed cleanly without damage to the aluminum oxide.

Figure 2 shows a transmission electron microscope picture of a model supported rhodium catalyst surface.[5] The large grains are due to the polycrystalline nature of the aluminum electrode. The catalyst particles are the small dots that speckle the picture. The average thickness used in the figure is 4 Å of rhodium. In this case the layer agglomerated into particles

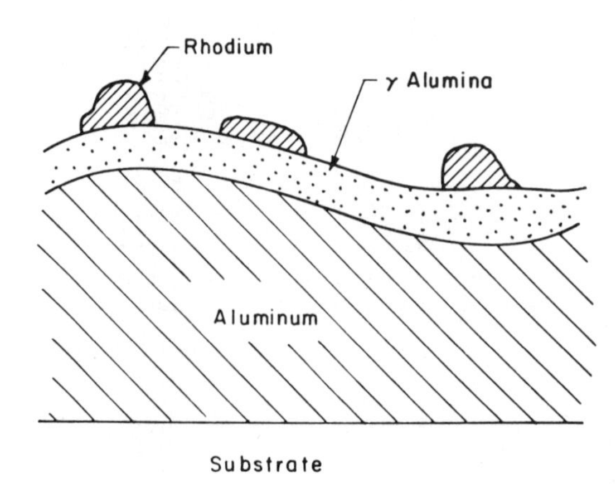

Figure 1. A schematic view of a model supported metal catalyst that can be studied with tunneling spectroscopy. The supported particles are formed from depositing metal vapor onto the oxide surface in high vacuum. The amount of metal deposited is generally measured with a quartz crystal microbalance and reported as an average coverage. Reproduced from reference 3.

Figure 2. Rhodium particles are shown in an electron micrograph of a specially prepared sample. The aluminum, evaporatively deposited and oxidized as in the preparation of a tunneling junction, is supported by a carbon-coated nickel grid. The relatively large crystallites are in the aluminum metal film. The oxide on the aluminum is too thin to be seen. The rhodium particles appear as small hemispheres with a typical diameter of two or three nanometers. Reproduced from reference 5.

with diameters on the order of 20 Å. The good dispersion and relative uniformity of the particles make this a good model system.

The supported transition metals used in this work are generally evaporated from pure material wound on tungsten filaments. Does tungsten contaminate the deposited particles? This question can be addressed by Auger spectroscopy. Figure 3 shows the results of Dubois *et al.*[3] showing that rhodium is deposited without resolvable tungsten contamination. Simi-

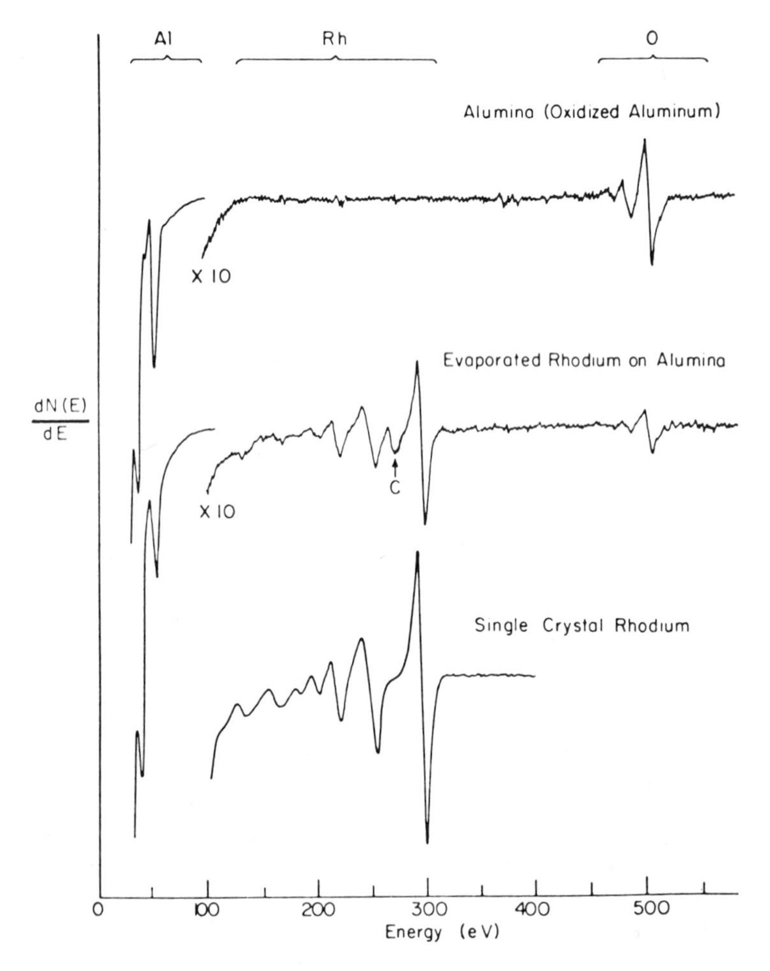

Figure 3. Auger electron spectra of alumina, alumina with rhodium particles (evaporated from tungsten) and single-crystal rhodium. No contamination of the rhodium particles by tungsten is observed for the evaporated rhodium. Reproduced from reference 3.

lar preliminary experiments with platinum and iron also show no tungsten contamination.

The chemisorption of gases on the particles is straightforward in principle. The particles are exposed to a known concentration of vapor for a measured time period. In practice it is often complicated by the effects of residual gases present in any vacuum system. Ultra-high vacuum techniques can greatly reduce this problem, but careful monitoring of the vacuum with a gas analyzer is necessary for good reproducibility of results over a series of experiments.

The final step to form the model catalyst is the deposition of the top metal electrode. This electrode is essential to form a tunneling junction; it is also the major source of problems in producing good models of real catalysts. In an ideal experiment the top electrode would be deposited only after the chemistry of interest has occurred and the surface cooled to 4.2 K. In practice this is extremely difficult. The effects of the top electrode on this work are easily observable; the broadening and downshifting of modes with large dipole derivatives,[5] the increased thermal stability of species covered,[6] and the diffusion barrier it presents to further adsorption are three examples.[7] Does this electrode prevent successful modeling of real catalysts? No, it does not. Comparison of the results obtained to date with tunneling spectroscopy with that of other techniques[3] shows that much can be learned with tunneling about supported metal particles. It is expected that tunneling will continue to provide unique information that is complementary to other techniques (see Chapter 6).

3. Experimental Results

3.1. Carbon Monoxide on Rhodium

3.1.1. Chemisorption of CO on Rhodium

Figure 4 shows the vibrations of CO chemisorbed on a model supported rhodium catalyst.[5] This spectrum results from a saturation exposure of CO on 20–30 Å particles formed at room temperature. There are three species of CO on the surface with overlapping modes. Identification of the species is difficult without first isolating them. Figure 5 shows spectra obtained by Klein *et al.*[8] in which the peak positions are shifted by isotopic substitution and by changing particle size. In general the peak positions are a function of many parameters: particle size, substrate temperature, CO exposure, background gases, and the top electrode metal are some of them.

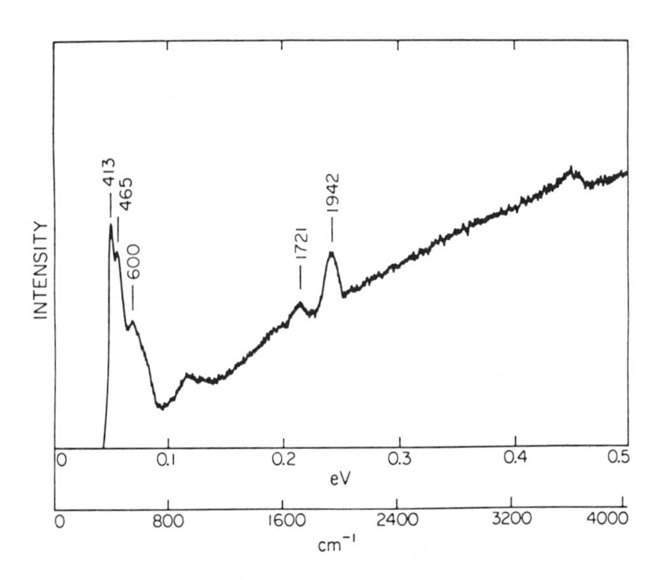

Figure 4. Differential spectrum of CO chemisorbed on alumina supported rhodium particles. Peaks labeled are those due to CO. The small peak at 116 meV is due to the alumina support and the peak near 445 meV is due the OH groups in the junction. Peak positions are not corrected for shifts due to the top lead electrode. Peak positions are found to vary with both rhodium coverage and CO exposure. Reproduced from reference 5.

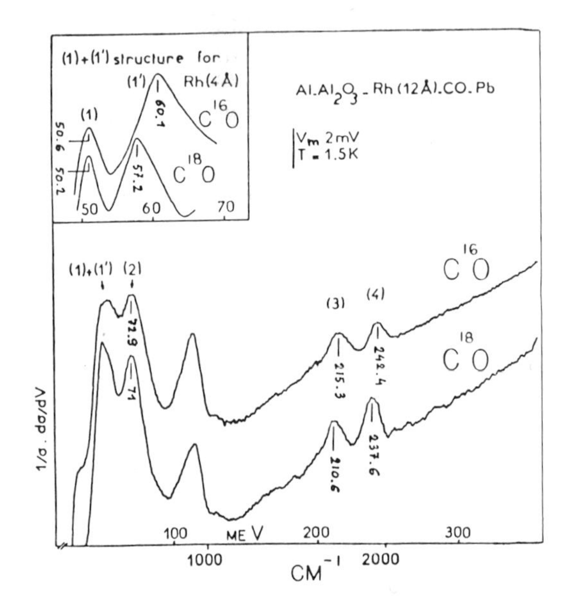

Figure 5. Tunneling spectra of two isotopes of CO chemisorbed on alumina supported rhodium (of average thickness 1.2 nm). Peaks (1) and (1′) are resolved only with a third or less of this rhodium coverage (see insert). The peak at 115 meV is due to the alumina support. Reproduced from reference 8.

Figure 6 shows CO chemisorbed on magnesium oxide supported rhodium particles obtained by de Cheveigné.[9] Again the vibrations of the CO are easily observed, but the peak position measurements are complicated by the background structure. To obtain reproducible peak position measurements a method of subtracting the background structure is invaluable.

Figure 7 shows six differential spectra in which the background structure has been subtracted.[10] Spectrum (a) shows a single species of CO chemisorbed on supported rhodium. This species is preferentially formed at high rhodium dispersions and saturation CO exposures. Measurements of the peak shifts with isotopes identify the peak near 52 meV as a bending mode, the peak near 70 meV as a stretching mode, and the peak near 243 meV as the CO stretching mode. Reference to the infrared literature identifies this linear species as a dicarbonyl, with two CO molecules adsorbed per rhodium atom.[5] Spectra (b–e) show the effects of reducing the particle dispersion and decreasing the CO exposure. The peaks of (a) are reduced as a new species is formed. Spectrum (f) results from heating the surface in vacuum after junction completion to 450 K. The CO chemisorbed on the rhodium particles is now almost entirely in a second linear form.

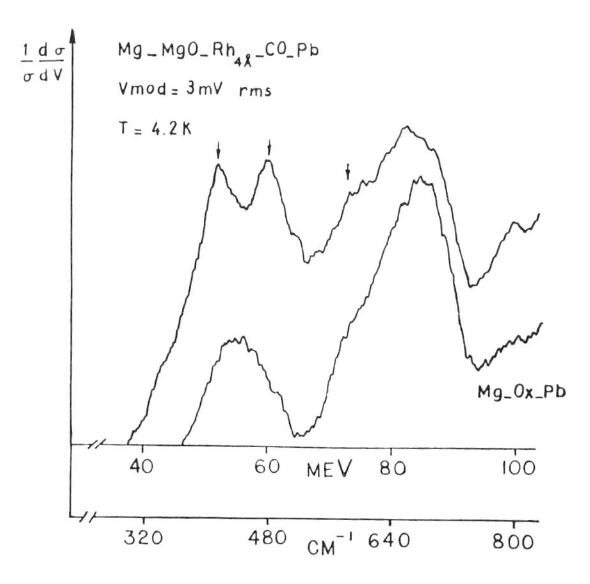

Figure 6. Tunneling spectra for CO chemisorbed on rhodium particles supported on magnesium oxide. The low-frequency modes of chemisorbed CO are again observed. The rhodium coverage is equivalent to a 0.4-nm uniform layer. The rhodium is not uniform, but forms small particles on the magnesium oxide surface. The lower spectrum shows the background structure. Reproduced from reference 9.

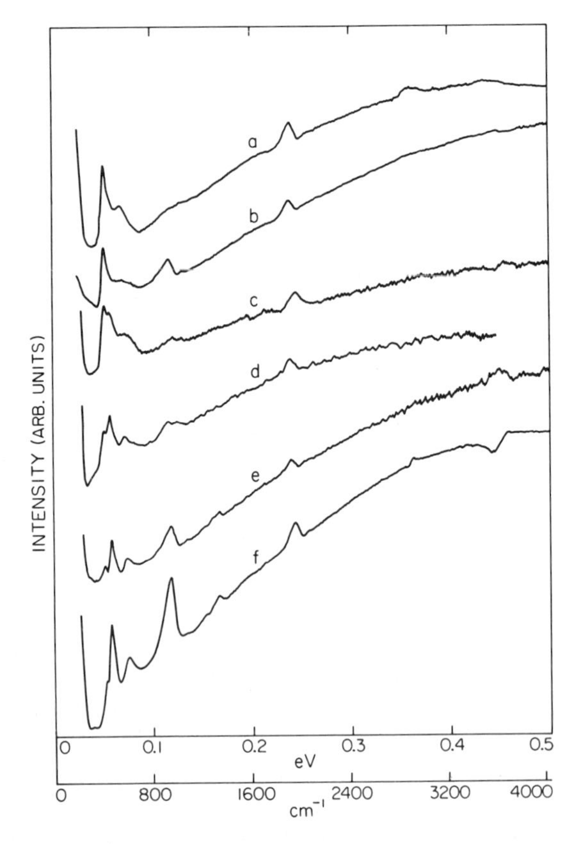

Figure 7. Differential spectra for CO chemisorbed on alumina supported rhodium particles. Spectrum (a) shows three modes due to linearly adsorbed CO; a bending vibration near 52 meV, a stretching vibration near 70 meV, and the CO stretching mode near 242 meV. This species has been identified as due to two CO molecules adsorbed per rhodium atom. Spectra (b, c, d, e) show the range of spectra obtained as the rhodium-to-CO ratio; substrate temperature and particle size are varied to produce a second linear species. Spectrum (f) has a bending mode near 58 meV and a low-energy stretching mode near 73 meV assigned to a linear RhCO species.

Measurements with isotopes identify the peak near 58 meV as a bending vibration and the peak near 73 meV as a stretching vibration. The positions of these modes are sensitive to the junction history. It is identified as a linear species with one CO molecule per rhodium atom. The variability of the mode frequencies of this species may reflect the coadsorption of different species. At any rate, the coadsorption of a second CO molecule is known to produce the shifts from 58 to 52 and 73 to 70 meV.

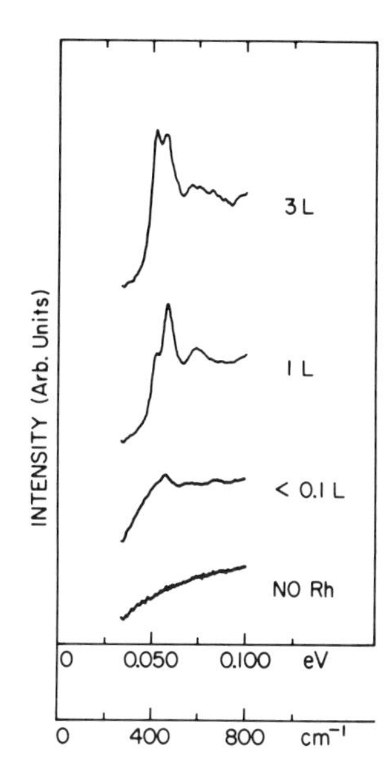

Figure 8. The low-energy region of the model alumina supported rhodium catalyst spectrum is shown for three different exposures to CO. The rhodium is in approximately 3-nm-diam particles. For CO exposures of one Langmiur (10^{-6} Torr sec $= 1.33 \times 10^{-4}$ Pa s or less, a peak near 58 meV due to RhCO dominates. For higher CO exposures a peak near 52 meV due to two CO molecules per rhodium atom rapidly becomes the most intense. Reproduced from reference 11.

Figure 8 shows the low-frequency region of CO on rhodium/alumina as a function of CO exposure.[11] For exposure to one Langmuir (10^{-6} Torr sec $= 1.33 \times 10^{-4}$ Pa sec) or less of CO the species formed is that of a single CO per rhodium atom. With increasing exposures the dominant species becomes the two-CO-per-rhodium-atom species. Note that the two stretching modes near 70 and 73 are not resolved when both are present. Resolution of the stretching modes also becomes difficult for low rhodium coverages and low dispersion.

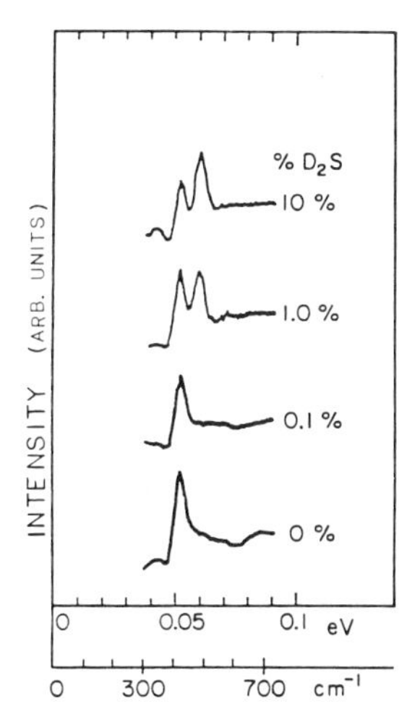

Figure 9. The low-energy region of the model alumina supported rhodium catalyst spectrum is shown for equal exposures to four mixtures of D_2S and CO. The rhodium is highly dispersed so that exposures to 100% CO forms the dicarbonyl species with a bending mode near 52 meV. With increasing D_2S exposure a second peak near 60 meV also forms due to the adsorption of only one CO molecule per rhodium atom. Reproduced from reference 11.

Figure 9 shows the coadsorption of CO and D_2S on rhodium–alumina as a function of the D_2S concentration.[11] At 0% D_2S the species formed is the dicarbonyl. With increasing D_2S the formation of the dicarbonyl is inhibited as the single CO per rhodium atom appears. The frequency of the bending vibration is also upshifted nearly 2 meV to 60 meV from its position observed without sulfur present. This suggests that the mechanism of sulfur poisoning on supported rhodium catalysts is to convert the more reactive dicarbonyl species into the less reactive linear species by blocking the adsorption of the second CO molecule. The expected rhodium–sulfur mode was not resolved in this work; this suggests it will be a weak and/or broad mode. The rhodium–carbon stretching mode (also broad and weak) is not resolved in this work with low rhodium coverage and low dispersion.

3.1.2. Hydrogenation of CO on Rhodium

One important reason for studying carbon monoxide adsorbed on transition metals is to understand the catalytic hydrogenation of carbon monoxide to form hydrocarbons, including synthetic fuels. Figure 10 shows tunneling spectra for CO on alumina supported rhodium before and after heating in the presence of hydrogen.[12] Before heating, the rhodium surface held three species of chemisorbed CO as shown in the upper spectrum. After heating to 420 K only one species remains, the linear Rh(CO). The other species have desorbed and/or reacted to form a hydrocarbon as seen by the growth of a CH stretching mode near 362 meV. Figure 11 shows the CH deformation region for this hydrocarbon with three isotopes of CO.[12] The spectra are the same for $^{12}C^{16}O$ and $^{12}C^{18}O$. This indicates that the hydrocarbon that is forming does not include the oxygen from the carbon monoxide in its structure. In contrast the upper spectrum which started with $^{13}C^{16}O$ shows definite shifts, showing that the hydrocarbon originated from the carbon monoxide; the hydrocarbon was not an impurity.

The identification of this hydrocarbon was challenging. There were no vibrational spectra for model hydrocarbons on rhodium. The best set of model compounds we could find were halogenated hydrocarbons, hydrocarbons bound to chlorine and bromine atoms. At first this may seem a poor choice for model compounds, but it turns out to be adequate. The stretching force constant in a carbon–rhodium bond and a carbon–halogen bond turn out to be nearly the same. Of course the mass must be corrected for, but this can be done by interpolating between results for iodine and bromine since rhodium is intermediate in mass between that of iodine and bromine.

Figure 12 shows three possibilities for what the unknown could be. The position of the unknown peaks are numbered (1–6) if observed directly and lettered (A, B) if obscured by other modes present in Figure 10. The position

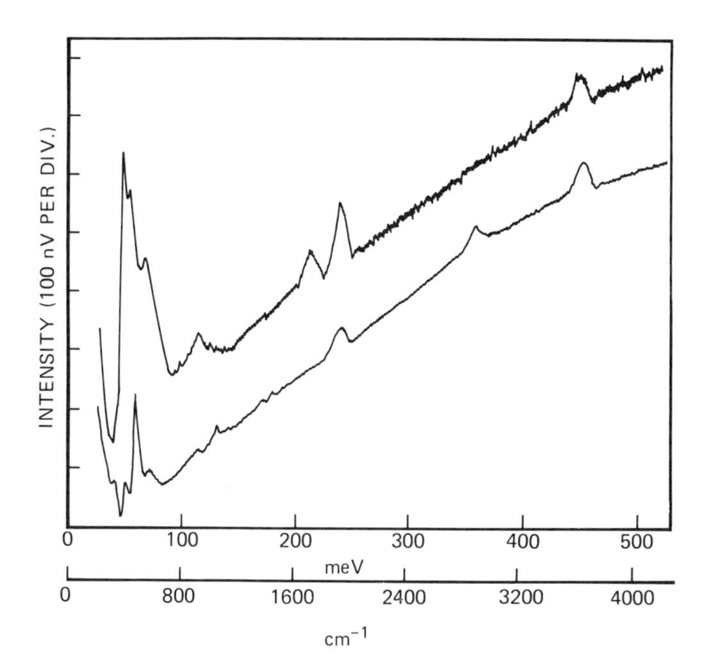

Figure 10. Differential spectra of CO chemisorbed on alumina supported rhodium particles before (upper spectrum) and after (lower spectrum) heating to 420 K in hydrogen. One of the three species of chemisorbed CO remains after heating and can be seen near 58, 73, and 242 meV. The other two species react and/or desorb while producing hydrocarbons on the rhodium particles as seen by the growth of a CH stretching mode near 362 meV. Reproduced from reference 12.

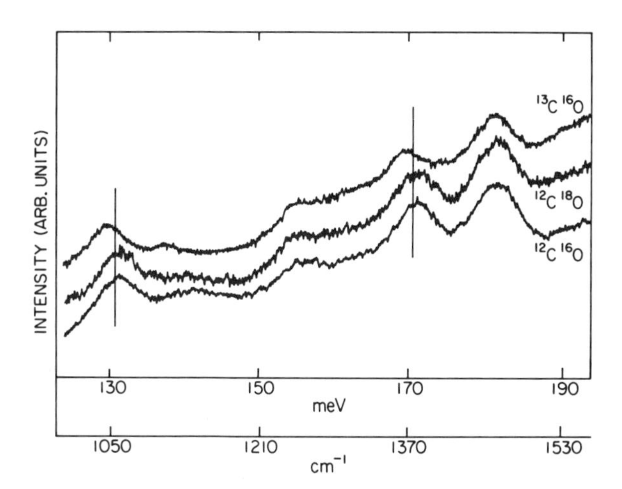

Figure 11. Differential spectra resulting from the use of three isotopes of CO to form hydrocarbons on the rhodium particles by heating in hydrogen. In the region from 125 to 190 meV are the C–H deformation modes. No modes are seen to shift with heavy oxygen, but several modes are seen to shift with heavy carbon. This suggests that the hydrocarbon is formed from the carbon but not the oxygen of the chemisorbed CO. Reproduced from reference 12.

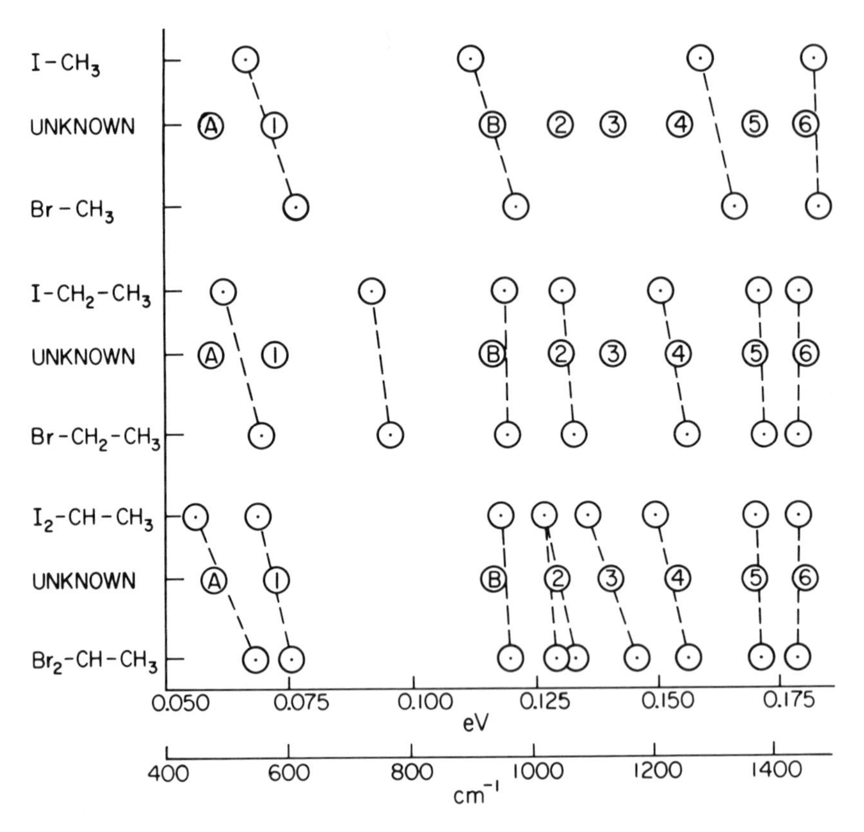

Figure 12. Comparisons are shown between mode positions of the hydrocarbon formed in the tunneling junctions and mode positions of known compounds. The vertical axis is scaled by the inverse square root of the mass of bromine, rhodium, and iodine. It is also displaced for each set of compounds. Modes labeled 1 to 6 of the hydrocarbon are observed directly. The measurement of the positions of modes labeled A and B is difficult due to overlap with other modes present. From the agreement with the two ethylidene species (and other tests) the hydrocarbon is identified as a μ-ethylidene species. Reproduced from reference 12.

of the halogenated compound's peaks are shown by circles with dots in them.[12] The position of the unknown between the iodine and the bromine is scaled according to the square root of the mass of the rhodium. Note that the agreement between the unknown peaks and the peaks for a methyl group is poor. The unknown has peaks where the methyl group does not have peaks. The addition of a CH_2 group increases the number of modes to rough agreement but the positions are still poor; in particular, the unknown has peaks where the CH_2CH_3 does not have peaks and CH_2CH_3 has peaks where the unknown does not have peaks. The ethylidene species $CHCH_3$

agrees both in number and position remarkably well with the unknown. This agreement, coupled with the isotope shift measurements, and other attempted fits to the data, leads to the conclusion that the CO was hydrogenated to form a stable ethylidene species.[12] This is the first observation, to my knowledge, of the formation of this species from the hydrogenation of CO on any supported metal catalyst.

3.2. Carbon Monoxide on Iron

Figure 13 shows the differential tunneling spectra of CO chemisorbed on highly dispersed iron particles on alumina.[13] Figure 14 shows the low-energy region with three isotopes of CO. As mentioned previously with rhodium, the isotope shifts of the modes allow identification as to type.[13] The two lowest modes shift more with heavy carbon than with heavy oxygen and are bending modes. The mode near 70 meV shifts more with heavy oxygen than with heavy carbon and is a stretching mode. Linearly adsorbed CO has two low-frequency bending modes that are expected to be degenerate if there is axial symmetry. The broken degeneracy of the two bending modes indicates that the CO is adsorbed at a site of twofold symmetry in a predominately linear fashion.[13] One way that this can be accomplished is with a CO molecule bound strongly to one iron atom and weakly to a second atom. More work is needed to determine the nature of this degeneracy-lifting bond. Possibilities include interactions between either the carbon or oxygen and the iron or lead surfaces.

The identification of the lower-frequency modes as bending modes is in marked contrast to the situation for iron pentacarbonyl in which most of

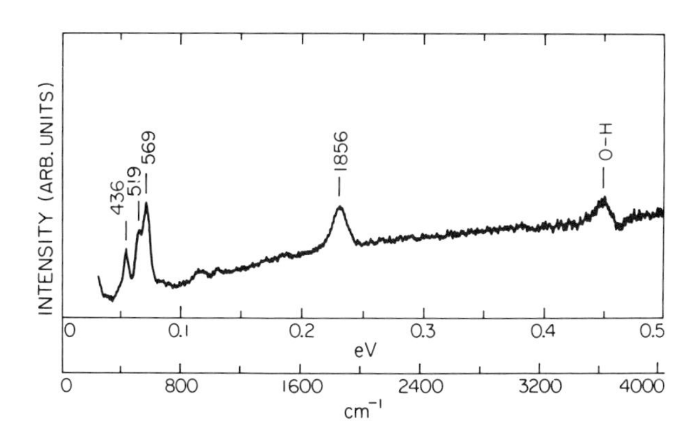

Figure 13. Differential spectrum of CO chemisorbed on alumina supported iron. Peaks labeled by wave number are the modes studied with isotopes. Peak positions have not been corrected for shifts due to the top electrode. Reproduced from reference 13.

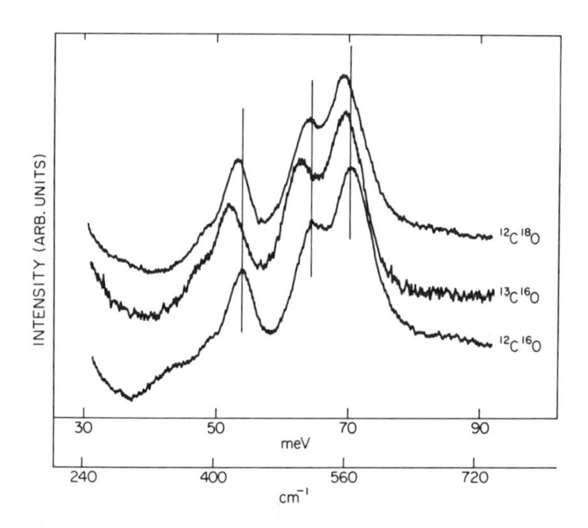

Figure 14. Differential spectra of the low-frequency region for three isotopes of CO chemisorbed on alumina supported iron. Bending modes have a larger shift for heavy carbon, while stretching modes have a larger shift for heavy oxygen. Reproduced from reference 13.

the bending modes are at higher energies than the stretching vibrations. This apparent contradiction can be resolved in the work on substituted iron carbonyls; as the number of CO molecules per iron atom decreases, the stretching modes increase in energy while the bending modes decrease in energy. In fact, extrapolation of the trends puts the modes in the range that are observed with tunneling. This illustrates the problems in using carbonyl data to directly assign the low-energy modes of species on supported metal catalysts.

Figure 15 shows the chemisorption of CO on alumina supported iron with six different top metal electrodes. The spectra were taken by Bayman *et al.*[14] The peak positions of the modes are seen to be dependent on the metal used. The relative intensities of the bending and stretching modes in the low-energy region reverse for silver and gold electrodes compared to lead or thallium. Indium and tin top electrodes appear to give a new species with only two low-frequency modes. One of these, near 60 meV, is halfway between the two bending modes seen in Figures 13 and 14. This mode is predicted to be a bending mode; this can be checked with isotopes. The remaining mode near 68 meV would be the stretching mode, and the overall peak shapes are similar to those of linear CO on rhodium. Clearly the top metal electrode can have strong effects on the spectra obtained, particularly for small molecules with large dipole derivatives such as CO. Further work with different top metals is clearly desirable. Until the tunneling environ-

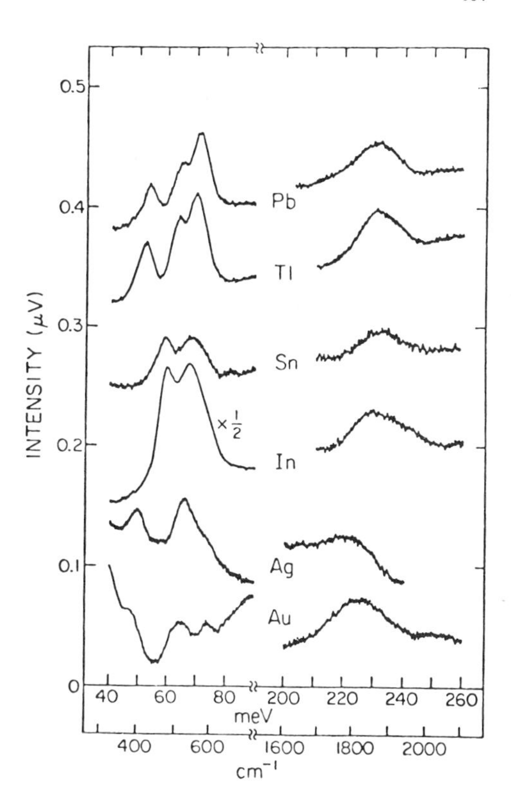

Figure 15. The top electrode has been found to have a large effect on the modes of CO chemisorbed on alumina supported iron. It is found that the frequencies observed for small molecules with large dipole derivatives such as CO depend on the nature of the top electrode. Here it is shown that the CO stretching mode shifts most with silver and gold top electrodes. Silver and gold also reverse the intensities of the low-frequency bending and stretching modes. Junctions with indium and tin top electrodes produce a different species with modes near 60 and 68 meV. Reproduced from reference 14.

ment is well characterized the direct comparison of tunneling frequencies (such as the relatively high metal–carbon stretching frequencies) to those found with other spectroscopies should be done with caution.

Figure 16 shows the attempt to hydrogenate the chemisorbed CO as in the rhodium experiments. Some hydrocarbon is seen to form but in very small amounts. Sintering of the supported iron particles causes a rapid rise in the background structure. The need for highly dispersed iron to avoid this large background structure prevents the hydrogenation of enough CO for identification at present.

3.3. Carbon Monoxide and Hydrogen on Nickel

Figure 17 shows the chemisorption of CO on alumina supported nickel particles.[6] Spectrum (a) shows a single species that is the most thermally stable, desorbing near room temperature. Spectra (b–d) were taken from junctions formed at successively lower temperatures, and each contains a

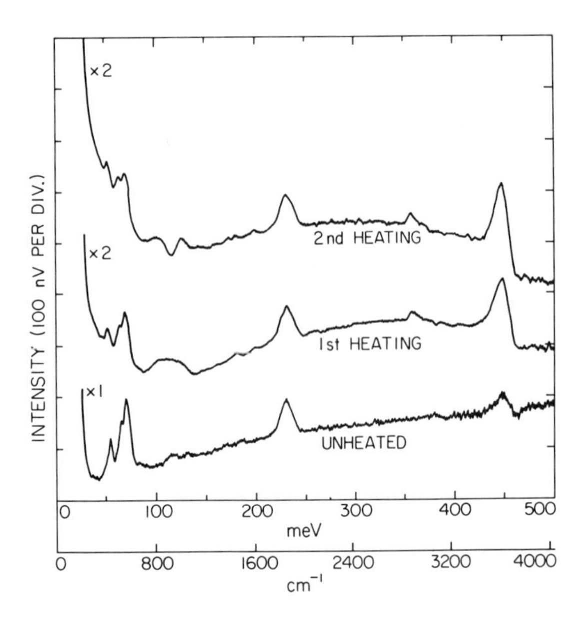

Figure 16. Differential spectra of CO chemisorbed on alumina supported iron that has been heated in hydrogen and CO. Very little hydrocarbon is seen to form in the junction; thus the measured intensities are too small to allow assignment at present. Sintering of the particles causes a rapid rise in the background structure at low frequencies.

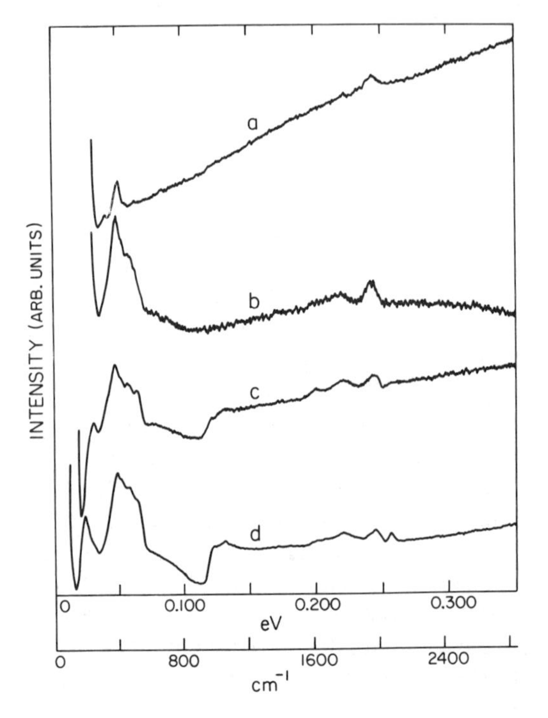

Figure 17. Differential spectra of CO chemisorbed on alumina supported nickel particles. Spectra (a) contains a single species that is the most stable. Spectra (b), (c), and (d) each contain an additional species of adsorbed CO. The relative ratios of species formed depend on the substrate temperature, particle size, and amount of CO present. Reproduced from reference 6.

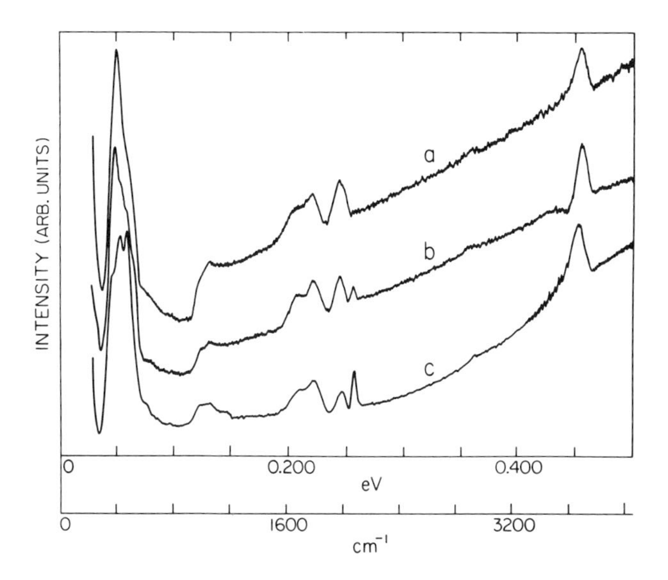

Figure 18. Differential spectra of CO on alumina supported nickel that show the desorption of a linearly adsorbed species with increasing substrate temperature. This species has modes at 45.5, 59.5, and 256.5 meV. The relatively small width of the CO stretching mode of this species suggests that it has a small dipole derivative and is relatively unshifted by the top lead electrode. Peak positions have not been corrected for top electrode shifts. Reproduced from reference 6.

new species of chemisorbed CO. Figure 18 shows the desorption of the least stable species. It has modes at 45.5, 59.5, and 256.5 meV. The CO stretching mode of this species is relatively narrow when compared to the other three species. This implies that it has a smaller dipole derivative and is less perturbed by the top electrode. It is predicted to be a linear species, with the mode at 45.5 meV a bending vibration and the mode at 59.5 meV a stretching vibration. This can be tested by experiments with isotopes. Figures 19 and 20 show the desorption of CO on nickel that occurs when a completed junction is heated in vacuum. The surface retains some CO at 350 K; junctions formed at 350 K contain no CO. This implies that the presence of the top electrode lends thermal stability the adsorbed CO.

Figure 21 shows the coadsorption of H_2 and CO on alumina supported nickel particles.[6] The presence of molecular hydrogen is seen directly near 510 meV. While tunneling spectroscopy is favored with good sensitivity for hydrogen, dissociated hydrogen as Ni–H has not been resolved. Hydrogen on the surface is seen to promote the formation of OH groups. This transfer of hydrogen gas to the surface as OH groups is thought to be a mechanism

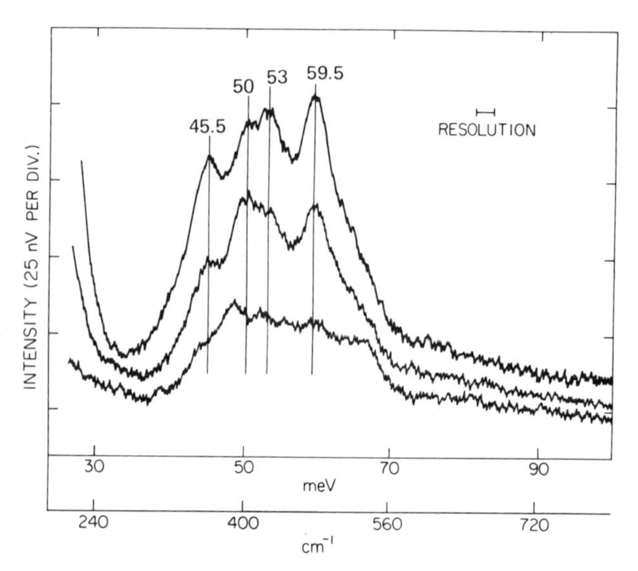

Figure 19. Differential spectra showing the desorption of all four species of CO on alumina supported nickel in the low-frequency region. Many overlapping modes are observed from 43 to 67 meV. It is found that the presence of the top electrode gives some thermal stability to the adsorbed CO. Peak positions have not been corrected for shifts due to the top lead electrode. Reproduced from reference 6.

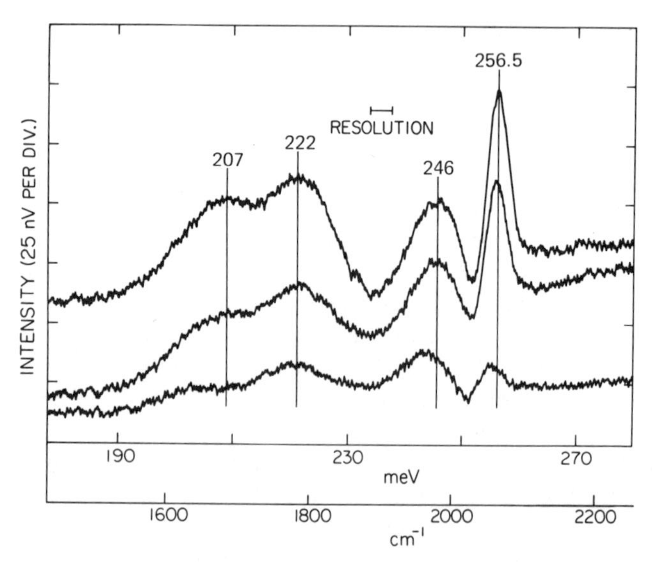

Figure 20. Differential spectra of CO chemisorbed on alumina supported nickel. The spectra show the decomposition of the CO as a completed junction is heated in vacuum. The frequencies of the CO stretching modes are seen to downshift with the desorption of CO. Indicated resolution is 3.9 meV. Peak positions have not been corrected for shifts due to the top lead electrode. Reproduced from reference 6.

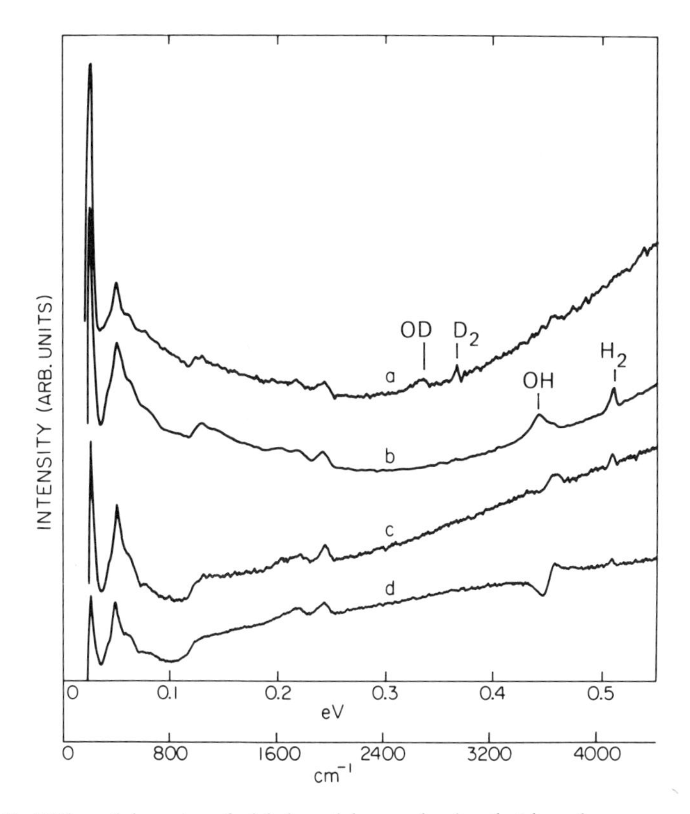

Figure 21. Differential spectra of nickel particles on alumina that have been exposed to both hydrogen and CO. Hydrogen is seen to coadsorb with the CO on the particles. The presence of hydrogen also is seen to cause the formation of OH groups. This transfer of hydrogen to the surface has been proposed as a mechanism for the "hydrogen spillover" seen on supported metal catalysts. Reproduced from reference 6.

of "hydrogen spillover," a well-documented phenomenom of supported metal catalysts.[6]

The attempt to hydrogenate the chemisorbed CO is shown in Figure 22. Very little surface hydrocarbon is seen to form. This lack of surface hydrocarbon may be an indication of nickel's tendency to produce primarily methane. Larger hydrocarbons such as the ethylidene found on rhodium probably do not form on nickel. The lack of observed hydrocarbons does not rule out the formation of methyl or methylene groups as these species are not expected to be thermodynamically stable under the reaction conditions used.[6]

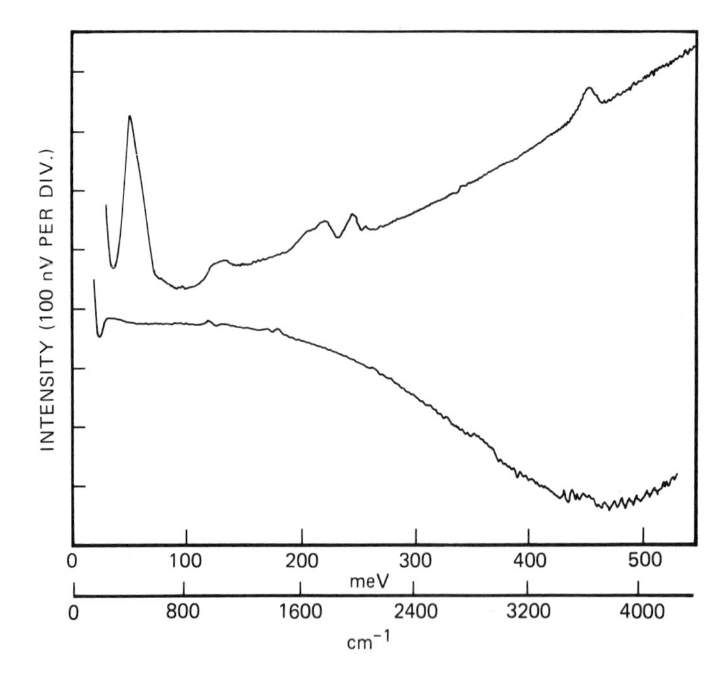

Figure 22. Differential spectra of CO on alumina supported nickel shown before (upper spectrum) and after (lower spectrum) heating in hydrogen. Very little surface hydrocarbon is seen to form. This result may reflect the selectivity of nickel catalysts for methanation. See reference 6.

3.4. Carbon Monoxide on Cobalt

The spectrum of CO on cobalt–alumina is shown in Figure 23. The cobalt is highly dispersed on the alumina and given a saturation exposure to CO. Under these conditions rhodium, iron, and nickel appear to give a single linear species. It is to be expected that cobalt, which is adjacent to these three metals in the Periodic Table, will do the same. Figure 24 checks this assumption with isotopes. Indeed, the lowest mode shifts as a linear bending mode and the higher mode shifts as a linear stretching mode. This "family" of linear species is listed in Table 1. A definite regularity can be seen in the frequency shifts between metals. It is tempting to predict the frequencies of linear CO on ruthenium.

Figure 25 shows the effect of increasing the amount of cobalt on the alumina surface. The frequencies begin to shift and new peaks rise as new species of chemisorbed CO appear. It is expected that the CO that has the lower CO stretching frequency is multiply bonded to the metal surface. This type of species, also seen on rhodium and nickel, has not yet been isolated

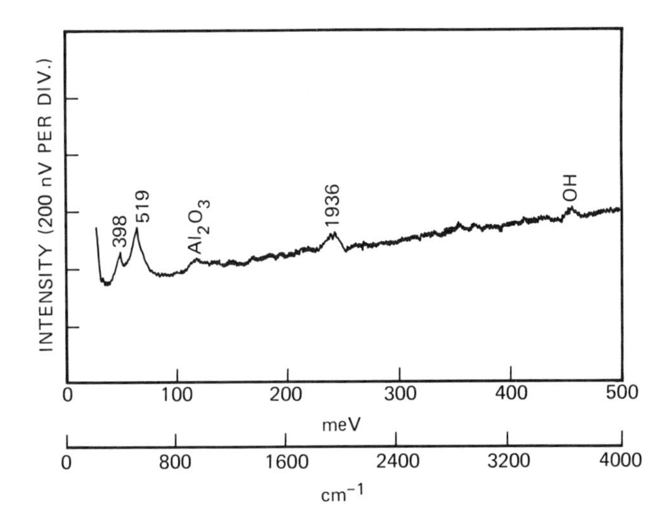

Figure 23. Differential spectrum of CO chemisorbed on alumina supported cobalt. The cobalt is highly dispersed and has been exposed to a saturation coverage of CO. The modes labeled were studied with isotopes. Peak positions have not been corrected for shifts due to the top lead electrode.

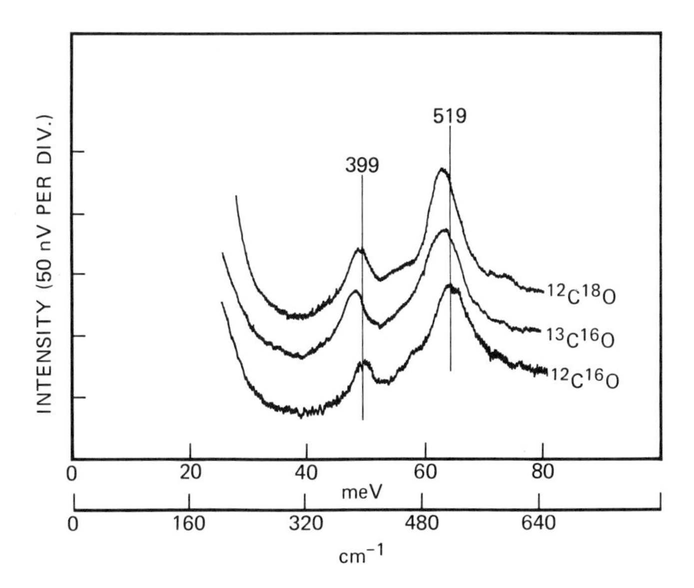

Figure 24. Differential spectra of CO on alumina supported cobalt showing the low-frequency region with three isotopes of CO. For highly dispersed cobalt two modes are seen; one shifts more with heavy carbon than with heavy oxygen, the other has approximately equal shifts for both heavy isotopes. The species is identified as linearly adsorbed CO, with a bending mode at 49.5 meV and a stretching mode at 64.3 meV. Peak positions have not been corrected for shifts due to the top lead electrode.

Table 1. Frequencies of Linear M–C–O (in meV)[a]

Metal	Mode type: MCO bending	MC stretching	CO stretching
Iron	60	68	230
Cobalt	49.4	64.3	238
Nickel	45.5	59.5	256.5
Ruthenium	(?)	(?)	(?)
Rhodium (I)	58	73	240
(II)	52	70	242

[a] Frequencies have not been corrected for top electrode shifts. Positions shown are averages; exact positions can and do vary.

for identification as to bonding type. Tunneling measurements have the potential, as yet unrealized, to unambiguously determine the nature of the bonding of this type of CO.

When CO on cobalt–alumina is heated hydrocarbons are readily formed. Figure 26 shows a representative junction that has been heated in hydrogen. The hydrocarbon formed is different from that formed on rhodium. It has a vibrational mode near 200 meV that can be expected to contain oxygen. Future experiments can be expected to identify this new species.

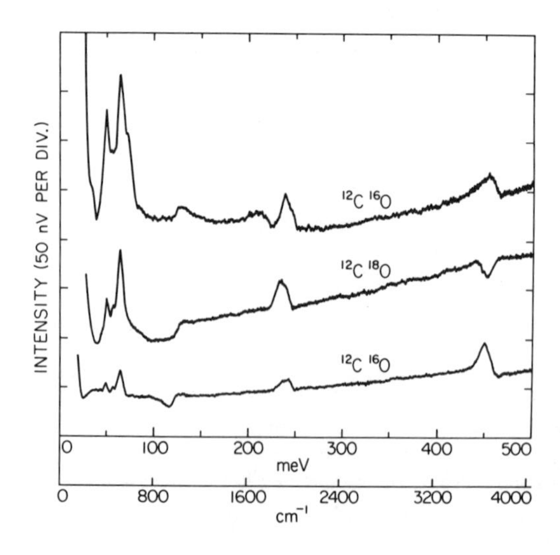

Figure 25. Differential spectra of CO chemisorbed on alumina supported cobalt particles. The spectra are for saturation exposures to CO (middle trace uses $C^{18}O$). The spectra show the effect of increasing the amount of cobalt evaporated onto the alumina. Mode positions are seen to shift and new species are formed. It is interesting (but hardly surprising) that the species formed on cobalt seem representative of the metals near it in the Periodic Table of the elements.

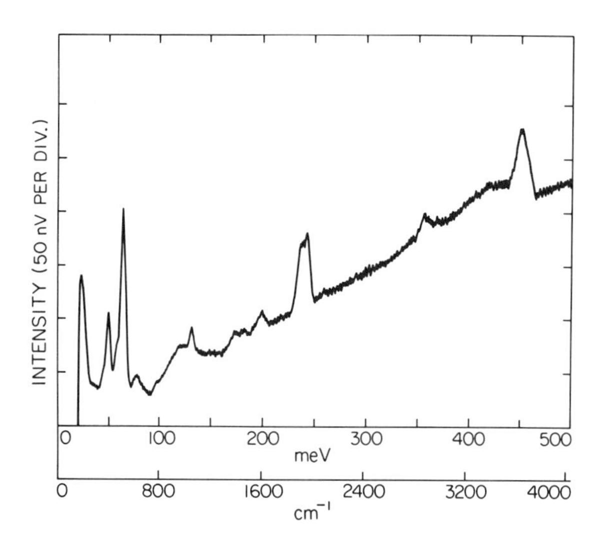

Figure 26. Differential spectrum of CO on alumina supported cobalt that has been heated in hydrogen and CO to 435 K. Hydrocarbons are seen to form in the junctions in small amounts. From the modes visible it is clear that this is a different species than the ethylidene formed in junctions that contain rhodium particles.

3.5. Ethanol on Silver

Figure 27 shows the CH stretching region for the chemisorption of ethanol on alumina supported silver particles.[15] This adsorption is representative of the first part of the oxidation of alcohols to aldehydes on supported silver catalysts. In this work by Evans *et al.*[15] the average particle diameter was determined with transmission electron microscopy, and the adsorption of ethanol was studied as a function of silver coverage. It was expected that ethanol would bond to the silver in much the same way it does to alumina by forming an ethoxide. Indeed, this was found to be the way it adsorbed. The frequencies for the ethoxide on silver were found to be slightly upshifted when compared to those on alumina.

4. Future Areas of Study

4.1. Acetylene on Palladium

Encouraging results have been found by Hansma and co-workers for the adsorption of small hydrocarbons on palladium and platinum. Figure 28

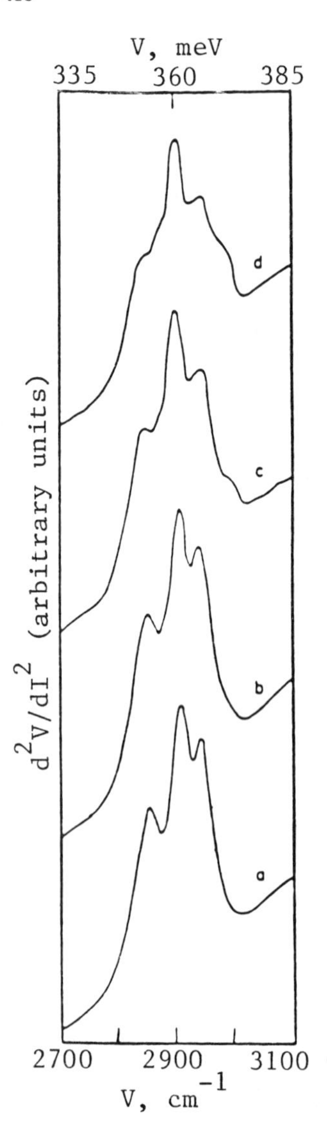

Figure 27. Spectra of ethanol chemisorbed on alumina supported silver for statistical Ag coverages of (a) 0, (b) 2.5, (c) 5.0, (d) 10.0 Å. The ethanol adsorbed as an ethoxide with mode positions slightly higher in frequency than that for ethanol on alumina alone. Reproduced from reference 15.

shows the chemisorption of acetylene on alumina supported palladium. The lower trace shows the species that results from adsorption at room temperature. The upper trace shows the results of heating the junction in hydrogen. The surface hydrocarbon is seen to react and form a new species that has a lower CH stretching frequency. Such reactions are of immense commercial importance. The ability of the noble metals to alter the structure of hydrocarbons by assisting the addition and removal of C–H and C–C

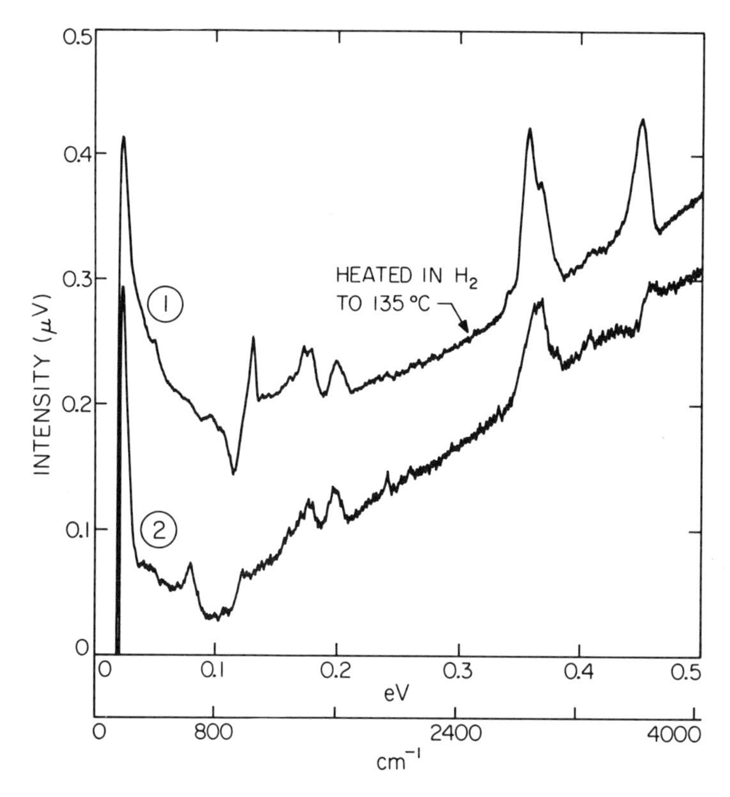

Figure 28. Differential spectra of acetylene chemisorbed on alumina supported palladium particles. Spectra are shown before and after heating in hydrogen. It is seen that the surface hydrocarbon is hydrogenated to form a second species after heating. Spectra were taken by Paul Hansma.

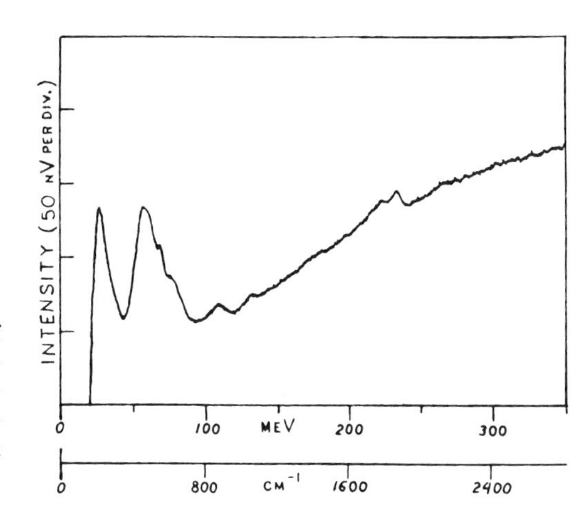

Figure 29. Differential spectrum of CO chemisorbed on alumina supported ruthenium particles. Preliminary work with ruthenium suggests that some CO dissociates on the surface at room temperature.

bonds is well exploited by industry. The different types of reaction intermediates, however, still remain to be observed experimentally. It can be expected that this and future work will be able to see and identify these important surface hydrocarbons.

4.2. Carbon Monoxide on Ruthenium

Have you guessed the frequencies of linear CO on ruthenium (via Table 1)? If you have, try to pick them out of Figure 29. This preliminary differential spectrum of CO chemisorbed on alumina supported ruthenium suggests that the correct answer will be forthcoming.

4.3. Carbon Monoxide on Platinum

Figure 30 is a preliminary spectrum of CO on alumina supported platinum. The spectrum is interesting in that this may be the first isolated bridging species with an approximately 80-meV "asymmetric stretch" mode of the M–C–M bond structure. The large structure near 117 meV is most likely the alumina support.

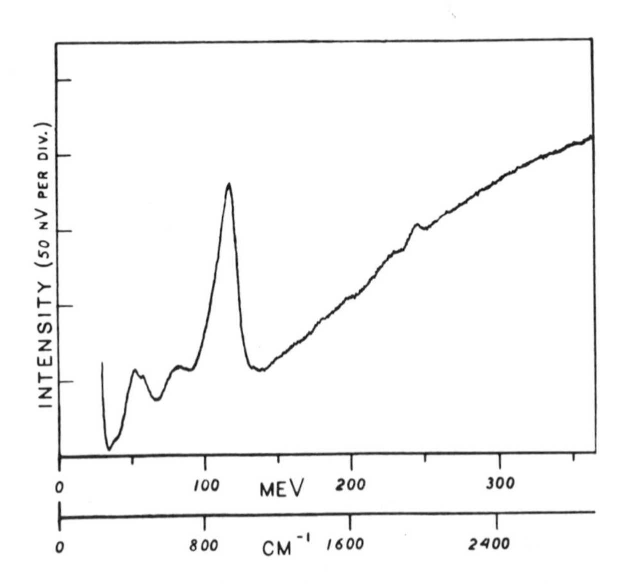

Figure 30. Differential spectrum of CO chemisorbed on alumina supported platinum particles. The large structure near 117 meV is due to the alumina support. This species is possibly a bridging CO species; the mode near 83 meV is in the expected frequency range of a Pt–C–Pt "asymmetric-stretching" vibration.

4.4. Other Molecules; Other Reactions

The future of tunneling spectroscopy in studying supported metal catalysts will involve the study of an ever-growing number of molecules (such as NO, C_xH_y, and $C_xH_yO_z$) on more metals under both oxidizing and reducing conditions. For example, the formation of NH_3 from N_2 and H_2 on iron oxide on alumina should be a fruitful research project. Another promising experiment is to look at the effect of a small amount of a second metal on the surface chemistry. A small amount of manganese on rhodium, for instance, alters the bonding of CO. With an ever-growing body of knowledge will come an understanding of such troublesome questions as those posed by the top metal electrode. At present it is clear that much can be learned about supported metals with tunneling, and that there are no fundamental impediments to further progress.

4.5. Low-Temperature Adsorption

In all present work with supported metal particles the completed junctions are heated to room temperature prior to insertion in liquid helium.

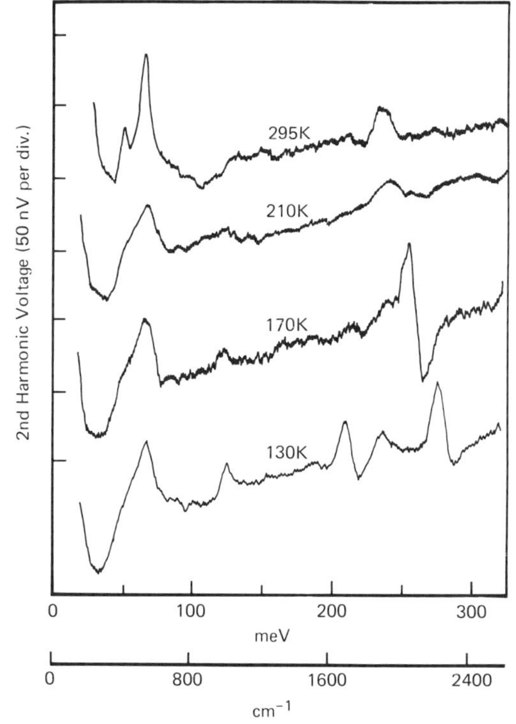

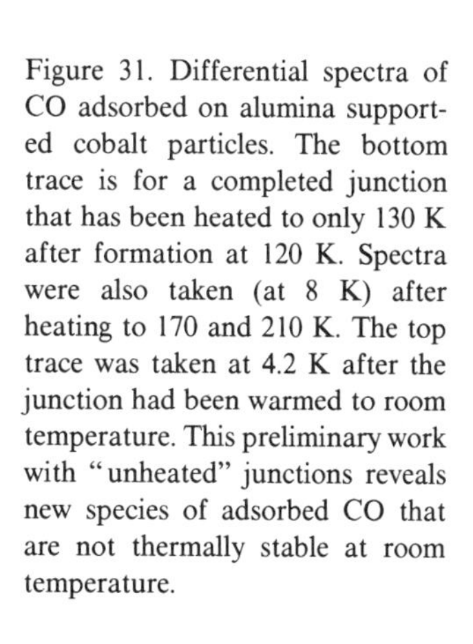
Figure 31. Differential spectra of CO adsorbed on alumina supported cobalt particles. The bottom trace is for a completed junction that has been heated to only 130 K after formation at 120 K. Spectra were also taken (at 8 K) after heating to 170 and 210 K. The top trace was taken at 4.2 K after the junction had been warmed to room temperature. This preliminary work with "unheated" junctions reveals new species of adsorbed CO that are not thermally stable at room temperature.

It is known that some of the surface species formed at low temperatures on these surfaces are not thermally stable at room temperature. While evaporation of the top lead electrode onto a liquid-nitrogen-cooled surface allows the observation of some of these species (as in the case of CO on nickel particles), further development of tunneling spectroscopy will allow the observation of more of these thermally unstable species. Figure 31 shows preliminary spectra of CO chemisorbed on supported cobalt particles. The spectra are for highly dispersed cobalt particles formed in vacuum on alumina at 120 K. The particles where subsequently exposed to approximately three monolayers of CO. The top lead electrode was also formed at 120 K. The completed junction was then warmed only to 130 K before a spectrum was taken (bottom trace) at 8 K. The junction was cycled to 170 and 210 K with spectra taken following each heating prior to warming to room temperature. A final spectrum (top trace) was then run at 4.2 K. This final spectrum is that found previously for CO on cobalt; the 130 and 170 K spectra contain new modes due to thermally unstable (at room temperature) CO species. It is to be expected that future work of this type will greatly extend the number and type of surface species observed on supported metal catalysts with tunneling spectroscopy.

5. Conclusions

(1) Tunneling spectroscopy can successfully model supported metal catalysts. Model catalysts have been made with eight metals and two supports; more can be expected.

(2) The chemisorption and reactions of small molecules can be studied over the full vibrational spectrum with excellent mode resolution.

(3) Tunneling spectra allow the unambiguous assignment of vibrational modes through the use of isotopes.

(4) The frequencies observed for chemisorbed CO are functions of many variables such as particle size, surface coverage, substrate temperature, and top electrode metal.

(5) Despite the problems and uncertainties created by the top metal electrode, much useful information can be obtained on the surface chemistry of supported metals. One example is the observation of D_2S blocking the formation of the dicarbonyl species on supported rhodium particles.

(6) Tunneling spectroscopy gives information not presently available with other techniques. For example, the identification of the ethylidene intermediate on supported rhodium particles has not been reported elsewhere.

(7) Linearly adsorbed CO typically has a bending mode around 45 to 60 meV and a stretching mode about 10 to 15 meV higher. This is in reverse

order from what might have been predicted from metal carbonyl data with higher CO-to-metal ratios.

(8) Tunneling spectroscopy can be used to study adsorbed species on alumina and alumina supported metal particles that are not thermally stable at room temperature.

References

1. G. A. Somorjai, Catalysis on the atomic scale, *Catal. Rev.* **18**, 173–197 (1978).
2. P. K. Hansma, W. C. Kaska, and R. M. Laine, Inelastic electron tunneling spectroscopy of carbon monoxide chemisorbed on alumina supported transition metals, *J. Am. Chem. Soc.* **98**, 6064–6065 (1976).
3. L. H. Dubois, P. K. Hansma, and G. A. Somorjai, The application of high-resolution electron energy loss spectroscopy to the study of model supported metal catalysts, *Appl. Surf. Sci.* **6**, 173–184 (1980).
4. P. K. Hansma, D. A. Hickson, and J. A. Schwarz, Chemisorption and catalysis on oxidized aluminum metal, *J. Catal.* **48**, 237–242 (1977).
5. R. M. Kroeker, W. C. Kaska, and P. K. Hansma, How carbon monoxide bonds to alumina-supported rhodium particles *J. Catal.* **57**, 72–79 (1979).
6. R. M. Kroeker, W. C. Kaska, and P. K. Hansma, Vibrational spectra of carbon monoxide chemisorbed on alumina-supported nickel particles: A tunneling spectroscopy study, *J. Chem. Phys.* **74**, 732–736 (1981).
7. R. C. Jaclevic and M. R. Gaerttner, Inelastic electron tunneling spectroscopy experiments on external doping of tunnel junctions by an infusion technique, *Appl. Surf. Sci.* **1**, 479–502 (1978).
8. J. Klein, A. Léger, S. de Cheveigné, C. Guinet, M. Belin, and D. Defourneau, An inelastic electron tunneling spectroscopy study of the adsorption of CO on Rh, *Surf. Sci.* **82**, L288–L292 (1979).
9. S. de Cheveigné, Thesis at the University of Paris VII, 1980.
10. S. Colley and P. K. Hansma, Bridge for differential tunneling spectroscopy, *Rev. Sci. Instrum.* **48**, 1192–1195 (1977).
11. R. M. Kroeker, W. C. Kaska, and P. K. Hansma, Sulfur modifies the chemisorption of carbon monoxide on rhodium/alumina model catalysts, *J. Catal.* **63**, 487–490 (1980).
12. R. M. Kroeker, W. C. Kaska, and P. K. Hansma, Formation of hydrocarbons from carbon monoxide on rhodium/alumina model catalysts, *J. Catal.* **61**, 87–95 (1980).
13. R. M. Kroeker, W. C. Kaska, and P. K. Hansma, Low-energy vibrational modes of carbon monoxide on iron, *J. Chem. Phys.* **72**, 4845–4852 (1980).
14. A. Bayman, W. C. Kaska, and P. K. Hansma, *Phys. Rev. B*, **25**, 2449 (1981).
15. H. E. Evans, W. M. Bowser, and W. H. Weinberg, The adsorption of ethanol on silver clusters supported on alumina, *Surf. Sci.* **85**, L497–L502 (1979).

14

Computer-Assisted Determination of Peak Profiles, Intensities, and Positions

J. G. Adler

1. Introduction

The essential prerequisite for determination of peak profiles and intensities is *calibrated* tunneling spectra. The first sections of this chapter will introduce a simple calibration scheme which can be implemented with any type of measurement system assisted by a minicomputer, microcomputer, or sophisticated calculator. The approach will be of a tutorial nature so that the experimenter can obtain the knowledge necessary to interface his own measurement system for use with a computer-controlled data acquisition system. Practical aspects of calibration, shortcuts, and tricks will be discussed in detail along with software–hardware tradeoffs. Following the details of calibration, the subtraction of backgrounds to IET peaks will be discussed in detail along with peak profiles in both the normal and superconducting states, from which peak intensities are easily determined. The use of *d*ifferential *i*nelastic *e*lectron *t*unneling *s*pectra (DIETS) will be briefly mentioned.

In Section 2.1 a brief review of the salient features of modulation spectroscopy is provided. With the exception of two early experimental techniques[1,2] all subsequent instrumentation in the literature is based on

J. G. Adler • Department of Physics, University of Alberta, Edmonton, Alberta T6G 2J1, Canada.

harmonic detection. Although for overall spectra (0–500 meV or higher) bridges provide no particular advantage, they are extremely useful and often essential for the precise determination of weak single peak profiles and hence much of the data used to illustrate this chapter was obtained using bridge techniques.

2. Measurement of Tunneling Conductance and Its Derivatives

2.1. Modulation Spectroscopy

One of the most effective ways to study nonlinear phenomena is by modulating one variable with a small sinusoidal signal and studying the device response to this modulation: a schematic of this technique is shown in Figure 1. In this example x is modulated with a signal $\delta \cos \omega t$ where the amplitude δ is small. The response is given by

$$\begin{aligned} y(x,t) &= y(x_0) + \left(\frac{dy}{dx}\right)_{x_0} \delta \cos \omega t + \frac{1}{2}\left(\frac{d^2y}{dx^2}\right)_{x_0} \delta^2 \cos^2 \omega t + \cdots \\ &= y(x_0) + \left(\frac{dy}{dx}\right)_{x_0} \delta \cos \omega t + \frac{1}{4}\left(\frac{d^2y}{dx^2}\right)_{x_0} \delta^2 (1 + \cos 2\omega t) + \cdots \end{aligned} \tag{1}$$

The representation in Figure 1 which maps x into y shows clearly how a pure sinusoid is distorted by the nonlinearity of $y(x)$. Both the magnitude of

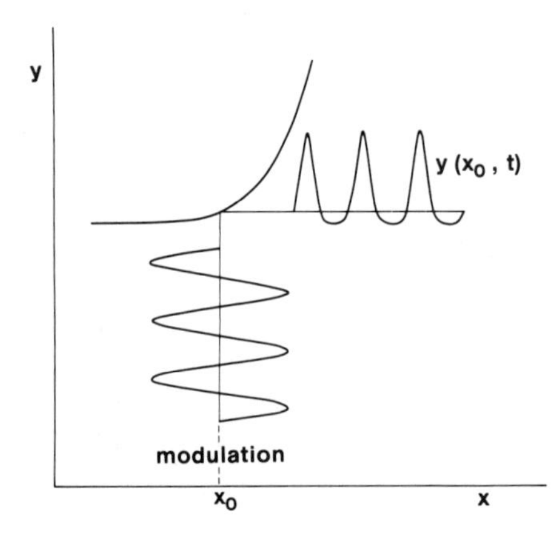

Figure 1. Distortion of a modulation signal $\delta \cos \omega t$ applied at x_0 into a complicated periodic function $y(x_0, t)$, by a nonlinear function $y(x)$.

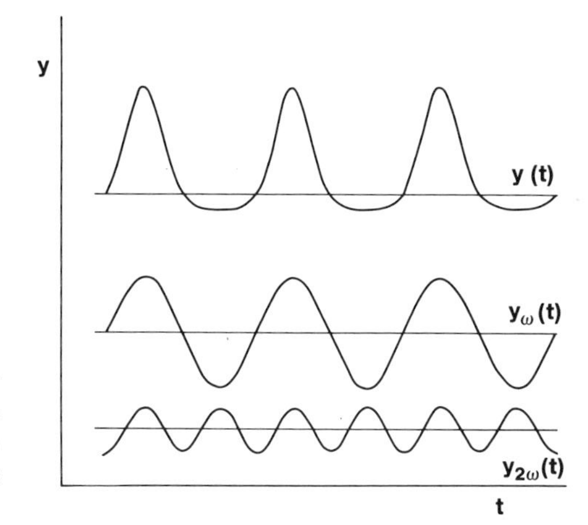

Figure 2. Illustration of the Fourier decomposition of $y(x_0, t)$ shown in Figure 1 into its fundamental $y_\omega(t)$ and second-harmonic contributions $y_{2\omega}(t)$.

the fundamental and its second harmonic are sketched in Figure 2. Examination of the analytic form of Eq. (1) shows that the slope dy/dx, and its derivative d^2y/dx^2, can be expressed as functions of ω and 2ω. The use of a small modulating signal at frequency ω along with a synchronous measurement of the nonlinear device response, $y(x, t)$ at both ω and 2ω (from which the derivatives dy/dx and d^2y/dx^2 may be obtained) over a range of x values will be referred to as modulation spectroscopy or harmonic detection.

Let us now consider how we apply such a technique explicitly to the measurements in inelastic electron tunneling spectroscopy. In electron tunneling the quantities of interest are $I(V)$, $\sigma = dI/dV$, and $d\sigma/dV = d^2I/dV^2$. To measure these quantities directly requires a constant voltage load line. In practice this type of load line requires feedback circuitry which is technically difficult and uses active devices which in turn introduce noise to the measurement. Constant current load lines on the other hand can be passively implemented by using high impedance sources. In the latter case dV/dI, the dynamic resistance and its derivative d^2V/dI^2 are observed. The second derivative can readily be converted using

$$\frac{d\sigma}{dV} = \frac{d^2I}{dV^2} = -\left(\frac{dI}{dV}\right)^3 \frac{d^2V}{dI^2} = -\sigma^3 \frac{d^2V}{dI^2} \qquad (2)$$

Except for tunneling experiments on junctions involving two superconductors in bias regions below 2 mV (where V is not a single-valued function of I) there is no need for measurements with a constant voltage load line. Implementation of a constant current load line is easy both in the case of direct measurements and bridge techniques. In many practical situations the true load line is somewhere between constant current and constant voltage (measurement of d^nV/dI^n or measurement of d^nI/dV^n. The calibration scheme discussed below is computer assisted and can obtain $\sigma = dI/dV$ and $(d\sigma/dV) = d^2I/dV^2$ easily independent of the nature of the load line in the circuit or whether the circuit be a bridge, direct measurement, or a feedback loop.

2.2. A Survey of Measuring Circuits

Much of the circuitry used in modern IETS owes its development to the revival of tunneling measurements motivated by Giaever's work on superconductors.[1] All circuits in the literature since 1960 except for two[2,3] use standard modulation techniques mentioned above. One rather unique circuit due to Thomas and Klein[2] obtains the second derivative information by monitoring the rate at which a servo potentiometer balances a bridge. References 2–24 give a survey of the measurement literature, which contains a vast variety of circuits. In this chapter we will discuss a method of calibrating both the first and second derivatives of the current voltage characteristics which can be applied to most all of these circuits with one exception.[17] As mentioned earlier, overall spectra can be measured with any of these circuits, but a detailed study of peak profiles is best carried out by a bridge circuit. To illustrate the calibration technique with real data the bridge system in the author's lab will be used. Indeed, it is in the realm of studying small individual peaks and their profiles that bridges really become useful. When one studies a small nonlinear region most of the variations of interest comprise only a small part of the total signal across either the junction or its load resistor. Thus in order to amplify the interesting part of the signal sufficiently to study it, a large amount of junction signal must be canceled or bucked off. This must be done prior to amplification because of the limited dynamic range of amplifiers. A bridge provides an excellent way to achieve such a cancellation. In particular, for the study of a small peak, the bridge must be balanced near the center of the peak and thus cancel the large fundamental component that might otherwise saturate the preamplifiers. A typical example of such a *short sweep* is shown in Figure 3.

One problem which can plague high-resolution measurements is external noise. Regardless of circuitry used, lead dress and shielding of circuitry is essential. Unfortunately in our modern environment one runs into all kinds of trouble both from other laboratory equipment as well as such amenities as microwave ovens, paging systems, etc. Local environment will dictate the amount of shielding from external sources necessary, which in some locations might necessitate a shielded room. Certainly, good grounding techniques are essential and these lend themselves primarily to trial-

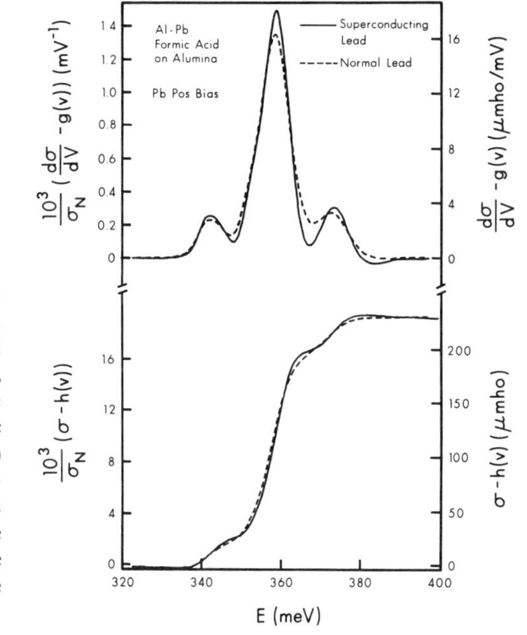

Figure 3. Calibration of an Al–alumina–formic-acid–Pb junction with a positive bias applied to the Al film; (a) the output of the 2ω lock-in, (b) the harmonic distortion, (c) the result of subtracting the harmonic distortion from the 2ω signal, (d) calibrated $d\sigma/dV$ found by multiplying (c) by $\sigma^3(V)$ and η_4, (e) same as curve (d) but with $\eta = \text{const}$, and the insert is the conductance. Figure from Magno and Adler (reference 24).

and-error solutions. When linking analog and digital equipment, however, attention must be given to the grounding scheme, often separating the analog and digital grounds. In terms of the analog equipment used the most critical component is the preamplifier used to detect $Y_{2\omega}$ in Figure 2. Other than this restriction most commercial phase sensitive detectors (lock-in amplifiers) may be used. In fact, to measure y_ω, simple cheap noncommercial lock-in amplifiers can readily be used.

The resolution obtained in most spectra relies on the use of a long time constant for measuring $y_{2\omega}$. Such a long time constant will enhance the signal-to-noise level of $y_{2\omega}$. This requires long sweeps, often 500 mV/hr or slower. Alternatively, many fast sweeps using a short time constant[18] can be averaged by a multichannel analyzer. Most of the circuits in the literature operate at relatively low frequencies, 500–1000 Hz, but some[10, 18] operate at 50 kHz, the only limitation being that of lead and junction capacitance. Junction capacitances are below 1 μF and typically 10 nF. Junction resistances can range from below 10 to over 1000 Ω, 50 Ω being ideal. Very low junction resistances cause heating at high biases while high resistance junctions tend to pick up noise. Finally, although harmonic distortion of the oscillator used is often a problem in modulation spectroscopy, the calibration scheme below circumvents this problem.

2.3. Calibration of Tunnel Conductance and Its Derivatives

For purposes of discussion in this section we will assume that the measuring system has two lock-in amplifiers. One of these operates at the modulation frequency ω and has an output Y_ω while the other operates at 2ω and provides an output voltage $Y_{2\omega}$. The details of the measurement system are not critical. We further assume that the lock-in outputs and the junction bias are digitized and fed to a minicomputer, microcomputer, or calculator; methods for possible interfacing will be detailed in Section 3.

The principle of the calibration scheme relies on substituting a resistance decade box for the junction and calibrating the system prior to each sweep over the resistance range of the junction in the region of the sweep. We shall assume constant current load line apparatus, thus the calibrated tunnel spectra will be given in Eq. (2). If the system had a constant voltage load line the calibration scheme would be the same but a simpler relation could be used. The calibration will be illustrated by a junction whose spectrum appears in Figure 4. These data were obtained using a bridge system[16] and are chosen to illustrate among other things the effect harmonic distortion has on such measurements. The junction conductance is shown in the inset on the upper right-hand corner and corresponds

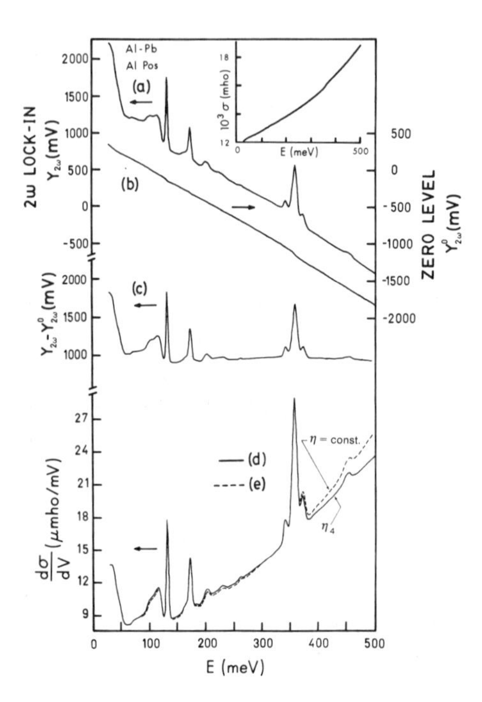

Figure 4. "Short sweep" for a hydrocarbon stretching peak in both the normal and superconducting state. The second derivative $d\sigma/dV = d^2I/dV^2$ has had a smooth polynomial background $g(V)$ subtracted from it and the conductance σ has had a smooth elastic background $h(V)$ subtracted from it. Such short sweeps can provide very high resolution containing as many as 2500 points per sweep. The superconducting curves have been shifted to allow for the effect of the energy gap. From Magno and Adler (reference 28).

to a resistance variation between 85 and 55 Ω. After adjusting the lock-in gains, etc. so that Y_ω and $Y_{2\omega}$ will use as much of the dynamic range of the amplifiers without saturating them, the junction is then replaced by a calibration resistance decade box and say 9 5-Ω steps from 50 to 90 Ω are used to obtain a least-squares fit to

$$Y_\omega = \sum_{n=0}^{N} c_n R_{\text{cal}}^n \tag{3}$$

where R_{cal} is the decade box resistance and Y_ω the fundamental lock-in output. A linear fit ($N=1$) is sufficient unless either the junction or the system response is very nonlinear; then a quadratic fit might be required. Even though the resistance box is a linear element there will be an output $Y_{2\omega}^0$ (zero level) because of harmonic distortion in the modulation signal; furthermore, if a bridge is used such distortion will vary with bridge balance[16, 24] and lock-in phase setting. The zero level and its variation with respect to R_{cal} can again be obtained using a least-squares fit

$$Y_{2\omega}^0 = \alpha_0 + \alpha_1 R_{\text{cal}} + a_2 R_{\text{cal}}^2 \tag{4}$$

The quadratic term can usually be ignored. Even in nonbridge systems α_1 can be nonzero owing to load line changes. Curve (b) in Figure 3 shows the harmonic distortion seen in a bridge which is driven by a source having about 1% 2ω content in the oscillation signal. The second-harmonic signal, $Y_{2\omega}$, at the bridge output due to the junction is shown in curve (a). The difference between curves (a) and (d) illustrates how much $Y_{2\omega}$ may differ from $d\sigma/dV$. Since both Y_ω and $Y_{2\omega}$ are measured simultaneously as a function of junction bias V one can obtain $d\sigma/dV$ from $Y_{2\omega}$ using the relation

$$\frac{d\sigma}{dV} = \eta_m(V)\sigma^3(V)\left[Y_{2\omega}(V) - Y_{2\omega}^0(V)\right] \tag{5}$$

where η_m is a Chebyshev polynomial of order m.[16, 24] Since $\sigma(V)$ can be obtained from $Y_\omega(V)$ using values of c_n in Eq. (3) and $Y_{2\omega}^0$ is known from Eq. (4), a least-squares technique can be used to find η_m using the integral equation

$$\sigma(V) - \sigma(V_a) = \int_{V_a}^{V} \eta_m(x)\sigma^3(x)\left[Y_{2\omega}(x) - Y_{2\omega}^0(x)\right] dx \tag{6}$$

Usually $m=4$ is used; a typical result is shown by curve 3(d); had η been assumed constant in this example the erroneous dashed curve 3(e) would have been obtained. It is convenient to interpolate $Y_{\omega}(V)$ and $Y_{2\omega}(V)$ onto a fixed grid of bias values V, spaced, say, 0.1 or 0.2 mV apart. Such equal-increment data facilitates computation, and the calculation of calibrated σ and $d\sigma/dV$ requires less than one minute for 2501 interpolated points.[16, 24] A test setup using a smart terminal (Tektronix 4052) and carrying out the interpolation *on-line* requires about 6 min at the end of the sweep to provide completely calibrated σ and $d\sigma/dV$ for 2501 points. This latter setup was easy to program and debug as it was done in BASIC.

3. Interfacing with a Computer

3.1. General Considerations

The considerations of this section are in two parts: the conversion of the analog signals to digital form and the transmission of such to the computer or calculator. Detailed considerations cannot be given here since much depends on what the individual has available in his own laboratory and perhaps more significantly, new and cheaper microprocessor controlled equipment appearing on the market daily. As time passes the methods available for interfacing become simpler to implement as more instruments adapt and thus help establish standards. One example of this is the IEEE-488 Bus[25] structure which is becoming commonplace, and although it has many limitations, the convenience of off-the-shelf interfacing usually far outweighs this. Various subsets of this standard appear under trade names such as GPIB, ASCII Bus, IEEE bus, HPIB, etc. In spite of minor variations in implementation of this standard, the instruments and computers (or calculators) are plug compatible with this interface. For the novice starting from scratch this is by far the easiest (not necessarily the cheapest) interfacing scheme.

3.2. Analog-to-Digital Conversion

There are a myriad of ways of achieving this function but in principle it requires a scanner (analog multiplexer) and an analog-to-digital (A/D) converter. This enables conversion of the triplet of signals: junction bias V, and the two lock-in outputs Y_{ω} and $Y_{2\omega}$ to be obtained in sequence. In practice the actual implementation of this function is changing rapidly with the decrease in price of digital and microprocessor circuitry. One can in some cases obtain three digital voltmeters (IEEE-488 compatible) for not much more in cost than a voltmeter and scanner. In order to obtain good resolution, a means of signal averaging is necessary; this may be carried out using a lock-in amplifier with a long, say 3-sec, time constant for $Y_{2\omega}$ or alternately a very short time constant along with a signal averager and 1000 or so sweeps.[18] Where higher resolution is required it is usually convenient to set up the instrumentation so its dynamic range is limited to the single peak to be studied for its line shape. For example, Figure 4 shows a high-resolution sweep of a C–H stretching peak; such a sweep (320 to 400 meV) takes about half an hour. In any case the use of a single sweep with a long time constant does not require as fast (more expensive) A/D converters as many averaged sweeps. The rapidly changing availability of new equipment makes specific recommendations regarding instrumentation impossible.

3.3. Digital Data Transmission from Analog Instrumentation

Four types of interfacing will be briefly considered here which may or may not be compatible with the controller (calculator, minicomputer, or microprocessor) available. They must also be interfaceable with instrumentation used, and will be presented in order of ease of implementation although even this may vary from system to system. The first of these and most recent is the "IEEE 488 Standard Digital Interface for Programmable Instrumentation.[11,25] It is probably the most expensive of the four options dealt with, but perhaps the most standardized and is increasingly more available both in instrumentation and among controllers. The second type of interfacing described is known as a *b*inary *c*oded *d*ecimal (BCD) interface; this type of instrumentation has been around for at least a quarter century and finds its commonality only in the method of coding the information. Actual implementation and voltage levels vary greatly among instruments. The third type of transmission to be discussed is the bit serial (RS232-C) interface which originated with the teletype. Finally, data transfer via 8 or 16 parallel lines with two control lines (strobe and flag) is mentioned. This may require some programming and circuitry to be provided by the user, but in the end the user will control things in the way he or she desires them.

3.3.1. IEEE 488 Standard Interface

By far the easiest to use, this interface can provide bidirectional, asynchronous communication (data, control, and interface management) with over a dozen different instruments at one time. Although this standard allows more than one controller on the line, in practice almost no commercial controllers implement this function (i.e., one controller only is allowed on most systems). This is, however, of no consequence for IETS experiments. The standardization is so complete that cabling and plugs are standard and if you are familiar with the interface you can use instruments and controllers off the shelf, connect them, and be measuring (not necessarily computing) within minutes. For our measurements one would use a scanner (with three channels: bias, $Y\omega$, and $Y_{2\omega}$) and a digital voltmeter both 488 compatible or three separate bus compatible voltmeters. The interconnect cabling has eight bidirectional data lines which transfer the information (1 byte). This information may be in any binary form; however, in most instruments this information consists of seven-bit ASCII (*A*merican *s*tandard *c*ode for *i*nformation *i*nterchange) characters. There are three lines which provide handshake control, i.e., bus timing. The final five lines provide interface management, that is they control which device on the line carries out a particular function at any given time. Thus the cable contains 16 lines in all. Speed of operation of such a bus is very system dependent but data transfers often take place at 1–3 kbyte/sec.

One very special advantage of this type of bus is that many storage devices (magnetic tapes and disks) are available for use on the bus. It is thus possible to interconnect the instrumentation, controller, and data storage using the same bus. Such a system can very quickly be put together from a series of building blocks. In the author's laboratory a Fluke scanner and voltmeter, a Tektronix 4052 terminal, and a Kennedy magnetic tape drive were interfaced to a bridge within an hour, and all the software to obtain calibrated IET spectra debugged and running in less than a week.[32] In the case of data storage using such a system it is worthwhile to binary code the output data on the bus (say a 4-byte word) since this may use less storage space on the magnetic tape as well as speeding up bus operation. This latter suggestion may require some software experience and may not be possible on some systems. The use of ASCII, on the other hand, is simple and straightforward.

3.3.2. The BCD Interface

This type of interface is available on many voltmeters and scanners and costs about the same, or is somewhat cheaper than, the IEEE 488 bus. It is, however, not nearly as convenient. Each decimal digit uses four lines in order to transfer it in binary coded decimal form. For example, a given reading having a sign, 5 decimal digits, and an exponent would require 28 lines. For the same information with six significant figures, 32 lines are needed. At least one other line between the voltmeter and controller is needed to send an end of conversion (EOC) command to the controller when the voltmeter data are ready to be transferred. The controller may have a read only memory (ROM) to translate BCD to its own binary representation, or the writer may simply convert BCD to decimal by means of a look-up table. Such a table is a series of logical comparisons of the four bits/line with the BCD equivalent of digits 0 to 9. Comparison is made cyclically until the right digit is found. This look-up can be carried out very rapidly on most controllers. Although BCD can also be used for the data storage, one usually has to build a multiplexing system, which does not make it worthwhile. It is easiest to use the IEEE 488 bus discussed above or the parallel data transfer method discussed at the end of this section.

3.3.3. The Serial Interface (RS-232C)

Unless the available instrumentation already has this interface, and no further funds can be spent, this is not the interface of choice. The data are usually sent as a string of ASCII characters (seven bits each) plus one bit for parity. Most systems also include one start bit and one or two (usually two) stop bits. These eleven (or 10) bits constitute a character. The input or output line level (high or low; one or zero) is compared to a clock signal and the data transmission is controlled by the clock. The system can be implemented with only three lines: input data, output data, and clock pulses. If only data transmission from the voltmeter to the controller is desired then the two lines, one data and a clock, will suffice. In some systems one can implement the software by treating the voltmeter as an external teletype, and in such instances programming is simplified. The only advantage of setting up such a system *ab initio* these days is if it is already there. One instance where such an interface may be desired is if the data were to be transmitted from the experiment via telephone line to the controller.

3.3.4. The Parallel Interface

This type of interface is usually 8, 16, or 32 lines in parallel plus one line from the controller to send out a pulse ("strobe") and a line for the controller to accept a pulse ("flag"). These lines are usually transistor–transistor–logic compatible (TTL) and in newer equipment they may be bidirectional (tristate). The most versatile data handling can be achieved with this type of interface. The software is usually written using ASSEMBLER language but in some newer controllers it may be done in BASIC. Fast data rates are achieved by the use of binary data. Such a binary representation can be very compact, using four bytes (32 bits) to represent any number which might be desired; this is a great advantage in data storage (Section 5.3). Some of the cheap home computers provide interfaces of this type under such names as

"S-100 bus." The standardization among manufacturers is not yet very good for these devices so the user must be familiar with hardware (*caveat emptor*).

In all instrumentation, much can be learned from a detailed study of advertising specifications of reputable manufacturers. Final choices must, however, be highly individual.

4. Peak Profile Determination

4.1. General Remarks

In order to obtain accurate peak profiles one must consider not only the effect of the elastic background on the spectra but also modulation and thermal smearing and the effect of such junction electrode factors as superconductivity. In fact, superconductivity, which aids resolution by replacing a fuzzy Fermi surface with a sharp BCS singularity, makes determination of the background more difficult.

The superconductivity of the lead cover electrode in an Al–alumina–formic-acid–Pb junction is shown in Figure 5. The top curve illustrates an

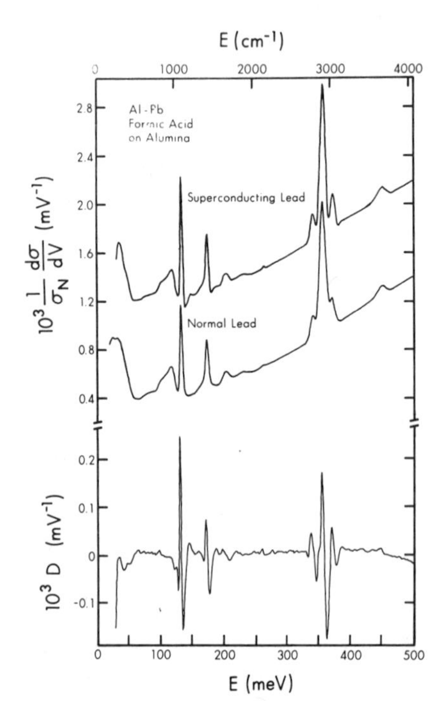

Figure 5. Comparison of $(1/\sigma_N)\, d\sigma/dV$ with the lead electrode normal and superconducting (curve offset vertically by 0.8 mV^{-1}); D is the difference between the top two curves, and σ_N is the conductance at zero bias. From Magno and Adler (reference 28).

undershoot (at biases above the peak) and peak sharpening obtained when the Pb electrode is superconducting. In the curve the superconducting data have been shifted down in energy to allow for the superconducting energy gap of Pb. The lowest curve shows the difference between the normal and superconducting spectra. A description of the effect of superconductivity on IET line shapes was given by Lambe and Jaklevic,(10) the effect of modulation was treated by Klein *et al.*,(26) and mention of superconducting electrode metal phonons was made by Mikkor and Vassel.(27) A detailed experimental comparison on a single vibrational mode taking the above effects into account was made by Magno and Adler.(28)

This section consists of two parts: the first deals with the expected line shape taking into account experimental factors which affect it, e.g., modulation, temperature, superconductivity; the second discusses ways of determining the background spectrum in order to determine the line profile and peak intensity.

4.2. Factors Affecting Peak Profile

This section discusses in a general way the factors which affect the peak profiles and considers a number of techniques for obtaining the background to an IET spectrum. The correct determination of this background is crucial in determining both the line shape and the integrated peak intensity. Throughout this section it is assumed that data are calibrated and in a form easily digestible by a computer or calculator, i.e., magnetic or paper tape or floppy disk. Often in tunneling a superconducting electrode is used, in particular a superconducting Pb counterelectrode provides handsome and enhanced spectra as shown in Figure 5. The superconducting counterelectrode produces a sharper (narrower and larger amplitude) line, an undershoot at energies just above the line, and finally metal phonon bumps within this undershoot. The two sharp peaks near 131 and 172 meV in Figure 5 illustrate this point; the phonon structure within the undershoot is illustrated in the upper solid curve of Figure 6. In the case of this superconducting data there is often a tendency to simply draw a background curve tangent to the minima of the curve as shown in Figure 7. A careful study indicates that such a background may give rise to errors as large as 30% in the integrated intensity.(28)

Consider first the undershoot in the superconducting structure; this was mentioned by Lambe and Jaklevic,(10) observed in detail by Klein *et al.*,(26) and more recently calculated by Kirtley and Hansma,(2) who failed to observe it. The phonon structure within the undershoot was first mentioned in the literature by Mikkor and Vassell(27) and later studied in detail by Magno and Adler.(28) All the above effects will be treated together here

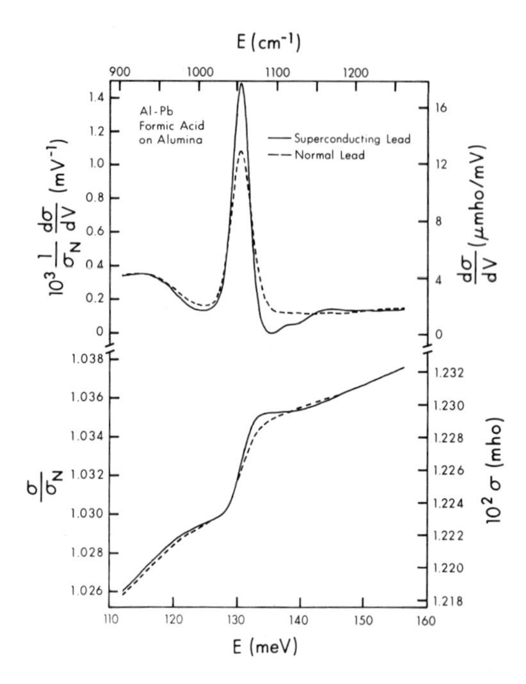

Figure 6. The normalized conductance and its derivative with the lead normal and superconducting for the peak near 131 meV. The superconducting data were shifted down by 0.84 meV, and σ_N is the conductance at zero bias. From Magno and Adler (reference 28).

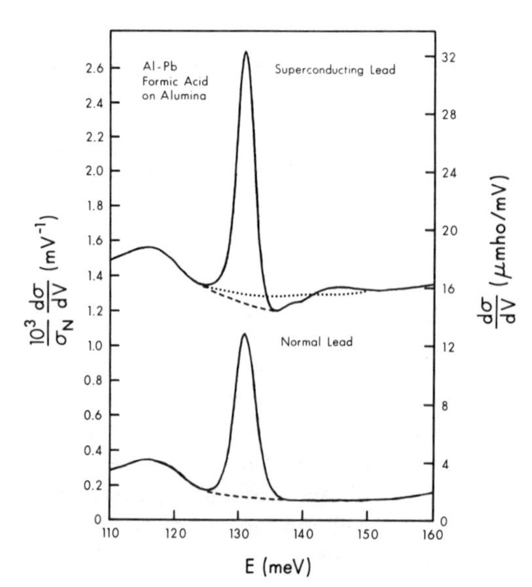

Figure 7. Choices of background curves in the normal and superconducting state. In the superconducting case the dotted background is correct as discussed in the text.

in a phenomenological way using a probe function[28] which can be obtained from the low bias spectrum of the tunnel junction that is being measured.

Consider the simple theoretical expression[10] for the increase in tunnel current due to an *inelastic* process

$$I_i \propto \int_0^\infty G(\omega) \int_{-\infty}^\infty N_1(E) N_2(E+eV-\hbar\omega) f(E) \times [1-f(E+eV-\hbar\omega)]\, dE\, d\omega \tag{7}$$

where $G(\omega)$ is the spectral distribution of the inelastic interaction, N_1 and N_2 are the electrode densities of states, and $f(E)$ the Fermi function. One can simplify (7) by differentiating it

$$\frac{d}{dV} I_i(V) \propto \int_0^\infty G(\omega) P(eV-\hbar\omega)\, d\omega \tag{8}$$

In this new expression we have introduced a probe function $P(eV-\hbar\omega)$. This probe function contains within it the superconducting density of states which in the case of strong coupling materials also contain phonon structure. For weak coupling superconductors such as aluminum the probe function is simple, the BCS density of states and phonon structure negligible. If *both electrodes are normal metals* and their densities of states are slowly varying, the N_1 and N_2 may be taken out of the integration (7) and one has

$$P_n(eV-\hbar\omega) = N_1 N_2 \frac{d}{dV} \int_{-\infty}^\infty f(E)[1-f(E+eV-\hbar\omega)]\, dE \tag{9}$$

which can be evaluated numerically. One final item that contributes distortion of the line shape is the modulation smearing[26] which must be included

$$\frac{d^2 I}{dV^2} = \frac{2m^2}{3\pi} \int_{-1}^{1} \left(\frac{d^2 I_i}{dV^2} \right)_{V+m\cos\nu t} (1-\cos^2\nu t)\, d(\cos\nu t) \tag{10}$$

where d^2I/dV^2 represents the measured inelastic structure, m is the modulation amplitude, and ν the modulation frequency.

The above expressions can be understood physically by inserting a delta function $\delta(\omega-\omega_0)$ for $G(\omega)$ in (8). In this case dI_i/dV is simply proportional to the probe function $P(eV-\hbar\omega_0)$. For two normal-metal electrodes this leads to a thermally smeared step in the tunnel conductance at $eV=\hbar\omega_0$ and hence a bell-shaped curve[10] $5.4k_BT$ wide in d^2I/dV^2. In the case of a superconducting electrode the step near $\hbar\omega_0 + \Delta$ is proportional to the junction conductance near the gap edge. Since this superconducting step is steeper (than when the electrodes are both normal) the peak in d^2I/dV^2 is both narrower and higher than in the case of normal electrodes. The rapid decrease in conductance at energies slightly above the superconducting energy gap is responsible for the undershoot in d^2I/dV^2 at energies beyond the peak. The peaks are of course found at higher energies in the case of superconducting electrodes than normal ones since the electrons decay to the edge of the gap rather than the Fermi surfacc. A detailed study of peak position has been made by Kirtley and Hansma,[29] who show that the shift ranges in position from Δ in the limit of no modulation to about $\Delta/2$ when the modulation amplitude is about 2Δ. They also noted that the shift is temperature dependent and decreases as the width of $G(\omega)$ approaches Δ.

In order to illustrate the method of obtaining a peak profile we will consider a single sharp isolated peak (131 meV) in Figure 5. This peak has been replotted in the upper half of Figure 6. All these data are at 4.2 K and superconductivity was quenched in a 5 kOe field to obtain the normal data. The superconducting data have been shifted down in energy[29] to take into account the Pb energy gap.

The important quantity in IETS is the increase in conductance due to a new inelastic channel. The lower half of Figure 6 shows these steps to be the same size in the normal and superconducting cases. In many cases, however, the size of these steps is tiny and thus in most experiments we measure $d\sigma/dV = d^2I/dV^2$ where the change is observed as a peak and we can evaluate the intensity

$$F(V) = \int \left[\frac{1}{\sigma_N} \left(\frac{d\sigma}{dV} \right) - g(V) \right] dV \tag{11}$$

where σ is the junction conductance, $g(V)$ a smooth background curve, and σ_N a value at which the data are normalized [e.g., $\sigma_N = \sigma(V=0)$ in the normal state and $\sigma_N = (V = V_n)$, where V_n is any bias value beyond phonon effects in the case of a superconducting electrode; for Pb we often choose $V_n = 30$ mV]. Note that by normalizing the spectra using σ_N we eliminate differences between junctions of different tunnel conductance. The spectrum $d\sigma/dV = d^2I/dV^2$ is given in μmho/mV while the normalized spectrum $(1/\sigma_N)\, d\sigma/dV$ is simply in units of mV^{-1}. Figure 7 illustrates that the choice in smooth background $g(V)$ is not unique. For the normal state data and background shown one obtains an intensity of 0.40×10^{-2}. For the superconducting data there are at least two choices of background shown in Figure 7. Using the dashed curve and integrating over the same energy range (124 to 136 meV) as used for the normal data we obtain 0.52×10^{-2} or an intensity which is 30% larger than in the normal case. On the other hand, using the dotted background and integrating over the larger region (124 to 150 meV) which includes the undershoot, one obtains 0.40×10^{-2}, consistent with the normal state data. Our criterion then, as to having obtained the correct background, is that we get consistent results regardless of the state (normal or superconducting) of the Pb electrode in the Al–oxide–formic-acid–Pb junction used in this example.

One can check that the correct choice of background has been used, in yet another way. One of the complicating factors in these illustrations, Figures 5–7, arises from the fact that the junction is Al–oxide–formic-acid–Pb and thus has two insulating layers to tunnel through: the oxide and the chemisorbed formate ions. This leads to an Al–O stretching peak at 118 meV near the C–H bend being studied at 130 meV. In order to deal quantitatively with such situations it is useful to study two junctions both sharing the same oxide layer, but with only one of these twin junctions having been doped (in this case with formic acid). Calibrated second derivatives allow one to obtain a difference spectrum of these twins. This type of spectrum is shown in Figure 8 under the symbol D and such

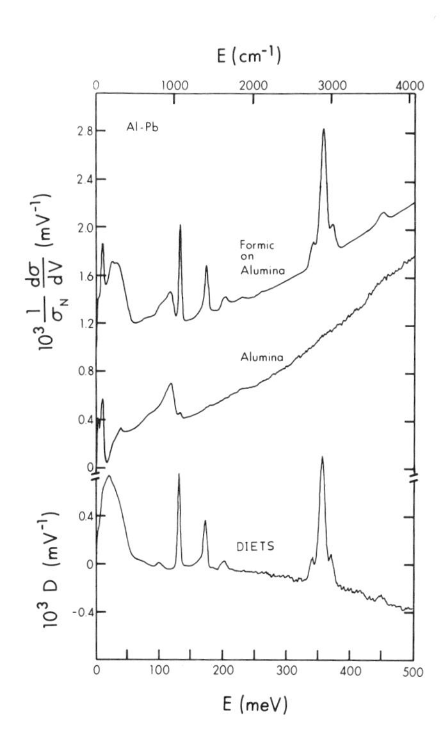

Figure 8. Spectra for a pair of twin junctions, one of which is Al–alumina–formic-acid–Pb and its control Al–alumina–Pb junction. The difference or DIETS spectra appears below.

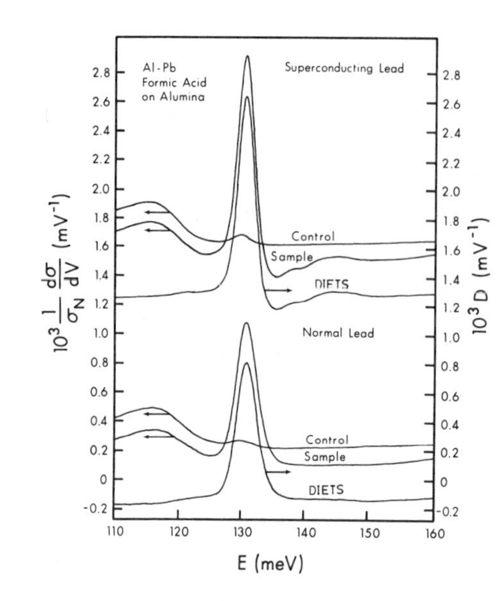

Figure 9. The control junction, sample, and difference curve are shown as an example of using DIETS to remove the alumina peak near 115 meV. The superconducting lead curves have been offset vertically by 1.4 units. Note complete absence of 118-meV peak (Al–O stretch) in DIETS curves. From Magno and Adler (reference 28).

*d*ifferential *i*nelastic *e*lectron *t*unneling *s*pectra will be referred to as DIETS. This DIETS curve D is simply the difference in spectra between the formic acid doped alumina junction and the control (undoped) alumina junction.

An illustration of the use of DIETS with both normal and superconducting electrodes is shown in Figure 9. Unfortunately (but often typically) the evaporator contains a minute amount of formic acid contamination whose integrated intensity is about 0.01×10^{-2}. Note that the DIETS curves (D in Figure 8) show no structure due to the aluminum oxide stretch near 118 meV. Using a smooth polynomial for $g(V)$ on the two DIETS curves one obtains a value in intensity of 0.39×10^{-2} for the normal state curve and 0.41×10^{-2} for the superconducting curve, both of which compare well to the 0.40×10^{-2} obtained from Figure 7.

One can now make a comparison of theoretical to experimental line profiles. Line shapes have been studied with reference to peak position by Kirtley and Hansma[29] using Gaussian, Lorentzian, and delta function lines without essential change in peak shift. In considering line profiles, Magno and Adler[28] found somewhat better fit using a Lorentzian line shape, and their data appear in Figure 10. The background used to obtain these fits is that shown by the dotted line in Figure 7. A Lorentzian spectral distribution

$$G(\omega)=\frac{A}{(\omega-\omega_0)^2+(\frac{1}{2}\Gamma)^2} \tag{12}$$

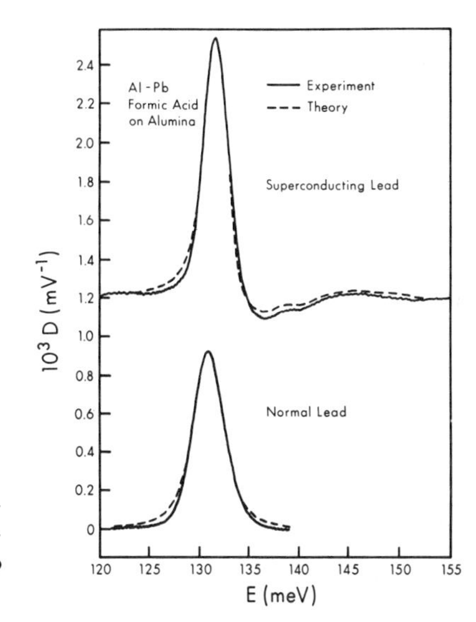

Figure 10. Comparison of theoretical and experimental line shapes for normal and superconducting Pb cover electrodes, from Magno and Adler (reference 28).

was used with $\hbar\omega_0 = 130.8$ meV and $\frac{1}{2}\hbar\Gamma = 1$ meV. The peak amplitude A was chosen in each case to provide the theoretical peak with the same height as the experimental one. Rather than calculate the theoretical superconducting density of states including Pb phonon bumps, experimental conductivity data at bias below 25 meV (above this, Pb multiphonon processes are negligible) was used for the probe function $P(eV - \hbar\omega)$. As seen in Figure 10 excellent agreement exists between theory and experiment. This is very striking in the bumps due to the superconducting Pb phonons and their harmonics manifested within the undershoot and above near 145 meV. In the peak profile calculations the Lorentzian line shape was clearly superior to a Gaussian function.

In conclusion it can be said that it is more difficult to determine peak background in the case of superconducting electrodes. Failure to choose the proper background of a narrow peak may lead to errors of 30% in intensity. To obtain good line profiles greater line separation is necessary in the case of superconducting electrodes than normal ones. For good resolution a 3-meV-wide peak requires a sweep of almost 10 meV on either side, while if an electrode is superconducting, data up to about 25 meV above the peak are required. The effect of such a strong coupling superconductor on line shape is vividly illustrated by the DIETS curve in Figure 11.

4.3. Peak Profiles of Junctions with Composite Barriers

Often in IETS one is confronted with the situation where one studies some substance chemisorbed on an oxide barrier rather than on the metal

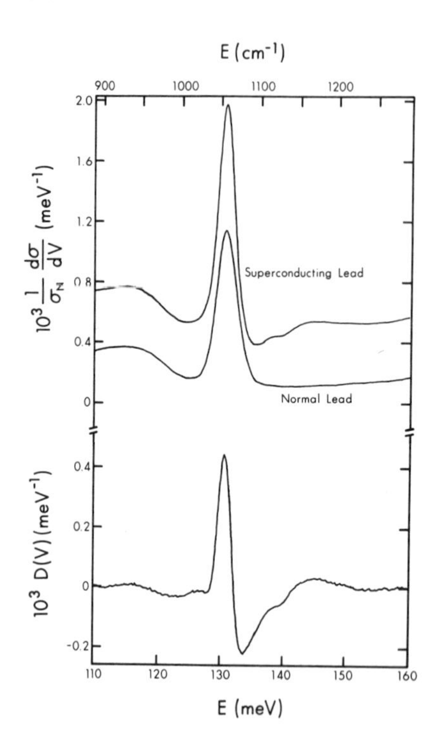

Figure 11. Illustration of superconducting phonon effects in the undershoot by use of the DIETS curve $D(V)$.

electrode itself. As previously mentioned, in such cases the spectrum of interest is the difference spectrum between an undoped oxide control junction and the doped junction. Very often junctions with aluminum oxide are used because of ease of preparation, and much data on the dopant material in the region below 130 meV is lost unless the difference spectrum is examined. An analog method has been devised by Colley and Hansma[17]: this method requires two closely matched resistance tunnel junctions, one in each arm of a bridge circuit. Such matched junctions are sometimes difficult to produce with different barriers. On the other hand, if accurately calibrated spectra exist, then the difference spectra (DIETS) can be computed directly as shown in Figure 8. The additional advantage of this second technique is that changes taking place in a given junction as a function of time can be compared. Perhaps one of the best illustrations of the use of DIETS with composite barriers appears in Figures 12 and 13. In Figure 12 the low-energy spectrum of formic acid chemisorbed on alumina is shown along with the spectrum of a control junction of undoped alumina. The DIETS curve shows the structure arising from the chemisorption process. Similar comparisons are shown for chemisorption of formic acid on both alumina and aluminum[30] in Figure 13.

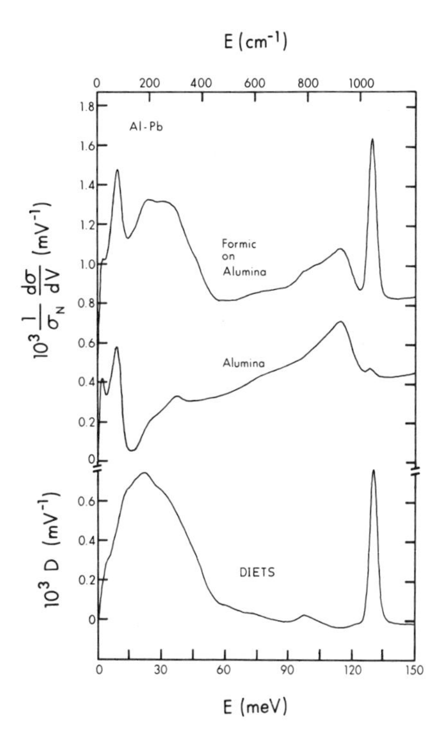

Figure 12. Low-energy DIETS spectra for formic acid chemisorbed on alumina.

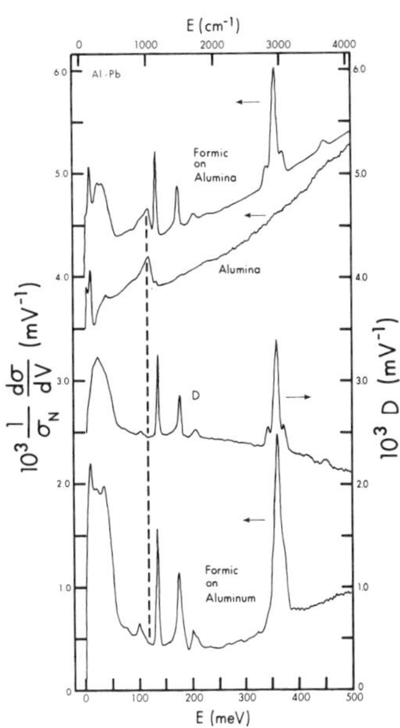

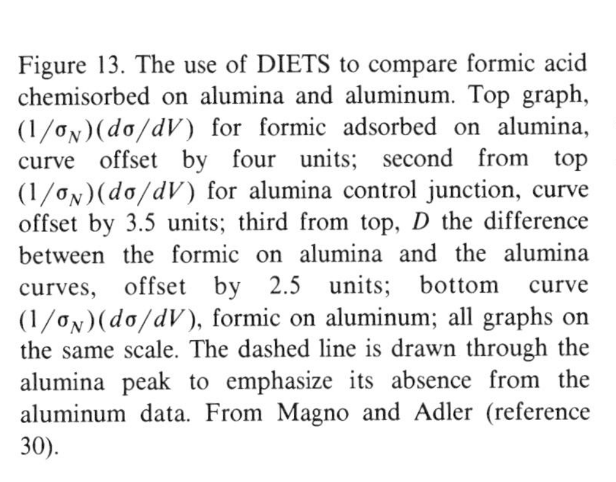
Figure 13. The use of DIETS to compare formic acid chemisorbed on alumina and aluminum. Top graph, $(1/\sigma_N)(d\sigma/dV)$ for formic adsorbed on alumina, curve offset by four units; second from top $(1/\sigma_N)(d\sigma/dV)$ for alumina control junction, curve offset by 3.5 units; third from top, D the difference between the formic on alumina and the alumina curves, offset by 2.5 units; bottom curve $(1/\sigma_N)(d\sigma/dV)$, formic on aluminum; all graphs on the same scale. The dashed line is drawn through the alumina peak to emphasize its absence from the aluminum data. From Magno and Adler (reference 30).

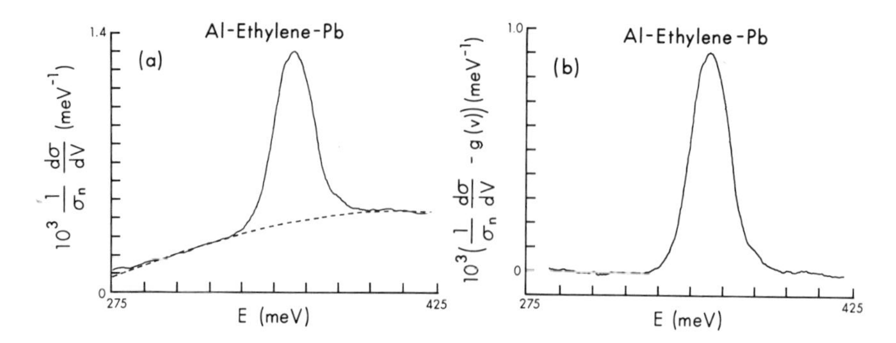

Figure 14. A C–H stretching peak showing the smooth background function $g(V)$ dashed curve is seen in figure (a). Curve (b) shows the integrand [Eq. (11)] used to obtain the intensity $F(V)$. From Adler, Konkin, and Magno (reference 31).

4.4. Peak Intensities

Once tunneling data are available in digital form in a computer then intensities may readily be computed. First a smooth background polynomial $g(V)$ [Eq. (11)] is chosen. Illustrations of such polynomials appear as dashed lines in Figures 14 and 15. Subtraction and subsequent integration yields

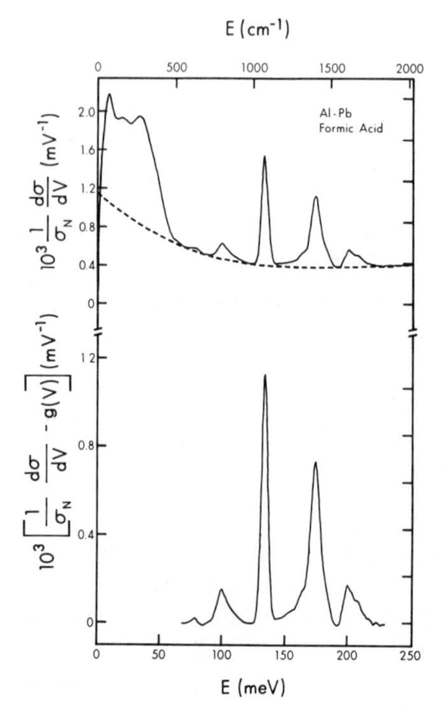

Figure 15. Formic acid chemisorbed on alumina taken with a 1-kOe field to suppress the superconductivity of the Pb and thus the undershoot is shown allowing one smooth polynomial $g(V)$ to be used as background to several peaks.

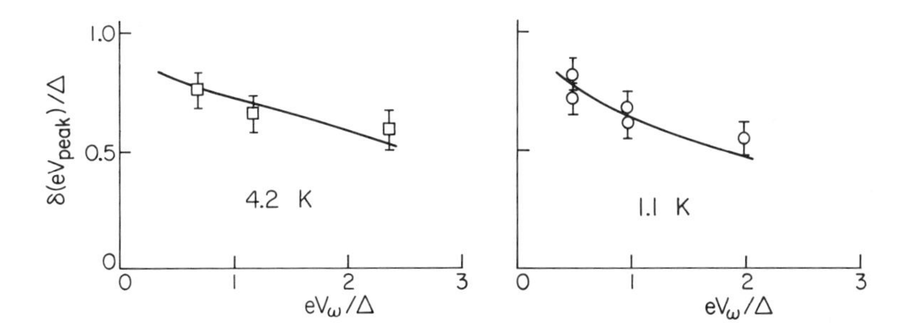

Figure 16. Peak positions as a function of modulation amplitude and energy gap for junctions with one electrode superconducting. From Kirtley and Hansma (reference 29).

the peak intensities. The care and attention required with superconducting data was illustrated in Section 4.2. For example, after subtracting off the proper choice of background $g(V)$ in Figure 7 and integrating over the range 124–150 meV for the superconducting data, one obtains a peak intensity of 0.40×10^{-2} identical to that obtained with a normal Pb electrode. This example also illustrates how much easier it is to choose the background in the normal state when the peaks are sharp (i.e., those which would provide a large undershoot if one electrode were superconducting).

4.5. Peak Positions

Peak positions may easily be obtained either from digital spectra or from a graphical representation thereof. The one problem which does arise again is if one of the junction electrodes is superconducting; the peaks will then be shifted to higher energy. Such shifts have been studied in detail by Kirtley and Hansma,[29] who found it to vary between Δ and $\frac{1}{2}\Delta$, where Δ is the superconducting energy gap. Figure 16 shows their results as a function of modulation amplitude V_ω. Their work also demonstrates the effect of different cover electrodes on peak positions.

5. Data Handling

5.1. General Comments

Because of rapid advances in computer and peripheral hardware it is difficult to give detailed advice. What follows is the result of more than a decade's experience with both minicomputer based and *smart terminal* (graphic terminal cum calculator, e.g., Tektronix 4052) based hardware, and the use of ASSEMBLER, FORTRAN, and BASIC languages for programming.

There are a myriad of tradeoffs involving speed and convenience vs cost. While most of the modern semiconductor circuitry is reliable, mechanical systems from keyboards to tape drives are much less so. It is in this latter area that cost and reliability go hand in hand. This is an important consideration in a laboratory situation where several people may use equipment, yet only one or none are capable of servicing the hardware. It may be tempting to set up with the cheapest home computer system and in fact it can work well, but *only if the user is willing to shoulder time-consuming servicing and maintenance himself*. For example, the keyboard on a Radio Shack system is not as durable as that on a Tektronix terminal.

Similar types of tradeoffs exist at the software end of things. Initial programming and debugging are easiest in a high-level language such as BASIC. However, the step-by-step execution of such programs is often very time consuming. For example, the calibration of σ and $d\sigma/dV$ discussed in Section 2.3 takes 23 sec in a computer using compiled ASSEMBLER and FORTRAN programs compared to 6 min 50 sec in BASIC on another machine. The computer-based system took several months to evolve and the BASIC system only a few days. Cost and availability of equipment locally along with servicing and help for a novice must also be considered.

For the above reasons this section will only discuss in a general way the type of programs found useful in IETS in the author's laboratory and finally give some opinions on data storage.

5.2. Data Calibration

The first illustration in this section describes a minicomputer-based data acquisition system using an integrating digital voltmeter and scanner.[16] The junction is swept and data taken at approximately equal time intervals although the junction bias is not necessarily a linear function of time. Y_ω and $Y_{2\omega}$ are continuously plotted on a CRT display during sweep. A computer program is used to acquire the data and simultaneously calibrate the conductance [using C_n in Eq. (3)] and to subtract the zero level $Y_{2\omega}^0$ [Eq. (4)]. The ensuing 2000 to 3500 triplets of points $(V, \sigma, Y_{2\omega} - Y_{2\omega}^0)$ are stored in the computer memory (if sufficient) or on magnetic tape. Upon completion of the sweep the computer calls another program which interpolates the data using a Lagrange interpolation, and results in up to 2501 points at equally spaced bias increments. Retaining data in memory (if space is available), the interpolation then calls an integration program[16, 24] which solves Eq. (6), the integral equation, and provides calibrated conductance and spectral data. The computation time can be very short, taking as little as 23 sec if the data remains in memory while programs are loaded in and out. If the data are input and output to nine-track tape between programs this increases to between 1 and 2 min.

Optimal processing speed is very dependent on the operating system *and* the user's familiarity with it. The system described was developed over many years[13, 16, 24] and is a mix of ASSEMBLY and FORTRAN programming using a minicomputer. If fast execution (points kept in memory during execution without input/output of data) is desired, such programming can be carried out in most mini- or microcomputers having access to compiler languages (ASSEMBLY, FORTRAN, ALGOL, PASCAL, etc.) and 64 kbytes of memory (this will handle about 3600 points). This same programming can be carried out in a computer having 16 kbytes but would require additional input/output such as magnetic tape or floppy disk. In the latter case speed is input/output rate dependent. All fast efficient systems require compiler-based languages and hence some patience in implementation, but once debugged are trouble free and a joy to use.

With the proliferation of small computers and calculators with display screens, similar programs can be implemented using higher-level languages such as BASIC or PASCAL. These systems usually do not allow access to compiler-based languages. These machines with their interpreted languages are slow and not very efficient because of their statement-by-statement program execution. They are, however, easy to implement and debug for the first time, because their statement-by-statement execution allows mistakes to be corrected on the spot. A similar program to the one described above was implemented using a Tektronix 4052 terminal and a voltmeter scanner combination interconnected by an IEEE-488 bus and basic programming. This system took about 7 min after the sweep to complete calibration[32] and display calibrated results on the screen. This latter system was set up and debugged in a day or two.

5.3. Data Storage

Computer-assisted data are like a two-edged sword: on the one hand it is very easy to review old data and analyze them, and on the other, the amount of data proliferates very rapidly. In fact, when compacted in binary form the average annual rate of data growth is about 3 miles of magnetic tape per annum. These high rates of data growth suggest that some thought has to be given to data storage.

At present one can consider three basic means of data storage: flexible disk ("floppy"), magnetic cartridge (cassette), or reel-to-reel magnetic tape.

Perhaps the quickest access to the data can be gained by the use of floppy disks. This rapid access must be traded for limited capacity and in some cases durability. Most current floppy disks only allow about a dozen sweeps to be stored. Magnetic cartridges[33] have capacities about half that of floppy disks but are slower since they are sequential access devices, being a ribbon of tape, while the floppy may be randomly accessed by sector and track since it is basically a rotating wheel. Reel-to-reel magnetic tape is available in length from 100 to 2400 feet and can easily store 200 sweeps. Access time near the start of the tape is rapid but not near the end. Commonly, tape drives are available with speeds between 12.5 and 75 in./sec.

Perhaps a more significant consideration when planning the data storage is the method of encoding. The simplest and most universal systems to interface (e.g., IEEE-488 bus) use ASCII. This is very convenient in terms of interfacing as was discussed in Section 3.3 but is not very compact in that one byte (8 bits) is used per character. Typically such systems use 12 and usually more bytes per number stored. On the other hand, with some extra work, numbers can often be packed in a binary format in which 4 bytes provide more than adequate precision and dynamic range. Compactness of storage significantly decreases access time. For long-term storage reel-to-reel tape is still the best compromise. As technology evolves, however, the price, access time, and storage density compromises may change for the various recording media.

5.4. Data Analysis

As was mentioned previously, the data are monitored on a CRT screen when recorded. In the case of the author's laboratory this has usually been a Tektronix 4010 or a 4052 terminal. These devices are blessed with the ability to access any point on the screen with a movable cross hair or pointer. Such a pointer is manipulated by an X and a Y potentiometer. A series of programs can be developed in which the user can readily analyze data in an interactive way. A partial description of some programs is given below.

A program that is often useful is one that enables one to plot (display on screen and if desired make a hard copy) the positive, negative, odd, or even part of either the conductance σ or its derivative $d\sigma/dV$, such plots can be done by normalizing by given conductance σ_N as shown in the previous illustrations. One can also normalize $d\sigma/dV$ by the conductance at each bias point and obtain the logarithmic derivative (Figure 17) $1/\sigma(V)(d\sigma/dV)=(d/dV)(\ln\sigma)$. Another program does a least-squares fit to the conductance to obtain barrier heights and thicknesses.[34,35] A program enables one to locate the pointer in a spectral peak and read its value in eV or cm^{-1} as well as the value of $d\sigma/dV$ at the peak. Other programs enable a smooth polynomial $g(V)$ [Eq. (11)] to be generated as the background to any peak (e.g., Figures 7, 14, and 15). The integrated intensity can

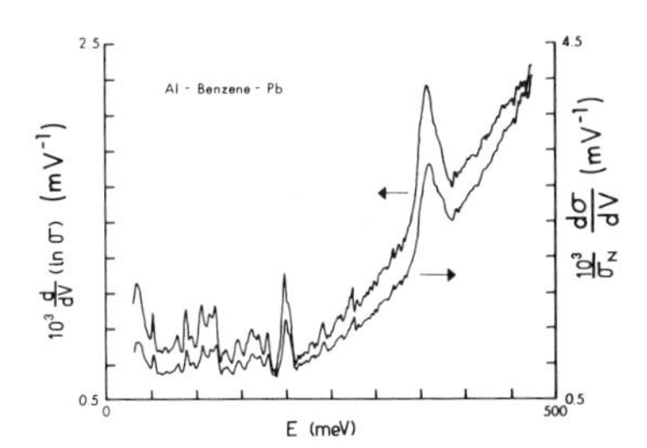

Figure 17. Comparison of a normalized second derivative $(1/\sigma_N)\,d\sigma/dV$ with a logarithmic derivative as discussed in text.

then be calculated within seconds for any peak. Other routines allow for difference spectra to be obtained, not only between different junctions (Figures 8, 12, 13) but also between different data for the same junction(36, 37) (Figures 5 and 11). Theoretical and experimental line profiles (Figure 10) can also be generated in this fashion.

References

1. I. Giaever, Energy gap in superconductors measured by electron tunneling, *Phys. Rev. Lett.* **5**, 147–148 (1960).
2. D. E. Thomas and J. M. Klein, Tunneling current structure resolution by differentiation, *Rev. Sci. Instrum.* **34**, 920–924 (1963).
3. J. S. Rogers, J. G. Adler, and S. B. Woods, Apparatus for measuring characteristics of superconducting tunnel junctions, *Rev. Sci. Instrum.* **35**, 208–213 (1964).
4. Ivar Giaever and Karl Megerle, Study of superconductors by electron tunneling, *Phys. Rev.* **122**, 1101–1111 (1961).
5. I. Giaever, H. R. Hart, Jr., and K. Megerle, Tunneling into superconductors at temperature below 1 K, *Phys. Rev.* **126**, 941–948 (1962).
6. William R. Patterson and J. Shewchun, Alternate approach to the resolution of tunneling current structure by differentiation, *Rev. Sci. Instrum.* **35**, 1704–1707 (1964).
7. D. E. Thomas and J. M. Rowell, Low level second harmonic detection system, *Rev. Sci. Instrum.* **36**, 1301–1305 (1965).
8. J. G. Adler and J. E. Jackson, System for observing small nonlinearities in tunnel junctions, *Rev. Sci. Instrum.* **37**, 1049–1054 (1966).
9. A. Gaudefroy-Demonbynes, E. Guyon, A. Martinet, and J. Sanchez, Dérivées premières et secondes de la caractéristique d'une jonction tunnel, *Rev. Phys. Appl.* **1**, 18–22 (1966).
10. J. Lambe and R. C. Jaklevic, Molecular vibration spectra by inelastic electron tunneling, *Phys. Rev.* **165**, 821–832 (1968).
11. Andrew Longacre, Jr., Biasing circuitry for tunnel junctions, *Rev. Sci. Instrum.* **41**, 448–449 (1970).
12. J. S. Rogers, Conductance bridge for electron tunneling measurements, *Rev. Sci. Instrum.* **41**, 1184–1186 (1970).
13. J. G. Adler, T. T. Chen, and J. Straus, High-resolution electron tunneling spectroscopy, *Rev. Sci. Instrum.* **42**, 362–368 (1971).
14. B. L. Blackford, Low impedance supply for tunnel junctions, *Rev. Sci. Instrum.* **42**, 1198–1202 (1971).

15. A. F. Hebard and P. W. Shumate, A new approach to high-resolution measurements of structure in superconducting tunneling currents, *Rev. Sci. Instrum.* **45**, 529–533 (1974).
16. J. G. Adler and J. Straus, Application of minicomputers in high-resolution electron tunneling, *Rev. Sci. Instrum.* **46**, 158–163 (1975).
17. S. Colley and P. Hansma, Bridge for differential tunneling spectroscopy, *Rev. Sci. Instrum.* **48**, 1192–1195 (1977).
18. R. C. Jaklevic and M. R. Gaerttner, Inelastic electron tunneling spectroscopy. Experiments on external doping of tunnel junctions by an infusion technique, *Appl. Surf. Sci.* **1**, 479–502 (1978).
19. Ursula Mazur and K. W. Hipps, An inelastic electron tunneling spectroscopy study of the adsorption of NCS, OCN, and CN from water solution by Al_2O_3, *J. Phys. Chem.* **83**, 2773–2777 (1979).
20. M. V. Moody, J. L. Paterson, and R. L. Ciali, High-resolution dc-voltage-biased ac conductance bridge for tunnel junction measurements, *Rev. Sci. Instrum.* **50**, 903–908 (1979).
21. L. D. Flesner and A. H. Silver, Improved method of measuring tunneling conductance, *Rev. Sci. Instrum.* **51**, 1411–1412 (1980).
22. Andrew A. Cederberg, Inelastic electron tunneling spectroscopy: Intensity as a function of surface coverage, *Surf. Sci.* **103**, 148–176 (1981).
23. A. B. Dargis, Digital inelastic electron tunneling spectrometer, *Rev. Sci. Instrum.* **52**, 46–51 (1981).
24. R. Magno and J. G. Adler, Data calibration in electron tunneling spectroscopy, *Rev. Sci. Instrum.* **52**, 217–223 (1981).
25. *IEEE Standard Digital Interface for Programmable Instrumentation* (IEEE Std 488-1975), The Institute of Electrical and Electronic Engineers, Inc., New York (1975).
26. J. Klein, A. Leger, M. Beli, D. Défourneau, and M. J. L. Sangster, Inelastic electron tunneling spectroscopy of metal–insulator–metal junctions, *Phys. Rev. B* **7**, 2336–2349 (1973).
27. M. Mikkor and W. C. Vassel, Phonon and plasmon interactions in metal–semiconductor tunneling junctions, *Phys. Rev. B* **2**, 1875–1887 (1970).
28. R. Magno and J. G. Adler, Intensity and lineshape measurements in inelastic electron tunneling spectroscopy, *J. Appl. Phys.* **49**, 5571–5575 (1978).
29. J. Kirtley and P. K. Hansma, Vibrational-mode shifts in inelastic electron tunneling spectroscopy: Effects due to superconductivity and surface interactions, *Phys. Rev. B* **13** 2910–2917 (1976).
30. R. Magno and J. G. Adler, A study of chemisorption of formic acid on different surfaces by electron tunneling, *J. Appl. Phys.* **49**, 4465–4467 (1978).
31. J. G. Adler, M. K. Konkin, and R. Magno, The technology of IETS, in *Inelastic Electron Tunneling Spectroscopy* (T. Wolfram, ed.) Springer-Verlag, Berlin (1978).
32. J. G. Adler, M. K. Konkin, D. P. Mullin, M. A. Ocampo, and J. Urias (to be published).
33. DC300A Data cartridge.
34. M. K. Konkin, R. Magno, and J. G. Adler, The use of barrier parameters for the characterization of electron tunneling conductance curves, *Solid State Commun.* **26**, 949–952 (1978).
35. R. Magno and J. G. Adler, The dependence of metal–insulator–metal conductance curves on chemisorbed ion concentration in the barrier, *Surf. Sci.* **78**, L250–L256 (1978).
36. M. K. Konkin and J. G. Adler, Annealing effects in tunnel junctions (thermal annealing), *J. Appl. Phys.* **50**, 8125–8128 (1979).
37. M. K. Konkin and J. G. Adler, Annealing effects in tunnel junctions (voltage annealing), *J. Appl. Phys.* **51**, 5450–5454 (1980).

15

Infusion Doping of Tunnel Junctions

R. C. Jaklevic

1. Introduction

1.1. Doping Requirements

One of the problems confronting the experimenter who wishes to study organic monolayers by electron tunneling is that of introducing organic molecules into the tunneling structure. The descriptive term "doping" will be used to describe this process. As can easily be seen from other chapters in this book and from a study of the literature in this field,[1-6] organic molecules with a wide variety of properties are studied by electron tunneling, and several doping techniques have evolved to accommodate them. To be successful the technique must achieve a stable coverage of one monolayer or less inside the tunnel junction. The doping procedure must not cause unwanted chemical changes in the deposit such as decomposition due to heating or reaction with the chamber walls. Impurities must be excluded and the deposited monolayer must be stable during the remaining fabrication steps. Hope has been placed in the possibility that tunneling spectroscopy will become a routine tool for analysis of monolayers of organics. Since taking the spectra from a prepared tunnel junction is relatively easy to do, it remains for junction fabrication and doping techniques to develop to

R. C. Jaklevic • Physics Department, Ford Scientific Laboratory, P.O. Box 2053 SL3012, Dearborn, Michigan 48121.

the degree that junctions can be prepared and doped easily and under a wide variety of conditions of interest to the surface chemist.

Early doping methods may be grouped under the general categories of either vapor or liquid doping. They have the common property that the molecular species are deposited on the oxidized aluminum strip as an intermediate step in junction fabrication. The final Pb overlay film is deposited by an additional vacuum evaporation step. More recently, an infusion method has been developed in which the tunnel junctions are processed to completion and stored until needed.[7,8] Organic molecules are later introduced into the junction by an infusion step. This chapter will be devoted to a description of the infusion method and experiments performed to understand the physical mechanism involved. Although it is usually better to refer to source articles for a detailed doping recipe, in this case the details are of interest. The phenomenon of infusion was not expected from previous knowledge of adsorbtion or diffusion kinetics and much remains to be learned about it. Also, the effect has implications beyond the simple goal of achieving a stable monolayer for the purpose of obtaining a tunneling spectrum. Most of the results here were obtained at the author's laboratory,[7,8] and those results obtained elsewhere will be noted in separate references.

1.2. Review of Other Doping Methods

1.2.1. Vapor Phase Doping

In vapor phase doping the aluminum surface, usually already oxidized, is exposed to the vapor of the molecules, often by simply admitting a small amount of the organic of interest directly into the vacuum chamber. If conditions are right, the required monolayer is deposited. It is especially well suited for molecules which have reasonable vapor pressure and form strong chemical bonds which firmly attach them to the surface. Examples of this are acids such as acetic, formic, etc., where molecular coverage limits at a single monolayer and thickness control is more or less automatic. For molecules with very low vapor pressure, deposition can be accomplished by evaporating from a heated source in a manner analogous to deposition of metal films. The molecules must be sufficiently stable to withstand the necessary heating, and positive control of the deposit is required to limit thickness to a single layer. Deposition must be done in a separate chamber if contamination of the vacuum system is to be avoided. At the other extreme are high-vapor-pressure organics with weak bonding strength, such as gases, solvents, and alcohols. These kinds of molecules desorb rapidly when exposed to the vacuum pumpdown needed for the final Pb overlay film deposition. For these, the substrate can be cooled to approximately 100 K and the molecules frozen in place for the final evaporation steps. This requires careful control of thickness which is usually difficult and inconvenient at these coverages and is especially susceptible to impurities. Vapor doping also has problems with those molecules which are chemically reactive or unstable when brought into contact with chamber walls or metal surfaces.

1.2.2. Liquid Phase Doping

In liquid phase doping the substrates with their oxidized aluminum films are removed from the vacuum chamber and the molecules are applied by direct deposition from a liquid. For example, a drop of a weak 0.1% solution may be applied to the substrate and the excess spun off leaving a very thin liquid layer. When the solvent evaporates away, a reasonably uniform molecular layer is left behind and junction fabrication is then completed. This method has proved convenient for low-vapor-pressure molecules or those that bond strongly to surfaces. For the former, thickness may be controlled by regulation of the solution concentration. The doping of large molecules, such as those of biological interest which tend to decompose if high-temperature evaporation is attempted, are suited to this method. It is also useful for working with very small numbers of molecules and avoids contamination of the vacuum chamber. A related scheme is to adsorb the molecules directly by dipping the substrates directly into the solution. This method is susceptible to airborne and solvent contamination but has been used successfully for a variety of molecular types.

1.2.3. Infusion Doping

The method of infusion doping offers a number of unique possibilities. These arise because junction fabrication is carried out to completion and, when required, doping takes place by infusing the molecules through the overlay metal electrode from an external source. Sample preparation is more efficient because the chores of doping and junction fabrication are separate. Doping is carried out in a single step and the junctions are immediately ready for measurement. Also a number of advantages are derived from the peculiar mechanism of infusion. For example, it turns out that there are molecules which can form stable layers by infusion but cannot be doped satisfactorily by the usual methods. Multiple doping of the same junction with different molecules can be done.

2. Experimental Description of Infusion

2.1. Junction and Film Preparation

The general techniques for junction preparation are well known,[1-6] so only those details important for infusion doping will be described. All films are prepared in organic free vacuum systems which are either ion or oil pumped. Low impurity levels are assured by operation of an electrical discharge in oxygen operated routinely during each run. Liquid-nitrogen cold traps are used in the oil pumped system. With a few exceptions the tunnel junctions used have Al base films and Pb overlay films. The Al is about 600 Å thickness and is deposited on glass or glazed ceramic substrates. The oxide film is produced on the Al by operation of an oxygen discharge for several minutes, enough to produce an initial tunneling resistance in the range 0.1–100 Ω. The lower values are not useful for tunneling but the infusion step causes a further increase in resistance bringing them to the required range of values. The presence of an initial oxide film is important, however, because intimate contact of the metal films precludes subsequent infusion. The overlay Pb film is deposited at a rate of about 30 Å/sec to a thickness of about 650 Å. This thickness was chosen after learning that infusion gets slower with increasing Pb thickness and that thinner films are quicker to deteriorate because of the degradation which accompanies infusion. Although there is some evidence that the Pb deposition rate affects infusion to some degree, the dependence is not critical for our vacuum conditions. Films made on both liquid-nitrogen-cooled and room temperature substrates have similar infusion behavior even though an electron microscope study shows that the former are free of holes or pores initially. The junctions are usually made in batches which are removed from the vacuum system, place in flame-cleaned petri dishes, and stored in clean dry air. The storage cabinet contains 5-Å molecular sieve material to remove moisture and a small amount of activated charcoal to remove organic material. Under these storage conditions, junctions remain stable in resistance and do not accumulate impurities. Experience has shown that ordinary laboratory air can cause oxide growth and organic contamination in a few hours.

2.2. Infusion Techniques

Infusion is carried out by exposing tunnel junctions to an atmosphere with a relative humidity (RH) in the range from 50% to 95%. The carrier gas may be air, nitrogen, oxygen, hydrogen, argon, or CO_2. For organic impurity control, all glassware is flame cleaned and the water is either triply distilled or singly distilled from water mixed with a small amount of potassium permanganate. Many of the initial experiments were done by simply placing the junctions inside a dish covered with a glass plate together with a few cm^3 of water. The junction remains there for several hours and the resistance is checked periodically until values in the range 30–500 Ω are attained. This range is usable in the tunneling spectrometer. Organic molecules of interest can be included by mixing a small amount into the water. In order to prevent dew formation and the damage it causes to Pb films, the covered dishes are placed in a dish of shallow water, which keeps them slightly cooler than ambient temperatures. This method is easy for exploratory work, but control of humidity and monitoring of resistance is cumbersome. Also impurities tend to accumulate in a static system.

A better way (Figure 1) is to create the desired RH by dividing the dry gas between two separate routes, one bubbling through pure water to achieve nearly 100% RH. The two flow rates are measured with precision flow meters and the gas trains are mixed and passed through the glass exposure chamber. The tunnel junction is mounted in the exposure chamber and is connected by electrical feed troughs to the external resistance measuring circuit. The water bubbler and exposure chambers are standard Pyrex gas washing flasks and all connecting tubing, valves, and check valves are either glass, Teflon, or polyethylene. Wire leads to the sample are placed

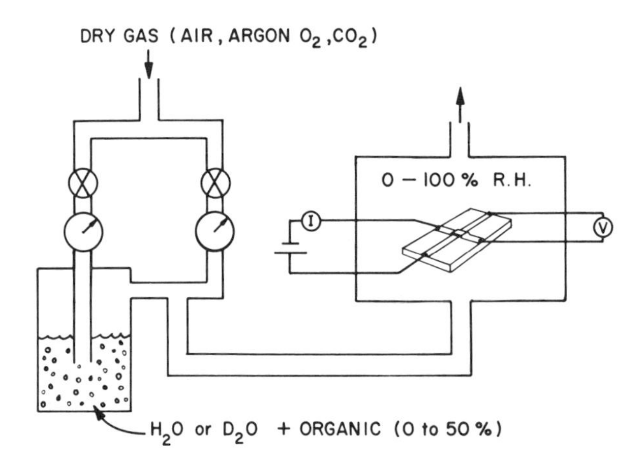

Figure 1. Schematic drawing of a flow-through system for infusion. The relative humidity (RH) can be controlled between 0% and 100% by control of the flow rates. While the tunnel junctions are exposed to the humid gas, the resistance is monitored. Organic molecules may be introduced by mixing with the water or by applying them to the outer surface of the overlay film of the tunnel junction. Impurity control is easier for this system than for static atmosphere exposure chambers.

downstream of the junction to avoid possible contamination by interactions of organic vapors with the metal surfaces. Usually argon or dry pure air is employed as the carrier gas although no effect due to various gases has been noted so far. Adjustment of the flow rates of the gas streams gives a RH in the exposure chamber adjustable from 0% to 100% which is stable over long periods of time and which can be changed with a time constant of about 10 sec. The latter is determined by the flow rate (about 5 l/min) and the volume of the system, which is about 1 l. Short time constants are valuable when quick changes in RH are needed at critical moments of the infusion process.

Organic molecules can be introduced by several convenient means. The main requirements are that the molecules be polar and be in contact with Pb overlay film simultaneously with the increased relative humidity. As a rule polar molecules are water soluble for low-molecular-weight substances. The easiest method is to add them directly to the water source. This works well for the high-vapor-pressure substances such as solvents, alcohols, and some acids and bases. The concentration required is determined by trial and error. Strong acids in a few percent (by volume) water solution produce very intense doping while stronger concentrations cause damage to the Pb film. A weak acid, like phenol, or one of the solvents requires higher concentrations to produce intense spectra. The exact composition of the vapor is not easy to calculate for liquid mixtures but at least it can be reproduced.

The molecules can be applied to the tunnel junction first and then infused later. One way is to coat the outside of the Pb overlay film with a layer of organic molecules beforehand and then expose the junction to humid gas. Molecules are applied by liquid doping but, in this case, the thickness of the organic layer may be many monolayers. Some molecules stick readily to the clean Pb film under simple exposure. After this organic outer layer is formed, the junction can be exposed to the infusion RH. Molecules migrate into the tunnel junction and form a stable monolayer at the tunneling interface. This technique is useful for strong adsorbers such as acids or low-vapor-pressure high-molecular-weight species.

2.3. Infusion Monitoring—Resistance and Capacitance

During infusion, the tunneling resistance grows to 100 Ω or more from initial values of 0.1–10 Ω. Resistances are monitored by periodic removal from the infusion chamber and measured on a low-voltage curve tracer or, for the case of the continuous gas flow infusion, constant monitoring is done by passing a small constant current (about one microampere) through the junction and measuring the voltage with a strip chart recorder. It is important that the voltage be limited to avoid junction damage.

When capacitance measurement is required, a small ac voltage is impressed at about 10 kHz and the ac current flowing through the junction is detected. This way the capacitance part of the impedance is found and the oxide thickness is calculated from the simple capacitance formula. This method is only useful if the resistance component of the impedance is 100 Ω or more.

3. Experiments Relating to Physical Mechanisms of Infusion

3.1. Resistance and Capacitance Behavior

The strip chart recorder provides a plot of resistance from 0.1 to 10^6 Ω versus time. There is a typical behavior of resistance during infusion as illustrated in Figure 2. For the first few minutes while the system is purged with dry gas the resistance remains stable. The RH is set to a certain value, say 80%, and resistance remains constant for a few more minutes and then begins to rise, slowly at first and then progressively faster. At any time the RH can be decreased to 0% and the resistance decreases almost immediately. This dropback occurs very quickly, in a few seconds, to values of one half to one tenth of the maximum. To attain useful tunneling resistance,

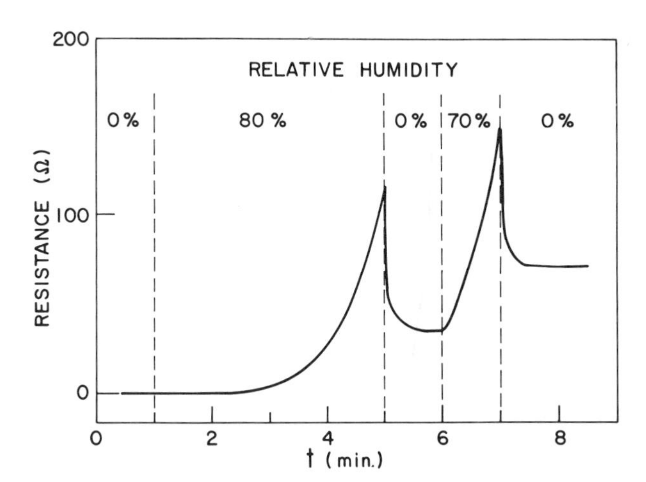

Figure 2. Resistance versus time behavior for a typical infusion of an Al–Al-oxide–Pb tunnel junction. Initial resistance is less than 1 Ω and is stable when stored at low RH. After exposure to about 80% RH there is an initial waiting time after which the resistance grows rapidly. The dropback in resistance each time the RH decreases is believed to accompany the deinfusion of water from inside the tunnel junction. Further infusion results in an immediate increase in resistance.

it is necessary to overshoot so that the value after dropback is in the range 25–100 Ω. The time required to reach a certain value for a fresh Pb film is a function of relative humidity and film thickness. Typically 15 min are required but the time can be much longer for very thick Pb. When stored in the dry gas the junctions remain stable indefinitely until needed for tunneling spectra measurement. If the laboratory air is humid, the transfer must be rapid to avoid additional infusion.

Capacitance data yield information about the growth of the effective insulator thickness and can be obtained after junctions have hundreds of ohms or more resistance. Observations show that infusion causes a growth of the oxide thickness and that there is a decrease in junction area due to pore formation, usually limited to 10% or less. The combination of capacitance and resistance data together with the tunneling spectra strongly suggest that most of the area of the junction is preserved during infusion in spite of evident change in the Pb film structure, and that this area is doped uniformly.

Experience has shown that the degree of infusion of the organic molecules cannot be determined from the resistance plots alone. The resistance increase signals a real growth of the oxide layer due to electrochemical processes similar to those which occur during liquid anodization of aluminum. If infusion is allowed to continue, resistances of several million ohms or more are attained. This limiting can be understood if it is supposed that the driving field for the process is supplied by the electric field in the oxide due to the difference in work function between Al and Pb. A thickness is reached where the internal fields are reduced to the degree that ion migration is no longer possible.

3.2. Film Porosity

The microstructure of the Pb film usually changes during infusion.[8, 9] Films which are known to be initially compact and smooth undergo structural changes accompanied by the development of a pore structure. Visually the films have a cloudy appearance if their sheet resistance increases by 10% or more. In extreme cases the films can be totally destroyed with exposure to high humidity and certain organic vapors. The existence of some type of pores after infusion can be inferred from the fact that, once a junction has undergone the resistance rise during the initial infusion, subsequent attempts do not require the initial time delay before the relatively rapid resistance rise begins. The junction appears to be "sensitized," and pore structure of the overlay films allows rapid infusion. Electron microscope examination of Pb films deposited on room-temperature substrates shows that a degree of porosity can exist before infusion[9] and a definite increase in pore size occurs during infusion. It must be emphasized that in many instances the amount of observable film damage or structural changes during exposure is almost unobservable but strong infusion occurs nonetheless. Therefore film damage and limitations caused by it at present are not viewed as fundamental problems. Initial results with Sn and Au overlay films indicate that damage during infusion is minimal for these metals.

3.3. Water Infusion and Organic Molecules

The tunneling spectra of junctions with infusion of pure water vapor was studied.[8, 9] When a D_2O source was employed, evidence of OD exchange with the initial OH was found. This is shown in Figure 3 where the spectra show that OH and OD groups can be exchanged during infusion. The initial concentration of OH and OD in these junctions is typical and is due to water (or heavy water) being present in a small amount in the electrical discharge during initial oxide growth. For these experiments, the initial resistances must be high enough that tunneling spectra could be run before infusion. Occasionally experiments like these failed to show complete exchange of the OH with infusing heavy water, or even no exchange. However, under the same conditions and for the same batch of samples, organic molecules would infuse readily. One possible explanation for this behavior may be understood from a model of the oxide deduced from tunneling experiments.[10] Hydroxyl groups are assigned to the interior as well as to the surface of the oxide. The surface groups would readily exchange with infusing water while the interior ones would not. The conditions determining which of the two types of hydroxyl groups are present depends on oxide growth methods, the amount of water vapor present during growth, and the temperature of the substrate during and after growth. Control over water vapor was marginal in our oxidation step and it remains to be seen if more careful oxidation procedures yield consistent and understandable hydroxyl exchange behavior.

As the data show, organic molecules are deposited inside the tunnel junction during infusion. Their mobility is promoted by the presence of the adsorbed water vapor. The rapid fallback of resistance after humidity drops suddenly indicates that some or all of the water present in the junction readily diffuses out at this time. Experiments also show that at the same time organic molecules remain inside the junction. Once doped and stored in dry air the organic tunneling spectrum remains stable for many months with no sign of loss of molecules. This is true even for molecules which cannot form a stable adsorbed layer at a free surface. The conditions which determine whether or not an organic layer is formed include their mobility

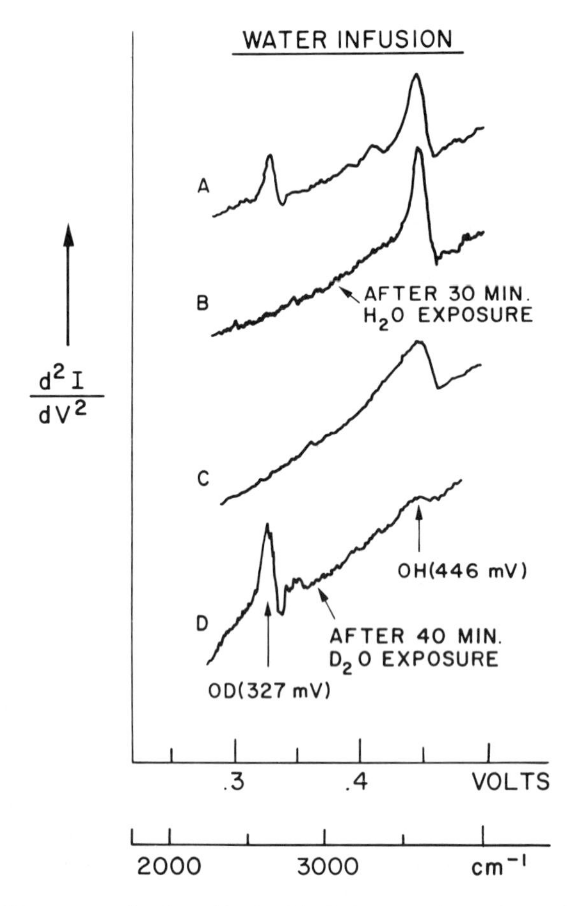

Figure 3. Tunneling spectra of a clean Al–Al-oxide–Pb junction shows the result of infusion of H_2O and D_2O. Curve A shows the initial OD peaks and curve B is from the same junction after H_2O infusion. Curve C is from a junction with no initial OD and curve D is from the same batch after D_2O infusion.

relative to water molecules, their relative concentration, and the possibility that they can form strong chemical bonds on the oxide surface. The last condition results in a stationary layer that cannot be removed by pure water infusion. It appears reasonable to say that the lack of success for infusion of molecules like CO_2, chloroform, or formaldehyde is at least partially a result of their having high mobility compared with water. They "deinfuse" with the excess water layer and hence are not trapped inside.

A simple experimental result illustrates that water affects the surface mobility of some adsorbed molecules. A diode was exposed to the pure, dry saturated vapor of deuterated formic acid and no effect on junction resistance or infusion spectra was observed. The sample was placed in storage

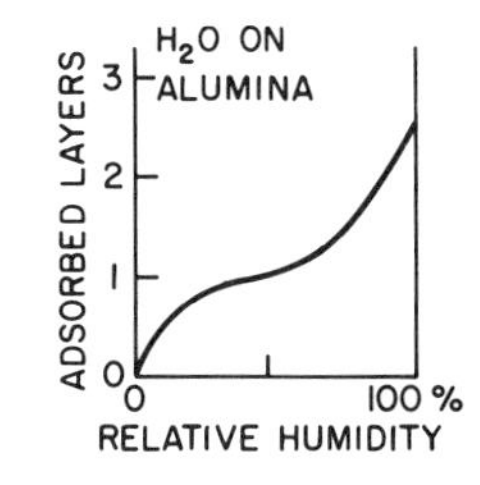

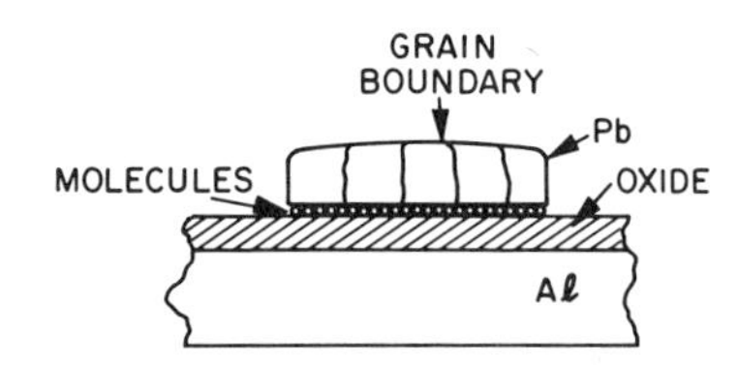

Figure 4. Schematic adsorbtion isotherm for water on alumina.[12] For large RH average coverage exceeds a single monolayer. Also shown is a cross section of a tunnel junction showing the grain boundaries in the Pb overlay film. During infusion this film develops a pore structure which allows penetration by the adsorbed water film along with organic molecules.

for one day and then infused with pure water vapor. A strong deuterated formic acid spectrum was obtained showing that the acid molecules would stick on the Pb overlay film but did not infuse into the junctions until water vapor was adsorbed. The major role played by the film of water is to drastically increase the mobility of the organic molecules. Similar behavior has been observed in ir experiments where fatty acids on glass are found to have an increased mobility in the presence of water vapor.[11]

The critical role of water vapor in the infusion process appears to be its ability to adsorb on the exposed surfaces of the Pb overlay films and penetrate to the Al oxide layer. Water is known to adsorb strongly on Al, glass, oxide, and organic surfaces. A typical behavior is shown in Figure 4 where it is seen that in the range 0% to 50% RH coverage increases from zero to one layer. The exact behavior above 50% depends on the type of surface and its microstructure. Highly porous structure can bind many monolayers.[12] These second and third layers of water are not bound as strongly as the initial ones, and molecules in these layers are quite mobile, behaving similarly to liquid water. In NMR experiments, motional narrowing of the water absorbtion peak is observed much like liquid water. Infusion occurs readily above 50% RH because the polar molecules are, in a sense, dissolved in a very thin stream of water adsorbed on the film surfaces. Penetration of the overlay films by water and organic molecules must progress along existing grain boundaries or pores which form during initial stages of infusion. At present this picture is qualitative and more microscopic details will be obtained as future experiments are done.

3.4. Masking Experiments

A series of experiments were performed which demonstrate that infusion takes place through the entire area of Pb overlay film rather than from the edges of the junction.[8] For this purpose a thick layer of Apiezon type-L vacuum grease was applied carefully over the junction to serve as an effective barrier to water or external organic molecules. The grease was removed by washing in pure benzene and the junctions were then run in the tunneling spectrometer. Results from a series of experiments where the edges of the junctions were masked showed that the dominant route for infusion was directly through the overlay film.

3.5. Sn and Au Overlay Films

Problems of stability of Pb films under infusion conditions have already been mentioned and some efforts have been made to find a better overlay film. With Al, Ag, Cr, and Cu no infusion could be produced. However, both Sn and Au electrodes show some success and warrant further study. With Sn at room temperature the film thickness must be above 2000 Å to achieve electrical continuity. If the substrates are cooled to 100 K or less Sn films of 500 Å or less can be made. Water, phenol, and several acids were successfully infused through Sn and presumably other molecules have a good chance for success. The spectra are comparable in intensity and behavior to Pb films and show very little sign of film degradation or structural changes. Apparently Sn films have an initial degree of porosity sufficient for successful infusion. The fact that Sn is not superconducting at 4 K ($T_C = 3.7$ K) creates some problems because the sheet resistance can contribute to voltage errors and broadening of spectral peaks. More work with Sn is needed to explore its potentials.

Another overlay film which has shown some promise is Au. It is continuous for films as thin as 100 Å. Usually, Au films made under good vacuum conditions show no infusion behavior, but films deposited under poor vacuum conditions in a utility-oil-pumped vacuum coating unit often show excellent infusion behavior. Success is not predictable, however, and attempts to produce Au overlay films by introducing various impurities into the clean vacuum system did not succeed. Apparently under certain unknown vacuum conditions Au films form with the required degree of porosity. Intense spectra of formic acid, propiolic acid, and phenol have been produced with Au films about 300 Å thick with no observed film deterioration. Work with Sn and Au is preliminary and the initial degree of success warrants further effort. Development of a stable and reliable overlay film for infusion would extend the scope of application of electron tunneling spectroscopy to the study of complex surface chemistry problems.

4. Examples of Molecules Infused

Infusion doping has been successful for a relatively large number of organic molecules.[8] In fact, the majority of molecules tried were doped successfully with a few exceptions, these being nonpolar molecules (benzene, CD_4, CH_4, CO_2, acetylene, C_2H_4, and napthalene) and some polar molecules (H_2S, CO, chloroform, formaldehyde, and SO_2). The spectra are comparable to or better than those obtained from junctions doped by other methods, when available for comparison. Therefore, no detailed spectral analyses will be given and only a few examples of several kinds of molecules will be given here. All spectra are taken at 4 K with a superconducting Pb electrode and the voltage scale has not been corrected for the Pb energy gap.

4.1. Acids and Bases

Organic acids are easy to infuse and reliably produce intense infusion spectra. Figures 5 and 6 show examples of these. For the carboxylic acids (formic and propiolic) the carboxyl group forms a strong bond to the oxide surface. The two peaks at 167 and 198 mV are due to the COO^- modes. Because of this bond, infusion cannot be reversed by exposure to pure water

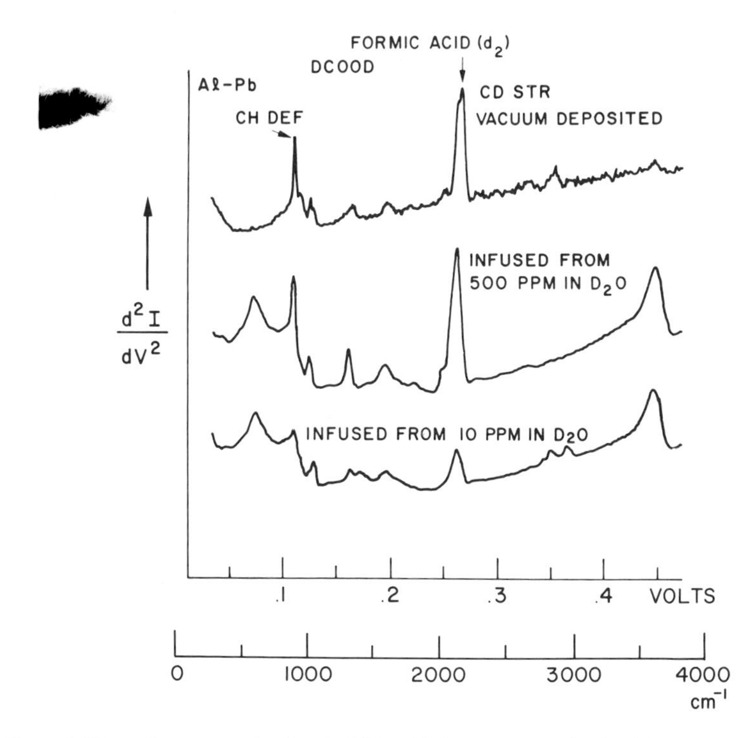

Figure 5. Tunneling spectra for formic (d_2) acid shows spectra obtained from vapor phase and infusion doping. The bottom curve demonstrates the high sensitivity for infusion doping to this molecule.

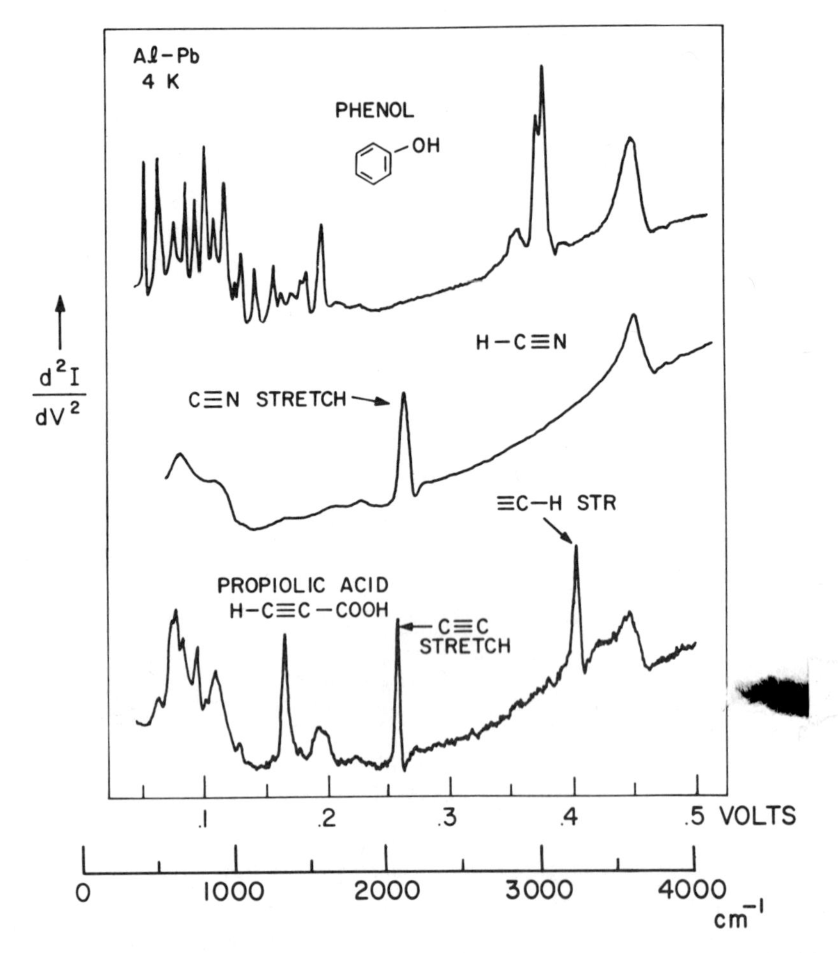

Figure 6. Tunneling spectra for phenol, HCN, and propiolic acid. All were obtained by infusion from water solutions.

vapor. The strong bonding means that infusion from very low concentrations of acids can produce strong spectra. As Figure 6 shows, a concentration of 10 ppm of deuterated formic acid in the infusing water source is easily detectable. Formic acid is a prevalent impurity often picked up inadvertently from laboratory air, tap water, or even singly distilled water. It has been suggested that it can be formed by reaction of CO_2 with H_2O during the process of infusion.[9]

Figure 6 shows examples of phenol, HCN, and propiolic acid. Phenol is a reliable infusing species but does not form a strong bond to the surface as inferred by the fact that it can readily be removed by infusion with pure water vapor. The HCN spectrum is also easy to obtain and is unique for its

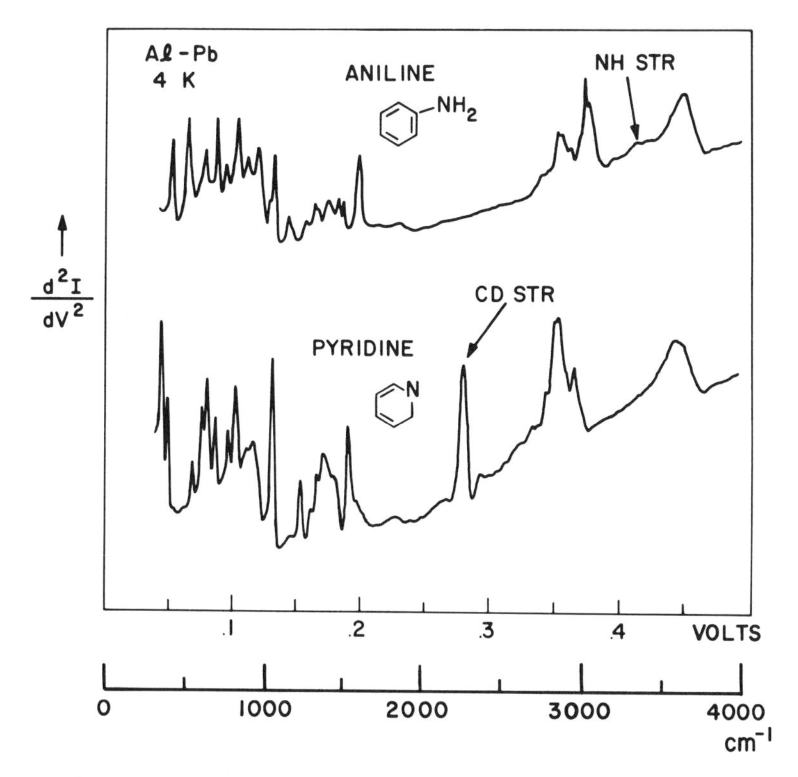

Figure 7. Tunneling spectra for aniline and pyridine. Both were obtained by infusion from water solutions.

simplicity: the single stretching peak at 260 mV corresponds to an adsorbed CN species. HCN is strongly adsorbed and therefore detectable from dilute sources. Propiolic acid adsorbs as a carboxylate ion (like formic acid) and produces a characteristic spectrum with an acetylene group triple bond. This molecule is an example where infusion doping succeeds when vapor doping does not. Attempts to dope from the vapor in the vacuum system give a complicated spectrum of reaction products produced when the propiolic acid molecules contact the chamber walls.

Figure 7 shows the spectra obtained for the weak bases aniline and pyridine. Both have been obtained by other doping methods. Pyridine does not form a stable adsorbed layer at room temperature and therefore cannot be doped by the vapor phase method at room temperature.

4.2. Solvents and Alcohols

This group of molecules is interesting for infusion because most of them cannot be doped by vapor phase doping at room temperature. They

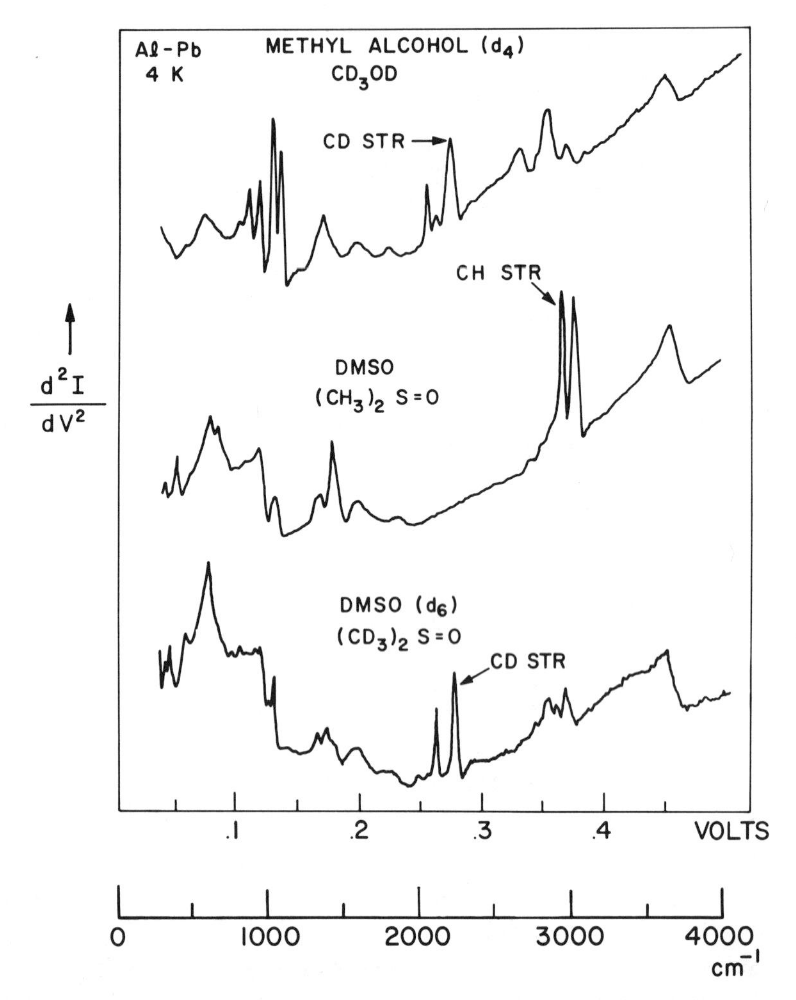

Figure 8. Tunneling spectra for methyl alcohol (d_4), and dimethylsulfoxide (DMSO). Both were obtained by infusion from water solutions.

tend to desorb rapidly during the vacuum pumpdown before the overlay Pb film is deposited. Figure 8 shows spectra for methyl alcohol and dimethylsulfoxide (DMSO). Acetone spectra (not shown) have also been obtained. All were doped from vapor produced by a water solution. Solvents tend to require a much higher concentration of molecules (50% by weight typically) because they do not form strong chemical bonds upon adsorbtion and probably tend to diffuse out along with the excess water.

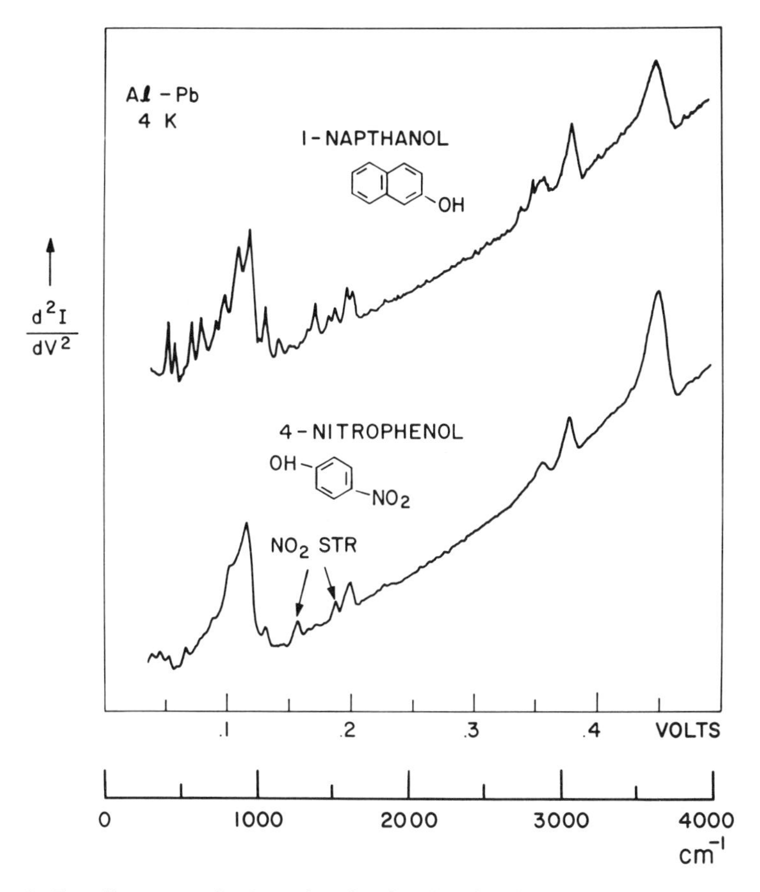

Figure 9. Tunneling spectra for 1-napthanol and 4-nitrophenol. Both were infused from solid films deposited on top of the overlay Pb.

4.3. Solid Phase Molecules

Molecules infused from the solid phase are shown in Figure 9. Both 1-napthanol and 4-nitrophenol are solids at room temperature and were infused by first depositing a thin layer of a 3% solution of chloroform on top of the Pb overlay and evaporating away the solvent. Subsequent infusion produced the spectra shown. The 1-napthanol is the largest molecule infused to date (largest dimension about 12 Å) and produces a spectrum very comparable to ir and Raman. The 4-nitrophenol is not as intense but clear indications of NO_2 group modes are present. 1-nitrophenol was also successfully infused by dipping the substrate directly into a weak solution (0.1%) with water.

5. Applications of Infusion

5.1. Hydrogenation and Deuteration of Propiolic Acid

These experiments investigate the behavior of adsorbed propiolic acid because it is found to be unstable with time at room temperature.[13] As Figure 10 shows, the spectrum transforms in about two weeks at room temperature or in about five minutes at 150°C while exposed only to dry air or helium gas. Study shows that no intermediate states are involved, the initial spectrum gradually dies while the new one grows. In the new one, the C≡C and ≡C—H peaks are gone and new peaks at 203 and 83 mV suggest that a C=C bond is present, indicating that a hydrogenation

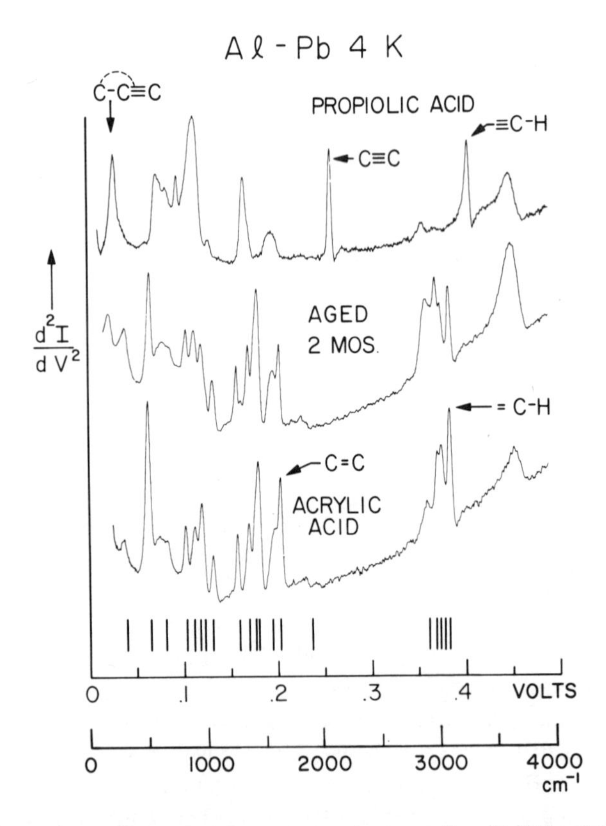

Figure 10. The top curve is the tunneling spectrum for propiolic acid (HC≡CCOOH) infused from a water solution. After aging (or heating), the spectrum transforms into the middle spectrum. This is compared with the spectrum of acrylic acid (H_2C=CHCOOH) infused from a water solution. The stick spectrum shows the ir and Raman frequencies for sodium acrylate.

reaction has occurred. The simplest possibility is that the acetylene carbon atoms have picked up two more hydrogen atoms to form an acrylate species. This is confirmed by the spectrum of Figure 10c for a junction which has been infused with acrylic acid from a 5% water solution. As can be seen, the spectra are virtually the same and compare well with peaks observed in the ir spectrum of sodium acrylate. Since the hydrogenation occurs without an external source of water vapor, the likely source for the hydrogen is the surface hydroxyl groups. The reaction is seen to be surface activated because a solution of propiolic acid and water remains stable for six months or more.

Another experiment[13] shows that the adsorbed propiolic species can interact with surface water or hydroxyl groups. Normal propiolic acid is infused in the presence of D_2O instead of H_2O. A partially deuterated molecular spectrum is obtained. Subsequent infusion with H_2O vapor restores most of the molecules to normal propiolic acid and still further exposure to D_2O vapor converts them back again. This is a surface-activated process because a 5% solution of normal propiolic acid in D_2O is not exchanged even after six months.

5.2. Solid-State Anodization of Aluminum

Measurements of resistance and capacitance both indicate that infusion results in an increase in the thickness of the oxide. The growth is accompanied by ion motion through oxide. This picture is consistent with the observed fact that oxide thickness tends to level off at a few million ohms and at oxide thicknesses of about 25 Å. In a series of experiments, it was found that further oxide growth occurs if a dc voltage is applied during infusion.[14] First, it is necessary to allow zero voltage infusion to proceed until the tunneling resistance has reached about one million ohms. This ensures that the electron tunneling current is reduced to low values. The aluminum is then made positive and the current is measured with time while the junctions are exposed to about 95% RH. With each increase in dc bias voltage, a dc current of about 10 μA (for our 4 mm^2 area junctions) is observed which gradually decays to less than one microamp. With each step increase in voltage the process repeats itself. This behavior is observed with Al electrodes for both Pb and Au overlay films and for a number of carrier gases including air, nitrogen, oxygen, and argon. The capacitance is measured after each anodizing step and a value for the thickness obtained. Data for thickness versus anodizing voltage are shown in Figure 11 where curves are shown for both Pb and Au overlay films. This behavior is similar to that observed with liquid anodization[14] where Al is placed in contact with an aqueous solution of weak acid. It is believed that hydroxyl ions are driven

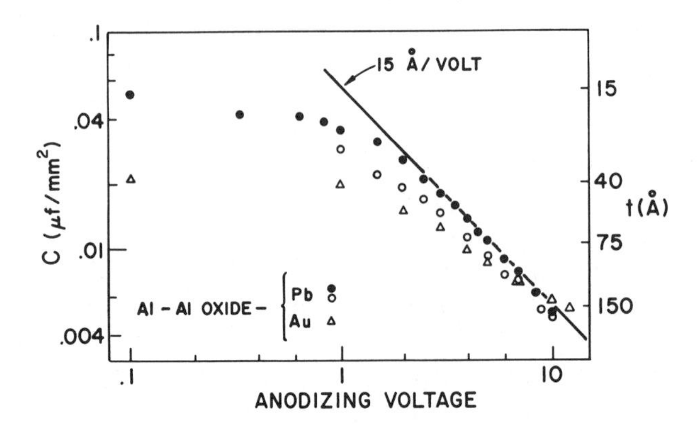

Figure 11. Variation of Al oxide thickness and capacitance with the applied dc voltage for overlay films of Pb (600 Å) and Au (100 Å) carried out at 95% RH and room temperature. The thickness scale was obtained assuming the dielectric constant of the oxide $K = 9$.

through the oxide by the applied field to the Al metal to react there to form the oxide. In fact a slope of 15 Å/V is observed, in close agreement with the present behavior. This supports the idea that the decaying current is ionic, probably hydroxyl ions, and is responsible for oxide growth. The slope and overall behavior of the curves does not depend upon the stepwise voltage increase taken between points. If the bias voltage is reversed, no oxide growth is observed.

Apparently a solid-state anodization is occurring whereby the role played by the electrolyte is assumed by the adsorbed layer of water. Solid-state anodization is of commercial importance in the growth of oxides in Ta capacitors. Anodizing behavior occurs when an electrically conducting manganese oxide layer is included in the Ta–Ta-oxide sandwich and serves as a source of oxidant to provide for "healing" of any defects in the insulating tantalum oxide. In experiments with this device it is interesting to note that ambient water vapor was also found to have large effect on the anodizing process.(16)

5.3. Other Applications

Several other experimental applications of infusion have been explored.(8) The passage of molecules through a plastic film was studied by coating the entire substrate and junction with a film of Formvar or collodian about 1 μm thick. This was done by dipping in a solution and evaporating away the solvent. Water vapor and organic molecules (phenol and formic acid) readily diffuse through these membranes in only a few seconds and cause infusion into the tunnel junctions. Evidently the high permeability of these species can be used to improve the selectivity and sensitivity to these molecules in a trace analysis experiment.

The potential for trace detection was demonstrated with formic acid in Figure 6. Similar results have been obtained by dipping the substrate directly into the aqueous solution. The water must be fairly clean since the survival rate of Pb films is short in impure water. Another example of trace analysis was the detection of HCN (approximately 100 ppm) in a gas stream emitted by automotive catalyst operated under abnormal conditions.

Another interesting area of experimentation is suggested by the expectation that there must be an observable emf during the infusion process which may be detected by measurement of a current in the external circuit. The buildup of an oriented dipole layer in the junction region, the reorientation of an existing layer of hydroxyl ions, and the migration of ionic species through the oxide all should be accompanied by the passage of a current. Charge rearrangements must occur to restore equilibrium between the two metal electrodes whenever the dipole layer is changed. Measurement of these currents, which are closely analogous to those produced in electrochemical cells, should be informative and correlate with details of infusion. It is difficult to predict the magnitude of the currents because they are transient and, for low resistance junctions, tunneling electrons will tend to short circuit the external current path. Preliminary measurements found several microamps of current (peak) and often complicated transient behavior. Large currents which tend to mask the effect are often generated at pressure or solder contact joints between the wire leads and the substrate films. These also depend upon the presence of water vapor and are undoubtedly electrochemical in origin. They can be avoided by using fired silver paste electrode with solder contacts. Monitoring of tunneling junction currents during infusion and correlating them with other junction parameters and the tunneling spectra may yield interesting information relating to oxide growth, chemical reactions, or corrosion processes.

6. Conclusions

(1) Infusion doping has been shown to be a useful and promising way to provide for the doping of junctions for tunneling spectroscopy and in addition offers unique experimental possibilities. For example, it can be used successfully with molecules which are difficult at best to dope by other methods.

(2) Batch processing provides a convenience and versatility which allows for working with molecules with a variety of properties and in interesting environments.

(3) Simultaneous and sequential doping make it possible to observe chemical reactions inside the junction.

(4) So far, only polar molecules have been used, but these include a wide spectrum of molecular types. There is reason to believe that other solvents can be found to play the role of the infusion medium and thus open the door to include nonpolar species as well.

(5) Problems relating to success rate and overlay film reliability are dependent upon the production of films which have the required porosity and are at the same time resistant to chemical attack and physical degradation during infusion. Initial results with Au and Sn are promising and many other possibilities, such as alloys or composites, can be explored. Membrane overlays, for example, can play the dual role of molecular selection and protection of the metal film from attack.

(6) Future developments will require much work, but will be a part of the gradual development of electron tunneling into a tool for surface analysis and chemistry.

References

1. J. Lambe and R. C. Jaklevic, Molecular vibration spectra by inelastic electron tunneling, *Phys. Rev.* **165**, 821–832 (1968).
2. D. G. Walmsley, in *Inelastic electron tunneling spectroscopy*, *Vibrational Properties of Adsorbates*, Springer-Verlag, Berlin (1980).
3. P. K. Hansma, Inelastic electron tunneling, *Phys. Rep.* **30C**, 146–206 (1977).
4. *Inelastic Electron Tunneling Spectroscopy*, (T. Wolfram, ed.), Springer-Verlag, Berlin (1978).
5. R. G. Keil, T. P. Graham, and K. P. Roenker, Inelastic electron tunneling spectroscopy, *App. Spectrosc.* **30**, 1–19 (1976).
6. W. H. Weinberg, Inelastic electron tunneling spectroscopy, *Ann. Rev. Phys. Chem.* **29**, 115–139 (1978).
7. R. C. Jaklevic and M. R. Gaerttner, Electron tunneling spectroscopy—External doping with organic molecules, *Appl. Phys. Lett.* **30**, 646–648 (1977).
8. R. C. Jaklevic and M. R. Gaerttner, Inelastic electron tunneling spectroscopy. Experiments on external doping of tunnel junctions by an infusion technique, *Appl. Surf. Sci.* **1**, 479–502 (1978).
9. W. J. Nelson, D. G. Walmsley, and J. M. Bell, Resistance and transmission electron micrography studies of the infusion doping of tunnel junction, *Thin Solid Films* **79**, 229–334 (1981).
10. W. M. Bowser and W. H. Weinberg, The nature of the oxide barrier in inelastic electron tunneling spectroscopy, *Surf. Sci.* **64**, 377–392 (1977).
11. R. L. Yang, J. B. Fenn, and G. L. Haller, Surface diffusion of stearic acid on aluminum oxide, *Am. Inst. Chem. Eng. J.* **20**, 735–742 (1974).
12. S. J. Gregg and K. S. Sing, *Adsorption*, *Surface and Porosity*, Academic Press, London (1967).
13. R. C. Jaklevic, Hydrogenation and deuteration of adsorbed propiolic acid as observed by electron tunneling, *Appl. Surf. Sci.* **4**, 174–182 (1980).

14. R. C. Jaklevic, Solid-state anodization of aluminum by vapor infusion, *J. Electrochem. Soc.* **126**, 1548–1550 (1979).
15. L. Young, *Anodic Oxide Films*, Academic Press, London (1961).
16. D. M. Smyth, Solid-state anodic oxidation of tantalum, *J. Electrochem. Soc.* **113**, 19–24 (1966).

16

Vibrational Spectroscopy of Subnanogram Samples with Tunneling Spectroscopy

P. K. Hansma and H. G. Hansma

As mentioned in Chapter 1, sample preparation is the most challenging part of tunneling spectroscopy. In sample preparation, the most critical step is doping the junction. Thus, it is understandable that a great deal of time and energy have gone into development of improved techniques for doping. The three general categories of techniques that have been discussed thus far in the book are (1) vapor doping, (2) liquid phase doping, and (3) infusion doping. In this chapter we discuss a relatively new technique for liquid phase doping in which a very small drop of dopant solution is placed on the tunneling barrier and evaporated to dryness. This simple new technique has the advantage that it requires orders of magnitude less of the doped compound than other techniques.

The first step in making the tunnel junctions is to evaporate aluminum through a stainless-steel mask into strips on glass slides. These aluminum strips are oxidized by exposure to oxygen and air. The chemicals are doped onto the strips using a small platinum loop as shown in Figure 1. Figure 1c is a scanning electron micrograph showing the 127-μm loop of platinum wire above an aluminum strip that is 100 μm wide and 0.08 μm thick. The platinum loop is held in the tripod just above the surface of the 1×3-in.

P. K. Hansma • Department of Physics, University of California, Santa Barbara, California 93106. **H. G. Hansma** • Department of Biological Sciences, University of California, Santa Barbara, California 93106.

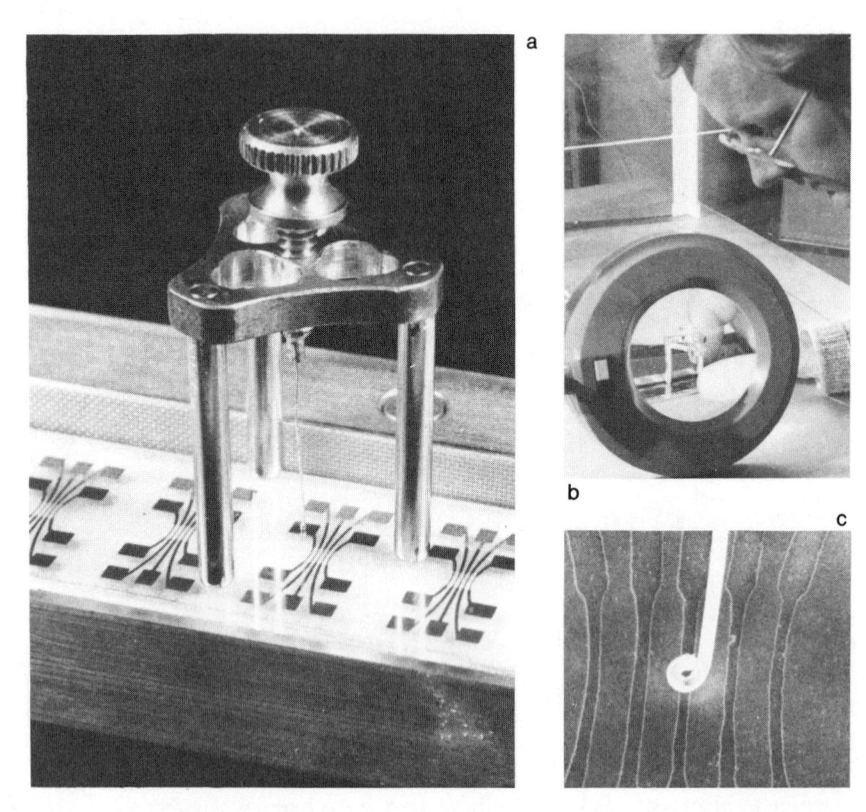

Figure 1. Apparatus and technique for applying nanoliter samples. (a) A platinum loop is held in a tripod that holds it just above the glass slide to avoid damaging the oxidized aluminum strips on the slide. (b) The loop is positioned by pushing the back two legs of the tripod against a guide rail. The loop is dipped into a microliter of stock solution at one end of the slide, then the tripod is tilted and slid on its back two legs to a position above the strip, as observed through the magnifying glass. When the third leg of the tripod is lowered onto the slide, the loop transfers a few nanoliters of liquid onto the strip. (c) A scanning electron micrograph of a loop and oxidized aluminum strips, 100 μm wide.

glass slide (Figure 1a). The height is adjusted with a brass thumb screw. In operation, the tripod is grasped as shown in Figure 1b and moved to the end of the slide where it is dipped into a drop of stock solution containing the chemical of interest. Typical size of the stock solution drop is 1 μl; the typical solution concentration is 0.1 g/l. Thus, the stock solution contains on the order of 1/10 μg of solute. The loop is raised from the stock solution by lifting the front leg of the tripod and then is moved to the junction area. The two back legs of the tripod are pressed against an indexing strip to determine the front–back position. The lateral position is determined by eye through the magnifying glass. When the loop is approximately over the junction area, the front leg of the tripod is lowered to the slide. The drop is thus transferred from the platinum loop to the junction area.

Typically, the five junctions in a set (Figure 1c) are doped with different dilutions of the stock solution by adding water to the drop of stock solution between successive pickups with the platinum loop. Finally, after all the desired junctions are doped, a drop of india ink is placed at each end of the slide with the doper. These two drops of india ink determine the line on which the dopant drops have been placed. They are used for orienting the slide over the mask when it is put back in the evaporation chamber for adding cross strips of lead metal (100 μm wide and 0.2 μm thick). The junctions are formed at the intersections of the lead and oxidized aluminum strips.

There are a few subtle differences between the experimental techniques for this subnanogram doping and the techniques described in Chapter 1: (1) It is valuable to use very small junction areas and very small dopant drops. Larger junction areas and dopant drops have given, in our limited experience, less reproducible results. The masks for the small metal strips can be chemically etched to order by several firms in thin stainless-steel sheets. (2) It is easy to get too high a junction resistance for these small junctions. Consequently steps must be taken to keep the resistance low. Rather than venting the chamber to pure oxygen, we vent to just 0.1 Torr of pure oxygen and then vent the rest of the chamber with nitrogen or argon (venting to pure nitrogen gave unreproducible results). After the chamber is vented, we carry the slides to the doping apparatus (it takes about 20 sec) and then at the doping apparatus maintain a gentle flow of nitrogen over the slides. (3) A convenient feature is that the platinum loop can be easily cleaned between dopants by simply holding it in the flame of a bunsen burner. It heats white hot and any organic material is burned off within a fraction of a second.

The completed junctions are treated in the same way as described in Chapter 1.

Since the concentrations of the dopant solutions are known, the amount of material transferred to the junctions can be determined if the drop size is known. The drop size can be measured by using drops of a radioactive solution. For example, we used a ^{14}C-glycine solution with an activity of approximately 500 μCi/ml. We made nine drops with the loop, cut the slides with a glass cutter to separate the drops, put each into a scintillation vial with scintillation cocktail,* and counted the vials. Standards were made by diluting the stock solution 5 μl to 500 μl and then counting 1 μl and 5 μl

*Two hundred and fifty milliliters of Triton X-114, 750 ml xylene, 3 g PPO, and 0.2 g POPOP (see reference 1).

of this dilute solution. The measured drop volumes were 4 ± 1 nl for the loop shown in Figure 1, and 7 ± 2 nl for the (similar) loop used for making the samples of Figures 2 and 3.

Figure 2 shows spectra obtained for benzoic acid solutions obtained by successive dilution of a 1-μl drop of 0.1 g/l stock solution. The top two traces were from separate junctions to illustrate the reproducibility of the technique. Note that the intensity of the benzoic acid peaks (e.g., near 400 and 1600 cm^{-1}) saturates near 7×10^{-10} g.

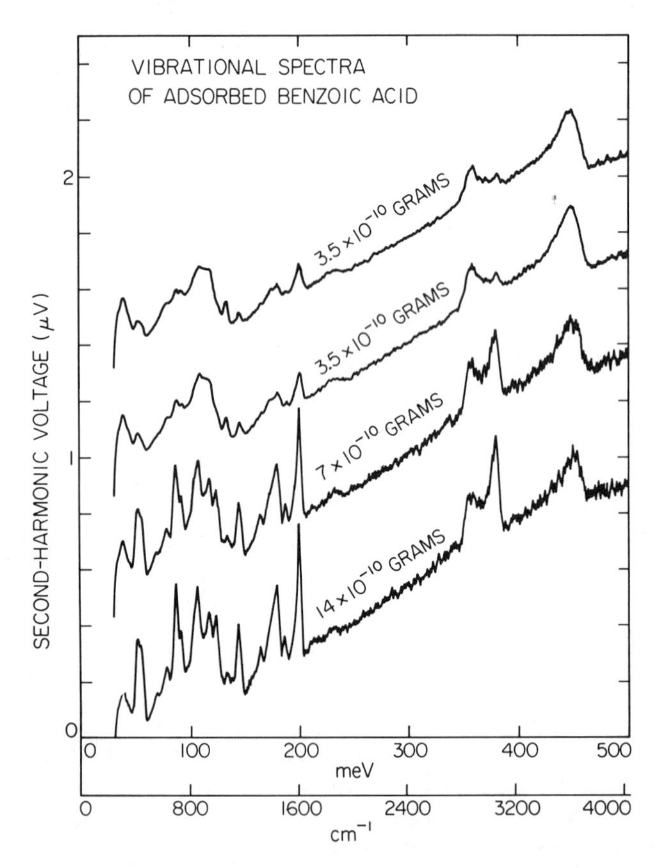

Figure 2. The doping apparatus of Figure 1 was used to apply 7-nl drops of 0.05, 0.05, 0.1, and 0.2 g/l solutions of benzoic acid in water to tunnel junctions. In addition to characteristic background peaks (e.g., near 900 and 3600 cm^{-1}), there are peaks due to adsorbed benzoic acid (e.g., near 400 and 1600 cm^{-1}) that increase with solution concentration. Note the reproducibility shown by the two top curves and the saturation of peak intensities shown by the two lower curves.

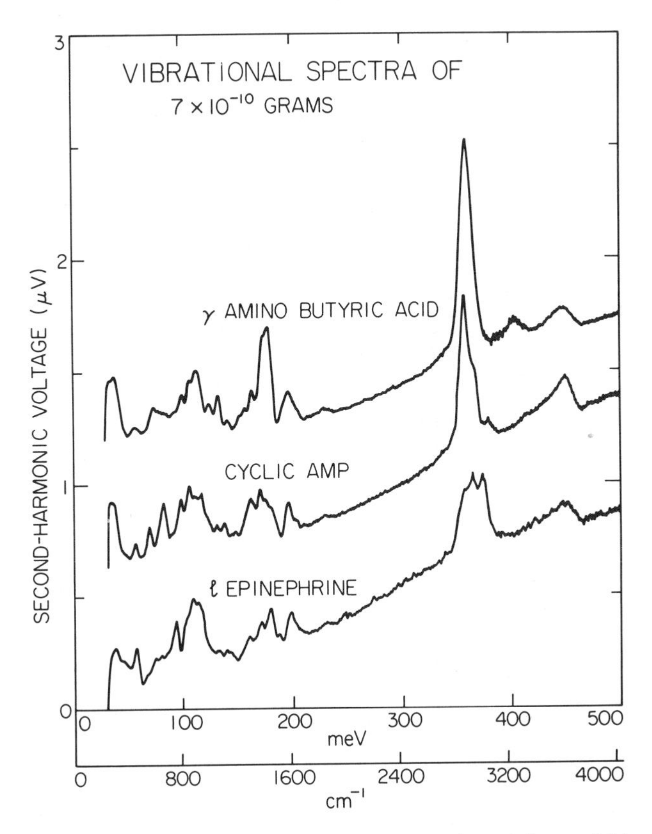

Figure 3. The doping apparatus of Figure 1 was used to apply 7-nl drops of 0.1 g/l aqueous solutions of an amino acid (upper trace), a nucleotide (center trace), and a hormone (lower trace). The hormone solution had 0.0005 *N* HCl added to promote solubility and stability. This quantity of material produced a saturation of peak intensities for these compounds as shown in Figure 2 for benzoic acid.

Figure 3 shows the vibrational spectra of 7×10^{-10} g of compounds of current interest in biochemistry. They illustrate that the technique can be applied to amino acids, nucleotides, and hormones. These compounds are also of interest as neurotransmitters.

Vibrational spectra of hundreds of other biochemicals have been previously obtained with a spin-doping technique.[2, 3] It should be possible to use the current microdoping technique with these and related biochemicals whenever one wants to use only a microliter or less of the stock solution.

It should be emphasized that these spectra of subnanogram quantities are not the minimum quantities that can be detected. In fact, quantities below a nanogram saturate the vibrational spectra. Previous evidence suggests that of order 1/100 of these saturation amounts can be detected in careful experiments.[4] Further, the loop used for these experiments made drops of order 460 μm in diameter. Though this made positioning easy in a 100×100-μm junction because of the large overlap, it could be reduced. For example, a 200-μm-diameter drop could deposit the same surface coverage with 1/5 the material. Thus, the sensitivity limit of the technique is probably below 10 pg.

A real question is how useful is this? There are a number of problems that prevent it from being a great boon to biologists and analytic chemists, at least at present: (1) The apparatus for doing tunneling spectroscopy is not available commercially. (2) There are no collections of tunneling spectra analogous to the huge collections of infrared or Raman spectra, though the number of published tunneling spectra has been growing rapidly. (3) Typical molecules react with the aluminum oxide to form surface adsorbed species. Though these surface adsorbed species are interesting in their own right, as previous chapters have shown, they are not the original parent compound, and any attempt to identify the original parent compound from the surface adsorbed species is by inference. Even if a huge collection of spectra was available, it is possible that more than one starting compound would give the same adsorbed species (for example, benzoyl chloride,[5] benzaldehyde,[6] and benzoic acid[7] all give adsorbed benzoate ions). (4) The range for quantitation is very small. Spectra saturate at approximately one monolayer and become undetectable for less than 100th of a monolayer. Ideally, one would like a much wider range over which quantitative results could be obtained. Since there is no guarantee that the tunneling peak height is a linear function of the dopant concentrations, standard curves would have to be plotted as well. (5) Vibrational spectroscopy, in general, is not good at distinguishing low levels of one compound in the presence of others. Assay techniques such as radioimmunoassay that can detect minute amounts of biochemicals in the presence of large amounts of buffers and other interfering chemicals are very much to be preferred in this regard.

For these and perhaps for other reasons, it seems unlikely that this technique will evolve into a general-purpose method for analyzing trace quantities of biochemicals. This kind of prediction is, of course, dangerous, because it does remain true that despite all of its limitations, tunneling does have sensitivity to orders of magnitude less material than any other technique for vibrational spectroscopy known at present. As discussed above, the minimum quantities detectable are probably below 10 pg (e.g., for benzoic acid, 10 pg corresponds to 5×10^{10} molecules, 1/10 picomole or one monolayer over a square less than 1/10 mm on a side).

Acknowledgments

We thank the Office of Naval Research for partial support of this work.

References

1. L. E. Anderson and W. O. McClure, An improved scintillation cocktail of high-solubilizing power, *Anal. Biochem.* **51**, 173–179 (1973).
2. P. K. Hansma and R. V. Coleman, Spectroscopy of biological compounds with inelastic electron tunneling, *Science* **184**, 1369–1371 (1974).
3. M. G. Simonsen, R. V. Coleman, and P. K. Hansma, High-resolution inelastic tunneling spectroscopy of macromolecules and adsorbed species with liquid-phase doping, *J. Chem. Phys.* **61**, 3789–3799 (1974).
4. R. M. Kroeker and P. K. Hansma, A measurement of the sensitivity of inelastic electron tunneling spectroscopy, *Surf. Sci.* **67**, 362–366 (1977).
5. D. G. Walmsley, I. W. N. McMorris, and N. M. D. Brown, Inelastic electron tunneling spectroscopy evidence for a surface chemical reaction, *Solid State Commun.* **16**, 663–665 (1975).
6. M. G. Simonsen, R. V. Coleman, and P. K. Hansma, High-resolution inelastic tunneling spectroscopy of macromolecules and adsorbed species with liquid-phase doping, *J. Chem. Phys.* **61**, 3789–3799 (1974).
7. J. Langan and P. K. Hansma, Can the concentration of surface species be measured with inelastic electron tunneling? *Surf. Sci.* **52**, 211–216 (1975).

Index

Acetaldehyde, 320
  spectrum, 321
Acetate
  orientation, 19
  spectrum, 19
Acetic acid, 318-320, 344-347
  deuterated, 320
  spectra, 319
Acetic anhydride, 320
  spectrum, 321
Acetone, spectrum, 247
Acetophenone
  comparison to L-phenylalanine, 205
  ring modes, 205
  spectrum, 204
Acetylacetone, 346
  as iron complex: *see* Iron complexes
  spectra on Al vs Mg, 353
Acetyl chloride, 320
  spectrum, 321
Acetylene on Pd, 415-418
  spectra, 417
Acid: *see* name of acid; Brønsted acid; Lewis acid
Acridine
  inhibition of corrosion, 304-306
  orientation on oxide, 305-306
  spectrum, 305
Acrylic acid, 326-327
  spectrum, 326
Acrylic acid, spectrum, 468
Adenine
  mode identifications, 213-214
  spectrum, 213
Adenine derivatives
  5′-AMP, 5′-ADP, 5′-ATP, 3′5′:cAMP, Apa, 216
  spectra of low-lying modes, 217
Adsorption
  of infused water on alumina, isotherm, 451
  low temperature, 419-427
Ag: *see* Silver
$\alpha^2F(\omega)$, 76, 88-95, 102, 103, 106
Alumina: *see* Aluminum oxide
Aluminum
  AES spectrum, 157
  corrosion of: *see* Corrosion of aluminum
  oxidation of, 157, 160-162
Aluminum oxide
  AES spectrum, 157
  Auger spectrum, 388
  electron microscopy, 386
  ELS spectrum, 162
  growth during infusion, 458
  preparation of, 29-30
  properties of, 155, 183-185
  Si compounds on, 248-249
  solid state anodization of, 469-470
  spectra, 248, 312, 315
  structure, 313
  thickness in junction, 43
  UPS spectrum, 160
Amino acids
  as dipolar ions, 204
  spectra, 203-207, 479
  surface adsorption, 204
γ-Amino butyric acid, spectrum, 479
Amplifier, lock in, 36-37
Analog to digital conversion, 430-431
Aniline, spectrum, 465
Arnold theory, 97-99
Asymmetry, opposite bias, arguments for long-range interactions, 66-68
Atomic radius of top metal electrode, effect on spectra, 23-25

Au: *See* Gold complexes
Auger electron spectroscopy (AES), 155-158
  spectra, 157, 388

Barrier, composite, SiO/AlO, 248-249
  See also: Oxide; Aluminum oxide; Insulator
Barrier potential
  assumptions, 55
  calculations, 45-47
Benzaldehyde, 170, 338
  spectra, 171
  spectra on Al vs Mg, 344
Benzene
  adsorption of, 180
  inelastic neutron scattering spectrum, 180-181
  reaction with Zirconium complex, 373-374
Benzoate ions
  effect of top metal electrode, 22-24
  electron-induced damage, 278
Benzoic acid
  configuration on oxide, 331
  p-deutero, spectra, 13, 14
  spectra of different concentrations, 478
  spectrum, 11
  See also: Benzoate ions
Benzyl alcohol
  spectra on Al vs Mg, 349
  spectrum, 333
Benzylamine, spectrum, 333
BET surface area, 167, 183, 184-185
Biochemical applications, 201-226, 479-480
Bottom metal electrode, preparation, 28
Bridge techniques, 424-428
p-Bromophenol, spectrum, 332
Brønsted acid, bonding to oxide, 326, 335-337, 350
1,3-Butanediol, spectra with different doping techniques, 31-32
Butyric acid C_4, 321
  spectrum, 322

Capacitance of junctions, behavior during infusion, 457
Caprylic acid C_8, 321
  spectrum, 322
Carbon monoxide
  on cobalt, spectra, 412-414, 419
Carbon monoxide (*continued*)
  frequency table, 414
  on iron, spectra, 405-408
  on iron particles, 24-25, 405-407
  on nickel, spectra, 407-412
  orientation on surface, 64-66
  on palladium, spectrum, 417
  on platinum, spectrum, 418
  on rhodium, 182-194, 397-402
    high resolution electron energy loss spectroscopy, 188-191
    infrared spectroscopy, 185-187, 189, 192, 193
    nuclear magnetic resonance spectroscopy, 191-192
    sample preparation, 182-185
    spectra, 187-189, 398-403
    ultraviolet photoelectron spectroscopy, 193-194
  on ruthenium, 418
    infrared spectrum, 172
    spectrum, 417
  in short-range interactions theory, 62
  spectra with different top metal electrodes, 24-25, 407
Carbon tetrachloride
  corrosion of aluminum, 293-296
  peak assignments, 296
  spectrum, 295
Carboxylate group
  modes, 322-323, 344, 347, 349
  peak intensities, 57
  spectra, 18, 19
Catalysis, 241-243, 359-421
  See also: Hydrogenation
Catalyst
  impregnation, 241-243
  supported cluster compound, 359, 391
  supported metal, 393-421
Cations, effects on ferrocyanide peaks, 262-263
C—H bond
  bending peak, 439
  peak intensities in theory and experiment, 57
  stretching peak, 428, 431
  stretch modes, intensity in C_n acids vs n, 325
Charge transfer
  in electronic transitions of inorganic ions, 238-240
  in redox reactions, 263-266

Chemisorption
of carbon monoxide, 182-194, 170-172
of carboxylic acids, 17
of hydrogen, 167, 177-181
See also: name of compound
p-Chlorophenol, spectrum, 332
Cluster compounds
experimental procedures for doping, 245-247, 361, 362
reaction with oxides, 229-230, 359, 360
See also: Cobalt complexes; Gold complexes; Iron complexes; Ruthenium Complexes; Silver complexes; and Zirconium complexes
Cn compounds: *see* name of compound
CO: *see* Carbon monoxide
Co: *see* Cobalt complexes
Cobalt, CO on, 412-415, 419-420
Cobalt complexes:
$Co(gly)_2$, spectrum, pH effects, 258-259
$[Co(gly)_2(NO_2)_2]^-$, spectrum, 257-258
$[Co(NCS)_4]^{-2}$, spectrum, 259-261
Coordination compounds, 229-267, 359-391
See also: name of metal
Complex, 229-267, 359-391
See also: name of metal
Computer
for data acquisition, 430-434
for data analysis, 448-449
for data calibration, 428, 446-447
for data handling, 445-446
for data storage, 447-448
Computer interface
BCD, 431-433
IEEE 488 bus, 431-447
parallel, 431-434
RS232-C, 431-433
Conductance, of junctions, 43
Corrosion inhibitors for aluminum, 292-293, 296-299, 304-307
acridine, 304-306
in chlorinated solvents, 291
formamide, 296-299
in hydrochloric acid, 304-307
mechanism, 291, 299
orientation of surface species, 306-307
surface species, 291-292
thiourea, 306-307
Corrosion of aluminum, 287-291, 293-296, 300-304
by carbon tetrachloride, 290, 293-296
by chlorinated hydrocarbons, 290-291
Corrosion of aluminum (*continued*)
effect of water, 288-289
in hydrochloric acid, 304
mechanism, 289-290
by organics, 288-290
surface species, 291-292
by tricholoroethane, 290-291
by tricholoroethylene, 300-304
Counterions, effects on ferrocyanide peaks, 262-263
5′-CMP (cytidine monophosphate)
effect of substrate cooling, 222
spectra after irradiation, 225
spectrum, 212
surface orientation of, 222
Current–voltage curves
structure in, 43-44
theory and experiment, 55
Cyanide: *see* Iron complexes; Silver complexes; Ruthenium complexes
Cyclic AMP, spectrum, 479
1,3,5-Cycloheptatriene, spectrum, 329
Cyclohexadiene, spectrum, 374
Cyclohexene, spectra, 329, 374
L-cysteine, SH stretching mode of, 208
spectrum, 206
Cytidine monophosphate: *see* 5′-CMP
Cytosine, spectrum, 212

Desorption, 408-410
Deuterated: *see* name of parent compound
Deuterium, (OD) OH exchange, 459-460
Diborane, reaction with alumina, 365-366
Dielectric constant, in adsorption of inorganic ions, 245-247
Differential IETS, 10, 439-440, 442-443
Diffusion barrier, 96
Dimethyl acrylic acid, spectrum, 326
Dimethyl sulfoxide, spectrum, 464
deuterated, spectrum, 464
Dipeptide, spectra, 205, 221
Diphenylamine, spectrum, 337
Dipole derivatives, 57, 58
effect of metal overlayer, 61
from ir data, 59, 60
for methyl group, 59
from vapor state data, 61
Dipoles
interaction of tunneling electrons, 18
orientation theory, 55-57
DMSO: *see* Dimethyl sulfoxide
DNA, calf thymus, comparison of IETS modes to Raman modes, 220

Doping
contamination during, 313
onto cooled substrate, 17
drop size, 477-478
effects of pH and vapor pressure, 245-246
infusion, 451-472
of inorganic ions, 250-252
from liquid source, 453, 475-481
methods, 30-32, 451-472
nanoliter doping, 475-481
from vapor source, 331, 353, 452-453
See also: Infusion doping
D-ribose
modes, 214
spectrum, 216
Dynamic resistance, 425

Elastic tunneling
and barrier models, 55
calculating currents, 44-47
defined, 43
Eliashberg equations, 91
Electrode: *see* Top metal electrode; Bottom metal electrode; specific metals
Electron energy loss spectroscopy (ELS), 161-162
See also: High resolution electron energy loss spectroscopy
Electron-irradiation
assumptions, 275
of benzoate, 280
spectra, 279
damage cross sections, 276
experiments, 273-275
exposure rate effects, 275
of fructose, spectra, 274
of 2,4-hexadienoate, 280
spectra, 279
of hexanoate, 280
spectra, 279
specimen heating, 275
suggestions for future, 280-282
Electron microscopy
scanning electron microscopy (SEM), 167
transmission electron microscopy (TEM), 166-167, 183-184, 394
Electron–phonon interactions, 71-106
coupling, 82-83, 89
as nuisance, 8-10
Electronic transitions, 109-118
in inorganic ions, 238-241

l-Epinephrine, spectrum, 479
Erbium, Er_2O_3 spectrum, 112
Ethanol, spectra, 247, 344
on Ag, spectrum, 408
Ethanolamine, spectrum, 342
Ethylene glycol, spectrum, 342
Ethylene oxidation, spectrum, 159
Ethylene, spectra, 369, 444
Ethylidene, 402-405
Esaki diode, 72, 74, 77-79
Evaporator, vacuum: *see* Vacuum evaporator
Extended x-ray absorption fine structure (EXAFS), 163-164

Fast modes, 133-135
See also: Polaritons, surface
Fe: *see* Iron complexes
Ferricyanide: *see* Iron complexes, $[Fe(CN)_6]^{-3}$
Ferrocyanide: *see* Iron complexes, $[Fe(CN)_6]^{-4}$
Forbidden transitions
electronic, of inorganic ions, 238-241
vibrational, of inorganic ions, 231-238
Formamide
inhibition of corrosion, 296-299
spectra, 297, 298
Formic acid, 344-347
deuterated, spectrum, 463
spectra, 427, 436, 439, 443, 444
spectra on Al vs Mg, 345
vapor vs infusion doping, spectra, 463
Frequencies, causes of variability, 400
with desorption, 410, 419
with temperature, 408-410
β-D-fructose, electron-irradiated spectra, 274
Functional groups, table of infrared absorption bands, 2-3

Gap function, 76, 94
Gas phase doping: *see* Doping
Glass
cleaning, 27
as junction substrate, 26-27
Glycinates, spectrum, 257-259
Glycine, spectra, 203, 257-258
5′-GMP (guanosine monophosphate), spectrum, 215

Gold
effect on CO spectrum, 407
See also: Gold complexes; Top metal electrode
Gold complexes
$[Au(CN)_2]^-$, 246
redox reactions, spectra, 264-266
$[Au(CN)_4]^-$, redox reactions, spectra, 264-266

Harmonic detection, 424-425
Harmonic distortion, 427
HCN, spectrum, 464
Hemoglobin, spectrum, 208
1-Heptene, 326, 328
spectrum, 326
1-Heptyne
configurations on oxide, 329-330
spectra on Al vs Mg, 352-353
spectrum, 335
2,4-Hexadienoate, electron-induced damage, 278
Hexanoate, electron-induced damage, 278
Hexanoic acid, 278
1-Hexene, 334-335
spectrum, 334
1-Hexyne
configurations on oxide, 335-336
spectrum, 335
3-Hexyne
configurations on oxide, 335-336
spectra on Al vs Mg, 351-353
spectrum, 328
High resolution electron energy loss spectroscopy (ELS), 177-179
of carbon monoxide, 188-191
spectra, 190
table, 169
Holmium, Ho_2O_3, 112, 240-241
Hormone, spectrum, 471
Hydrocarbons
unsaturated, spectra, 326, 331
vibrational frequencies depend on top metal electrode, 22, 24-25
See also: name of compound
Hydrogen
and acetylene, 415-418
and CO, 407-411
ELS spectra, 178
inelastic neutron scattering spectrum, 181
Hydrogenation of CO
on Co, 414-415
spectrum, 415
on Ni, 411-412
spectrum, 412
on Rh, 402-405
spectrum, 403
Hydrogen chemisorption, 167
surface area of metal, 177-181
Hydrogen cyanide, spectrum, 464
Hydrogen sulfide, deuterated, and CO, 401, 402
Hydroxyl group
effect of top metal electrode, 21-24
orientation, 50
theory and experiment, 55
See also: OH(OD)

Indium
effect on CO spectrum, 407
See also: Top metal electrode
Induced dipoles, in tunneling theory, 50-51, 60-61
Inelastic neutron scattering 179-181
table, 169
Inelastic tunneling
and barrier models, 55
calculations, 47-68
defined, 43
Inelastic tunneling theories
complex model, 52-69
difficulties with simple model, 51
simple model, 47-51
Infrared spectroscopy
for assigning partial charges, 57
carboxylate vibrational peaks compared to tunneling, 18
comparison to other spectroscopies (table), 169
mode intensities relative to tunneling, 17, 57
peak intensities compared to tunneling, 50
reflection–absorption infrared (RAIR) spectroscopy, 169-173, 188
table of absorption bands, 2-3
tables of modes, 189, 193
transmission infrared spectroscopy, 168-170, 185-187
See also: specific compounds
Infusing doping, 451-472
mechanism, 461-462

Infusion doping (*continued*)
method, 454-457
organic polar molecules, 462-468
porosity of metal overlay film, 462, 458-459
resistance vs time, 457-458
role of water vapor, 459-461
solid phase, 467
Inorganic ions, 229-267
In situ, 419-420
Insulator, thickness in junction, 43
See also: Aluminum oxide; Barrier
Interactions
predictions from opposite bias conductance step ratios, 67-68
short vs long range, 66-68
Interactions, electron–phonon: *see* Electron–phonon interactions
Interactions, long-range electron–molecule: *see* Long-range interactions
Image dipole, effect on vibrational frequencies, 23-24
Impregnation catalysts, 241-243
Iodine, EXAFS spectrum, 164
Ionic motion
during infusion, 469
emf generated during infusion, 471
solid state anodization of Al, 469-470
Iron complexes
$Fe(acetylacetone)_3$, 259
$[Fe(CN)_6]^{-3}$, redox reactions, spectrum, 263-265
$[Fe(CN)_5NO]^{-2}$, spectrum, 260-261
$[Fe(CN)_6]^{-4}$, 230, 237, 239, 260-261
spectrum, 249
tunneling, Raman, ir spectra, 252-256
$Fe_3(CO)_{12}$
crystal structure, 384
spectra, 385
table of mode assignments, 386
Iron particles, CO, on, 405-407
Irradiation, electron: *see* Electron-irradiation
Irradiation, UV: *see* UV radiation damage
IR spectroscopy: *see* Infrared spectroscopy
Isocyanate, 229-230
spectra, peak assignments, 250-252
Isotopic substitutions
^{13}C in CO spectra, 402, 403, 413
of ferrocyanide, 254-256
^2H and CO spectra, 409
spectrum, 410
Isotopic substitutions (*continued*)
^{18}O in CO spectra, 398, 402, 403, 413, 414
of OD and OH, 459-460
of Si–H, 248-249

Junctions
conductance vs applied bias, 11
cutting apart, 34-35
fabrication, 28-33, 454
idealized view, 5-6
light-emitting: *see* Light-emitting junctions
measuring $I-V$ characteristics, 36-38
storage, 33, 446
various sizes, 34
water analogy, 4-5
See also: Bottom metal electrode; Capacitance; Doping of samples; Resistance; Sample holder; Sample preparation; Top metal electrode

Lauraldehyde, 321
spectrum, 323
Lead
effect on CO spectrum, 407
normal vs superconducting, 434, 436, 441-442, 445
See also: Top metal electrode
Lewis acid, bonding to oxide, 328, 334-337, 350
Light-emitting junctions
efficiency, 122, 148-151
frequency spectrum, 139
with holographic gratings, 142-145
with periodic array structures, 150
with roughened electrodes, 138-142
small-particle junctions, 146-147
three sample geometries, 124-125
Liquid doping: *see* Doping
Localized electromagnetic excitations, 146-150
Lock-in amplifier, use of, 36-37
Long range electron–molecule interactions, 62
calculations, 47-61
intrinsic dipole potential, 57
models, 44
Low-energy electron diffraction (LEED), 165-166

Low-energy vibrational modes
effect of top metal electrode, 24-25
in junctions with iron particles, 24-25, 405-407
Low-temperature adsorption, 419-420

Magnesium oxide, 344-346
complex with Si, 248-249
LEED spectra, 165-166
spectra of compounds on, 347-355
spectrum, 345
spectrum of deuterated, 345
McMillan–Rowell inversion, 91-95
Metal: *see* Bottom metal electrode; Top metal electrode; $M(XY)_6$; name of metal
Metal carbonyls: *see* Iron complexes; Ruthenium complexes; Zirconium complexes
See also: Carbon monoxide
Metal complexes, 229-267, 359-388
See also: name of metal
Metal films: *see* Bottom metal electrode; Top metal electrode
Metal particles, 394
Metal phonons, cause structure in IV curves, 43-44
Metal tris diimine, 240
Methanol
deuterated, spectrum, 466
spectrum, 247
Methyl alcohol: *see* Methanol
Methyl sulfonate
orientation on oxide, 19, 59-61
peaks, theory and experiment, 60, 61
spectrum, 19
MgO: *see* Magnesium oxide
Microscope slides
cleaning, 27
as junction substrate, 26-27
Model catalysts, 363-397
Modulation spectroscopy, 424-426
Modulation voltage broadening of peaks, 15-17
$M(XY)_6$, 230, 235-239, 247, 253-256, 261-265

1-Napthanol, spectrum, 467
[NCO]: *see* Isocyanate $(OCN)^-$
[NCS]: *see* Thiocyanate
Neurotransmitter, spectra, 479
NH–rhodanine–merocyanine, spectrum, 115
Nickel, CO on, 407-412
Nickel complex:
$Ni(gly)_2$, spectra, 257-258
p-Nitrophenol, spectra, 332, 467
Nitroprusside: *see* Iron complexes
Normal metal phonons, 86-88, 104, 105
NS interface, 97-98, 99
Nuclear magnetic resonance (NMR) spectroscopy, 191-192
Nucleic acids
dominant modes and resolution, 219
spectra of DNA and RNA, 219
Nucleosides, spectra for adenine derivatives, 217
Nucleotides, 214-221
dominant vibrational modes of, 214
preparation of junctions, 214
spectra, 210-217, 471
surface adsorption effects, 221-223
surface adsorption of $-PO_3^{2-}$ moiety, 216
surface orientation effects, 221-223
vibrational modes due to sugar, 214
vibrations of phosphomonester group, 215

1-Octene, 326-328
spectrum, 326
[OCN]: *see* Isocyanate
(OD)OH
exchange during infusion doping, 459-460
spectrum, 460
See also: Isotopic substitutions
Orientation
of acridine, 305-306
of dipoles, 50
of hydrocarbons, 328-330
of methyl sulfonate, 19, 59-61
of molecules, 44
of organic acids, 18-19, 326, 331
on surface, 18-19
theory of effect on tunneling intensities, 55-57
of thiourea, 306-307
See also: Selection rules
Oxidation
of aluminum, methods, 29-30
of inorganic ions, 263-266
See also: Aluminum oxide

Oxide: *see* Aluminum oxide; Barrier; Magnesium oxide
Oxide phonons and structure in IV curves, 43-44
Oxygen, AES spectrum, 157

Pair potential, 76, 91, 94
Palladium acetylene on, 415-418
Palmitaldehyde C_{16}, 321
  spectrum, 323
Partial charge
  model for surface bonding, 57
  in tunneling theory, 5[illegible]3
Pb: *see* Lead
Peak assignments: *see* name of compound
Peak intensity, definition, 438
Peak positions
  measuring, 38
  shifts, 16, 438, 440, 445
Peak profile
  determination, 434-443
  modulation and thermal smearing, 434-438
  normal vs superconducting electrodes, 427, 428, 434, 440, 444
Peak shifts, 16, 438, 440, 445
Peak widths, 14-17
Pentacene, spectrum, 114
2,4-Pentanedione, RAIR spectrum, 173
Peptides, resolution of, 209
pH dependence
  of ion absorption, 244-246
  of metal glycinates, 257-259
  of redox reactions, 264
  spectra, 246, 257, 258, 264
Phenols
  spectra, 332, 464
  spectra on Al vs Mg, 348
l-Phenylalanine, spectra, 204, 207
2-Phenylethylamine, spectrum, 337
Phonon
  assisted tunneling, 72-74, 77-86
  spectrum: *see* $\alpha^2F(\omega)$
  structure, 43-44, 435
Phthalocyanines, spectrum, 115
Plasmons: *see* Polaritons
Plastic film, diffusion of organic molecules through during infusion, 470-471
Platinum, acetylene on, 418
Point dipole, in inelastic tunneling theory, 47-50
Polaritons, surface
  in Al–I–Ag tunnel junction, 131-136
  at a dielectric-metal interface, 126-130
  fast mode, 133-135
  localized by microstructures, 146, 150
  probability of excitation by electrons, 123
  probability of radiative decay, 150
  slow mode, 132-134
Polyamino acids
  L-tryptophyl-L-phenylalanine spectrum, 205
  substrate cooling effects on, spectra, 221
Polyuridylic acid, spectrum, 218
Potassium: *see* associated anion
Potential energy, equations for inorganic ions, 232-234
Preparation of junctions, 26-33
  See also: Doping
Probability current, 83
L-Proline, spectrum, 207
1,3-Propane diamine, spectrum, 338
1,3-Propanediol, spectrum, 338
Propiolic acid, 326
  effect of aging, spectrum, 468
  spectra, 326, 464, 468
Propionic acid C_3, 321, 344-348
  spectra of Al vs Mg, 347
  spectrum, 322
Propylene, spectrum, 371
Proteins, resolution of, 209
Proximity effect, 96
Proximity electron tunneling spectroscopy (PETS), 95-106
Proximity junction, 96-99
Purine bases, 209-214
  mechanism of bonding, 209
  spectrum, 213
Pyridine
  adsorption of, 174-175, 177
  deuterated, ELS spectra, 177
  Raman spectra, 174-175
  spectrum, 465
Pyrimidine bases, 209-214
  mechanism of bonding, 209
  ring deformation modes, 210
  ring modes, 210
  spectra, 210-212

Quantum size effect oscillations, 188
Quartz balance, 394

Radiative decay, 150
  See also: Polaritons, surface
Radius, atomic: *see* Atomic radius
Raman spectroscopy, 171, 174-176
  active modes, 50-51
  compared to other spectroscopies (table), 169
  mode intensities relative to tunneling, 17, 57
  surface enhanced (SERS), 175-176
  See also: specific compounds
Rare earth oxides, 111-113
  spectra, 112
Reduction of inorganic ions, 263-266
Reflection–absorption infrared spectroscopy: *see* Infrared spectroscopy
Renormalization function $Z(w)$, 91-92, 94
Reproducibility, nanoliter doping, 478
Residual gas analyzer, 397
Resistance of tunnel junctions, 33
  during infusion, 457-458
  optimum resistance ratio, 36
Resolution, 14-17
Rhodium, 397-405
  See also: Carbon monoxide on rhodium
Rhodium complex
  $[RhCl(CO_2]_2$
    crystal structure, 379
    spectra, 380, 381
    table of mode positions, 383
Ruthenium, 418
  spectrum with CO, 172, 417
Ruthenium complexes:
  $[Ru(CN)_6]^{-4}$, 237
  $Ru_3(CO)_{12}$
    crystal structure, 375
    spectrum on alumina, 376
    table of mode assignments, 378

Sample doping: *see* Doping; Infusion doping
Sample holder, 35
Sample preparation
  techniques, 26-33
  See also: Doping; Junctions
Scanning electron microscopy: *see* Electron microscopy, scanning
Schottky barrier, 79-82, 86
$[SCN]^-$: *see* Thiocyanate
Secondary ion mass spectrometry (SIMS), 158-159
  spectrum, 159
Selection rules, 17-20
  symmetry, 54-57
  symmetry predicted tunneling peak intensities, 57
  table, 235
  See also: Orientation
Self-energy effect, 82
Semiconductor phonons, 72-74, 77-86
Sensitivity
  maximum, 480
  nanoliter doping, 477, 480
  of tunneling spectroscopy, 10-14, 477-480
L-Serine, spectrum, 206
Short-range interaction
  between tunneling electron and carbon monoxide, 62-68
  not considered in tunneling, 44
Si: *see* Silicon complexes
Signal-to-noise ratio of spectra, 16-17
Silicon complex, SiH, spectra from SiO evaporation, 248
Silver, 407
  ethanol on, 415
  iodine absorption on, SEXAFS, 164
  spectrum, 416
  See also: Top metal electrode
Silver complex, AgI, EXAFS spectrum, 164
Sintering of particles, 408
Slides, microscope, 26-27
Slow modes, 132-134
  See also: Polaritons, surface
Sodium: *see* associated anion
Spectroscopy: *see* Auger electron spectroscopy; Electron energy loss spectroscopy; High resolution electron energy loss spectroscopy; Inelastic neutron scattering; Infrared spectroscopy; Nuclear magnetic resonance spectroscopy; Raman spectroscopy; Reflection–absorption infrared spectroscopy; Secondary ion mass spectroscopy; Ultraviolet photoelectron spectroscopy; Vibrational spectroscopy; X-ray photoelectron spectroscopy
Spectrum: *see* name of compound
Stearaldehyde C_{18}, 321
  spectrum, 324

Strong coupling superconductors, 91-95
Substrate cooling, effects on spectra, 221-223
Substrates for junctions, 26-27
Superconducting lead energy gap, use in characterizing junctions, 36
Superconductivity
causes structure in IV curves, 43-44
effect on vibrational frequencies of benzoate, 22-23
and phonons, 89
Superconductor phonons, 99-104
Surface analytical techniques, 154-182
Surface area, measured by gas adsorption, 167
Surface concentration
determinations of, 10-12
monolayer, 10
and peak intensities, 49-50, 477-478
related to solution concentration, 12
Surface enhanced Raman spectroscopy: *see* Raman spectroscopy
Surface extended x-ray absorption fine structure (SEXAFS), 163-164
Surface plasmons: *see* Polaritons, surface
Surface polaritons: *see* Polaritons, surface
Surface reactions: *see* name of compound
Symmetry coordinates, 53
tunneling theory, 58-59

TCNQ, 263
Temperature broadening of peaks, 15-17
Tetracyanin, 113-114
spectrum, 113
Tetramethylammonium ion, spectrum, 259-260
Thallium
effect on CO spectrum, 23-25, 399
See also: Top metal electrode
Thermal broadening of peaks, 15-17
need for cryogenic temperatures, 52
Thiocyanate, spectra and peak assignments, 249-251, 259-261
Thiourea
inhibition of corrosion, 306-307
orientation on oxide, 306-307
Threshold spectroscopy, 77-88
Thymine, spectrum, 211
Tin, effect on CO spectrum, 23-25, 399
See also: Top metal electrode
Top electrode: *see* Top metal electrode
Top metal electrode, 20-25
effect of atomic radius on spectra, 23-25, 399
fabrication, 32-33
porosity of Pb, 458-459
redox of inorganic ions, 265-266
types useful for infusion, 462, 469-470
See also: specific metals
Trace time of spectra, 16-17
Transfer Hamiltonian
for calculating elastic tunneling current, 45-47
disadvantages, 47
in inelastic tunneling, 47-51, 52-54
Transmission electron microscopy: *see* Electron microscopy, transmission
Transmission infrared spectroscopy: *see* Infrared spectroscopy
Transition metals: *see* Top metal electrode; name of metal
Tricholoroethane, corrosion of aluminum, 290-291
Trichloroethylene
chemisorption on aluminum oxide, 303
corrosion of aluminum, 300-304
peak assignments, 302
spectrum, 301
Trifluoroacetic acid, 320
spectrum, 319
Tungsten, contamination by, 396
Tunneling currents, calculating, 44-47
Tunneling density of states, 72, 89-91
Tunneling spectrum
apparatus for measuring, 36-37
calibrating *x*-axis, 38
differential, 10
high energy vibrations, 109-118
low energy vibrations, 10
reveals vibrational energies, 7
See also: name of compound
Tunnel junction: *see* Junctions

Ultraviolet photoelectron spectroscopy (UPS), 159-161, 193-194
spectra, 194
Ultraviolet radiation damage: *see* UV radiation damage
5′-UPM (uridine monophosphate), spectrum, 210
Unsaturated acids, spectra, 326
Unsaturated hydrocarbons, 326-330

Uracil, spectrum, 210
UV radiation damage of nucleotides, 223-226
  method of UV exposure, 224
  primary mode damage, 224
  UV damage spectra of 5′-CMP, 225

Vacuum evaporator, for making junctions, 27-28
Vapor doping: *see* Doping
Vibrational spectroscopy
  characteristic mode positions (table), 2-3
  comparisons of various techniques (table), 169
  See also: High resolution electron energy loss spectroscopy; Inelastic neutron scattering; Infrared spectroscopy; Selection rules; Raman spectroscopy.

Water, deuterated, spectrum, 460
Water, spectra, 247, 452
Whiskey, Bushmill's spectrum, 341
WKB approximation, 84

Xenocyanin, 113-114
  spectrum, 114
X-ray photoelectron spectroscopy (XPS), 155
  spectra, 156

Zirconium complex, $Zr(BH_4)_4$
  spectra, 338, 367
    with acetylene, 372
    with cyclohexene, cyclohexadiene, and benzene, 374
    with ethylene, 369
    with propylene, 371
  tables of mode assignments, 364, 370

Conversion Table between Millielectron Volts and Wave Numbers

meV	cm^{-1}	meV	cm^{-1}	meV	cm^{-1}	meV	cm^{-1}	meV	cm^{-1}
1	8	51	411	101	815	151	1218	201	1621
2	16	52	419	102	823	152	1226	202	1629
3	24	53	427	103	831	153	1234	203	1637
4	32	54	436	104	839	154	1242	204	1645
5	40	55	444	105	847	155	1250	205	1653
6	48	56	452	106	855	156	1258	206	1661
7	56	57	460	107	863	157	1266	207	1670
8	65	58	468	108	871	158	1274	208	1678
9	73	59	476	109	879	159	1282	209	1686
10	81	60	484	110	887	160	1290	210	1694
11	89	61	492	111	895	161	1299	211	1702
12	97	62	500	112	903	162	1307	212	1710
13	105	63	508	113	911	163	1315	213	1718
14	113	64	516	114	919	164	1323	214	1726
15	121	65	524	115	928	165	1331	215	1734
16	129	66	532	116	936	166	1339	216	1742
17	137	67	540	117	944	167	1347	217	1750
18	145	68	548	118	952	168	1355	218	1758
19	153	69	557	119	960	169	1363	219	1766
20	161	70	565	120	968	170	1371	220	1774
21	169	71	573	121	976	171	1379	221	1782
22	177	72	581	122	984	172	1387	222	1791
23	186	73	589	123	992	173	1395	223	1799
24	194	74	597	124	1000	174	1403	224	1807
25	202	75	605	125	1008	175	1411	225	1815
26	210	76	613	126	1016	176	1420	226	1823
27	218	77	621	127	1024	177	1428	227	1831
28	226	78	629	128	1032	178	1436	228	1839
29	234	79	637	129	1040	179	1444	229	1847
30	242	80	645	130	1049	180	1452	230	1855
31	250	81	653	131	1057	181	1460	231	1863
32	258	82	661	132	1065	182	1468	232	1871
33	266	83	669	133	1073	183	1476	233	1879
34	274	84	677	134	1081	184	1484	234	1887
35	282	85	686	135	1089	185	1492	235	1895
36	290	86	694	136	1097	186	1500	236	1903
37	298	87	702	137	1105	187	1508	237	1912
38	306	88	710	138	1113	188	1516	238	1920
39	315	89	718	139	1121	189	1524	239	1928
40	323	90	726	140	1129	190	1532	240	1936
41	331	91	734	141	1137	191	1541	241	1944
42	339	92	742	142	1145	192	1549	242	1952
43	347	93	750	143	1153	193	1557	243	1960
44	355	94	758	144	1161	194	1565	244	1968
45	363	95	766	145	1169	195	1573	245	1976
46	371	96	774	146	1178	196	1581	246	1984
47	379	97	782	147	1186	197	1589	247	1992
48	387	98	790	148	1194	198	1597	248	2000
49	395	99	798	149	1202	199	1605	249	2008
50	403	100	807	150	1210	200	1613	250	2016

meV	cm^{-1}	meV	cm^{-1}	meV	cm^{-1}	meV	cm^{-1}	meV	cm^{-1}
251	2024	301	2428	35				451	3638
252	2032	302	2436	35[illegible]				452	3646
253	2041	303	2444	35[illegible]				453	3654
254	2049	304	2452	35[illegible]				454	3662
255	2057	305	2460	35[illegible]		[illegible]	[illegible]	455	3670
256	2065	306	2468	356	2871	406	3275	456	3678
257	2073	307	2476	357	2879	407	3283	457	3686
258	2081	308	2484	358	2887	408	3291	458	3694
259	2089	309	2492	359	2896	409	3299	459	3702
260	2097	310	2500	360	2904	410	3307	460	3710
261	2105	311	2508	361	2912	411	3315	461	3718
262	2113	312	2516	362	2920	412	3323	462	3726
263	2121	313	2524	363	2928	413	3331	463	3734
264	2129	314	2533	364	2936	414	3339	464	3742
265	2137	315	2541	365	2944	415	3347	465	3750
266	2145	316	2549	366	2952	416	3355	466	3759
267	2153	317	2557	367	2960	417	3363	467	3767
268	2162	318	2565	368	2968	418	3371	468	3775
269	2170	319	2573	369	2976	419	3379	469	3783
270	2178	320	2581	370	2984	420	3387	470	3791
271	2186	321	2589	371	2992	421	3396	471	3799
272	2194	322	2597	372	3000	422	3404	472	3807
273	2202	323	2605	373	3008	423	3412	473	3815
274	2210	324	2613	374	3016	424	3420	474	3823
275	2218	325	2621	375	3025	425	3428	475	3831
276	2226	326	2629	376	3033	426	3436	476	3839
277	2234	327	2637	377	3041	427	3444	477	3847
278	2242	328	2645	378	3049	428	3452	478	3855
279	2250	329	2654	379	3057	429	3460	479	3863
280	2258	330	2662	380	3065	430	3468	480	3871
281	2266	331	2670	381	3073	431	3476	481	3879
282	2274	332	2678	382	3081	432	3484	482	3888
283	2283	333	2686	383	3089	433	3492	483	3896
284	2291	334	2694	384	3097	434	3500	484	3904
285	2299	335	2702	385	3105	435	3508	485	3912
286	2307	336	2710	386	3113	436	3517	486	3920
287	2315	337	2718	387	3121	437	3525	487	3928
288	2323	338	2726	388	3129	438	3533	488	3936
289	2331	339	2734	389	3137	439	3541	489	3944
290	2339	340	2742	390	3146	440	3549	490	3952
291	2347	341	2750	391	3154	441	3557	491	3960
292	2355	342	2758	392	3162	442	3565	492	3968
293	2363	343	2766	393	3170	443	3573	493	3976
294	2371	344	2775	394	3178	444	3581	494	3984
295	2379	345	2783	395	3186	445	3589	495	3992
296	2387	346	2791	396	3194	446	3597	496	4000
297	2395	347	2799	397	3202	447	3605	497	4009
298	2404	348	2807	398	3210	448	3613	498	4017
299	2412	349	2815	399	3218	449	3621	499	4025
300	2420	350	2823	400	3226	450	3629	500	4033

1-MONTH